OXFORD PAPERBACK REFERENCE

A Dictionary of **Statistics**

Graham Upton is Professor of Applied Statistics in the Department of Mathematical Sciences at the University of Essex. His more than 100 publications include seven books, four co-authored with Ian Cook and published by Oxford: *Understanding Statistics, Introducing Statistics,* and *Statistics S1 and S2.*

Ian Cook was a lecturer in mathematics at the Universities of Hull and Essex. He was also a Chief Examiner in Mathematics at A Level for 30 years. He is co-author with Graham Upton of *Understanding Statistics, Introducing Statistics,* and *Statistics S1 and S2.*

SEE WEB LINKS

To find recommended web links for this and many other Oxford reference titles, visit http://global.oup.com/booksites/reference/ when you see this sign.

A Dictionary of
Statistics

THIRD EDITION

GRAHAM UPTON
and IAN COOK

Great Clarendon Street, Oxford, OX2 6DP,
United Kingdom

Oxford University Press is a department of the University of Oxford.
It furthers the University's objective of excellence in research, scholarship,
and education by publishing worldwide. Oxford is a registered trade mark of
Oxford University Press in the UK and in certain other countries

First published 2002
Second edition 2006
Second edition revised 2008
Third edition 2014

Impression: 1

British Library Cataloguing in Publication Data

Data available

ISBN 978-0-19-967918-8

Printed in Great Britain by
Clays Ltd, St Ives plc

Contents

Preface to Third Edition vi

Preface to Second Edition vi

Preface to First Edition vii

Abbreviations in Biographies viii

Dictionary 1

Appendices 468

I Statistical Notation 468

II Mathematical Notation 469

III Greek Letters 470

IV Upper-Tail Percentage Points for the Standard Normal Distribution 471

V The Standard Normal Distribution Function 472

VI Percentage Points for the *t*-Distribution 474

VII Percentage Points for the *F*-Distribution 475

VIII Percentage Points for the Chi-Squared Distribution 478

IX Pseudo-Random Numbers 480

X Selected Landmarks in the Development of Statistics 481

XI Honours and Awards 483

XII Further Reference 486

Preface to the Third Edition

For this edition we have added nearly 250 new entries. The majority refer to the application of statistical methods in the fields of machine learning, genetics, and medical statistics. We have also taken the opportunity to add more than 200 new web links, and to reorganize some material to improve its accessibility. Examples of the latter include the explanations of sample mean, sample variance, and their population equivalents. The previously minor entry on the two-by-two table is now greatly enhanced by the collation of related entries and relevant new material to provide a comprehensive account of this frequently encountered topic. In this edition we have also included about 250 extra web links. For the most part these links provide extensions to our short biographies and usually include at least one picture of the individual concerned.

GRAHAM UPTON
IAN COOK

COLCHESTER
DECEMBER 2013

Preface to Second Edition

For this second edition we have added more than 500 entries, often reflecting the terms used in standard statistical computer packages, bringing the total to more than 2000. We have undertaken major revisions of topics (such as multiple comparison tests and tests of equality of variance) with the aim of drawing together descriptions of related single entries into larger multiple entries that help the reader to appreciate better the similarities and differences between alternatives. In addition to updating the existing biographies we have added nearly 60 further biographies, bringing the total to more than 200. We have replaced the previous list of addresses of societies by their websites (reflecting the evolution in methods of communication). A new addition is a list of the major prize-winners in Statistics.

We are very grateful to Paul Baxter, Michael Burch, Andrew Colman, Tim Liao, and David Shrewsbury for highlighting problems in the first edition, to Robert Cummings, Tim Earl, and Katrina Morrow for supplying diagrams, and to eminent statisticians worldwide for help with their own biographies. We are particularly grateful to Nick Cox for his numerous excellent suggestions.

GRAHAM UPTON
IAN COOK

COLCHESTER
DECEMBER 2005

Preface to First Edition

This dictionary has two aims: (1) to provide a satisfactory amount of accurate information about subjects of interest to the user, and (2) to induce the user to read about topics other than those of immediate concern. The achievement of the second aim has (we hope) been effected by a deliberate breadth in the dictionary's scope. As well as entries on statistical topics, there are entries on related topics in mathematics, operational research, and probability.

In deciding on the topics for inclusion, we have had to think of the probable users. Many of the people using this dictionary as an *aide-mémoire* in Statistics will be those who are meeting the subject for the first time, as students at school or university. Another large group of readers will be specialists in other subjects who have found the need to analyse their own data and have then encountered the gobbledegook associated with computer packages. Our selection of topics has been made with all of these people in mind.

Dictionaries vary widely in style. This became very apparent to us once we started on this project. We have taken the view that if a reader needs to look up an 'elementary' topic then that reader may well need a rather long explanation, possibly with an example. Conversely, a reader looking up an 'advanced' topic will be a reader who already has a deep statistical knowledge and needs rather less help.

We have included approximately 150 short biographies. The criterion for inclusion has been that the individual concerned has made an important contribution to the development of the subject of Statistics, or that the individual's name forms part of the title of a topic (or both). We have not restricted ourselves to the dead, and would be pleased to hear from any (of the living) who feel that our entry is a misrepresentation of their career. There will be surprise omissions as well as surprise inclusions, and we will be pleased to receive nominations for future inclusion. Certainly, no omission should be regarded as a comment on the achievements of that individual.

The dictionary concludes with a glossary, tables, a brief overview of the history of statistics and suggestions for further reading. We hope these will prove useful.

We have tried to be consistent in our presentation and accurate in what we present. However, we have feet of clay and we therefore welcome any correspondence that may improve future editions.

GU/ITC

COLCHESTER
DECEMBER 2001

Abbreviations in Biographies

Abbreviations that appear as dictionary entries are excluded from this list.

AAAS	American Academy of Arts and Science
AAS	Australian Academy of Sciences
AMS	American Mathematical Society
ANU	Australian National University
ASQ	American Society for Quality
Caltech	California Institute of Technology
CSIRO	Commonwealth Scientific and Industrial Research Organization
ETH	Eidgenössische Technische Hochschule
FRS	Fellow of the Royal Society
FRSC	Fellow of the Royal Society of Canada
FRSE	Fellow of the Royal Society of Edinburgh
FRSNZ	Fellow of the Royal Society of New Zealand
FSU	Florida State University
GWU	George Washington University
IC	Imperial College, London
IoS	Institute of Statisticians
IMA	Institute of Mathematics and its Applications
ISU	Iowa State University
KCL	King's College, London
LMS	London Mathematical Society
LSE	London School of Economics and Political Science
LSHTM	London School of Hygiene and Tropical Medicine
MIT	Massachusetts Institute of Technology
NAE	National Academy of Engineering
NAS	National Academy of Sciences
NYU	New York University
ORS	Operational Research Society
OSU	Ohio State University
QMUL	Queen Mary, University of London
SUNY	State University of New York
U	'University' or 'the University of'
UCB	University of California at Berkeley
UCL	University College, London
UCLA	University of California at Los Angeles
UCSD	University of California at San Diego
UNC	University of North Carolina at Chapel Hill

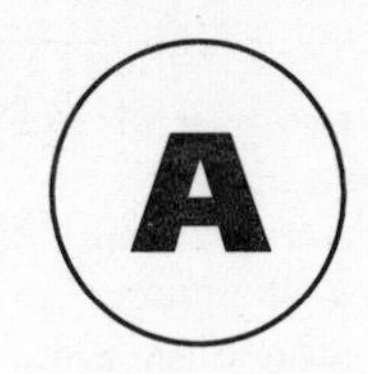

Aalen model A *linear regression model describing the manner in which the *expected value of a response variable, $Z(t)$, depends on explanatory variables $X_1(t), \ldots, X_k(t)$ with time-dependent coefficients $\beta_0(t), \beta_1(t), \ldots, \beta_k(t)$:

$$\mathrm{E}[Z(t)] = \beta_0(t) + \beta_1(t)X_1(t) + \cdots + \beta_k(t)X_k(t).$$

Abbe, Ernst Carl (1840–1905; b. Eisenach, Germany; d. Jena, Germany) German mathematician and physicist. Abbe's father was a book printer and factory worker and his childhood was one of privation. Nevertheless, he studied at U Jena and U Göttingen, receiving his PhD in 1861. In 1863 he was appointed to a lectureship at Jena on the basis of a dissertation that, in effect, derived the *chi-squared distribution. Following an approach from Carl Zeiss, most of his subsequent work was concerned with optics and astronomy. A lunar crater is named after him, also a minor planet, and several schools in Germany.

SEE WEB LINKS

- Fuller biography and photograph.

abscissa *See* CARTESIAN COORDINATES.

absolute difference The *absolute value of the difference between two numbers. *See also* MEAN ABSOLUTE DEVIATION.

absolute error The *absolute value of the difference between (commonly) an *observation and the value predicted by, or estimated from, some *model.

absolute value (modulus) The value of a number disregarding its sign. Denoted by a pair of '|' signs: thus the modulus of -2.5 is $|-2.5| = 2.5$.

absorbing barrier (absorbing state) *See* MARKOV PROCESS.

acceptable quality level *See* ACCEPTANCE SAMPLING.

acceptable risk In the context of a medical treatment this describes a situation in which the expected benefits outweigh the potential hazards of the treatment.

a

acceptance region The set of values of the *statistic, in a *hypothesis test, which lead to acceptance of the *null hypothesis.

acceptance–rejection algorithm A method for generating values of a *continuous random variable for use in a *simulation. Suppose that the random variable X, which takes values in the interval (a, b), has *probability density function f. Denote the maximum value of $f(x)$ by M. Let u and v be two random numbers *uniformly distributed in the interval (0, 1). Write $r = a + (b - a)u$ and $s = Mv$, so that r and s are uniformly distributed on (a, b) and $(0, M)$, respectively. Calculate $f(r)$. If $f(r) > s$ then r is accepted as a value of X. Otherwise, it is rejected and a new pair of values is taken for u and v.

SEE WEB LINKS

- R code and demonstration.

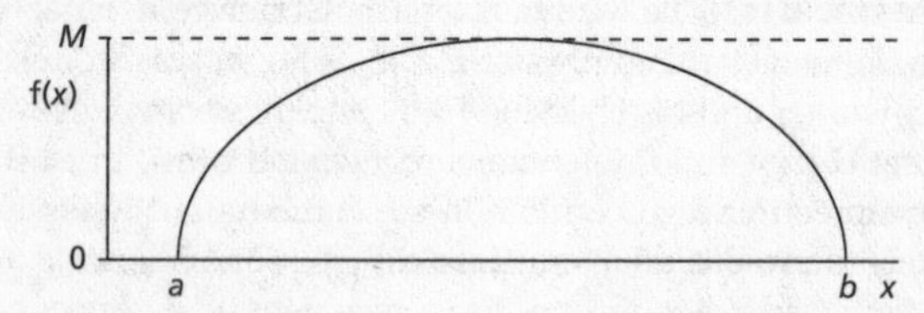

Acceptance–rejection algorithm. Uniform random numbers are generated in the intervals (a, b) and $(0, M)$. If the point generated lies between the graph of $f(x)$ and the x-axis, then the value of X is accepted.

acceptance sampling A method of *quality control. A random *sample is taken from a *batch of output and the decision to accept or reject the batch is based on either the number of *defectives in the sample (**inspection by attributes**) or on some summary *statistic such as the *sample mean (**inspection by variables**).

In the case of inspection by attributes, the *probability of accepting a batch is a function of the proportion, p, of defectives in the batch. Any acceptance sampling scheme that does not sample 100% of a batch will lead to the occasional rejection of batches with very low proportions of defectives (the **producer's risk**), and to the occasional acceptance of batches with very high proportions of defectives (the **consumer's risk**).

As p increases, so the probability that a batch will be rejected increases. The **lot tolerance percent defective** (**LTPD**) is the value of p (expressed as a percentage) that the sampling scheme would expect to reject on a given proportion (usually 90%) of occasions.

The maximum proportion of defectives that is regarded as desirable by the consumer is called the **acceptable quality level** (**AQL**). The graph relating the probability of acceptance to p is called the **operating characteristic curve (OC-curve)**. The average quality level of the items in the batches released after inspection is the **average outgoing quality (AOQ)**. The AOQ is usually calculated under the assumption that defective items found during testing will be replaced before the batch is released. The AOQ has minima at $p=0$ and $p=1$ and its maximum is termed the **average outgoing quality limit (AOQL)**.

Suppose that the proportion of defectives remains constant from batch to batch. Eventually a batch will be rejected. The **average run length (ARL)** is the average number of batches inspected up to and including the one that is rejected. *See* QUALITY CONTROL.

accessible *See* MARKOV PROCESS.

ACF *See* AUTOCORRELATION.

acquiescence bias The tendency of an interviewee to agree with the questioner. For example, if the question is 'Did you vote in the last election?' then the number replying 'Yes' is likely to provide an overestimate of the proportion who voted. On the other hand, if the question had been 'Did you abstain in the last election?', then this would also be likely to provide an overestimate.

action line *See* QUALITY CONTROL.

actuarial statistics The branch of Statistics concerned with insurance and loss, including reinsurance, *ruin theory, and *run-off triangles.

AdaBoost *See* BOOSTING.

adaptive sampling A method of *sequential sampling in which the later sampling procedure is affected by the earlier results. Examples include the comparison of rival treatments in *clinical trials and the use of *kriging in the exploration for oil.

addition law for probabilities (law of total probability) Law stating that if two events (*see* SAMPLE SPACE) A and B are *mutually exclusive then

$$\mathrm{P}(A \cup B) = \mathrm{P}(A) + \mathrm{P}(B).$$

For example, the *probability that, when a normal six-sided die is rolled, it shows a multiple of 3 is

$$\frac{1}{3} = \frac{1}{6} + \frac{1}{6} = \mathrm{P}(\text{shows } 3) + \mathrm{P}(\text{shows } 6).$$

The generalization to a list $A_1, A_2, \ldots$ of mutually exclusive events is the **law of total probability**:

a

$$P(A_1 \cup A_2 \cup \cdots) = \Sigma P(A_j) = P(A_1) + P(A_2) + \cdots.$$

An equivalent form is: If an event A is the union of mutually exclusive events $A_1, A_2, \ldots,$ then

$$P(A) = \Sigma P(A_j) = P(A_1) + P(A_2) + \cdots.$$

Another form of the law is: If the events $B_1, B_2, \ldots$ are mutually exclusive and *exhaustive, then

$$P(A) = \Sigma P(A \cap B_j).$$

This follows from the previous form by taking $A_j = A \cap B_j$, for all j, and noting that, in this case, $A = A_1 \cup A_2 \cup \cdots$, and that the events $A \cap B_1$, $A \cap B_2, \ldots$ are mutually exclusive.
See also UNION, INTERSECTION.

additive model A *model in which the combined effect of the explanatory variables (and their *interaction) is equal to the sum of their separate effects.

ADF test *See* DICKEY–FULLER TEST.

adjacent; adjacency matrix *See* GRAPH.

adjusted R^2 *See* ANOVA.

admissibility A term (introduced by *Wald in 1939) used in *statistical inference in several contexts. A procedure is admissible if there is no alternative procedure that performs at least as well under all circumstances and performs better under some circumstances.

age-specific failure rate *Alternative name for* HAZARD RATE.

age-specific rate When a rate, such as a *birth rate, *incidence rate, or *mortality rate is calculated for individuals of a specified age (or age range) then the rate is described as being age-specific.

agglomerative clustering methods Methods for grouping *multivariate data into clusters (*see* CLUSTER ANALYSIS). Suppose there are n data items. The agglomerative clustering methods start by regarding these as n separate clusters of size 1. The two clusters judged closest together (on some criterion) are then merged to reduce the number of clusters to $(n-1)$. This procedure could be continued until all the items would be collected into a single cluster.

The three simplest criteria are as follows. In **single linkage clustering** the distance between two clusters is defined as the least distance between an item in one cluster and an item in the other cluster. In **complete linkage clustering**, by contrast, the distance between two clusters is defined as the greatest distance between an item in one cluster and an

a

item in the other cluster. As a compromise, **group-average clustering** uses the average of the distances between every member of one cluster and every member of the other cluster. The process of agglomeration is often represented using a *dendrogram. *See also* DISTANCE MEASURE; WARD'S METHOD.

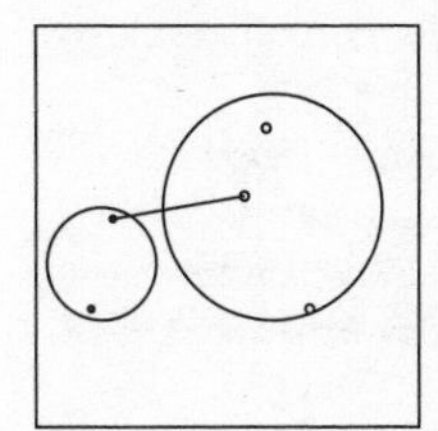
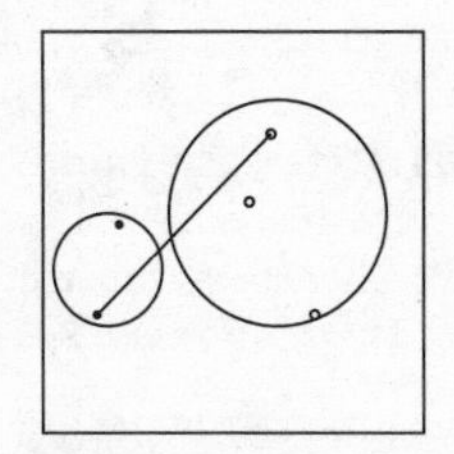
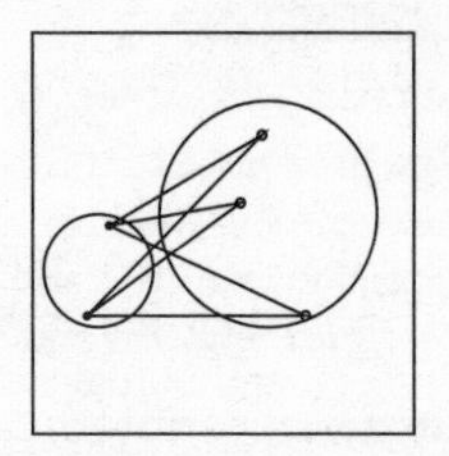

Single linkage distance Complete linkage distance Group distances to be averaged

Agglomerative clustering methods. Examples of the distance definitions used in clustering.

AH *Abbreviation for* ALTERNATIVE HYPOTHESIS.

AI *Abbreviation for* ARTIFICIAL INTELLIGENCE.

AIC (Akaike's information criterion) *See* MODEL SELECTION PROCEDURE.

AID *Abbreviation for* AUTOMATIC INTERACTION DETECTOR.

Ajne test *See* CIRCULAR UNIFORM DISTRIBUTION.

Akaike, Hirotugu (1927–2009; b. Fujinomiya, Japan; d. Tsukuba, Japan) Japanese statistician and mathematician. On graduating from U Tokyo in 1952, Akaike joined the staff of the *Institute of Statistical Mathematics, becoming its Director General between 1986 and 1994. He was awarded the Kyoto Prize (Japan's highest award for global achievement) in 2006.

SEE WEB LINKS

- Memorial website.

Akaike's information criterion *See* MODEL SELECTION PROCEDURE.

aleatory variable *Alternative term for* RANDOM VARIABLE.

algorithm A procedure consisting of a series of steps, often repetitive, for solving a problem.

alias *See* FACTORIAL DESIGN; FACTORIAL EXPERIMENT.

Allen, Sir Roy George Douglas (1906–83; b. Stoke-on-Trent, England; d. London, England) English economist and statistician. He

a

studied mathematics at Cambridge U, graduating in 1927. In 1928 he joined the faculty of the LSE where he spent his entire career and was appointed Professor of Statistics in 1944. He was President of the *Econometric Society in 1951 and of the *RSS in 1969, being awarded the latter's *Guy Medal in Gold in 1978. He was knighted in 1966.

SEE WEB LINKS

- Biography and photograph.

Allgemeines Statistisches Archiv The journal of the *Deutsche Statistische Gesellschaft. First published in 1890, it was the journal of the Society between 1911 and 2006, when it was superseded by **AStA Advances in Statistical Analysis* and **AStA Wirtschafts- und Sozialstatistisches Archiv.*

allocation problem *See* ASSIGNMENT PROBLEM.

allometry The study of the interdependence of size and shape in living organisms.

Almon model *See* DISTRIBUTED LAGS MODEL.

alpha (α) The *probability, in a *hypothesis test, of rejecting the *null hypothesis when it is, in fact, true. Usually called the **significance level**.

alternating renewal process A *renewal process in which the process alternates between states A and B. Let the average length of a period in state A be μ_A and the average length of a period in state B be μ_B. The long-run proportion of the time that the system is in state A is

$$\frac{\mu_A}{\mu_A + \mu_B}.$$

An alternating renewal process is a special case of a *semi-Markov process. When the states are 'working' and 'under repair', the *probability that the system is working at time t is called the **availability**.

alternative hypothesis The hypothesis, in a *hypothesis test, that will be accepted if the *null hypothesis is rejected. The term was introduced by *Neyman and Egon *Pearson in 1933.

American Society for Quality (ASQ) An organization founded in 1946 with the aim of furthering quality in all its aspects. Since 1959 it has published the journal **Technometrics* in collaboration with the *American Statistical Association. It awards the *Shewhart Medal annually.

SEE WEB LINKS

- Society home page.

American Statistical Association (ASA) A scientific and educational society founded in 1839 in Boston, Massachusetts. It is the second oldest professional society in the USA. Its current headquarters are in Alexandria, Virginia. The Association has nearly 18000 members. It publishes seventeen journals, including the **Journal of the American Statistical Association* and the **American Statistician*. Since 1965 the *Wilks Medal has been presented annually for distinguished contributions to Statistics.

SEE WEB LINKS

- Association home page.

American Statistician A quarterly journal published by the *American Statistical Association, concentrating on statistical methodology. It was first published in 1947.

SEE WEB LINKS

- Journal home page.

analysis of covariance *See* ANOVA.

analysis of variance *See* ANOVA.

ancillary statistic In the context of *estimation of an unknown *parameter, a statistic whose value provides information incidental to the estimation process. The most usual case is where the *sample size is not fixed. For example, we might wish to know the proportion of a sweet pea mixture that has red flowers. We plant N seeds, but our estimate will be based on the ancillary statistic n, the number of seeds that germinate and produce flowers. The term was introduced by Sir Ronald *Fisher in 1925.

ANCOVA *See* ANOVA.

Anderson, Theodore Wilbur (1918– ; b. Minneapolis, MN) American mathematical statistician who specialized in the analysis of *multivariate data. A graduate of Northwestern U, Anderson (supervised by *Wilks) gained his PhD at Princeton U in 1945. In 1946 he joined the staff at Columbia U, moving in 1967 to Stanford U. He was Editor of the **Annals of Mathematical Statistics* 1950–2, and President of the *IMS in 1963. He was the IMS's *Wald Lecturer in 1982 and the *COPSS *Fisher Lecturer in 1985. He was the *ASA's *Wilks Award winner in 1988. Anderson was still producing single-authored papers in 2012 at the age of 93.

SEE WEB LINKS

- Fuller biography, interview and photographs.

Anderson–Darling test A general test, published in 1952, that compares the fit of the observed *cumulative distribution function with that expected. It was derived by *Anderson and David A. Darling as a modification of the *Cramér–von Mises test. The test statistic A^2 is given by

a

$$A^2 = -\frac{1}{n}\sum_{j=1}^{n}(2j-1)[\ln\{F(x_{(j)})\} + \ln\{1-F(x_{(j)})\}] - n,$$

where F is the hypothesized cumulative distribution function, n is the *sample size, and $x_{(j)}$ is the jth ordered *observation ($x_{(1)} \le x_{(2)} \le \cdots \le x_{(n)}$). The statistic can also be used to test for *normal distribution and *exponential distribution with unknown *parameters estimated by their sample equivalents. In some cases, as shown in the following table, an adjusted test statistic is required.

	TEST STATISTIC	UPPER TAIL PROBABILITY			
		0.10	0.05	0.025	0.01
Specified distribution	A^2	1.933	2.492	3.070	3.857
Normal, estimated mean ($n > 20$)	A^2	0.894	1.087	1.285	1.551
Normal, estimated variance ($n > 20$)	A^2	1.743	2.308	2.898	3.702
Normal, estimated mean and variance	$A^2(1+\frac{3}{4n}+\frac{9}{4n^2})$	0.631	0.752	0.873	1.035
Exponential, estimated mean	$A^2(1+\frac{3}{10n})$	1.062	1.321	1.591	1.959

Andrews plot A plot suggested in 1972 by the American David Andrews as an alternative method to *Chernoff faces for representing *multivariate data in two dimensions. The plot can help to identify *outliers and to establish similarities within groups of data items. For an item with values given by the row *vector ($a\ b\ c\ d\ e \ldots$), the function $x(t)$ is defined as

$$x(t) = \frac{a}{\sqrt{2}} + b\sin(t) + c\cos(t) + d\sin(2t) + e\cos(2t) + \cdots.$$

The resulting graph is drawn for ($-\pi \le t \le \pi$), for each data item. The form of the graph is dependent upon the ordering of the characteristics $a, b, \ldots$; the usual advice is to order the characteristics in declining order of their (supposed) importance.

angular histogram *Alternative name for* circular histogram (*see* CYCLIC DATA).

angular uniform distribution *Alternative name for* CIRCULAR UNIFORM DISTRIBUTION.

anisotropic; anisotropy *See* ISOTROPY.

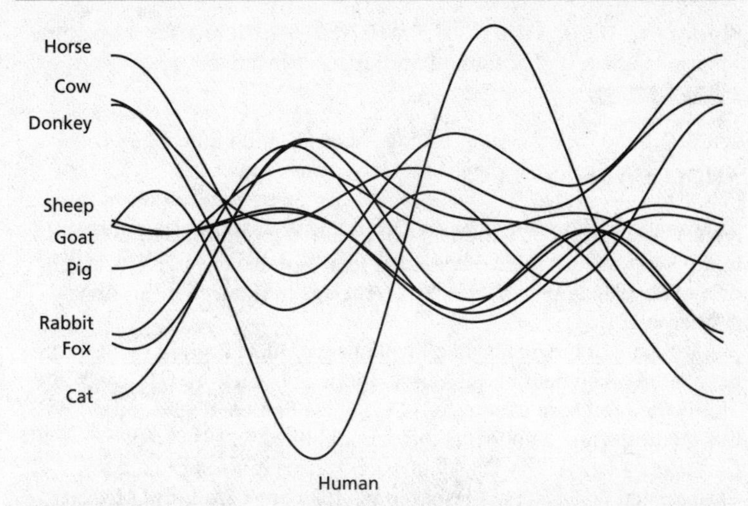

Andrews plot. This plot compares humans with nine familiar animal species, using five characteristics (body weight, brain weight, hours of sleep, lifespan, and gestation). It appears that humankind is a race apart and that it is difficult to tell the sheep from the goats.

Annals of Applied Probability A bi-monthly publication of the *Institute of Mathematical Statistics. It was first published in 1973.

SEE WEB LINKS

- Journal home page.

Annals of Applied Statistics A quarterly publication of the *Institute of Mathematical Statistics. It was first published in 2007.

SEE WEB LINKS

- Journal home page.

Annals of Mathematical Statistics The single journal of the *Institute of Mathematical Statistics from 1930 to 1973, before its subdivision into three parts.

Annals of Probability A bi-monthly publication of the *Institute of Mathematical Statistics. It was first published in 1973.

SEE WEB LINKS

- Journal home page.

Annals of Statistics A bi-monthly publication of the *Institute of Mathematical Statistics. It was first published in 1973.

SEE WEB LINKS

- Journal home page.

Annals of the Institute of Statistical Mathematics A quarterly English language publication of this Japanese institute.

SEE WEB LINKS

- Journal home page.

ANOCOVA *See* ANOVA.

ANOVA (analysis of variance) Abbreviation for **analysis of variance**, it refers to the attribution of variation in a *variable to variations in one or more explanatory variables. The term was introduced by Sir Ronald *Fisher in 1918.

A measure of the total variability in a set of *data is given by the sum of squared differences of the *observations from their overall *mean. This is the **total sum of squares** (*TSS*). It is often possible to subdivide this quantity into components that are identified with different causes of variation. The full subdivision is usually set out in a table, as suggested by Sir Ronald Fisher in his 1925 book *Statistical Methods for Research Workers*. Each row of the table is concerned with one or more of the components of the observed variation. The columns of the table usually include the **sum of squares** (*SS*), the corresponding number of *degrees of freedom (v), and their ratio, the **mean square** ($= SS/v$).

After the contributions of all the specified sources of variation have been determined, the remainder, often called the **residual sum of squares** (*RSS*) or **error sum of squares,** is attributed to *random variation. The mean square corresponding to *RSS* is often used as the yardstick for assessing the importance of the specified sources of variation. One method involves comparing ratios of mean squares with the *critical values of an *F-distribution.

As an example, suppose that four varieties of tomatoes are grown in three grow-bags giving the yields (in g) shown below. The explanatory variables are the grow-bags and the varieties.

T_1	T_3	T_4	T_2
1890	1740	1620	1970

Bag 1

T_2	T_1	T_3	T_4
1850	1760	1800	1890

Bag 2

T_3	T_1	T_2	T_4
1810	1910	2100	1710

Bag 3

The following ANOVA table results:

SOURCE OF VARIATION	DEGREES OF FREEDOM	SUM OF SQUARES	MEAN SQUARE
Differences between grow-bags	2	12 950	6 475
Differences between varieties	3	93 425	31 142
Residual	6	72 250	12 042
Total	11	178 625	

Since the mean square for varieties is much greater than that for grow-bags we can conclude that differences between varieties are more important. However, the residual sum of squares amounts to nearly half the total sum of squares, indicating that there are major unexplained sources of variation.

The proportion of variation explained by the model is R^2 given by

$$R^2 = 1 - \frac{RSS}{TSS},$$

which is sometimes called the **squared multiple correlation** or the **coefficient of determination**. In the example $R^2 = 0.60$. The quantity $1 - R^2$ is the **coefficient of alienation**. In the case of a linear relation between two variables, R^2 is the square of the sample *correlation coefficient.

A quantity often calculated is the **adjusted R^2** in which, in the previous formula, RSS is replaced by $RSS/(n-p-1)$ and TSS is replaced by $TSS/(n-1)$, where n is the number of observations and p is the number of *parameters in the model. In the example, $n=12$, $p=5$ and the adjusted R^2 is 0.26.

In 1989, Sir David *Cox and E. Joyce Snell introduced a more general definition based on the *likelihood. For a sample of size n, let $\mathrm{L}(\hat{\boldsymbol{\beta}})$ denote the maximum of the likelihood based on some proposed *model involving the *parameters $\boldsymbol{\beta}$, and let $\mathrm{L}(\mathbf{0})$ denote the likelihood when all parameters are set to zero. The **Cox–Snell R^2** is

$$1 - \left(\frac{\mathrm{L}(\mathbf{0})}{\mathrm{L}(\hat{\boldsymbol{\beta}})}\right)^{\frac{2}{n}}.$$

A preferable statistic, in the sense that all values between 0 and 1 can be achieved, is obtained by dividing the Cox–Snell R^2 by its maximum value $(1 - \mathrm{L}(\mathbf{0})^{\frac{2}{n}})$. This variation was suggested in 1991 for a discrete model by the Dutch medical statistician Nico J. D. Nagelkerke and is now described as **Nagelkerke's R^2**.

The proportion of the total sum of squares that is attributed to any particular source of variation is referred to as **eta-squared** (η^2). Thus, the difference between grow-bags has $\eta^2 = \frac{12\,950}{178\,625} = 0.072$. A related statistic is **partial eta-squared**, which is the ratio of the sum of squares of interest to the total of that sum of squares and the residual sum of squares. Thus, for grow-bags, the partial eta-squared is $\frac{12\,950}{12\,950+72\,250} = 0.152$.

Often, particular comparisons between the treatments are of interest. These are referred to as **contrasts**. The set of contrasts consisting of, say, (i) a comparison of treatment 1 with the average effect of treatments 2 to t, (ii) a comparison of treatment 2 with the average effect of treatments 3 to t, etc. are called **Helmert contrasts**. Both these contrasts and those corresponding to *orthogonal polynomials lead to a diagonal *variance–covariance matrix.

In an ANOVA analysis each explanatory variable takes one of a small number of values. If, instead, some explanatory variables are continuous in nature (*see* CONTINUOUS VARIABLE), then the resulting models are called **ANOCOVA** models (also **ANCOVA** or **analysis of covariance** models). ANOVA can also be thought of as *multiple regression using only *dummy variables. *See also* EXPERIMENTAL DESIGN.

Ansari–Bradley test *See* TEST FOR EQUALITY OF SCALE.

Anscombe, Francis John (1918–2001; b. Hove, England; d. New Haven, CT) English statistician. Graduating from Cambridge U in 1939, Anscombe worked during the Second World War for the Ministry of Supply. His tasks included a mathematical solution for firing rockets during D-Day. In 1948 he joined the faculty at Cambridge U, where, on behalf of the Fitzwilliam Museum, he purchased a work by the then unknown painter Francis Bacon. When the Museum decided it was too modern, he retained it and subsequently sold it to pay for his four children's education. He was keen that his PhD students should be involved with real problems. Both *Deming and *Tukey were fond of quoting his maxim that it is better to 'realize what the problem really is, and solve that problem as well as we can, instead of inventing a substitute problem that can be solved exactly, but is irrelevant'. In 1956 he was recruited to a chair at Princeton U by Tukey who 'wanted someone to talk to, not at'. In 1963 he moved to Yale U, where he founded its Department of Statistics. He was the *COPSS *Fisher Lecturer in 1982.

SEE WEB LINKS

- Obituary and photograph.

Anscombe residual An alternative to the usual residual (*see* REGRESSION DIAGNOSTICS) which *Anscombe proposed in 1953 for cases where the *random errors in a general *regression model do not have a *normal distribution. The idea is to produce quantities that do have near-normal distributions, and the form for the residual depends upon the error distribution assumed.

In the case of a *Poisson distribution, an Anscombe residual is given by

$$r = \frac{3(y^{\frac{2}{3}} - \hat{y}^{\frac{2}{3}})}{2\hat{y}^{\frac{1}{6}}},$$

where y and $\hat{y}$ are, respectively, observed and *fitted values. In the case of a *gamma distribution the formula becomes

$$r = \frac{3(y^{\frac{1}{3}} - \hat{y}^{\frac{1}{3}})}{\hat{y}^{\frac{1}{3}}},$$

and in the case of an *inverse normal distribution the formula is

$$r = \frac{\ln y - \ln \hat{y}}{\sqrt{\hat{y}}}.$$

Anscombe's regression data Artificial *data created by *Anscombe to illustrate (*see diagram overleaf*) the necessity for studying residuals. Each of the four data sets has the same fitted *regression line, $y = 3 + 0.5x$, and the same summary *ANOVA table (with the same regression sum of squares, total sum of squares, and value for R^2).

x	y	x	y	x	y	x	y
10	8.04	10	9.14	10	7.46	8	6.58
8	6.95	8	8.14	8	6.77	8	5.76
13	7.58	13	8.74	13	12.74	8	7.71
9	8.81	9	8.77	9	7.11	8	8.84
11	8.33	11	9.26	11	7.81	8	8.47
14	9.96	14	8.10	14	8.84	8	7.04
6	7.24	6	6.13	6	6.08	8	5.25
4	4.26	4	3.10	4	5.39	19	12.50
12	10.84	12	9.13	12	8.15	8	5.56
7	4.82	7	7.26	7	6.42	8	7.91
5	5.68	5	4.74	5	5.73	8	6.89

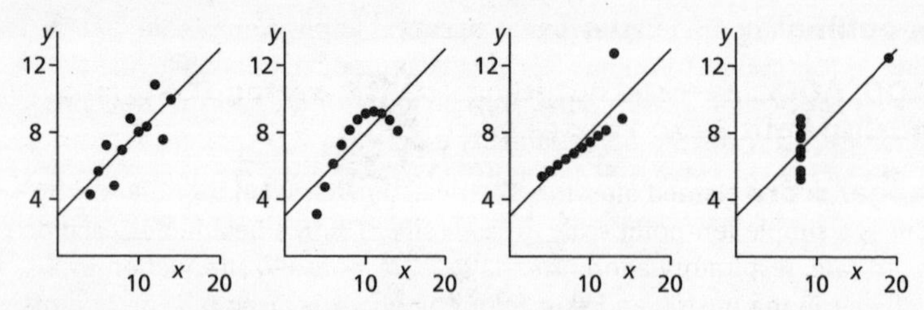

Anscombe's regression data. Each data set has the same mean and variance for *x*, the same mean and variance for *y*, the same fitted regression line, and the same residual sum of squares—Anscombe created these data sets to emphasize the need for the statistician to look carefully at data.

ant colony optimization An *optimization procedure that seeks to mimic an ant's apparent ability to find the shortest distance between two points. The ant's choices are based on the quantities of pheromones left by previous ants. These build up faster on shorter routes. The computer version similarly leaves markers behind to guide subsequent choices.

antecedent variable An explanatory variable, referring to an earlier time point, used as part of an explanation of the variation in a **consequent** (dependent) **variable**. *See also* REGRESSION.

antedependence The dependence of the value of a quantity at time t on its value at one or more previous times. *See also* AUTOCORRELATION.

antithesis (plural: antitheses) *See* ANTITHETIC VARIABLE.

antithetic variable A variable used in a *variance reduction technique for *simulation. It refers to the use of simulated values that are not *independent but are negatively correlated.

Suppose, for example, that we wish to estimate the expected value of the *random variable X, and that we simulate a value x using the function g of the *pseudo-random numbers $u_1, u_2, \ldots, u_n$ ($0 \le u_j \le 1$, for all j):

$$x = \mathrm{g}(u_1, u_2, \ldots, u_n).$$

The antithetic variable procedure makes use of the **antitheses** of the random numbers, namely $(1-u_1), (1-u_2), \ldots, (1-u_n)$, to form x' given by

$$x' = \mathrm{g}(1-u_1, 1-u_2, \ldots, 1-u_n).$$

Write X' as the corresponding random variable. It can be shown that the *variance of $\frac{1}{2}(X+X')$ is never greater than half the variance of X. There is therefore a gain in precision as well as an increase in efficiency, since each pseudo-random number is used twice.

A-optimality *See* EXPERIMENTAL DESIGN.

AOQ; AOQL; average outgoing quality; average outgoing quality limit *See* ACCEPTANCE SAMPLING.

Apgar score Named after the American paediatrician Virginia Apgar, this is a simple ten-point scale for assessing a baby's health. For each of heart rate, respiratory effort, muscle tone, response to stimulation by a catheter in the nostril, and skin colour, the baby is given 0, 1, or 2 points. An overall score of ten suggests a bonny bouncing baby.

Applied Statistics *See* *JOURNAL OF THE ROYAL STATISTICAL SOCIETY*.

AQL *See* ACCEPTANCE SAMPLING.

Aranda-Ordaz, Francisco Javier (1951–91; b. México City; d. México City) Mexican statistician. Aranda-Ordaz obtained his BS and MS degrees at Universidad Nacional Autónoma de México (UNAM) before going to IC where the supervisor of his PhD (awarded in 1981) was Sir David *Cox. He returned to Mexico, joining the faculty at UNAM, where he was Professor of Statistics at the time of his early death.

Aranda-Ordaz transformations Transformations, for a *proportion, p, suggested in 1981 by *Aranda-Ordaz. The transformations have the form ($\alpha > 0$)

$$y = \ln\left[\frac{1}{\alpha}\{(1-p)^{-\alpha} - 1\}\right].$$

The case $\alpha = 1$ corresponds to the *logistic transformation.

arbitrage A sure-win betting scheme. If there are h horses in a race, with the odds quoted for horse j being o_j to 1 against that horse winning, then an arbitrage is possible if $\delta > 0$, where

$$\delta = 1 - \sum_{j=1}^{h} \frac{1}{1+o_j}.$$

If such a set of odds exists, then it is referred to as a **Dutch book**. In this case a win of N (ignoring betting costs) is guaranteed by backing each horse, with x_j, the bet on horse j, being given by

$$x_j = \frac{N}{\delta(1+o_j)}.$$

As a simple example, suppose that there are two horses in a race, each with odds of two to one against. A punter placing 1 cent on each (total cost

2 cents) would receive back 2 cents, together with the winning stake—for an assured gain of 1 cent.

Arbuthnot, John (1667–1735; b. Inverbervie, Scotland; d. London, England) Scottish mathematician and royal physician. In 1692 the first English work on *probability was Arbuthnot's translation of *De Ratiociniis in Ludo Aleae* (*Calculation in Games of Chance*) by the Dutchman Christiaan *Huygens. After graduating in medicine at St Andrews U in 1696, Arbuthnot moved to London. In 1700 his *Essay on the Usefulness of Mathematical Learning* was published. In 1704 he was elected FRS. In the following year he was appointed physician to Queen Anne. His 1710 paper on the imbalance between male and female births may be regarded as the first application of probability to *social statistics.

SEE WEB LINKS

- Fuller biography and portrait.

arc *See* GRAPH; NETWORK.

archetypal analysis A procedure for expressing *multivariate data as a *linear combination of a few representative members (**archetypes**) of that data set.

ARCH model Abbreviation for AutoRegressive Conditional Heteroskedasticity model. This type of *model is used in finance to model changes over time in asset prices which are typically *heteroskedastic.

arc-sine law *See* RANDOM WALK.

arc-sine transformation For a *random variable *X*, having a *binomial distribution with *parameters *n* and *p*, a transformation that *stabilizes the *variance and produces a variable with an approximate *normal distribution whose variance is almost independent of *p* is

$$\sin^{-1}\left(\sqrt{\frac{X}{n}}\right).$$

An improvement suggested by *Anscombe in 1948 is

$$\sin^{-1}\left\{\sqrt{\left(X+\frac{3}{8}\right)\Big/\left(n+\frac{3}{4}\right)}\right\}.$$

Both transformed variables have *expected value and variance given approximately by $\sin^{-1}(\sqrt{p})$ and $1/(4n)$, respectively. *See also* FREEMAN–TUKEY TRANSFORMATION.

arc-sinh transformation *See* NEGATIVE BINOMIAL DISTRIBUTION.

ARE *Abbreviation for* ASYMPTOTIC RELATIVE EFFICIENCY.

areal coordinates Relative to a given triangle of reference $A_1 A_2 A_3$ the areal coordinates of a point P inside, or on the boundary of, the triangle are (x_1, x_2, x_3), where x_1, x_2, x_3 are the areas of triangles PA_2A_3, PA_3A_1, PA_1A_2, respectively. In the usual case, where the triangle of reference is

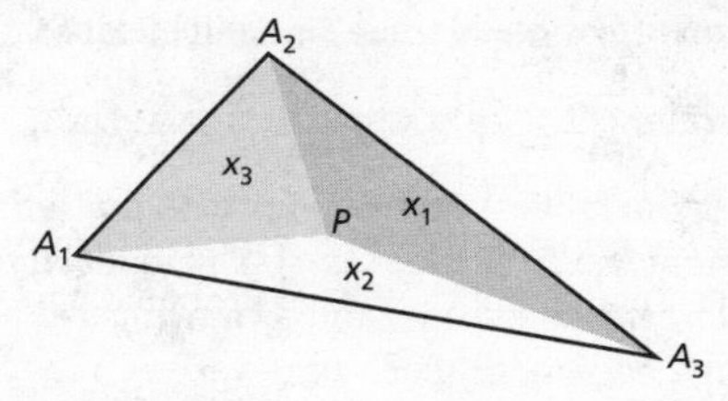

Areal coordinates. The location of a point is specified by the areas of the component regions.

equilateral, the coordinates are called *barycentric coordinates. The idea can be generalized to a tetrahedron of reference, using volumes, and to a *simplex of reference.

area sampling Sampling with primary sampling units (*see* CLUSTER SAMPLING) that are non-overlapping regions of the earth's surface. Examples of such regions are parishes, counties, fields, political constituencies. Often the sampling procedure involves the use of aerial photographs.

Arfwedson distribution A *distribution, presented in 1951 by the Swede, Gerhard Arfwedson, concerned with the case where each *observation takes one of k equally likely values. Let M be the *random variable denoting the number of values that do not occur in a *sample of size n. The distribution of M is given by

$$\mathrm{P}(M=m)=\binom{k}{m}\sum_{j=0}^{k-m}(-1)^j\binom{k-m}{j}\left(\frac{k-m-j}{k}\right)^n,\quad m=0,1,\ldots,(k-1).$$

The distribution has also been called the **coupon-collecting distribution**, since one application is to find the probability that a person having n randomly selected coupons (which might be cigarette cards, plastic toys from cereal packets, etc.) will have at least one of each of the k equally likely varieties.

ARIMA model A *model for a *time series which resembles an *ARMA model except that it is presumed that the time series has a steady underlying trend (*see* MOVING AVERAGE). The model therefore works with the differences between the successive observed values, instead of the values themselves. To retrieve the original data from the differences requires a form of integration and the model is therefore called an **autoregressive integrated moving average** model. A model incorporating additional space-time variation is called a **STARIMA model**.

arithmetic mean *Alternative name for* SAMPLE MEAN.

ARL (average run length) *See* ACCEPTANCE SAMPLING.

ARMA model A *model for a *time series with no trend (*see* MOVING AVERAGE; the constant *mean is taken as 0). The model incorporates terms from both an *autoregressive model and a *moving average model. A model incorporating additional space-time variation is called a **STARMA model**. For an equivalent model for time series with a trend, *see* ARIMA MODEL.

Armitage, Peter (1924– ; b. Huddersfield, England) English medical statistician. Armitage began studying mathematics at Cambridge U in 1941, but, in 1943, he was recruited into the Ministry of Supply, encountering practical statistics whilst working alongside *Barnard. In 1946 he completed his studies at Cambridge, joining the faculty at LSHTM in 1947. In 1976 he succeeded *Bartlett as Professor of Biomathematics at Oxford U. He was President of the *IBS in 1972, editing the Society's journal **Biometrics* from 1980 to 1984, and becoming an Honorary Life Member of the Society in 1998. He was President of the *RSS in 1982, and is one of only three individuals (the others being *Plackett and *Durbin) to have a complete set of *Guy Medals (Bronze in 1962, Silver in 1978, Gold in 1990). He was President of the *ISCB in 1990.

SEE WEB LINKS

- Biography and photograph.

Armitage–Doll model A model proposed by *Armitage and *Doll to describe the process of a malignant cancer. They proposed that the disease could be regarded as having n stages, with the time between moving from one stage to the next having an exponential distribution. This model is a pure birth process (*see* BIRTH-AND-DEATH PROCESS) with death as an absorbing state (*see* MARKOV PROCESS).

AR model *Alternative name for* AUTOREGRESSIVE MODEL.

artificial intelligence Research in which the aim is to construct an intelligent computing machine, including both hardware and software, that can tackle problems that usually require human intelligence. Such

a

problems include *expert systems, the playing of games, and language translation. A major success is the chess-playing machine Deep Blue, although this machine does not emulate human analysis of chess, but uses vast computing power at each stage to consider millions of possible subsequent positions. In 1950, *Turing proposed a test for intelligence in a computing machine: an observer, posing questions via a keyboard and obtaining answers on a monitor, must be convinced that a human, rather than a machine, is responding to the questions.

artificial neural network (artificial neural net) A computing system built from a large number of simple processing elements dealing individually with parts of a large problem. The net may have several layers of processing elements and the processing elements are massively interconnected. Adaptively adjusted weights are applied to the inputs and to the connections between the processing elements. There are many applications, including, for example, speech and pattern recognition, oil and gas exploration, financial forecasting, and health care cost reduction.

ASA *Abbreviation for* AMERICAN STATISTICAL ASSOCIATION.

ASN (average sample number) *See* SEQUENTIAL SAMPLING.

ASQ *Abbreviation for* AMERICAN SOCIETY FOR QUALITY.

assignment problem (allocation problem) A *linear programming problem in which the aim is to allocate individuals to tasks in a manner that optimizes their overall effectiveness.

association Two *variables are associated if they are not independent, i.e. if the value of one variable affects the value, or the *distribution of the values, of the other. Thus, for a human population, height and weight are associated, and so are actual skin-colour and ethnicity. In the case of numerical variables an appropriate measure is the *correlation coefficient. In the case of *ordinal variables and *categorical variables, an alternative **measure of association** is required.

*Yule used the term 'association' in his 1900 paper that proposed a measure suitable for the case of two variables each having two categories.

For two categorical variables (A and B having, respectively, J and K categories) a measure with a probabilistic interpretation is **Goodman and Kruskal's lambda** (λ) suggested by *Goodman and *Kruskal in 1954. Suppose that we are asked to guess the category of B for the next *observation. An intelligent guess would be the category that was the commonest so far (with *marginal total equal to f_{0m}, say). Judging by the past data, the *probability that our guess will be correct will be P, given by $P=f_{0m}/f_{00}$, where $f_{00}=\Sigma_j \Sigma_k f_{jk}$, $f_{j0}=\Sigma_k f_{jk}$, $f_{0k}=\Sigma_j f_{jk}$ and f_{jk} is the frequency of the (j, k) category combination. Suppose now that the next observation belongs to category j of variable A. Our best guess for the category of B now corresponds to the maximum of $f_{j1}, f_{j2}, \ldots, f_{jK}$. The revised probability of our being correct will be estimated by

$$\frac{\max_k(f_{jk})}{f_{j0}}.$$

Since the estimated probability of the next observation belonging to category j of A is estimated as f_{j0}/f_{00}, the estimated probability of being correct taking into account the category of A for the next observation is P_A, given by

$$P_A = \frac{\sum_{j=1}^{J} \max_k(f_{jk})}{f_{00}}.$$

The statistic λ, given by

$$\lambda = \frac{P_A - P}{1 - P},$$

is a measure of the *proportional reduction in error resulting from knowledge about A.

With *ordinal variables there are more appropriate measures that take account of the ordering. Let j and j' be two categories of one ordinal variable, A, and let k and k' be two categories of a second ordinal variable, B. The quantities S, D and T_B are defined in terms of pairs of observations, one belonging to cell (j, k) and the other to cell (j', k'):

$S=$ the total number of pairs for which, when $j>j'$, $k>k'$;
$D=$ the total number of pairs for which, when $j>j'$, $k<k'$;
$T_B=$ the total number of pairs for which, when $j>j'$, $k=k'$.

These quantities are calculated using every pair of observations. Two measures based on these statistics are **Goodman and Kruskal's gamma** (γ), proposed by Goodman and Kruskal in 1954, which is given by

$$\gamma = \frac{S - D}{S + D},$$

and **Somers's d_{BA}**, proposed by the sociologist Robert H. Somers in 1962, which is the preferred measure when B is believed to depend on A. The formula is

$$d_{BA} = \frac{S - D}{S + D + T_B}.$$

See also TWO-BY-TWO TABLE.

assumed mean *Alternative description of* WORKING MEAN.

AStA Advances in Statistical Analysis A quarterly publication of the *Deutsche Statistische Gesellschaft concentrating on statistical methods and applications and review articles. It was first published in 2007 when it superseded **Allgemeines Statistisches Archiv.*

AStA Wirtschafts- und Sozialstatistisches Archiv A quarterly publication of the *Deutsche Statistische Gesellschaft concentrating on economic and social statistics. It was first published in 2007 when it superseded **Allgemeines Statistisches Archiv.*

asymmetric distribution *See* SYMMETRIC DISTRIBUTION.

asymmetric matrix *See* MATRIX.

asymptote If part of a graph is unbounded and there is a fixed straight line l such that the distance from a point P on the graph to l tends to 0 as $OP \to \infty$, where O is the origin, then l is an asymptote to the curve. Alternatively, it is the limiting position of the tangent to the graph at P, as $OP \to \infty$. For example, the asymptotes of the graph of

$$y = 2x + 3 - \frac{1}{x-2}$$

are $y = 2x + 3$ and $x = 2$.

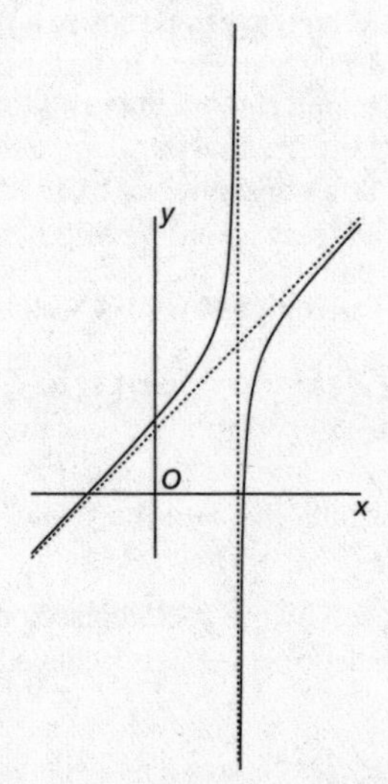

Asymptote. The dotted lines are the asymptotes (vertical, $x = 2$; oblique, $y = 2x + 3$) of the graph $y = 2x + 3 - 1/x - 2$.

asymptotically efficient estimator An *estimator whose efficiency tends to unity as the *sample size increases.

asymptotically unbiased estimator *See* ESTIMATOR.

asymptotic distribution The limiting *distribution of some *random variable as (usually) the sample size is increased. For example, under the conditions of the *central limit theorem, the asymptotic distribution of the *sample mean is a *normal distribution.

asymptotic normality The *distribution of a *statistic is said to be asymptotically normal if, as the *sample size increases, the distribution of the statistic approaches a *normal distribution.

asymptotic relative efficiency The limiting ratio of the *variances of two alternative *estimators of a *parameter as the *sample size increases.

attenuation The *sample correlation coefficient between two *variables that are subject to measurement error will be reduced by such errors—this is referred to as attenuation.

attribute A basic property of an *experimental unit whose value cannot be changed by the experimenter. For example, the experimenter can vary the amount of fertilizer in an experiment, but cannot change the composition of the soil in a field. *See also* ACCEPTANCE SAMPLING.

attrition A problem that affects most *longitudinal studies. For example, in a *panel study, some of the original participants will not be participating at the end of the study—as a consequence of death, emigration, or failure to provide a forwarding address when moving house.

augmented Dickey–Fuller test (ADF) *See* DICKEY-FULLER TEST.

Australian and New Zealand Journal of Statistics A quarterly journal, first published in 1998, formed by merging the *Australian Journal of Statistics* and the *New Zealand Statistician*. It is jointly published by the *Statistical Society of Australia Inc. and the *New Zealand Statistical Association.

SEE WEB LINKS

- Journal home page.

Australian Journal of Statistics The original journal published by the *Statistical Society of Australia Inc. and superseded in 1998 by the **Australian and New Zealand Journal of Statistics*.

autocorrelation (serial correlation) A measure of the linear relationship between two separate instances of the same *random variable, as distinct from the *population correlation coefficient, ρ, which refers to the linear relationship between two different random variables. As with ρ, the possible values lie between −1 and 1 inclusive, with unrelated instances having a theoretical autocorrelation of 0.

In the case of a *time series, autocorrelation measures the extent of the linear relation between values at time points that are a fixed interval (the **lag**) apart. Similarly, **spatial autocorrelation** quantifies the linear relationship between values at points in space that are a fixed distance apart (in any direction in the case of an isotropic process; *see* ISOTROPY). It is usually found that spatial autocorrelation is near 1 for points close together and decays to 0 as the distance increases—thus the daily rainfalls at the Lords and Oval cricket grounds in London will resemble each other closely, but will bear little or no resemblance to the rainfalls at the Kensington Oval in the West Indies. The phrase 'serial correlation' was introduced by *Yule in 1926, while 'autocorrelation' was first used in 1933 as a description of a (related) function used by *Wiener in 1926.

For a random variable X at time (or location) t, the *population **autocorrelation function (ACF)** for lag l, ρ_l, is given by

$$\rho_l = \frac{\mathrm{Cov}(X_t, X_{t+l})}{\mathrm{Var}(X_t)},$$

the **autocovariance function** for lag l divided by the *variance of X_t (which, for a *stationary process, is equal to that of X_{t+l}). At low lags autocorrelation is usually positive. It usually declines towards 0 as the lag increases. *See also* PARTIAL AUTOCORRELATION FUNCTION.

The **sample autocorrelation** for lag l, r_l, is given (for $l = 1, 2, \ldots, n-1$) for the sequence of n values $x_1, x_2, \ldots, x_n$ (ordered in space or time) by

$$r_l = \frac{n\sum_{t=1}^{n-l}(x_t - \bar{x})(x_{t+l} - \bar{x})}{(n-l)\sum_{t=1}^{n}(x_t - \bar{x})^2},$$

where $\bar{x}$ is the sample mean.

A plot of the variance of $1/\{2(X_t - X_{t+l})\}$ against l is called a **variogram** (or **semi-variogram**). The related plot of autocorrelation versus lag is called a **correlogram**, and a plot of the autocovariance against lag may be called a **covariogram**. *See also* SERIAL CORRELATION; PERIODOGRAM.

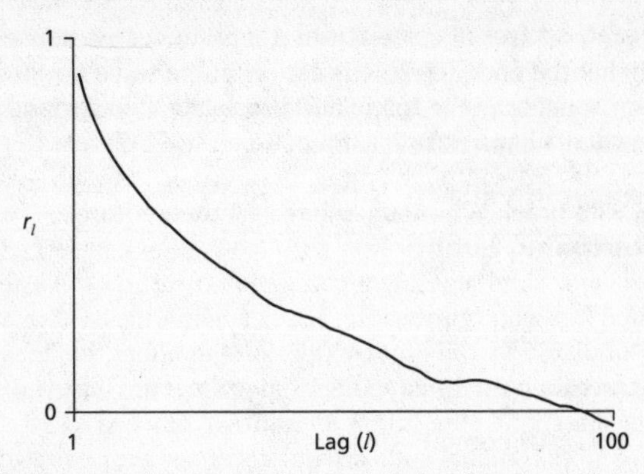

Autocorrelation. A correlogram showing a typical plot: the autocorrelation generally reduces as the lag increases.

autocovariance function *See* AUTOCORRELATION.

automatic interaction detector (AID) A forerunner of *CHAID as a method for finding relationships in *multivariate data.

autoregressive integrated moving average *See* ARIMA MODEL.

autoregressive model (AR model; auto-regressive process) A *model for a *time series having no trend (*see* MOVING AVERAGE; the constant *mean is taken as 0). Let $X_1, X_2, \ldots,$ be successive instances of the *random variable X, measured at regular intervals of time. Let ε_j be the random variable denoting the *random error at time j. A pth-order model relates the value at time j to the preceding p values by

$$X_j = \alpha_1 X_{j-1} + \alpha_2 X_{j-2} + \cdots + \alpha_p X_{j-p} + \varepsilon_j,$$

where $\alpha_1, \alpha_2, \ldots, \alpha_p$ are constants. Such a model is written in brief as AR(p). The AR(1) process is a Markov chain (*see* MARKOV PROCESS). Autoregressive models can also be expressed as *moving average models. Models combining both type of process include *ARMA models and *ARIMA models.

The **Yule–Walker equations**, introduced by *Yule in 1927 and Walker in 1931, relate $\alpha_1, \alpha_2, \ldots, \alpha_p$ to the population *autocorrelation values $\rho_1, \rho_2, \ldots, \rho_p$ by the p equations:

$$\rho_k = \alpha_1 \rho_{1-k} + \alpha_2 \rho_{2-k} + \cdots + \alpha_p \rho_{p-k}, k = 1, 2, \ldots$$

with $\rho_{-a} = \rho_a$ for all a, and with $\rho_0 = 1$.

availability *See* ALTERNATING RENEWAL PROCESS.

average For a set of data, a loosely used term for a *measure of location—either the *mode, or, in the case of numerical data, the *median or the *mean. Its meaning is often restricted to this last case. *See also* MOVING AVERAGE; WEIGHTED AVERAGE.

average outgoing quality (AOQ) *See* ACCEPTANCE SAMPLING.

average outgoing quality limit (AOQL) *See* ACCEPTANCE SAMPLING.

average run length (ARL) *See* ACCEPTANCE SAMPLING.

average sample number (ASN) *See* SEQUENTIAL SAMPLING.

axes *See* CARTESIAN COORDINATES.

axial data *See* CYCLIC DATA.

axial distribution A *distribution of an axis through a point. In two dimensions it is a distribution on a semicircle, while in three dimensions it is a distribution over a hemisphere. It is not possible to determine whether $\leftrightarrow$ points to the right or the left, but the inclination of its axis is apparent.

Babbage, Charles (1792–1871; b. London, England; d. London, England) English mathematician and inventor. He studied mathematics at Cambridge U where he was a co-founder of the 'Analytical Society' which advanced the cause of what is now the standard notation for differentiation. After graduation in 1814, he was elected FRS in 1816 and FRSE in 1820 (the year in which he was a co-founder of what is now the Royal Astronomical Society). He is known as the 'Father of Computing', having formulated the idea of a mechanical calculator during his student days. A first model was demonstrated in 1822, at which time he stated 'I wish to God these calculations had been executed by steam'. A working model of his second machine is at the Science Museum in London. At the time of his death, however, the notion was almost forgotten. From 1828 to 1839 he was Lucasian Professor of Mathematics at Cambridge U. The initial decision to found the *RSS was taken at Babbage's house in 1834. A crater on the Moon is named after him.

SEE WEB LINKS
- Fuller biography and portrait.

BACI design An *experimental design used in ecological and environmental studies. The design was introduced in the 1970s and makes use of *repeated measures. *Observations are taken at several time points Before some environmental change (e.g. the diversion of a river) and at several time points After the change. At each time point, observations are taken at two locations: one certainly unaffected by the change is termed the **Control location** and the other, potentially affected, is called the **Impact location**.

background variable An explanatory variable that can affect other (dependent) variables but cannot be affected by them. For example, one's schooling may affect one's subsequent career, but the reverse is unlikely to be true. *See also* REGRESSION.

back-to-back stem and leaf plot *See* STEM AND LEAF DIAGRAM.

backward elimination *See* STEPWISE PROCEDURE.

bagging *Bootstrap samples are taken from the original *regression or classification data, with the same *model selection procedure or *classification tree procedure being used on each sample, and the results are aggregated to produce predictions or classifications that have improved precision. *See also* BOOSTING; MACHINE LEARNING.

Bahadur, Raghu Raj (1924–1997; b. New Delhi, India; d. Chicago, IL) Indian mathematical statistician. A graduate of U Delhi, Bahadur gained his PhD from UNC in 1950. His career was split between the *Indian Statistical Institute and U Chicago. He was the 1974 *Wald Lecturer of the *IMS.

SEE WEB LINKS

- Curriculum vitae.

Bahadur efficiency A measure of the efficiency of one *estimator relative to another that is based on the *distribution functions of the estimators.

Balaam's design *See* CROSSOVER TRIAL.

balanced design An *experimental design in which the same number of *observations is taken for each experimental condition.

balanced incomplete block design An *experimental design in which t *treatments are compared in b *blocks of size k, where $k < t$, and b and k are chosen so that bk is a multiple of t. The balanced nature of the design is reflected in the fact that each treatment appears the same number (r) of times, and every pair of treatments appear together in a block on the same number (λ) of occasions. To satisfy these requirements,

$$bk = rt \text{ and } \lambda(t-1) = r(k-1).$$

An example with four treatments (A–D) and blocks of size 3 is shown below.

B	A	C

D	A	B

D	A	C

C	B	D

Banach, Stefan (1892–1945; b. Kraków, Poland; d. Lviv, Ukraine) Polish mathematician. As a baby he was given away by his parents and brought up by a laundress. At the age of fifteen he was earning a living by coaching in mathematics. In 1919 he was appointed as a lecturer at U Lviv, Ukraine, and was awarded his doctorate despite having no previous degree. He was imprisoned in Lviv during the Second World War and died shortly afterwards from his consequent ill-health.

SEE WEB LINKS

- Fuller biography and photograph.

Banach's matchbox problem A classic problem in *probability. A pipe smoker has two full boxes of n matches. When he wants a match, he is equally likely to select either box. The question concerns the number of matches left in the second box on the first occasion that he chooses a box and finds it empty. The probability that there are exactly k ($k = 0, 1, 2, \ldots, n$) matches left in the second box is

$$\binom{2n-k}{n}\left(\frac{1}{2}\right)^{2n-k}.$$

bandit problems Problems concerned with the determination of an optimal strategy. The bandit referred to is the 'one-armed bandit' otherwise known as a 'fruit machine'. For an actual machine in an amusement arcade the general advice would be not to play it, since it is the machine owner who will benefit in the long run. However, the term 'one-armed bandit' in statistics refers to the problem of deciding whether to 'play' when the expected pay-off may not be negative. Statisticians also consider k-armed bandits for which the question is 'Which of the k arms should be played?' One application of the resulting theory is to the medical problem of deciding which of a number of possible treatments should be given to a patient—here the pay-off is measured in terms of the patient's future health.

An optimal strategy is based on the **Gittins index**, which is defined as the maximum value, over all N, of the quantity

$$\frac{\sum_{t=1}^{N}\beta^{t-1}\mathrm{E}\{X(t)\}}{\sum_{t=1}^{N}\beta^{t-1}},$$

where $\mathrm{E}\{X(t)\}$ is the *expected value of the payout at the tth play of the bandit and β is the discount rate.

bandwidth *See* KERNEL METHOD.

bar chart; bar diagram; bar graph A diagram (*opposite*) for showing the *frequencies of a *variable that is *categorical or *discrete. The lengths of the bars are proportional to the frequencies. The widths of the bars should be equal. If the widths of the bars are negligible then the diagram may be called a **line graph**. Diagrams resembling bar charts were used in a theoretical context in the 14th century. *See also* COMPOUND BAR CHART; MULTIPLE BAR CHART.

Barnard, George Alfred (1915–2002; b. Walthamstow, England; d. Brightlingsea, England) English logician, with a special interest in *statistical inference, who was one of the first to advocate *Monte Carlo

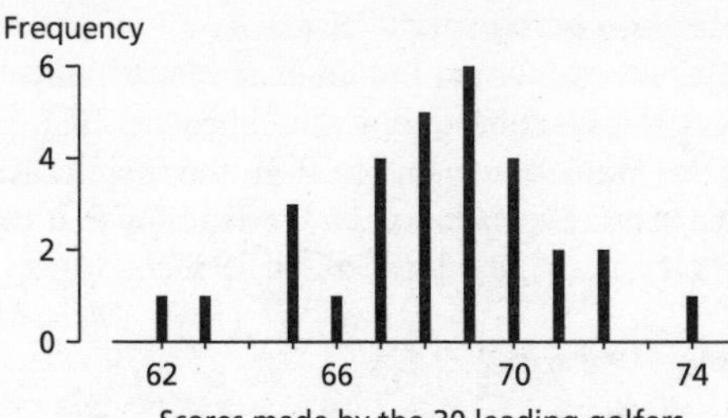

Scores made by the 30 leading golfers
in the final round of the 1992 Scottish Open

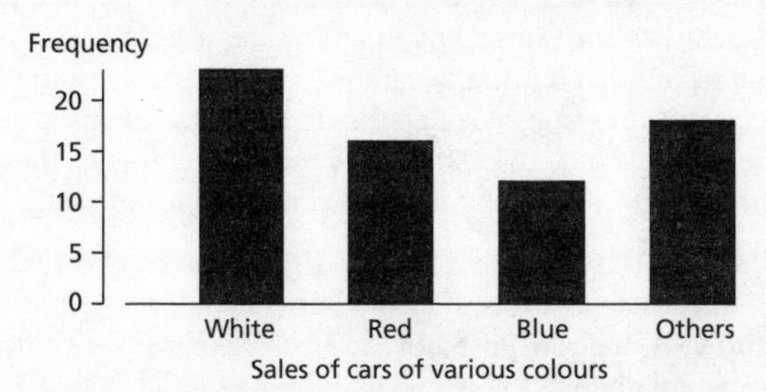

Sales of cars of various colours

Bar chart. The top diagram is a bar chart for a discrete variable (note that 0 need not appear); the bottom diagram illustrates data for a categorical variable. The choice of width for the bars is arbitrary, but the width should not be the same as that of the gap between successive bars.

---

tests and the *likelihood principle. Barnard, who graduated from Cambridge U in 1937, developed an interest in Statistics after being recruited to the Ministry of Supply in 1942. There he supervised the work of *Armitage, *Lindley, and *Plackett. In 1945 he joined the faculty at IC, moving in 1966 to U Essex as the founding Professor of Statistics. He was Chairman of the IoS in 1961, President of the ORS in 1963, President of the IMA in 1970, and President of the *RSS in 1972. He was the *COPSS *Fisher Lecturer in 1976. He received the *Guy Medal of the RSS in Silver in 1958 and the Guy Medal in Gold in 1975. He was made an Honorary Fellow of the RSS in 1993.

SEE WEB LINKS

- Fuller biography, interview, and photographs.

Bartlett, Maurice Stephenson (1910–2002; b. London, England; d. Exmouth, England) English statistician. Bartlett obtained his doctorate from Cambridge U in 1932 (supervised by *Wishart). After a year at UCL, he worked for ICI before joining the faculty at Cambridge U. During the Second World War he was seconded to the Ministry of Supply where he worked alongside *Anscombe, David *Kendall (his former research student), and *Moran. Appointed Professor of Mathematical Statistics at Manchester U in 1947, he subsequently held posts at UCL (1960) and Oxford U (1967). He was elected FRS in 1961 and was the *COPSS *Fisher Lecturer in 1964. He was President of the *RSS in 1966 and was awarded its *Guy Medal in Silver in 1952 and in Gold in 1969.

SEE WEB LINKS

- Fuller biography,interview, and photographs.

Bartlett–Lewis model A model used for describing clusters of events in time or space. The idea is that clusters of events are initiated at random points in space or time (according to a *Poisson process). Each of these events then results in a number of sub-events that are distributed in space or time according to some specified distribution. Initially used as a model to describe the incidence of failures in a production process, these models are now popular for describing the spatial occurrence of rainfall.

Bartlett's identity A *matrix identity (introduced by *Bartlett in 1951) used in *multivariate analysis. If **M** is a non-singular $n \times n$ matrix, **v** is an $n \times 1$ vector, $\mathbf{v}'$ is the transpose of **v**, and k is a constant such that $1 + k\mathbf{v}'\mathbf{M}^{-1}\mathbf{v} \neq 0$, then

$$(\mathbf{M} + k\mathbf{v}\mathbf{v}')^{-1} = \mathbf{M}^{-1} - \frac{k\mathbf{M}^{-1}\mathbf{v}\mathbf{v}'\mathbf{M}^{-1}}{1 + k\mathbf{v}'\mathbf{M}^{-1}\mathbf{v}}.$$

Bartlett's test for eigenvalues A *test used in *principal components analysis. The test is of the *null hypothesis that the smallest m eigenvalues of a $n \times n$ *variance-covariance matrix are equal to 0. Writing $k = n - m + 1$, the test statistic is

$$X^2 = \nu\left\{m \ln\left(\frac{1}{m}\sum_{j=k}^{n} \lambda_j\right) - \sum_{j=k}^{n} \ln(\lambda_j)\right\},$$

where ν is the number of *degrees of freedom associated with the variance-covariance matrix, and $\lambda_1, \lambda_2, \ldots, \lambda_n$ are the eigenvalues. Under the null hypothesis, X^2 has an approximate *chi-squared distribution with $\frac{1}{2}(m-1)(m-2)$ degrees of freedom.

Bartlett test *See* TEST FOR EQUALITY OF VARIANCE.

Bartlett window *See* PERIODOGRAM.

Barton–David test *See* TEST FOR EQUALITY OF SCALE.

barycentric coordinates Coordinates describing the position of a point in a triangle or *simplex. The position of any point P inside, or on the boundary of, a given triangle of reference $A_1A_2A_3$ can be specified by finding masses m_1, m_2, m_3 such that P is the centre of mass of particles of masses m_1, m_2, m_3 placed at A_1, A_2, A_3 respectively. In this case P is said to have barycentric coordinates (m_1, m_2, m_3). For an internal point of the triangle the masses are all positive. If $k > 0$ then (km_1, km_2, km_3) represents the same point, so barycentric coordinates are usually chosen so that $m_1 + m_2 + m_3 = 1$. With this convention, the barycentric coordinates of A_1 are $(1, 0, 0)$, and those of the midpoint of A_1A_2 are $(\frac{1}{2}, \frac{1}{2}, 0)$. Suppose a random trial can result in one of three possibilities with probabilities p_1, p_2, p_3, such that $p_1 + p_2 + p_3 = 1$, then allocation of these probabilities can be represented by a point inside, or on the boundary of, the triangle. The triangular display is sometimes referred to as a **ternary diagram**. The idea of barycentric coordinates can be generalized to the case of a tetrahedron in three dimensions or a simplex. *See also* AREAL COORDINATES.

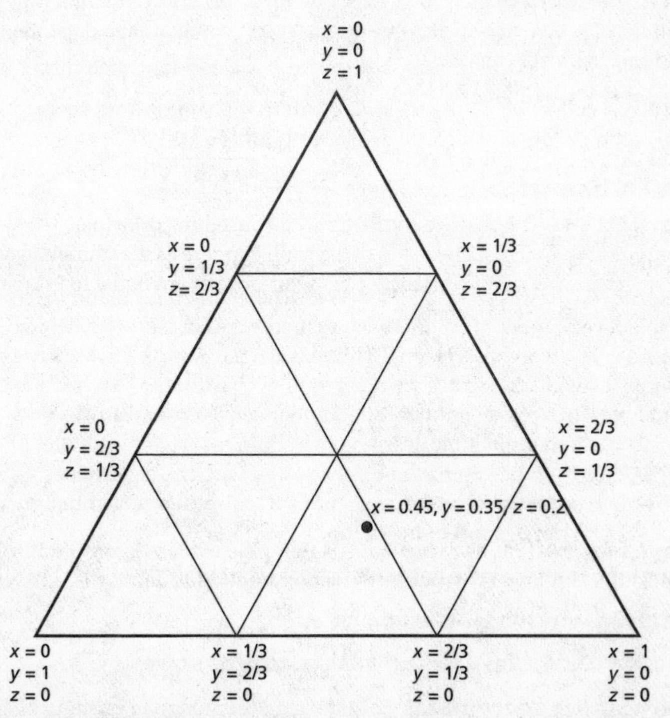

Barycentric coordinates. Illustrated for compositions of three classes having proportions *x*, *y*, and *z*. The case where the three classes are equally likely corresponds to the centroid of the equilateral triangle.

base period *See* INDEX NUMBER.

base-weighted index *See* INDEX NUMBER.

BASIC (Beginner's All-purpose Symbolic Instruction Code) An elementary computer programming language. There are many different 'dialects'.

SEE WEB LINKS
- The 1964 user's manual.

basic feasible solution *See* LINEAR PROGRAMMING.

Basu, Debabrata (1924–2001; b. Dhaka, Bangladesh; d. Calcutta, India) Bangladeshi statistician whose career was divided between India and the USA. Basu was a faculty member at FSU and the *Indian Statistical Institute where he obtained his PhD supervised by *Rao.

Basu theorems These theorems, due to *Basu, are concerned with the notions of sufficient statistics (*see* ESTIMATOR), *conditional independence, and *ancillary statistics.

batch A collection of items forming part of the output from some production process. An alternative term is **lot**.

bathtub curve *See* HAZARD RATE.

Bayes, Reverend Thomas (1701–61; b. London, England; d. Tunbridge Wells, England) Nonconformist minister in Tunbridge Wells, England. Bayes was elected FRS in 1742. The eponymous theorem has led to the development of an approach to Statistics that runs parallel to the methods of *hypothesis testing. This approach is referred to as *Bayesian inference and its advocates are referred to as *Bayesians. The theorem was contained in an essay not published until after Bayes's death and was largely ignored at the time.

SEE WEB LINKS
- Information and portrait.

Bayes factor (Bayes ratio) A measure of the evidence provided by the data, D, in favour of model M_1 as opposed to model M_2. The Bayes factor is B, given by

$$B = \frac{P(D \mid M_1)}{P(D \mid M_2)}.$$

The value of B may be assessed with the following table, derived from that suggested by *Jeffreys:

$2\ln(B)$	EVIDENCE IN FAVOUR OF M_1
<0	Negative
0–2.2	Not worth more than a bare mention
2.2–6	Positive
6–10	Strong
>10	Very strong

b

Bayesian A statistician who analyses data using the methods of *Bayesian inference. The term was used by Sir Ronald *Fisher in 1950.

Bayesian Analysis A quarterly journal published by the *International Society for Bayesian Analysis.

SEE WEB LINKS

- Journal home page.

Bayesian inference An approach concerned with the consequences of modifying our previous beliefs as a result of receiving new *data. By contrast with the 'classical' approach which begins with a *hypothesis test that proposes a specific value for an unknown *parameter, θ, Bayesian inference proposes a **prior distribution** (often simply called a prior), $p(\theta)$, for this parameter. Data $x_1, x_2, \ldots, x_n$ are collected and the *likelihood $f(x_1, x_2, \ldots, x_n|\theta)$ is calculated. *Bayes's theorem is now used to calculate the **posterior distribution**, $g(\theta|x_1, x_2, \ldots, x_n)$. The change from the prior to the posterior distribution reflects the information provided by the data about the parameter value. For any particular event the initial probability is described as the **prior probability** and the subsequent probability as the **posterior probability**.

If nothing is known about the value of a parameter, then a **non-informative prior** is used—typically, this is a *uniform distribution over the feasible set of values of the parameter. Another approach, the **empirical Bayes** method, utilizes the data to inform the prior distribution.

In a similar way, if nothing is known about the underlying distribution, then the **principle of indifference** effectively states that all possible values should be assigned the same probability of occurrence. This is also called the **principle of insufficient reason**.

Subjective probability measures the degree of belief an individual has in an uncertain proposition. This could form the basis for a prior distribution. Another term is **personal probability**, though this may be used to suggest that the person's selected probability is misguided.

Often, however, a more useful choice for the form of a prior distribution is a member of a family of distributions which is such that the posterior distribution is another member of that family, so that the effect of the data can be interpreted in terms of changes in parameter values. Such a prior is called a **conjugate prior**.

Sometimes useful information is available. For example, an appropriate prior for the amount taken by a supermarket on a Saturday might be a *normal distribution centred on the amount taken the previous Saturday. This would be an **informative prior**.

*Jeffreys argued that an appropriate prior should be unaffected by the way a model is expressed: this leads to the **Jeffreys prior** which is proportional to $\sqrt{\mathrm{I}(\theta)}$, where $\mathrm{I}(\theta)$ is the *Fisher information. Since it is only the relative sizes of the prior values that matter, those values need not sum or integrate to 1. Such a prior is called an **improper prior**.

In the **Dempster–Shafer theory of evidence** (suggested by *Dempster in 1967 and later developed by his research student, Glenn Shafer) the Bayesian approach is developed to handle events with imprecisely known probabilities. The theory uses concepts termed 'belief' and 'plausibility' as lower and upper bounds on event probabilities.

Bayesian information criterion *See* MODEL SELECTION PROCEDURE.

Bayes net; Bayes network *See* GRAPHICAL MODEL.

Bayes ratio *See* BAYES FACTOR.

Bayes's theorem A simple form of Bayes's theorem is

$$\mathrm{P}(A \mid B) = \frac{\mathrm{P}(A)\mathrm{P}(B \mid A)}{\mathrm{P}(A)\mathrm{P}(B \mid A) + \mathrm{P}(A')\mathrm{P}(B \mid A')},$$

where A' denotes the complementary event (*see* SAMPLE SPACE) to the event A. For example, suppose a man has two coins in his pocket. One is *unbiased, whereas the other is double headed. He takes one coin at random from his pocket and tosses it. Given that the coin falls heads, the *probability that it is the double-headed coin is

$$\frac{\frac{1}{2} \times 1}{(\frac{1}{2} \times 1) + (\frac{1}{2} \times \frac{1}{2})} = \frac{2}{3}.$$

The general form is

$$\mathrm{P}(A_j \mid B) = \frac{\mathrm{P}(A_j)\mathrm{P}(B \mid A_j)}{\Sigma \mathrm{P}(A_k)\mathrm{P}(B \mid A_k)},$$

where the events $A_1, A_2, \ldots$ are *mutually exclusive and *exhaustive.

Behrens–Fisher problem A problem concerned with the comparison of the *means of two *populations having *normal distributions with different *variances. The problem was first discussed by B. V. Behrens in

1929. Although Behrens's method of solution was unclear, his conclusions were confirmed by Sir Ronald *Fisher in 1935.

The *null hypothesis is that the populations have the same mean, and the suggested solution is an application of the *Welch statistic. In this case, the test statistic t is given by

$$t = (\bar{x}_1 - \bar{x}_2) \Big/ \sqrt{\frac{s_1^2}{n_1} + \frac{s_2^2}{n_2}},$$

where $\bar{x}_1$ and $\bar{x}_2$ are the sample means, n_1 and n_2 are the sample sizes, and s_1^2 and s_2^2 are the sample variances (using the divisors $n_1 - 1$ and $n_2 - 1$). If the populations do have the same mean then t is an observation from an approximate *t-distribution with v *degrees of freedom, where v is taken as the nearest integer to

$$(a+b)^2 \Big/ \left(\frac{a^2}{n_1 - 1} + \frac{b^2}{n_2 - 1}\right),$$

where $a = s_1^2 / n_1$ and $b = s_2^2 / n_2$. If $n_1 \leq n_2$, then the above formula ensures that $(n_1 - 1) \leq v \leq (n_1 + n_2 - 2)$.

If the populations can be assumed to have the same variance then the problem is simple and the standard t-test (*see* HYPOTHESIS TEST) is appropriate.

Belgian Statistical Society A society, founded in 1937, to provide a focus for statisticians in Belgium. The Society has more than three hundred members and publishes a regular newsletter.

SEE WEB LINKS

- Society home page.

bell-curve A curve that resembles the axial cross-section of a bell. One example is the graph of the *probability density of a *normal distribution. It is often assumed that every bell-curve must arise from a normal distribution, but this is not the case. For example, all the *t-distributions (including the *Cauchy distribution) also give rise to bell-curves. The term can also be used to describe a *histogram that approximates a bell-curve.

Bell–Doksum tests Tests corresponding to *Kruskal–Wallis tests, in which the *ranks used by those tests are replaced by corresponding *normal scores.

Bellman, Richard Ernest (1920–1984; b. Brooklyn, NY; d. Los Angeles, CA) American applied mathematician. Bellman obtained his BA from Brooklyn College in 1941 and his MA in mathematics from U Wisconsin-Madison in 1943. In 1946 he obtained his PhD at U Wisconsin and joined their faculty in 1946. In 1952 he joined the Rand

Corporation, introducing the concept of *dynamic programming in the following year. He moved to U South Carolina in 1965. Elected a fellow of AAAS in 1975, he was awarded the IEEE Medal of Honor in 1979.

SEE WEB LINKS

- Fuller biography and photograph.

Bellman–Harris process A *branching process starting with a single individual. At the end of that individual's life it gives rise to n further individuals. All the individuals have the same lifetime distribution and n is an observation from a specified probability distribution. The question of interest is the number of individuals in existence at a given timepoint. The process was introduced by *Bellman and Harris in 1952 and has been used in biology to model the growth of populations and the spread of epidemics.

Benders's decomposition A technique, introduced by the Dutchman Jacques Benders, that is used in *linear programming. The variables are first divided into groups **x** and **y** and the problem is reformulated as one of maximizing (or minimizing) a semi-linear objective function of the form $\mathbf{c}'\mathbf{x} + \mathrm{g}(\mathbf{y})$, subject to constraints such as $\mathbf{Ax} + \mathbf{h}(\mathbf{y}) \leq \mathbf{b}$, where g is a known function, **h** is a known vector-valued function, **A** is a known *matrix, **b** and **c** are known *vectors, and **c**′ is the transpose of **c**.

Benford's Law The first digits of numbers in a collection of related multi-digit numbers (commonly cited examples are tables of logarithms, populations of towns, newspaper circulations) are not equally likely. Instead, the probability that the first digit is D is

$$\log_{10}\left(1 + \frac{1}{D}\right).$$

Thus, in such collections, approximately 30% of numbers begin with the digit 1. *See also* ZIPF'S LAW.

Benjamini-Hochberg test *See* FALSE DISCOVERY RATE.

Berger, James Orvis (1950–   ; b. Minneapolis, MN) American mathematical statistician specializing in *Bayesian inference. Berger studied at Cornell U, obtaining his PhD in 1974 and joining the faculty at Purdue U. He was Editor of the **Annals of Statistics* from 1994 to 2000 and President of the *IMS in 1995. In 1997 he moved to Duke U. He was the *COPSS President's Award winner in 1985 and its *Fisher Lecturer in 2001. He was the IMS *Wald Lecturer in 2007. He was elected to the NAS in 2003.

SEE WEB LINKS

- Fuller biography, interview, and photographs.

Berkson, Joseph (1899–1982; b. New York City; d. Rochester, MN) American statistician and physician. Berkson obtained his BS from City College (now City U) New York in 1920, his AM from Columbia U in 1922, and his MD from Johns Hopkins U in 1927. From 1932 to 1964 he worked at the Mayo Clinic (a centre for medical education and research) and at U Minnesota. Berkson coined the word '*logit'. He was elected to the NAS in 1979.

Berkson's paradox Two *independent events become conditionally dependent if it is given that at least one of them occurs or does not occur. This has been expressed as the statement 'If you don't die of one thing then you will die of another'.

Bernoulli The quarterly journal of the *Bernoulli Society for Mathematical Statistics and Probability. It was first published in 1995.

SEE WEB LINKS

- Journal home page.

Bernoulli, Daniel (1700–82; b. Gröningen, Netherlands; d. Basel, Switzerland) Swiss mathematical physicist. Daniel Bernoulli is best known to statisticians for his solution of the *St Petersburg paradox posed by his cousin Nicolaus *Bernoulli. Daniel, a famous prodigy, was the son of Johann Bernoulli and the nephew of Jacob *Bernoulli, both being Professors of Mathematics. He gained a succession of degrees from U Basel whilst still a teenager: BA in philosophy and logic in 1715, MA in 1716, Doctor of Medicine in 1720. In 1724, whilst practising medicine in Venice, he applied mathematics to the design of an hour-glass for use at sea. This won him a prize and a post as Professor of Mathematics at St Petersburg U, where he worked on a number of *probability problems. He returned to U Basel in 1734, initially as Professor of Botany, then as Professor of Physiology and finally, in 1750, as Professor of Physics.

SEE WEB LINKS

- Fuller biography and portrait.

Bernoulli, Jacob (Jacques) (1654–1705; b. Basel, Switzerland; d. Basel, Switzerland) Swiss mathematician. Jacob Bernoulli is best known to statisticians for his *Ars Conjectandi* (*The Art of Conjecture*), a treatise on *probability, published posthumously in 1713, in which he derived the form of the *binomial distribution. Jacob was the uncle of Nicolaus *Bernoulli. At U Basel he studied philosophy and theology according to his parents' wishes, whilst studying mathematics and astronomy for his own satisfaction. After graduation Jacob travelled around Europe studying with fellow mathematicians. In 1683 he returned to U Basel to teach mechanics.

He studied and published in many areas of mathematics, including, in 1689, a statement of the *law of large numbers.

SEE WEB LINKS

- Fuller biography and portrait.

Bernoulli, Nicolaus (1687–1759; b. Basel, Switzerland; d. Basel, Switzerland) Swiss mathematician. Nicolaus Bernoulli was a prolific correspondent and poser of the *St Petersburg Paradox, solved by his cousin Daniel *Bernoulli. He was a nephew of Jacob *Bernoulli, who supervised his Master's degree in mathematics at U Basel. He was awarded his PhD at Basel in 1709 for a study of the application of probability theory to legal problems. In 1716 he was appointed to Galileo's chair at U Padua. In 1722 he left Italy and returned to U Basel, as Professor of Logic and later Professor of Law.

Bernoulli distribution The *distribution of a *discrete random variable taking two values, usually 0 and 1. An experiment or trial that has exactly two possible results, often classified as 'success' or 'failure', is called a **Bernoulli trial**. If the *probability of a success is p and the number of successes in a single experiment is the *random variable X, then X is a **Bernoulli variable** (also called a **binary variable**) and is said to have a Bernoulli distribution with *parameter p. The *mean of the distribution is p and the *variance is $p(1-p)$. The *probability function is given by

$$\mathrm{P}(X=1)=p, \qquad \mathrm{P}(X=0)=1-p.$$

A binomial variable (*see* BINOMIAL DISTRIBUTION) with parameters n and p is the number of successes in n *independent Bernoulli trials and may be regarded as the sum of n independent *observations of a Bernoulli variable with parameter p. The phrase 'Bernoullian trial' was used in a 1937 book on probability.

Bernoulli Society for Mathematical Statistics and Probability (BSMSP) Society founded in 1975 as an autonomous section of the *International Statistical Institute. Its object is the advancement, through international contacts, of the sciences of *probability (including the theory of *stochastic processes) and mathematical statistics and of their applications. The Society has more than 1000 members and publishes four issues of its journal **Bernoulli* per year.

SEE WEB LINKS

- Society home page.

Bernoulli trial; Bernoulli variable *See* BERNOULLI DISTRIBUTION.

Bernstein, Sergi Natanovich (1880–1968; b. Odessa, Ukraine; d. Moscow, Russia) Russian mathematician and probabilist. Bernstein obtained a doctorate from the Sorbonne U in Paris in 1904 and another from Kharkov U in 1913. From 1908 to 1933 he taught at Kharkov and subsequently at the Mathematical Institute in Moscow.

SEE WEB LINKS

- Photograph.

Bernstein inequality A stronger version of the *Chebyshev inequality. Let $X_1, X_2, \ldots, X_n$ be *independent *random variables, such that, for $j = 1, 2, \ldots, n$, X_j has *expected value 0 and $|X_j| \le M$. *Bernstein showed, in a 1926 paper, that, for all positive ε,

$$P\left(\left|\sum_{j=1}^{n} X_j\right| > \varepsilon\right) \le 2\exp\left\{-\frac{1}{2}\varepsilon^2 \Big/ \left(\sum_{j=1}^{n}\sigma_j^2 + M\varepsilon\right)\right\},$$

where $\mathrm{Var}(X_j) = \sigma_j^2$. *See also* CHEBYSHEV INEQUALITY; HÖLDER INEQUALITY; KOLMOGOROV INEQUALITY; MARKOV INEQUALITY; MINKOWSKI INEQUALITY.

Berry–Esséen theorem A theorem published in 1941 that concerns the *distribution of the *mean, $\bar{X}$, of n *independent identically distributed *random variables, $X_1, X_2, \ldots, X_n$. The theorem requires that the *expected values of X_j, X_j^2, and $|X_j|^3$ are finite and equal to 0, σ^2, and τ, respectively for all j. Denoting the distribution function of the *standard normal distribution by Φ, the theorem states that there is a value c for which

$$\left|P\left(\bar{X} < \frac{a\sigma}{\sqrt{n}}\right) - \Phi(a)\right| \le \frac{c\tau}{\sigma^3\sqrt{n}},$$

for all a. The value of c is known to be less than 0.475.

Bessel correction *See* SAMPLE VARIANCE.

best linear unbiased estimator *See* BLUE.

beta (β) The *probability, in a *hypothesis test concerning the value of a *parameter, of accepting the *null hypothesis when the *alternative hypothesis is, in fact, true. Also referred to as the probability of a Type II error. The value of β depends upon the true parameter value. The probability of rejecting the null hypothesis when the alternative hypothesis is true is the power of the test. Its value is $1 - \beta$.

beta-binomial distribution (Polya distribution) A *compound distribution that results from allowing the success *probability in a sequence of *Bernoulli trials to have a *beta distribution. If the *parameters of that distribution are α and β, then the probability of obtaining r successes in n trials is given by the beta-binomial distribution as

$$\binom{n}{r}\frac{\mathrm{B}(\alpha+r,\beta+n-r)}{\mathrm{B}(\alpha,\beta)}, \quad r=0,1,\ldots,n,$$

b

where B is the *beta function.

The distribution has *mean np and *variance

$$np(1-p)\frac{(n+\alpha+\beta)}{(1+\alpha+\beta)},$$

where

$$p=\frac{\alpha}{\alpha+\beta}.$$

beta distribution A distribution often used as a prior distribution for a *proportion. The *probability density function, for a *random variable X having a beta distribution is

$$\mathrm{f}(x)=\frac{1}{\mathrm{B}(\alpha,\beta)}x^{\alpha-1}(1-x)^{\beta-1}, \qquad 0<x<1,$$

where α and β are positive *parameters, and B is the *beta function. The name appears in a 1911 publication by *Gini. The distribution has *mean

$$\frac{\alpha}{\alpha+\beta},$$

and *variance

$$\frac{\alpha\beta}{(\alpha+\beta)^2(\alpha+\beta+1)}.$$

If both $\alpha>1$ and $\beta>1$, the distribution has *mode at

$$\frac{(\alpha-1)}{(\alpha+\beta-2)}.$$

If both $\alpha<1$ and $\beta<1$ then the distribution is U-shaped, whereas, if just one of α and β is <1, then the distribution is J-shaped.

If $Y_1, Y_2, \ldots, Y_k$ are *independent *random variables, with Y_j having a *chi-squared distribution with ν_j *degrees of freedom, then the ratio

$$\sum_{j=1}^{k-1} Y_j \Big/ \sum_{j=1}^{k} Y_j$$

has a beta distribution with $\alpha=\frac{1}{2}\sum_{j=1}^{k-1}\nu_j$ and $\beta=\frac{1}{2}\nu_k$.

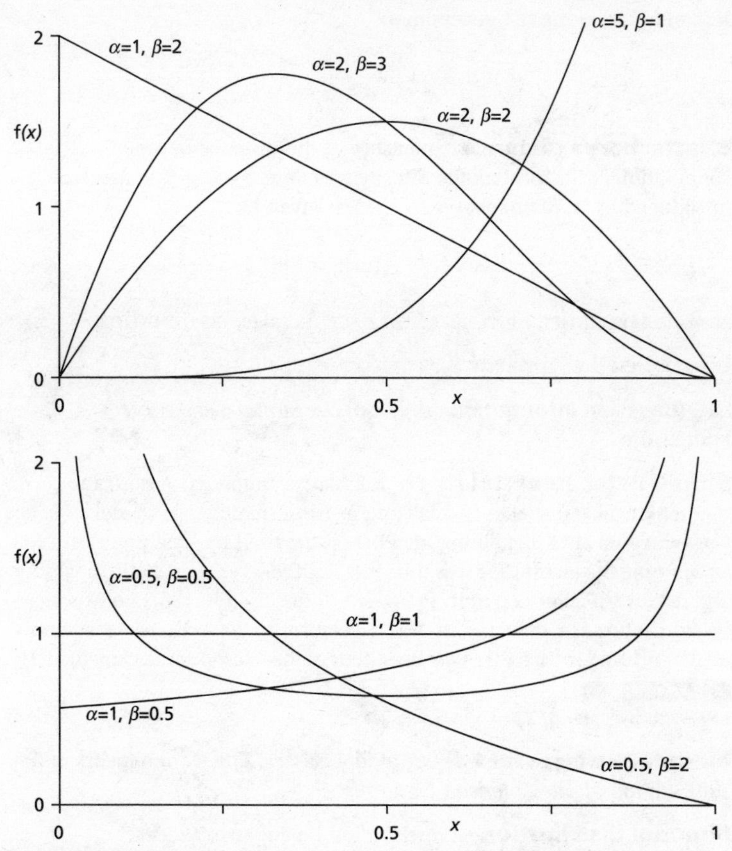

Beta distribution. The distribution has a variety of shapes, which depend on the values of α and β.

beta function The beta function B is given by

$$\mathrm{B}(a,b)=\int_0^1 t^{a-1}(1-t)^{b-1}\mathrm{d}t,$$

where $a>0$, $b>0$. The function is symmetric, i.e. $\mathrm{B}(a, b)=\mathrm{B}(b, a)$, for all a, b.

The beta function is related to the *gamma function Γ by

$$\mathrm{B}(a,b)=\frac{\Gamma(a)\Gamma(b)}{\Gamma(a+b)}.$$

If a and b are positive integers then

$$\mathrm{B}(a, b) = \frac{(a-1)!(b-1)!}{(a+b-1)!}.$$

Bhattacharya distance A measure of the distance between *populations with *probability density functions f and g. The measure, introduced by A. Bhattacharya in 1946, is given by

$$\cos^{-1}\left(\int_{-\infty}^{\infty} \{\mathrm{f}(x)\mathrm{g}(x)\}^{\frac{1}{2}} \mathrm{d}x\right).$$

See also HELLINGER DISTANCE; KULLBACK–LEIBLER INFORMATION.

bias; biased estimator *See* ESTIMATOR.

BIC (Bayesian information criterion) *See* MODEL SELECTION PROCEDURE.

Bickel, Peter John (1940– ; b. Bucharest, Romania) American mathematical statistician specializing in *semi-parametric models. Bickel was educated at UCB, gaining his PhD (supervised by *Lehmann) in 1963 and joining the faculty. He was the 1979 *COPSS President's Award winner and its 2013 *Fisher Lecturer. President of the *IMS in 1981, he was its *Wald Lecturer in 1980 and its *Rietz Lecturer in 2004. He was elected to AAAS and NAS in 1985. He was President of the *Bernoulli Society in 1991.

SEE WEB LINKS

- University website.

bimodal Having two *modes or modal classes. The word appears in the 1901 edition of *The American Naturalist.*

bimodal distribution A *distribution having two *modes. A distribution with two or more modes is called multimodal; one with a single mode is called *unimodal.

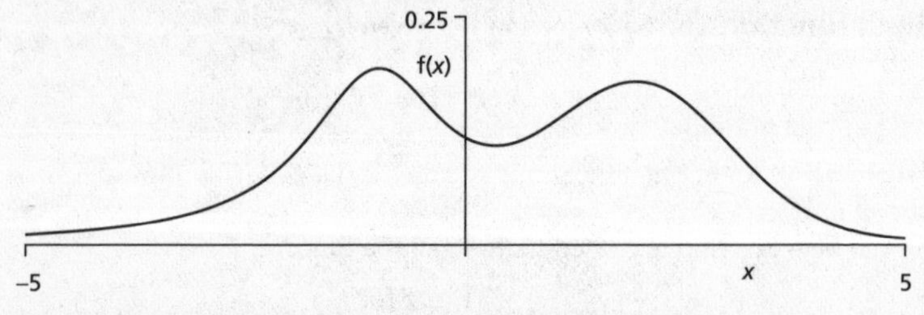

Bimodal distribution. The distribution illustrated is a mixture of two normal distributions.

binary Adjective describing a variable that can have only two possible values. The values are usually 0 and 1. The **binary system** is a system of counting using just these digits. In this system the decimal numbers 1, 2, 3, 4, . . . , 8, . . . , 16, . . . , 32, . . . , become 1, 10, 11, 100, . . . , 1 000, . . . , 10 000, . . . , 100 000, ...

binary regression *Regression in which the response variable can take only the values 0 and 1.

binary system *See* BINARY.

binary variable *See* BERNOULLI DISTRIBUTION.

Bingham distribution An *axial distribution for data on a hemisphere.

binomial coefficient The coefficient of a power of *x* in the *binomial expansion of $(1+x)^n$. The general binomial coefficient is defined, for any value of *n* and for any positive integer *r*, by

$$\binom{n}{r} = \frac{n \times (n-1) \times (n-2) \times (n-3) \times \cdots \times (n-r+1)}{1 \times 2 \times 3 \times \cdots \times r}.$$

By convention, $\binom{n}{0} = 1$. Suppose now that *n* is a non-negative integer. In this case $\binom{n}{r}$ is an integer and $\binom{n}{n} = 1$. Furthermore,

$$\binom{n}{r} = 0 \quad \text{for} \quad r > n.$$

If $r \leq n$ then $\binom{n}{r}$ is also written as ${}^nC_r$ (*see* COMBINATION) and

$$\binom{n}{r} = \binom{n}{n-r} = \frac{n!}{r!(n-r)!}.$$

The phrase 'binomial coefficient' appears in English in a mathematics text of 1876.

binomial distribution The distribution associated with the *random variable, *X*, defined as the number of 'successes' in *n* *independent trials each having the same *probability, *p*, of success. The random variable *X* is said to be a **binomial variable** and to have a binomial distribution with *parameters *n* and *p*. This is written as $X \sim \mathrm{B}(n, p)$. The *mean of this distribution is *np* and the *variance is $np(1-p)$. The *probability function is given by

$$\mathrm{P}(X=r) = \binom{n}{r} p^r (1-p)^{n-r}, \qquad r = 0, 1, \ldots, n.$$

The distribution takes its name from the fact that successive probabilities are the terms in the expansion in ascending powers of *p*, by the binomial theorem, of $(q+p)^n$, where $q=1-p$. The first published derivation of the distribution was by Jacob *Bernoulli in 1713.

As an example, suppose that a computer generates fifteen random integers between 0 and 9 inclusive. The number of these integers that are odd has a B(15, 0.5) distribution. The number that are non-zero has a B(15, 0.9) distribution, and the number that are greater than 7 has a B(15, 0.2) distribution. The diagram shows the graphs of the probability functions for these distributions.

If we note that $\mathrm{P}(X=0)=q^n$, successive probabilities can be calculated using the recurrence relation

$$\mathrm{P}(X=r)=\frac{n-r+1}{r}\times\frac{p}{q}\times \mathrm{P}(X=r-1).$$

If $(n+1)p$ is not an integer the graph is *unimodal, with *mode at the (integer) value of r such that and $(n+1)p-1$ and $(n+1)p$ are both modal values (as in the B(15, 0.5) case illustrated).

A binomial random variable with parameters n and p may be regarded as the sum of n independent observations of a Bernoulli variable (*see* BERNOULLI DISTRIBUTION) with parameter p. The sum of two independent binomial variables with parameters n_1, p and n_2, p, respectively, is also a binomial variable, with parameters (n_1+n_2), p.

For large values of np and nq the **normal approximation to the binomial distribution** may be used:

$$\mathrm{P}(X\le r)\approx\Phi(z),\qquad \text{where } z=\frac{r+\frac{1}{2}-np}{\sqrt{np(1-p)}},$$

and Φ is the *cumulative distribution function for a standard *normal variable (*see* NORMAL DISTRIBUTION). The '½' is a *continuity correction. The result, that a binomial distribution with p = ½ may be approximated by a normal distribution, underlies the derivation of the normal distribution by *de Moivre in 1733 and is sometimes referred to as the **de Moivre–Laplace theorem**. For large values of n and small values of p the **Poisson approximation to the binomial distribution** may be used:

$$\mathrm{P}(X=r)\approx\frac{(np)^r\mathrm{e}^{-np}}{r!}.$$

The word 'binomial' was used in its mathematical sense in a 1557 text entitled *The Whetstone of Witte* by Robert Recorde. The 'binomial distribution' was so named by *Yule in 1911.

SEE WEB LINKS

- Applet.

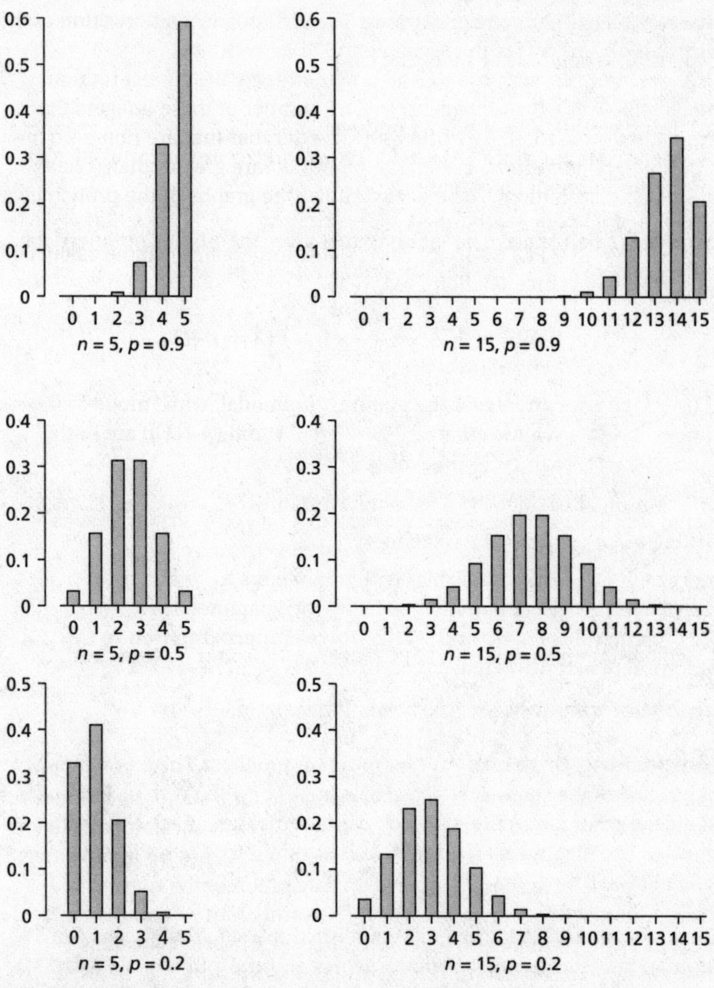

Binomial distribution. The distribution is *skewed to the right if $p > 0.5$ and to the left if $p < 0.5$. It is symmetric if $p = 0.5$.

binomial expansion The application of the *Maclaurin series to $(1+x)^n$, for any value of n:

$$(1+x)^n = 1 + nx + \frac{n(n-1)}{2}x^2 + \cdots = \sum_{r=0}^{\infty} \binom{n}{r} x^r,$$

where $\binom{n}{r}$ is a *binomial coefficient. The series converges and the expansion is valid in the following cases:

(i) for any value of n, if $-1 < x < 1$,
(ii) for any value of x, if n is a non-negative integer, in which case the series is finite, since $\binom{n}{r} = 0$ for $r > n$, and the expansion is that given in the *binomial theorem.

binomial theorem The theorem that gives the expansion of the nth power, where n is a non-negative integer, of a binomial:

$$(a+b)^n = a^n + \binom{n}{1}a^{n-1}b + \cdots + \binom{n}{r}a^{n-r}b^r + \cdots + \binom{n}{n-1}ab^{n-1} + b^n,$$

where

$$\binom{n}{r} = \binom{n}{n-r}.$$

The *binomial coefficient $\binom{n}{r}$ is often written as ${}^nC_r$ (*see* COMBINATION). Setting $a = b = 1$ gives the relation

$$\binom{n}{0} + \binom{n}{1} + \cdots + \binom{n}{n} = 2^n.$$

The theorem appears in the 1742 *Treatise of Fluxions* by *Maclaurin.

binomial variable *See* BINOMIAL DISTRIBUTION.

bin-packing problem An *optimization problem. There is a supply of bins, all of the same size. These bins are to be filled with collections of different numbers of items (each of the same size). Each collection must go in a single bin. The problem is to minimize the number of bins required.

Bioinformatics A journal, first published in 1985, that focuses on developments in genome *bioinformatics and computational biology. It now appears monthly.

SEE WEB LINKS

- Journal home page.

bioinformatics A subject concerned with extracting information from data arising from study of DNA, genomes, etc. The subject interweaves biology, computer science, and statistics.

Biometrics The quarterly journal of the *International Biometric Society. It was first published under that name in 1947, with Gertrude *Cox as the first Editor.

SEE WEB LINKS

- Journal home page.

Biometrika The first *Statistics journal to specialize in *biometry. The first issue of *Biometrika* appeared in 1901, with Karl *Pearson as Editor. Pearson remained as Editor until 1936. Subsequent editors include Egon *Pearson (1936–65) and Sir David *Cox (1965–91). It is published quarterly.

SEE WEB LINKS

- Journal home page.

biometry The measurement of quantities in the living world. The word is often used as a synonym for *biostatistics. 'Biometry' was the title of the first article (by *Galton) in the first volume of **Biometrika*.

Biostatistics A quarterly journal, first published in 2000, that concentrates on applications of Statistics to the study of human health and disease.

SEE WEB LINKS

- Journal home page.

biostatistics Statistics applied to the living world. It includes *demography, epidemiology, and clinical trials. Specialized measurement techniques include *capture–recapture methods and the analysis of *line transects.

biplot (Gabriel biplot) A diagram similar to a *scatter diagram that attempts to represent *observations having several coordinates on a diagram having (usually) just two coordinates. It was introduced by *Gabriel in 1971.

Birnbaum, Zygmunt William (1903–2000; b. Lviv, Ukraine; d. Seattle, WA) Polish statistician, who spent most of his career in the USA. Following parental wishes, Birnbaum studied law at U Lviv obtaining his MA in 1925. He then followed his own interests and studied mathematics, obtaining his PhD in 1929. After practising as an actuary for some years, his interests focused on Statistics. In 1937 he emigrated to the USA and, in 1939, he was appointed to the faculty of U Washington. He was President of the *IMS in 1964 and was presented with the *Wilks Award of the *ASA in 1984.

SEE WEB LINKS

- Fuller biography, interview, and photographs.

Birnbaum inequality For a standard normal random variable, Z, *Birnbaum showed in a 1942 paper that, for $z > 0$,

$$\mathrm{P}(Z > z) \geq \frac{1}{2\sqrt{2\pi}}(\sqrt{4+z^2} - z)\mathrm{e}^{-\frac{1}{2}z^2}.$$

b

Birnbaum–Saunders distribution A two-parameter *distribution originally proposed by *Birnbaum and Saunders in 1969 to describe the lifetime ($x > 0$) of a component liable to experience a fatigue failure:

$$\mathrm{f}(x) = \frac{1}{2\alpha\sqrt{2\pi\beta x}}\left(1 + \frac{\beta}{x}\right)\exp\left(-\frac{(x-\beta)^2}{2\beta\alpha^2 x}\right).$$

The *parameters α and β are positive. The *mean and *variance are, respectively, $\beta(1 + \frac{1}{2}\alpha^2)$ and $(\alpha\beta)^2(1 + \frac{5}{4}\alpha^2)$.

birth-and-death process A continuous-time *Markov process with states $\{0, 1, \ldots\}$ for which the only possible transitions from state n are to $n+1$ (a birth) or to $n-1$ (a death). If the *probability of a death is 0 then the process is a **pure birth process.**

birthday problem A well-known, but intriguing *probability problem. There are n people in a room. Assume that none is born on 29 February and that the remaining 365 days are all equally likely as birthdays. What is the smallest value of n for which the probability that at least two have the same birthday is greater than 0.5?

The answer is not 183, but 23. The *complementary event is that all n people have different birthdays. The probability of this, p_n, is

$$\frac{365}{365} \times \frac{364}{365} \times \cdots \times \frac{366-n}{365},$$

which reduces surprisingly quickly as n increases:

n	3	5	9	13	16	19	22	23	26	30	34	40	46
p_n	0.99	0.97	0.91	0.81	0.72	0.62	0.52	0.49	0.40	0.29	0.20	0.11	0.05

birth rate The number of births occurring in a stated population during the stated period of time, usually a year, as a proportion of the number in the stated population. A total or crude birth rate utilizes all births, usually expressed as births per 1000, whereas an *age-specific rate is usually reported on the basis of the number of births per 1000 persons in this age group. The birth rate may be standardized when comparing birth rates over time, or between countries, to take account of differences in the age structures of the populations.

biserial correlation A measure of the *association between a binary variable (*see* BERNOULLI DISTRIBUTION), X, taking values 0 and 1, and a *continuous random variable, Y. If it is assumed that for each value of X the *distribution of Y is a *normal distribution, with different *means but the same *variance, then an appropriate measure is the **point biserial correlation coefficient**. This is estimated from a sample as r_{pb} $(-1 \leq r_{pb} \leq 1)$, given by

$$r_{pb} = \frac{(\bar{y}_1 - \bar{y}_0)\sqrt{p(1-p)}}{s_y},$$

where $\bar{y}_1$ and $\bar{y}_0$ are the mean Y-values corresponding to the two values of X, s_y^2 is the *sample variance (using the $n-1$ divisor) of the combined set of n Y-values, and p is the proportion of X values equal to 1.

If it can be assumed that X is a dichotomous (*See* CATEGORICAL VARIABLE) representation of an underlying continuous random variable, W, with W and Y having a bivariate normal distribution (*see* MULTIVARIATE NORMAL DISTRIBUTION), then an appropriate measure is the **biserial correlation coefficient**. This is estimated as r_b, given by

$$r_b = \frac{\bar{y}_1 - \bar{y}_0}{s_y}\sqrt{\frac{p(1-p)}{u}} = \frac{r_{pb}}{\sqrt{u}},$$

where

$$u = \frac{1}{\sqrt{2\pi}} e^{-\frac{1}{2}h^2},$$

and h is the value defined by $P(Z \geq h) = p$, for a standard normal variable (*see* NORMAL DISTRIBUTION) Z.

bit A *binary digit. Computers store information in the form of bits. A sequence of (usually) eight bits is called a **byte**. The term 'bit' was coined by *Tukey in the 1940s; the term 'byte' by Dr Werner Buchholz of IBM in 1956.

bivariate data Data consisting of pairs of values (x_1, y_1), $(x_2, y_2), \ldots$ taken from a *bivariate distribution. The term 'bivariate' dates from an article published in 1920.

bivariate distribution When, as a result of an experiment, values for two *random variables X and Y are obtained, it is said that there is a bivariate distribution. In the case of a sample, with n pairs of values $(x_1, y_1), (x_2, y_2), \ldots, (x_n, y_n)$ being obtained, the methods of *correlation and *regression may be appropriate. Associated with the experiment is a **bivariate probability distribution**. In the case when X and Y are *discrete random variables, the distribution is specified by the **joint distribution** giving $P(X = x_j \;\&\; Y = y_k)$ for all values of j and k.

b

The **marginal distribution** of X is given by

$$P(X = x_j) = \sum_k P(X = x_j \,\&\, Y = y_k),$$

and the marginal distribution of Y is given by

$$P(Y = y_k) = \sum_j P(X = x_j \,\&\, Y = y_k).$$

If X and Y are *independent random variables then

$$P(X = x_j \,\&\, Y = y_k) = P(X = x_j) \times P(Y = y_k).$$

The *expected values and *variances of X and Y are given in the usual way from these marginal distributions. For example,

$$E(X) = \sum_j x_j P(X = x_j).$$

The **conditional distribution** of X, given that $Y = y_k$, is given by

$$P(X = x_j \mid Y = y_k) = \frac{P(X = x_j \,\&\, Y = y_k)}{P(Y = y_k)},$$

and the **conditional expectation** of X, given that $Y = y_k$, written as $E(X|Y = y_k)$, is defined in the usual way as

$$E(X \mid Y = y_k) = \sum_j x_j P(X = x_j \mid Y = y_k).$$

The conditional distribution of Y and the conditional expectation of Y, given that $X = x_j$, are defined similarly.

The **mutual information** of the variables X and Y is given by

$$I(X, Y) = \Sigma\Sigma_{jk} P(X = x_j \,\&\, Y = y_k) \log[P(X = x_j \,\&\, Y = y_k)/\{P(X = x_j)P(Y = y_k)\}].$$

The base of the logarithm is usually taken to be 2 (*compare* DIVERSITY INDEX).

In the case when X and Y are *continuous random variables, the distribution is specified by the **joint probability density function** $f(x, y)$ with the property that if R is any region of the (x, y) plane then

$$P[(X, Y) \in R] = \int\int_R f(x, y)\,dx\,dy.$$

The **marginal distribution** of X then has *probability density function (pdf) $f_x(x) = \int_{-\infty}^{\infty} f(x, y)dy$ and the marginal distribution of Y has pdf $f_y(y) = \int_{-\infty}^{\infty} f(x, y)dx$. If the two random variables are independent of one another then $f(x, y)$ is the product of the pdfs of the two marginal distributions. The expected values and variances of X and Y are given in the usual way from these marginal distributions. In this case the mutual information is given by

$$I(X, Y) = \int \int f(x, y) \log[f(x, y)/\{f_x(x)\ f_y(y)\}] dx dy.$$

The probability density function for the conditional distribution of X, given that $Y=y$, is

$$\frac{f(x, y)}{\int_{-\infty}^{\infty} f(x, y) dx},$$

and

$$E(X \mid Y = y) = \frac{\int_{-\infty}^{\infty} x f(x, y) dx}{\int_{-\infty}^{\infty} f(x, y) dx}.$$

See also MULTIVARIATE DISTRIBUTION.

bivariate normal distribution *See* MULTIVARIATE NORMAL DISTRIBUTION.

bivariate probability distribution *See* BIVARIATE DISTRIBUTION.

biweight function *See* *M*-ESTIMATE.

Black–Scholes model A financial model of the variations over time in (for example) the price of stocks on the stock market. The model, proposed in 1974, is based on the notion that the underlying price variations could be modelled as *Brownian motion.

black swan event An event that, given the current *model, is a surprise at the time of its occurrence and has a big impact, but is subsequently rationalized by a reformulation of the model (e.g. by assuming errors with a *heavy-tailed distribution in place of a *normal distribution).

Blackwell, David Harold (1919–2010; b. Centralia, IL; d. Berkeley, CA) American mathematical statistician. Blackwell was educated at U Illinois, gaining his PhD (supervised by *Doob) at the age of 22. In 1944 he joined the faculty of Howard U, where he published the result now called the *Rao–Blackwell theorem. In 1954 he moved to UCB. President of the *IMS in 1956, he was its *Rietz Lecturer in 1961 and its *Wald Lecturer in 1976. In 1986 he was the *COPSS *Fisher Lecturer. In 1965 he was elected to the NAS (the first African-American to be so honoured) and in 1969 to the AAAS. He was President of the *Bernoulli Society in 1975 and was made an Honorary Fellow of the *RSS in 1976.

SEE WEB LINKS

- Fuller biography, interview, and photographs.

Bland-Altman plot A type of plot originally suggested by *Tukey, and independently introduced by Altman and Bland in 1983 in a medical context. Measurements of a characteristic (e.g. blood pressure) are made independently by two observers (or two instruments) on a series of n individuals; denoting these measurements by $(x_{11}, x_{21}), \ldots (x_{1n}, x_{2n})$, the plot is a *scatter diagram with $(x_{1i} - x_{2i})$ giving the vertical coordinate and $(x_{1i} + x_{2i})/2$ giving the horizontal coordinate. The plot is also known as **Tukey's mean-difference plot** or **MD-plot**.

blinding A method of avoiding bias in the context of treating a disease. In a medical experiment the comparison of treatments could be biased if either the patient, the doctor administering the treatment, or the data analyst knew which treatment was allocated to which patient. If the patient (or the doctor, or the data analyst) is unaware of which treatment is being given, then they are said to be 'blind to' the treatment allocation process. If neither the patient nor the doctor is aware of the treatment allocation then the process is said to be **a double-blind trial**.

Typically the patient's permission will be required for participation in such a trial. Agreement by a patient to being treated in an unknown fashion may be difficult to obtain. In **Zelen's design**, the allocation of treatments to patients is at random, but the allocation is revealed to the patient so that the patient's decision concerning participation is fully informed.

In such an experiment, one treatment is often a dummy—for example, a pill of the same shape and flavour as the genuine pill, but with no active ingredients. This is called a **placebo**. In practice there is often a **placebo effect**, in which the treated person shows a benefit despite the treatment theoretically having no effect. The effect is a reflection of the psychological benefit of believing that one is being given an effective treatment.

The related **Hawthorne effect** recognizes that individuals behave differently when aware that they are being studied, whilst the **Pygmalion effect** recognizes that individuals are influenced by the expectations that others have of them.

block In the context of *experimental design, a homogeneous group of *experimental units.

block kriging *See* KRIGING.

BLUE (best linear unbiased estimator) An *estimator that is unbiased, is formed from a linear combination of the *observations, and has the smallest *variance of all such estimators.

BMDP (Biomedical Computer Programs) A computer package permitting many types of statistical analysis.

SEE WEB LINKS

- Computer package home page.

body mass index *See* QUETELET INDEX.

bomb packing A *probability fallacy. At a time when plane hijackers used the threat that they would explode a bomb in their luggage if the pilot did not do as they asked, it was suggested that one should pack a bomb in one's own luggage, since the probability of there being two bombs on the same aircraft was minimal. The logic fails, since the event of interest is not whether there is a bomb on the plane, but whether one of the other passengers has packed a bomb.

Bonferroni, Carlo Emilio (1892–1960; b. Bergamo, Italy; d. Florence, Italy) Italian mathematician specializing in the mathematics of finance. Bonferroni was educated at U Turin. He was appointed Professor at U Bari (1923) and then at U Florence (1933). He was an excellent musician and, in his younger days, a keen glacier climber.

SEE WEB LINKS

- Fuller biography and portrait.

Bonferroni inequality An inequality concerning the joint *probabilities of occurrence of combinations of events. Let $E_1, E_2, \ldots, E_m$ be m events, with E'_j denoting the *complementary event to the event E_j, for $j = 1, 2, \ldots, m$. *Bonferroni developed various bounds and the inequality that is most often cited is

$$1 - \{P(E'_1) + P(E'_2) + \cdots + P(E'_m)\} \leq P(E_1 \cap E_2 \cap \cdots \cap E_m).$$

The most usual context has the event E_j defined as 'hypothesis test j produces a non-significant result'.

Bonferroni *t*-test *See* MULTIPLE COMPARISON TEST.

Boole, George (1815–64; b. Lincoln, England; d. Ballintemple, Ireland) Self-taught English mathematician with a flair for languages. At school Boole excelled at Latin and by the age of 16 was an assistant school teacher, whilst studying mathematics for his own interest. To support his parents he opened his own school, continuing with his work in mathematics. This work became so well known that at the age of 35 he was appointed Professor of Mathematics at Queen's College, Cork. His *An Investigation into the Laws of Thought, on Which are Founded the Mathematical Theories of Logic and Probabilities*, in which he introduced what is now known as *Boolean algebra, was published in 1854. This algebra has found many applications, particularly in the design of computers. He was elected FRS in 1857.

SEE WEB LINKS

- Fuller biography and portrait.

Boolean algebra The algebra, developed by *Boole, of events (*see* SAMPLE SPACE), *unions, *intersections and *complementary events in a *sample space S. For any events A, B, C the following algebraic laws hold:

$$
\begin{array}{cc}
A \cup A = A, & A \cap A = A, \\
A \cup A' = S, & A \cap A' = \phi, \\
A \cup B = B \cup A, & A \cap B = B \cap A, \\
A \cup S = S, & A \cap S = A, \\
A \cup \phi = A, & A \cap \phi = \phi, \\
A \cup (B \cup C) = (A \cup B) \cup C, & A \cap (B \cap C) = (A \cap B) \cap C, \\
A \cup (B \cap C) = (A \cup B) \cap (A \cup C), & A \cap (B \cup C) = (A \cap B) \cup (A \cap C), \\
(A \cup B)' = A' \cap B', & (A \cap B)' = A' \cup B',
\end{array}
$$

where A' and B' are the complementary events of A and B respectively, and ϕ is the empty set. The last line of the above comprises the **de Morgan laws**. The penultimate line comprises the distributive laws and the antepenultimate line comprises the associative laws.

Boole inequality Let $E_1, E_2, \ldots, E_n$ be n events, with $P(E_j)$ denoting the *probability that event E_j occurs. Then the following inequality concerning the probability of the *union of events was demonstrated by *Boole in 1862:

$$P(E_1 \cup E_2 \cup \cdots \cup E_n) \leq \sum_{j=1}^{n} P(E_j).$$

boosting An iterative form of *bagging in which the *bootstrap sampling in subsequent iterations (*see* ITERATIVE ALGORITHM) is weighted to increase the *probability of including *data poorly predicted in the previous iteration. The *observations not included in each bootstrap sample are referred to as the **out-of-bag sample** and may be used to further improve accuracy. The most widely used boosting algorithm is **AdaBoost**. *See also* MACHINE LEARNING.

SEE WEB LINKS

- Applet for AdaBoost.

Booth, Charles (1840–1916; b. Liverpool, England; d. Thringstone, England) A wealthy English philanthropist and a pioneer of social statistics. In the 1880s Booth reported the findings of his investigation into the extent of London's poverty using colour-coded street maps. In 1892 he was President of the *Royal Statistical Society and was the first recipient of its *Guy Medal in Gold. In 1899 he was elected FRS.

bootstrap A *computer-intensive *resampling method for estimating the properties of a *distribution while making minimal assumptions. In this respect it resembles the *jackknife. The idea is simple. Suppose we have n *observations $x_1, x_2, \ldots, x_n$ from an unknown distribution. We assume that the *population being sampled has $\frac{1}{n}$ of its observations equal to x_1, $\frac{1}{n}$ equal to x_2, and so on. Using *pseudo-random numbers we now select m sets of n observations from this hypothetical distribution. If we are interested in, for example, the *median of the original distribution, then we can use the m 'sample' medians to give an overall estimate (*see* ESTIMATOR) and a *confidence interval for that estimate.

In practice bootstrap estimators are often slightly biased. However, this bias can be estimated—again using bootstrap methods. Estimation using this form of bias correction is referred to as the **double bootstrap**. The term 'bootstrap' was introduced by *Efron in 1979.

Borda scores A system proposed by the French mathematician, Jean Charles de Borda (1733–1799), for determining the outcome of a voting contest with n candidates. Under this system a voter awards one point to the least-favoured candidate, two points to their next choice, and so on, up to n points for their most favoured candidate. The winner is the candidate with the largest total number of points.

Borel–Cantelli lemma If $E_1, E_2, \ldots$ is an infinite sequence of *independent events with *probabilities $p_1, p_2, \ldots$, and if the sum of these probabilities is finite, then the probability that an infinite number of the events occur is zero.

Bortkiewicz, Ladislaus Josephowitsch von (1868–1931; b. St Petersburg, Russia; d. Berlin, Germany) Prussian economist and statistician. Bortkiewicz studied law as an undergraduate at U St Petersburg. His doctorate was obtained from U Göttingen in 1893. After a period in Strasbourg he returned to St Petersburg and it was there, in 1898, that he published a work entitled *The Law of Small Numbers* that dealt with properties of the *Poisson distribution and introduced the *Prussian horse-kick data that have become familiar to generations of students of Statistics. In 1901 he settled at U Berlin, becoming Professor of Statistics and Political Economy in 1920.

Bowley, Sir Arthur Lyon (1869–1957; b. Bristol, England; d. Haslemere, England) English social statistician. Bowley was a mathematics graduate at Cambridge U. In 1895 he joined LSE, being appointed their first Professor of Statistics in 1919. He was the author of the influential textbook *Elements of Statistics*, first published in 1901, with a seventh edition in 1937. In this book he wrote 'A knowledge of statistics is like a knowledge of a foreign language or of algebra; it may prove of use at any time under any circumstances'. He was President of the

*Econometric Society in 1938. In the same year he was President of the *RSS having been awarded its *Guy Medal in Silver in 1895 and in Gold in 1935.

SEE WEB LINKS

- Portrait.

Bowley coefficient of skewness *See* QUARTILE.

Bowley index *See* PRICE INDEX.

Box, George Edward Pelham (1919–2013; b. Gravesend, England; d. Madison,WI) English statistician who initially trained as a chemist. Married to Joan, the daughter of Sir Ronald *Fisher, Box's career was principally spent in the United States. During the Second World War, when determining the effects of poisonous gases, Box noted that the results varied and required statistical analysis. After the war, he studied mathematics and statistics at UCL, obtaining his PhD (supervised by Egon *Pearson and *Hartley) in 1952. He was appointed Professor of Statistics at U Wisconsin–Madison in 1960. In 1968 he was awarded the *Shewhart Medal of the *ASQ. He was President of the *ASA in 1978 having been presented with its *Wilks Award in 1972. He was the *COPSS *Fisher Lecturer in 1974, President of the *IMS in 1980, and its *Rietz Lecturer in 1981. He was elected FRS in 1985. In 1993 he was awarded the *Guy Medal in Gold of the *RSS and was made an Honorary Fellow of the Society in 1993. An oft-quoted statement attributed to Box is 'All models are wrong, some models are useful'.

SEE WEB LINKS

- Fuller biography, interview, and photographs.

box-and-whisker diagram *See* BOXPLOT.

Box–Behnken design *See* RESPONSE SURFACE.

Box–Cox transformation A transformation to normality suggested by *Box and Sir David *Cox in 1964. They proposed a family of transformations that might be used to convert a general set of n *observations into a set of n *independent observations from a *normal distribution with constant *variance. The transformation involves a *parameter λ that can be estimated from the *data using the method of *maximum likelihood. The transformed observation, $y^{(\lambda)}$, is related to the original observation, y, by

$$y^{(\lambda)} = \begin{cases} (y^{\lambda} - 1)/\lambda & \lambda \neq 0, \\ \ln y & \lambda = 0. \end{cases}$$

b

Box–Jenkins procedure A general strategy for the analysis of *time series based on the use of *ARIMA models or, for seasonal *data, *SARIMA models. The procedure was set out by *Box and *Jenkins in their 1970 book *Time Series Analysis: Forecasting and Control.* The first stage consists of removing trends (*see* MOVING AVERAGE) or *cycles from the data. An appropriate type of *model must then be identified and its *parameters estimated. The estimated model is then compared with the original data and adjustments are made if necessary.

Box–Ljung test A test of whether a proposed *ARMA model describes a *time series. Suppose that the model has p *parameters and the time series has n *observations. Let x_j and $\hat{x}_j$ be, respectively, the jth observed and fitted values, define a_j by $a_j = x_j - \hat{x}_j$ and define r_k by

$$r_k = \sum_{t=1}^{n-k} a_t a_{t+k} \Big/ \sum_{t=1}^{n} a_t^2.$$

The Box–Ljung statistic, Q, is given by

$$Q = n(n+2) \sum_{k=1}^{m} \frac{r_k^2}{(n-k)},$$

where m is arbitrary and very much smaller than n, but usually about 24. If the model is correct then Q is an observation from a *chi-squared distribution with $(m-p)$ *degrees of freedom.

Box–Muller transformation A procedure, suggested by *Box and Mervin Muller in 1958, for the *simulation of *observations from a *normal distribution. If u_1 and u_2 are two *independent observations from a continuous *uniform distribution on the interval (0, 1), then the quantities x and y, given by

$$x = \cos(2\pi u_1)\sqrt{-2\ln(u_2)}, \qquad y = \sin(2\pi u_1)\sqrt{-2\ln(u_2)},$$

where $2\pi u_1$ is taken to be in radians, are independent observations from a standard normal distribution.

boxplot (box-and-whisker diagram, box–whisker diagram) A graphical representation of numerical *data, introduced by *Tukey and based on the *five-number summary. The diagram has a scale in one direction only. A rectangular box is drawn, extending from the lower *quartile to the upper quartile, with the *median shown dividing the box. 'Whiskers' are then drawn extending from the end of the box to the greatest and least values. Multiple boxplots, arranged side by side, can be used for the comparison of several samples.

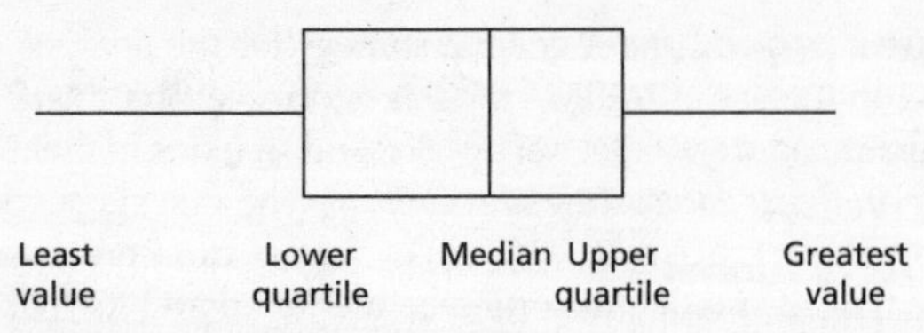

Boxplot. The basic boxplot illustrates five key values: the minimum value, the maximum value, the quartiles, and the median.

In **refined boxplots** the whiskers have a length not exceeding 1.5 × the interquartile range (*see* QUARTILE). Any values beyond the ends of the whiskers are shown individually as *outliers.

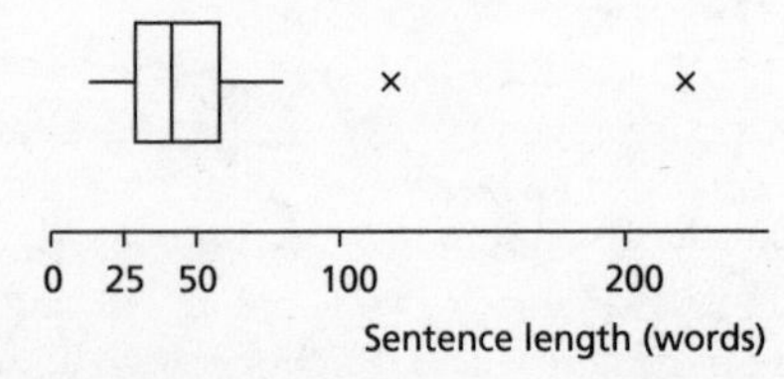

Refined boxplot. In this diagram, individual outliers are indicated. The data illustrated are the lengths of the first eighteen sentences of *A Tale of Two Cities* by Charles Dickens. The first sentence—'It was the best of times, it was the worst of times, . . . '—stretches to 118 words, and the fourteenth sentence has 221 words.

SEE WEB LINKS
- Applet.

Box–Tidwell transformation A general power transformation, introduced by *Box and Tidwell in 1962, in which the observation y is transformed to the value

$$y^{(p)} = \begin{cases} y^p & \text{if } p \neq 0, \\ \ln y & \text{if } p = 0, \end{cases}$$

where $y > 0$ and p has some specified value.

box–whisker diagram *See* BOXPLOT.

Bradley, Ralph Allen (1923–2001; b. Smith Falls, Ontario; d. Athens, GA) Canadian mathematical statistician. A graduate of Queen's U, Ontario in 1944, Bradley gained his PhD (supervised by *Hotelling) at UNC in 1949. After posts at Virginia Polytechnic Institute and FSU, he ended his career at U Georgia. Editor of *Biometrics* from 1957 to 1962, Bradley was President of the *IBS in 1965, and of the *ASA in 1981.

SEE WEB LINKS

- Fuller biography, interview, and photographs.

Bradley–Terry model A *model, proposed by *Bradley and Milton Terry in 1952, that describes the *probability that one treatment is preferred to another. An experiment compares t treatments, two at a time. When two treatments (j and k, say) are compared, the outcome is a statement of which of the two is preferred. The model specifies that there are constants $\pi_1, \pi_2, \ldots, \pi_t$, with $0 \leq \pi_s \leq 1$, for all s, and $\Sigma_{s=1}^{t} \pi_s = 1$, which are such that the probability that treatment j is preferred to treatment k is

$$\frac{\pi_j}{\pi_j + \pi_k}.$$

branch and bound A procedure, introduced in the early 1960s, for solving an integer programming (*see* LINEAR PROGRAMMING) problem. The problem is first solved ignoring the integer constraint. The solution obtained being noted, a variable is given an integer value either above or below the apparent maximum. Each resulting reduced problem is solved in its turn. The result is a branching tree reporting bounds on the optimum derived from problems related to the original.

branching process A *stochastic process with varying numbers of states. Consider a *population in which each individual has *probability p_k of providing k descendants in the next generation. Let X_j be the number of individuals in the jth generation. The sequence $X_1, X_2, \ldots,$ is a *Markov process known as a branching process. This formulation was originally suggested by *Galton and Henry *Watson in 1874 in the context of the extinction of surnames, and may be referred to as the **Galton–Watson model**. Extinction is certain if and only if

$$\sum_{k=0}^{\infty} kp_k \leq 1$$

and otherwise the probability of extinction is equal to P, the solution (between 0 and 1) of

$$P = \sum_{k=0}^{\infty} P^k p_k.$$

Bray–Curtis distance *See* DISTANCE MEASURE.

break-point model *See* LINEAR PIECEWISE REGRESSION MODEL.

Breslow, Norman Edward (1941– ; b. Minneapolis, MN) American biostatistician. Breslow gained his BA in mathematics at Reed College, Portland in 1962. He obtained his PhD in biostatistics from Stanford U in 1967 (supervised by *Efron) and then joined the faculty at U Washington. He was the *COPSS *Fisher Lecturer in 1995 and President of the *IBS in 2003. He was elected an Honorary Life Member of the IBS in 2006.

SEE WEB LINKS

- University website.

Brier score; Brier skill score *See* SKILL SCORE.

Brillinger, David Ross (1937– ; b. Toronto, Canada) Canadian statistician, specializing in *time series. Brillinger graduated in mathematics from U Toronto in 1959, gaining his PhD (supervised by Tukey) from Princeton U in 1961. After five years at LSE, Brillinger was appointed Professor of Statistics at UCB in 1969. He was the *Wald Lecturer of the *IMS in 1983 and its *Neyman Lecturer in 2005. He was the *COPSS *Fisher Lecturer in 1991, and was elected FRSC in 1985. Awarded the Gold Medal of the *SSC in 1992, he was its President in 2000. He was President of the *IMS in 1994 and of *TIES in 2006. In 2002 he was awarded the *Parzen Prize.

SEE WEB LINKS

- University website.

Brown–Forsythe test *See* TEST FOR EQUALITY OF VARIANCE.

Brownian motion A continuous-time version of the *random walk, named after Robert Brown (1773–1858), a Scottish botanist. In 1827, Brown noticed the erratic movement of pollen grains under water. The first explanation of this motion (in terms of the bombardment of the pollen by the surrounding water molecules) was given by Einstein in 1905. Norbert *Wiener provided a concise mathematical description in 1918 and the motion is also called a **Wiener process.**

Starting from the origin at time 0, the path of a particle is made up of independent increments (in d dimensions) in random directions which are such that its distance from the origin at time t is an *observation from a *normal distribution with *mean 0 and *variance proportional to t.

An alternative description is provided by assuming that it is the velocity rather than the position which is changing through time as a consequence of collisions and friction. This model is called the **Ornstein–Uhlenbeck process.**

Brownian motion. This simulation, in two dimensions, illustrates the typical mixture of apparently random wanderings mixed with apparently purposeful movement.

brushing An interactive method for the *exploratory data analysis of *multivariate data. In a simultaneous display of *scatter diagrams of the values of pairs of *variables, the highlighting ('brushing') of data points in one diagram results in the same data items being highlighted in the other diagrams. This helps in understanding relations between variables and in identifying *outliers.

BSMSP *Abbreviation for the* BERNOULLI SOCIETY FOR MATHEMATICAL STATISTICS AND PROBABILITY.

bubble plot A *scatter diagram (*see overleaf*) using circles as plotting symbols in which the areas of the circles indicate the values of a third *variable.

bubble sort A simple but not very efficient *algorithm for arranging a set of n numbers in order of increasing magnitude. The method starts with the right-hand pair of numbers, swapping if necessary so that the smaller number is on the left of the pair, and proceeds towards the left, making a total of $(n - 1)$ comparisons. At the end of this pass the smallest number is

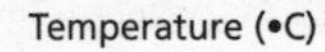

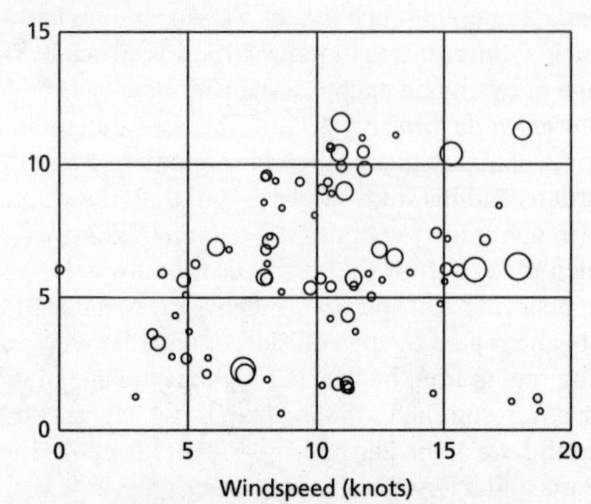

Bubble plot. The diagram displays the relation between windspeed and temperature near Bolton, England during March 2000. Each circle corresponds to an hour during which rain was recorded. The areas of the circles are proportional to the amount of rain.

---

at the left end of the line. The algorithm recommences with the right-hand pair of numbers and again proceeds towards the left. This time $(n - 2)$ comparisons are made and the pass ends with the second-smallest number in second position. The procedure is repeated until the complete ordering is achieved: this requires a total of $\frac{1}{2}n(n - 1)$ comparisons.

16		8		13		4
					×	
16		8		4		13
			×			
16		4		8		13
	×					
4		16		8		13
					○	
4		16		8		13
			×			
4		8		16		13
					×	
4		8		13		16

Bubble sort. Illustration of the simple neighbour-swapping algorithm. In the example × indicates a swap and ○ that no swap is required.

Buffon, Georges Louis Leclerc, Comte de (1707–88; b. Montbard, France; d. Paris, France) French aristocrat educated in law and medicine. Buffon's principal interest was in nature: he was struck by the diversity of life. At the age of twenty he encountered the *binomial theorem and his first work, *Sur le jeu de franc-carreau*, introduced differential and integral calculus into *probability theory. He was appointed keeper of the French botanical gardens and his study of plants and their development led to his publication (in about forty volumes) of *Histoire Naturelle* (*Natural History*), which tried to show the continuity of nature. He had no concept of evolution, believing that species are fixed. He speculated that the Earth might have been created by the collision of a comet with the Sun. Based on the cooling rate of iron, he proclaimed that the age of the Earth was 75 000 years; as a result the Catholic Church (which, at that time, claimed 6 000 years as the age of the Earth) ordered that his books should be burnt. A street in Paris and a lunar crater are named after him.

SEE WEB LINKS

- Commentary and portrait.

Buffon's needle problem A *probability problem posed by *Buffon in 1777. A needle of length l is to be randomly thrown on to a piece of paper that is covered by parallel lines that are a distance d apart; the problem is 'What is the probability that the needle crosses a line?' The answer is

$$\frac{2l}{\pi d}.$$

This provides an empirical method for the estimation of π (though many throws are needed if an accurate answer is required).

SEE WEB LINKS

- Animation.

Bulletin of the International Statistical Institute A journal, which first appeared in 1886, that reports the proceedings of the biennial sessions of the *International Statistical Institute.

SEE WEB LINKS

- Journal home page.

bump hunting A term apparently originated by experimental physicists and now used in *data mining. It refers to the problem of finding clustered values in *multivariate data.

Burman, John Peter (1924–98; b. London, England; d. Blackheath, England) English statistician. Burman was educated at Cambridge U and then spent two years working for the Ministry of Supply. He was the co-author with *Plackett of the 1946 paper that introduced the influential

Plackett and Burman design (*see* FACTORIAL DESIGN). His subsequent career was at the Bank of England, where he worked on the development of *ARMA models.

burn-in period *See* MARKOV CHAIN MONTE CARLO (MCMC) METHODS.

Burt, Sir Cyril Lodowicz (1883–1971; b.Stratford-on-Avon, England; d. London, England) English pioneer of educational psychology. At Oxford U he studied philosophy. In 1908, on graduating, he joined the faculty at U Liverpool, where he worked with delinquent boys in the docklands area. In 1913 he was appointed as the first ever educational psychologist for London County Council. He joined UCL in 1924 as Professor of Education, becoming Professor of Psychology in 1931. In 1942 he was President of the British Psychological Society and he was knighted in 1946. After his death, his work on identical twins was declared to be fraudulent, though this remains a subject of debate.

SEE WEB LINKS

- Fuller biography and portrait.

Burt table A symmetric table that is used in *correspondence analysis. It shows the *frequencies for all combinations of categories of pairs of *variables in a *data set. As an example, the Burt table shown displays simultaneous information on the occurrence of category combinations for the variables Age, Health, and Class.

		AGE		HEALTH		CLASS	
		YOUNG	OLD	ILL	WELL	MIDDLE	WORKING
AGE	YOUNG	64	0	9	55	47	17
	OLD	0	36	17	19	21	15
HEALTH	ILL	9	17	26	0	7	19
	WELL	55	19	0	74	61	13
CLASS	MIDDLE	47	21	7	61	68	0
	WORKING	17	15	19	13	0	32

byte *See* BIT.

C; C^{++} The programming language C^{++} is an advanced form of the language C, which originated at Bell Laboratories in the USA.

calibration The variables X and Y are related by a function f having an unknown *vector $\boldsymbol{\theta}$ of parameters. Values of Y are measured at known values of X. The measured values are denoted by y and x so that

$$y = \mathrm{f}(\boldsymbol{\theta}, x) + \varepsilon,$$

where ε is a measurement error. In the future it is desired to estimate the value of X (ξ, say) by measuring the value of Y:

$$y = \mathrm{f}(\boldsymbol{\theta}, \xi) + \varepsilon.$$

This is calibration. The problem is how best to allow for the uncertainty in the elements of $\boldsymbol{\theta}$ in the estimation of ξ.

Canadian Journal of Statistics The quarterly journal of the *Statistical Society of Canada. The first volume appeared in 1973. The journal includes articles written in either English or French.

SEE WEB LINKS

- Journal home page.

Canberra distance *See* DISTANCE MEASURE.

canonical correlation analysis; canonical variate analysis A method of assessing the relationship between two groups of *variables (for example, the relationship between three measures of a worker's ability and four measures of his or her performance). The method, suggested by *Hotelling in 1936, identifies the two *linear combinations of variables (one from each group) that have the maximum *correlation.

canonical form *See* EXPONENTIAL FAMILY.

cap A name for the ∩ sign denoting *intersection.

capability analysis A method determining the extent to which the long-term performance of an industrial process complies with engineering requirements or managerial goals. Often it is required that output should lie between upper and lower **specification limits** (*USL*; *LSL*). One measure of the extent to which this is achieved is the **potential capability**, C_p, given by

$$C_p = \frac{USL - LSL}{6\sigma},$$

C

where σ is the *standard deviation of the output *distribution. An equivalent statistic is C_r, the **capability ratio**, which is $1/C_p$.

capability ratio *See* CAPABILITY ANALYSIS.

capture–recapture method A method for estimating the size of a *population. The method is generally applied to living organisms, such as fish in a lake.

The simplest example proceeds as follows. A *sample of n_1 individuals is obtained from the population (for example, by catching fish in a net). Each of these individuals is now marked in some fashion (for example, by attaching a tag) and is returned to the population. Later, a second sample (of n_2 individuals) is taken from the population. Suppose m of these individuals are tagged. The sample proportion of tagged individuals is m/n_2, and the population proportion is n_1/N, so that an estimate of the population size, N, is $\hat{N}$, given by

$$\hat{N} = \frac{n_1 n_2}{m}.$$

This estimate is known as the **Petersen estimator** of the population size. An alternative that avoids the problem caused when m is zero is provided by the **Chapman estimator**:

$$\hat{N} = \frac{(n_1 + 1)(n_2 + 1)}{m + 1} - 1.$$

carry-over effect *See* CROSSOVER TRIAL.

CART (Classification and Regression Trees) A program for constructing a *classification tree.

Cartesian coordinates The usual system for identifying the location of a point in two, or more, dimensions. The position of a point *P* in a plane can be represented by a pair of numbers (x, y), relative to two **axes** *Ox* and *Oy* which are straight lines meeting at the **origin** *O* represented by (0, 0). In the usual case of rectangular (or perpendicular, or orthogonal) axes, the **abscissa** x is the perpendicular distance of *P* from *Oy* and the **ordinate** y is the perpendicular distance of *P* from *Ox*. The value of x is positive if *P* lies in the half-plane to the right of *O*, and is negative in the left-hand half-plane. Similarly, the value of y is positive in the upper half-plane and negative in the lower half-plane. The plane is therefore divided into four

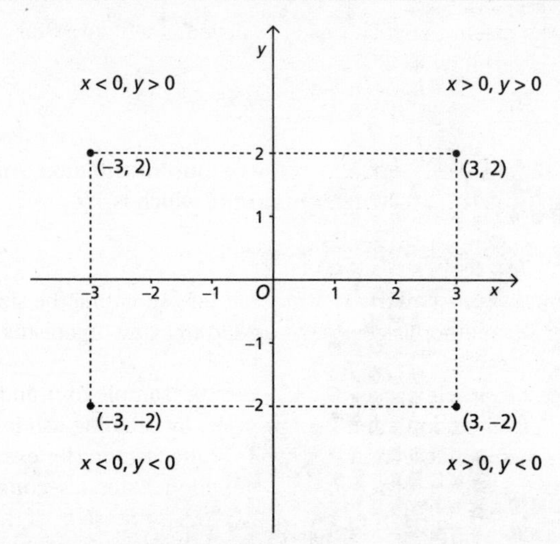

Cartesian coordinates. The axes meet at the origin O. The coordinates of a point are (x, y), where x and y are the corresponding signed distances along the two axes.

quadrants corresponding to the four combinations of coordinate signs. The units of distance in the directions Ox and Oy may be different and are usually indicated by numbers on the axes.

The term 'Cartesian' is derived from *Descartes who first introduced coordinates. The word 'coordinate' was introduced by *Leibniz in about 1693. The phrase 'Cartesian coordinates' was used in 1844.

The idea of coordinates can be generalized to three-dimensional space. The process can be reversed by considering any ordered set of three real numbers (x, y, z) as a point in three-dimensional space. This can be further generalized by considering the row *vector $(x_1\ x_2 \ldots x_n)$ to be the coordinates of a point in an n-dimensional space. This space is usually denoted by $\mathbb{R}^n$.

cartogram A distorted map in which regions are drawn not to an areal scale but to some other scale such as population. Some cartograms attempt to retain the shape of the geographical region they represent; others, such as the **rectangular cartogram**, use a single shape to represent all regions of equal importance. *See diagram overleaf.*

case *See* RETROSPECTIVE STUDY.

case-controlled study *See* RETROSPECTIVE STUDY.

casement plot A method for examining the relations between three *variables simultaneously. A casement plot consists of a side-by-side

C

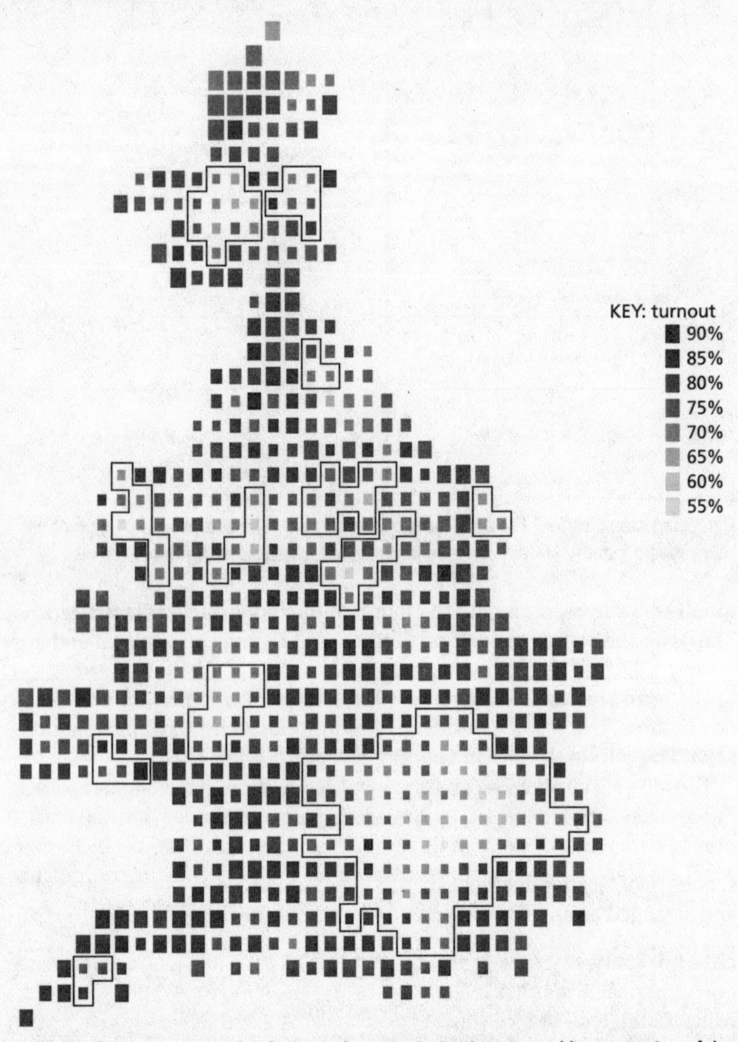

Rectangular cartogram. The diagram shows variation in turnout (the proportion of the electorate who actually cast their votes) in the British constituencies in the 1992 general election. Each constituency is represented by a rectangle whose area indicates the physical size of the constituency. The big conurbations (Greater London, Birmingham, Manchester, etc.) are indicated as outlined regions. Somewhat paradoxically, turnout is lower in city centres than in the rural constituencies.

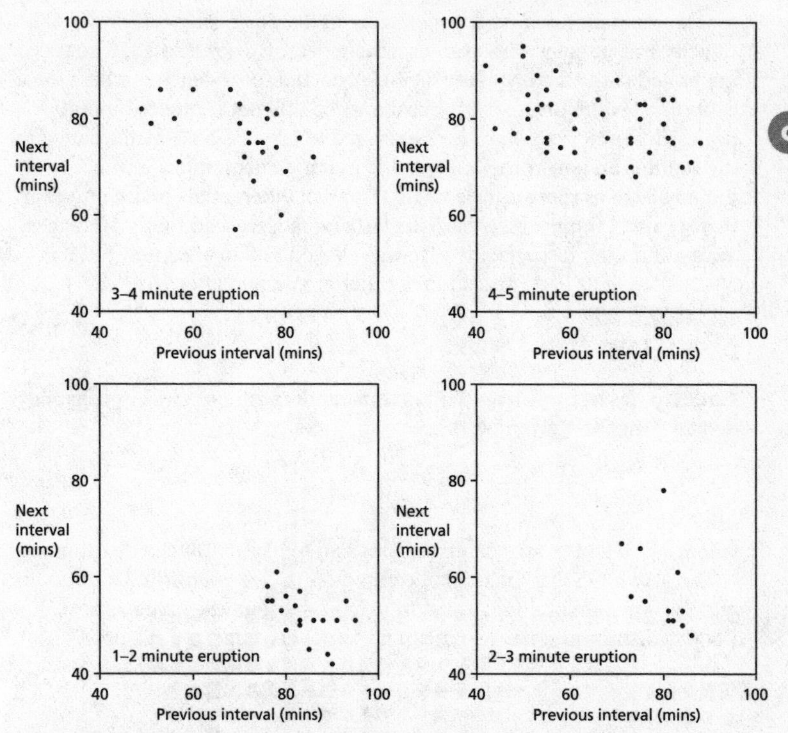

Casement plot. The relation between the time intervals before and after eruptions of *Old Faithful, shown separately for different lengths of the intervening eruption.

sequence of *scatter diagrams of X against Y, subdivided by the values of Z and arranged as an array resembling a casement window.

categorical variable (nominal variable) A *variable whose values are not numerical. Examples include gender (male, female), paint colour (red, white, blue), type of bird (duck, goose, owl). A variable with just two categories is said to be **dichotomous,** whereas one with more than two categories is described as **polytomous.** The corresponding nouns are **dichotomy** and **polytomy**.

cauchit link *See* GENERALIZED LINEAR MODEL.

Cauchy, Baron Augustin-Louis (1789–1857; b. Paris, France; d. Sceaux, France) French mathematician and engineer. After studying first mathematics and then engineering at the École Polytechnique, Cauchy was initially employed as an engineer. However in 1815 he was appointed to the faculty of the École Polytechnique to teach mathematical analysis. Despite his many publications, he was not promoted to full professor for 18 years because he refused to take the oath of allegiance to the republican government. He made major developments in the theories of determinants (*see* MATRIX), partial differential equations, group theory, and complex variables. His last words were said to be 'Men pass away—but their deeds abide'. He was elected FRS in 1832 and FRSE in 1845. A street in Paris and a lunar crater are named after him.

SEE WEB LINKS

- Fuller biography and portrait.

Cauchy distribution A *continuous random variable with *probability density function f given by

$$\mathrm{f}(x) = \frac{k}{\pi\{k^2 + (x-m)^2\}}, \qquad -\infty < x < \infty,$$

where $k > 0$ and m are *parameters, is said to have a Cauchy distribution.

The graph of f is a *bell-curve centred on m (*see opposite*). The *mode and the *median are both equal to m, and the *quartiles are $m \pm k$. A Cauchy distribution has no *mean or *variance, since, for example,

$$\int_{-\infty}^{\infty} \frac{kx}{\pi\{k^2 + (x-m)^2\}}\,\mathrm{d}x$$

does not exist.

The standard Cauchy distribution is given by $k=1$, $m=0$, and in this case the distribution is a *t-distribution, with one *degree of freedom.

Since the Cauchy distribution has neither a mean nor a variance, the *central limit theorem does not apply. Instead, any linear combination of Cauchy variables has a Cauchy distribution (so that the mean of a random sample of observations from a Cauchy distribution has a Cauchy distribution).

If X and Y have *independent standard normal distributions (*see* NORMAL DISTRIBUTION) then Y/X has a standard Cauchy distribution. Equivalently, if U has a *uniform continuous distribution on $-\frac{1}{2}\pi < u < \frac{1}{2}\pi$ then tan U has a standard Cauchy distribution. A geometrical representation of this is as follows. Let O be the origin of *Cartesian coordinates, and let A be the point (0, 1). If the random point P, with coordinates (X, 0), is such that the angle OAP (= U, say) has a uniform continuous distribution on $-\frac{1}{2}\pi < u < \frac{1}{2}\pi$, then X has a standard Cauchy distribution.

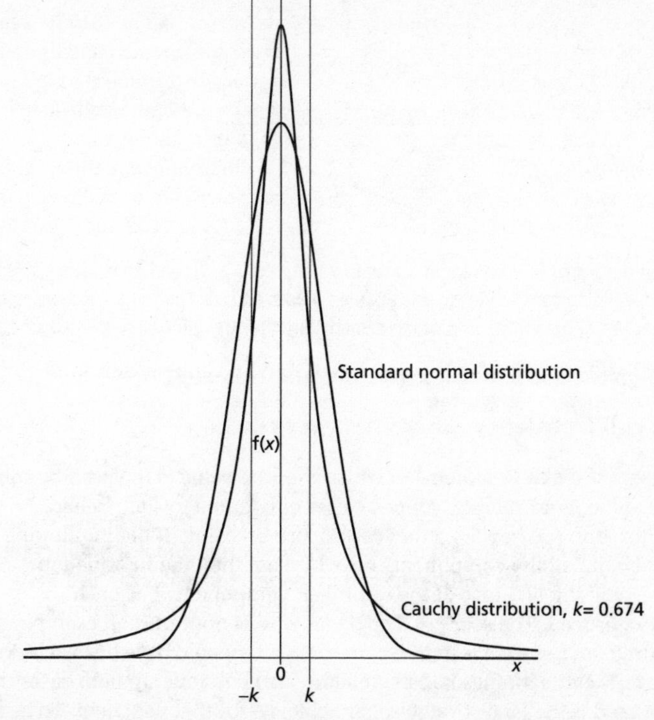

Cauchy distribution. The Cauchy distribution illustrated has $m = 0$ and $k = 0.674$. Also illustrated is the standard normal distribution. Both distributions have 25% of their area above 0.674 and 25% below -0.674. The fatter tails of the Cauchy distribution are apparent.

Cauchy–Schwarz inequality If X and Y are two *random variables then

$$\{E(XY)\}^2 \leq E(X^2) \times E(Y^2),$$

with equality if and only if $P(aX = bY) = 1$ for some real constants a and b, at least one of which is non-zero.

causal diagram A graphical representation (*see diagram overleaf*) of the relationships between *variables often with single-headed arrows from each explanatory variable (*see* REGRESSION) to each of the dependent variables that it affects, and double-headed arrows linking correlated explanatory variables.

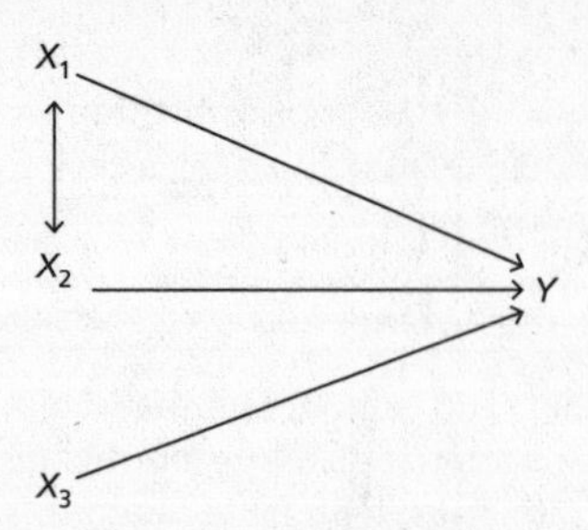

Causal diagram. The diagram indicates that variables X_1, X_2, and X_3 influence Y, with X_1 and X_2 being correlated with one another. Values such as *partial correlations might be attached to the arrows in order to demonstrate the strength of a relationship.

cdf *Abbreviation for* CUMULATIVE DISTRIBUTION FUNCTION.

cell; cell frequency *See* CONTINGENCY TABLE.

censored data Data items in which the true value is replaced by some other value. For example, suppose a set of components are being monitored to see how long they last before breaking. If the monitoring stops before all the components have broken, then the information concerning the lifetimes of the unbroken components has been **right-censored**. The score of a cricketer who is not out is an example of censored data, since it is not known what score would have been achieved if the cricketer's innings had been allowed to continue. In both cases the value used is the largest value so far achieved for that data item. To avoid bias, subsequent calculations should take account of the censoring.

censored regression models *Regression models in which the *data on the dependent variable may be regarded as being *censored data. The analysis of such models is also called **tobit analysis**.

census *Data for an entire *population—as opposed to a random *sample. In England the most famous early census is that of the 'Domesday Book'. The first modern census was the 1666 census of the 3215 inhabitants of New France (now Canada!). In Europe the first complete demographic census was that in Sweden in 1749. The first federal census in the USA took place in 1790, and the first complete demographic censuses in both Britain and France occurred in 1801. In both the United Kingdom and the USA, censuses now occur every ten years.

centile *See* PERCENTILE.

central composite design *See* RESPONSE SURFACE.

central limit theorem (clt) A theorem, proposed by *Laplace, explaining the importance of the *normal distribution in Statistics. Let X_1, $X_2, \ldots, X_n$ be *independent *random variables each having the same *distribution, with *mean μ and *variance σ^2. Let $\bar{X}$, given by

$$\bar{X} = \frac{1}{n}(X_1 + X_2 + \cdots + X_n),$$

denote the sample mean. The central limit theorem states that, for large n, the distribution of $\bar{X}$ is approximately a *normal distribution with mean μ and variance $\frac{1}{n}\sigma^2$. Thus, for a large random *sample of *observations from a distribution with mean μ and variance σ^2, the distribution of the sample mean is approximately normal with mean μ and variance $\frac{1}{n}\sigma^2$ and the distribution of the sample total is approximately normal with mean $n\mu$ and variance $n\sigma^2$. The phrase 'central limit theorem' appears in the title (in German) of an article by *Pólya published in 1920.

central moment *See* MOMENT.

central tendency (centrality) The tendency of quantitative *data to cluster around some central value. The central value is commonly estimated by the *mean, *median, or *mode, whereas the closeness with which the values surround the central value is commonly quantified using the *standard deviation or *variance. The phrase 'central tendency' was first used by the American psychologist E. L. Thorndike in 1905.

centred moving average *See* MOVING AVERAGE.

cgf *See* CUMULANT.

CHAID (Chi-square Automatic Interaction Detection) A program for constructing a *classification tree.

chain graph *See* GRAPHICAL MODEL.

change-point problem A problem concerned with the identification of the point in a *time series at which some change occurs. Suppose that *data are collected on one or more *variables at a series of time points in order to examine the properties of the variables and possible relationships between them. The time of collection appears to be incidental. However, frequently this turns out to be not quite true. For example, in one case the explanation for an apparent change in the mean of a variable was found to be attributed to a tall meter reader who looked down at a dial being replaced by a short one who looked up at the same dial. Another example concerns changes in the responses to a question in an annual survey that turn out to be a consequence of a change in the survey design. In these examples, the cause of the change and hence the time that the change took place have been identified. Often, however, though it is evident that a change has taken place, it is not clear at what precise point this occurred.

c

Chapman, Sydney (1888–1970; b. Eccles, England; d. Boulder, CO) English mathematician. In 1907, after graduating as an engineer at Manchester U, he studied mathematics at Cambridge U. In 1910 he became assistant to the Astronomer Royal, and this led to his research in geomagnetic theory. His subsequent academic career involved appointments at Manchester U (1919), IC (1924), and Oxford U (1946). He was elected FRS in 1919 and was awarded the Society's Copley Medal in 1964. He was President of the LMS in 1929 and awarded its de Morgan Medal in 1944. He was elected FRSE in 1953.

SEE WEB LINKS

- Fuller biography and photographs.

Chapman estimator *See* CAPTURE-RECAPTURE METHOD.

Chapman–Kolmogorov equation *See* MARKOV PROCESS.

characteristic equation *See* MATRIX.

characteristic function The (complex-valued) function $\mathrm{E}(\mathrm{e}^{itX})$, where X is a random variable, t is a real variable and $\mathrm{i} = \sqrt{-1}$. If X is continuous and has *probability density function f then the characteristic function is a multiple of the *Fourier transform of f. The characteristic function is closely related to the *moment-generating function and has the advantage of always existing. The first English use of the phrase 'characteristic function' was by *Kullback in 1934.

characteristic value; characteristic vector *See* MATRIX.

Chebyshev, Pafnuty Lvovich (1821–94; b. Okatovo, Russia; d. St Petersburg, Russia) Russian mathematician. Chebyshev spent most of his career at St Petersburg U, where he developed a mathematical school with an international reputation (*Markov was one of his students). He made major contributions in many branches of mathematics. To statisticians he is best remembered for his work on the weak law of large numbers (*see* LAWS OF LARGE NUMBERS) and the introduction of polynomials as a means of handling the fitting of a curvilinear regression model (*see* MULTIPLE REGRESSION MODEL). A lunar crater is named after him.

SEE WEB LINKS

- Chebyshev 190th anniversary stamp.

Chebyshev distance *See* DISTANCE MEASURE.

Chebyshev–Hermite polynomials Polynomials that play an important role in the *Edgeworth expansion and *Gram-Charlier expansion. The rth of these polynomials, $H_r(x)$, is the coefficient of $t^r/r!$ in the expansion of $\exp(tx - \frac{1}{2}t^2)$ and is given by

$$H_r(x) = x^r - \frac{r(r-1)}{2.1!}x^{r-2} + \frac{r(r-1)(r-2)(r-3)}{2^2.2!}x^{r-4} - \cdots.$$

Chebyshev inequality For a *random variable X with *expected value μ and *standard deviation σ,

$$P(|X - \mu| > k\sigma) \leq \frac{1}{k^2}$$

for all positive values of the constant k. *See also* BERNSTEIN INEQUALITY; HÖLDER INEQUALITY; KOLMOGOROV INEQUALITY; MARKOV INEQUALITY; MINKOWSKI INEQUALITY.

Chernoff, Herman (1923– ; b. New York City) American applied mathematician and statistician. Chernoff was educated at City College (now City U) NY and at Brown U, where he obtained his doctorate in 1948. After a period at U Illinois (Urbana) he moved to Stanford U in 1952. In 1974 he became Professor of Applied Mathematics at MIT and in 1985 moved to become Professor of Statistics at Harvard U. He now holds the title of Professor Emeritus at both MIT and Harvard U. His statistical interests include optimal *experimental design, pattern recognition, and *sequential sampling. In 1968 when he was President of the *IMS he was its *Wald Lecturer. In 1989 he was its *Rietz Lecturer. In 1975 he was the *COPSS *Fisher Lecturer and in 1987 he was the *ASA's *Wilks Award winner. He was the 2013 *Rao Prize recipient.

SEE WEB LINKS

- Fuller biography, interview, and photographs.

Chernoff faces An alternative to *Andrews plots as a method of giving a two-dimensional representation of *multivariate data. The shape and size of the various characteristics of a face are varied according to the values of the *variables concerned. *See diagram overleaf.*

SEE WEB LINKS

- Description of R functions for automated face drawing.

Chinese postman problem A *network problem that can be formulated as a *combinatorial optimization problem. A postman has to travel down every street in a town. The streets have different lengths. The problem is to devise a route that minimizes the distance the postman

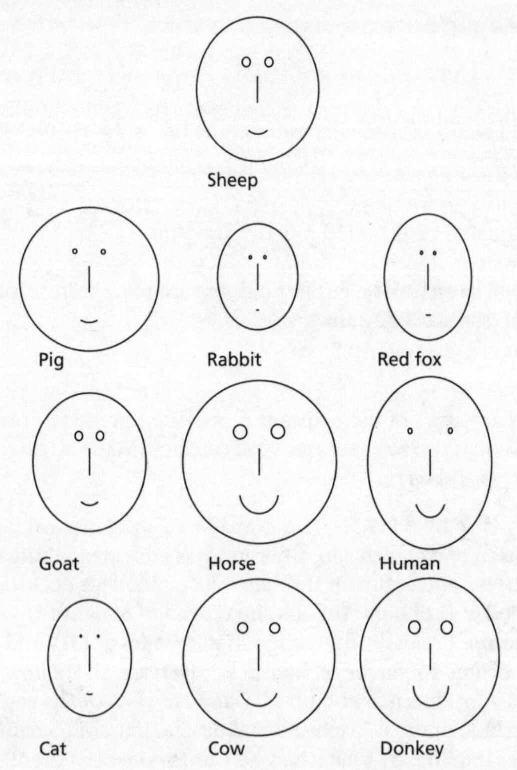

Chernoff faces. The horizontal scaling, vertical scaling, size of eyes, and shape of mouth are governed by the values of various attributes of the creatures being compared. Note the similarity between sheep and goat!

---

travels. The adjective 'Chinese' refers to the nationality of the original poser of the problem (Mei-Ku Kuan) and not to any unusual behaviour by Chinese postmen.

chi-squared distribution (χ^2) If $Z_1, Z_2, \ldots, Z_\nu$ are ν *independent standard normal variables (*see* NORMAL DISTRIBUTION), and if Y is defined by

$$Y = \sum_{j=1}^{\nu} Z_j^2,$$

then Y has a chi-squared distribution with ν *degrees of freedom (written as χ_ν^2). The *probability density function f is given by

$$f(y) = \frac{e^{-\frac{1}{2}y} y^{\left(\frac{1}{2}\nu - 1\right)}}{2^{\frac{1}{2}\nu}\Gamma\left(\frac{1}{2}\nu\right)}, \qquad y > 0,$$

where Γ is the *gamma function. The form of the distribution was first given by *Abbe in 1863 and was independently derived by *Helmert in 1875 and Karl *Pearson in 1900. It was Pearson who gave the distribution its current name.

The chi-squared distribution has *mean ν and *variance 2ν. For $\nu \leq 2$ the *mode is at 0; otherwise it is at $(\nu - 2)$. A chi-squared distribution is a special case of a *gamma distribution. The case $\nu = 2$ corresponds to the *exponential distribution. Percentage points for chi-squared distributions are given in Appendix VIII.

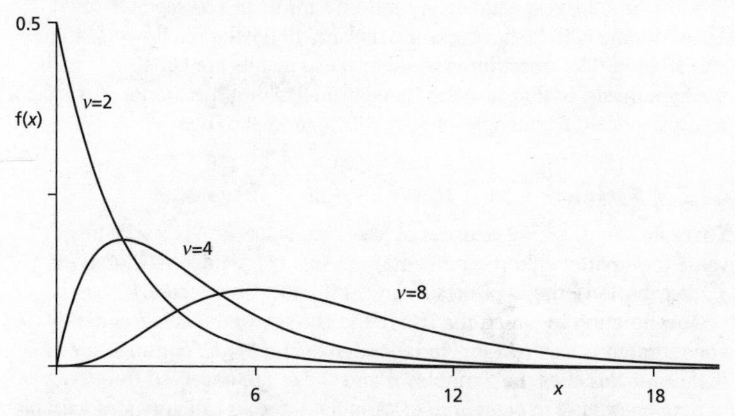

Chi-squared distribution. All chi-squared distributions have ranges from 0 to ∞. Their shape is determined by the value of ν. If $\nu > 2$ then the distribution has a mode at $(\nu - 2)$; otherwise the mode is at 0.

chi-squared test (Pearson goodness-of-fit test) A *goodness-of-fit test, introduced by Karl *Pearson in 1900, that is popular because of its simplicity. Let O denote the observed *frequency of an outcome in a *sample and let E denote the corresponding *expected frequency under some *model. The test statistic is X^2, defined by

$$X^2 = \Sigma r^2,$$

where the summation is over all the outcomes whose frequencies are being compared, and r, the **Pearson residual**, is defined by

$$r = \frac{O - E}{\sqrt{E}}.$$

It is important to note that the comparison involves frequencies and not proportions.

If there are m observed frequencies, and p *parameters have been estimated using these frequencies, then, if the model is correct, the observed value of X^2 will approximate an observation from a *chi-squared distribution with $(m-p-1)$ *degrees of freedom. In the case of a *continuous random variable, the frequencies will refer to ranges of values. For other random variables it is usual to combine together neighbouring rare values into a single category, since the chi-squared approximation fails if there are too many small expected frequencies.

As an example, suppose it is hypothesized that a type of sweet pea occurs in shades of white, red, pink, and blue, with proportions $\frac{1}{4}$, p, $(\frac{3}{4}-3p)$, and $2p$, respectively. A random sample of 120 seeds is sown. All germinate with 20 having white flowers, 10 having red flowers, 40 pink, and 50 blue. The question is whether these results are consistent with the hypothesis. In this case the *maximum likelihood estimate of p is 0.15, so the expected frequencies are 30, 18, 36, and 36. Thus

$$X^2=\frac{(20-30)^2}{30}+\frac{(10-18)^2}{18}+\frac{(40-36)^2}{36}+\frac{(50-36)^2}{36}=12.8.$$

There are $4-1-1=2$ degrees of freedom. Since 12.8 is a very large value (compared with the percentage points (Appendix VIII) of a χ^2_2-distribution), the hypothesis can confidently be rejected.

One situation in which the use of the chi-squared test is frequently encountered is as a test for *independence in a $J \times K$ *contingency table that cross-classifies the variables A and B. Let the observed frequency of data belonging to category j of variable A and to category k of variable B be f_{jk}. Write

$$f_{j0}=\sum_{k=1}^{K} f_{jk}, \qquad f_{0k}=\sum_{j=1}^{J} f_{ik}, \qquad f_{00}=\sum_{j=1}^{J}\sum_{k=1}^{K} f_{jk}.$$

Then, according to the *null hypothesis of independence, the expected frequency e_{jk} is given by

$$e_{jk}=\frac{f_{j0}f_{0k}}{f_{00}},$$

and the test statistic X^2 is given by

$$X^2=\sum_{j=1}^{J}\sum_{k=1}^{K}\frac{(f_{jk}-e_{jk})^2}{e_{jk}}.$$

If the null hypothesis of independence is correct, then the distribution of X^2 can be approximated by a chi-squared distribution with

$(J-1)(K-1)$ degrees of freedom. In the special case where $J=K=2$ (a two-by-two table), the chi-squared approximation is improved by using the Yates-corrected chi-squared test (*see* TWO-BY-TWO TABLE).

In 1946 *Cramér suggested that a measure of *association could be based on the value of X^2. This is **Cramér's** ***V***, given by

$$V=\sqrt{\frac{X^2}{Mf_{00}}},$$

where M is the smaller of $J-1$ and $K-1$.

chi-squared variable A *random variable having a *chi-squared distribution.

Cholesky, André-Louis (1875–1918; b. Montguyon, France; d. Bagneux, France) French mathematician and geodesist. Cholesky was trained at the École Polytechnique and entered the artillery branch of the French army. By 1905 he was attached to the Geodesic Section and spent the winter of 1907–8 mapping first Crete, then Algeria, Tunisia, and Morocco. He was killed in action in North France in the First World War.

SEE WEB LINKS

- Biography, photographs, etc.

Cholesky decomposition If **A** is a non-singular symmetric *matrix, then it can be expressed as the product **LL**′, where **L** is a lower triangular matrix and **L**′ is the transpose (*see* MATRIX) of **L**. The decomposition is useful in solving sets of linear equations and in matrix inversion.

Chow test The *null hypothesis is that the response variable **Y** is related to k explanatory variables by the *multiple regression model $\mathrm{E}(\mathbf{Y})=\mathbf{X}\boldsymbol{\beta}$, for which the residual sum of squares (*see* ANOVA) is R. The data consists of two sub-samples (1 and 2; of sizes n_1 and n_2) and the alternative hypothesis is that in sub-sample j ($j=1,2$) the model is $\mathrm{E}(\mathbf{Y})=\mathbf{X}\boldsymbol{\beta}_j$ with residual sum of squares R_j. The Chow test, introduced in 1960, assumes independent normal errors and has test statistic F, given by

$$F=\frac{(n_1+n_2-k)(R-R_1-R_2)}{k(R_1+R_2)}.$$

The test statistic is compared with an *F-distribution having k and (n_1+n_2-2k) *degrees of freedom.

chuck-a-luck A fairground gambling game. The player nominates a number from 1 to 6. Three fair dice are rolled. If the player's number appears then the player receives back his stake multiplied by $(1+k)$, where k is the number of dice displaying the nominated number. The average win per unit staked is 199/216.

circular data *See* CYCLIC DATA.

C

circular distribution A distribution suitable for *cyclic data. Examples include the *circular uniform distribution, the *von Mises distribution, and the *wrapped Cauchy distribution.

circular histogram *See* CYCLIC DATA.

circular mean *See* CYCLIC DATA.

circular normal distribution *See* VON MISES DISTRIBUTION.

circular uniform distribution The *uniform distribution for directional and *cyclic data. The *probability density function f is constant for directions (in *radians)

$$f(\theta) = \frac{1}{2\pi}, \qquad -\pi < \theta \leq \pi.$$

For a data set $\theta_1, \theta_2, \ldots, \theta_n$, a test of circular uniformity amounts to a test of the *null hypothesis, H_0, that there is no preferred direction, with the alternative being that the data come from some *unimodal circular distribution. The **Rayleigh test**, introduced by Lord *Rayleigh in 1880, uses the test statistic R^2 given by

$$R^2 = \left(\sum_{j=1}^{n} \sin\theta_j\right)^2 + \left(\sum_{j=1}^{n} \cos\theta_j\right)^2.$$

Under H_0 the approximate *probability of observing a value $\geq R^2$ is given by

$$\exp\left(\sqrt{1 + 4n + 4(n^2 - R^2)} - (2n + 1)\right).$$

For very large values of n, $T = 2R^2/n$ is an *observation from an approximate *chi-squared distribution with two *degrees of freedom.

As an alternative, the **Ajne test** (suggested by the Swede Björn Ajne in 1968) uses as its test statistic the number of observations contained within the semicircle for which this number is least.

city-block metric *See* DISTANCE MEASURE.

class; class boundary (class limit) When numerical *data are grouped into classes (e.g. $1 \leq x < 5$, $5 \leq x < 10, \ldots$), the values marking the limits of a class are called the lower and upper class boundaries. Some texts reserve the term 'class limit' for use with *continuous variables.

class frequency The number of *observations in a *class.

classification tree; classifier A rule for predicting the class of an object from the values of its predictor variables. Classification trees are much used in *data mining. They differ from *discriminant analysis in that judgements are reached by considering variables hierarchically rather than simultaneously. Classification tree programs include CART, CHAID, and QUEST. *See also* DECISION TREE.

class limit *See* CLASS.

clinical trials A carefully designed set of studies of the value of alternative treatments or procedures used for the prevention, diagnosis, or treatment of a disease. *See* BLINDING.

clinimetrics The study of rating scales and indexes for the description of clinical phenomena.

Clopper–Pearson method A method suggested by C. J. Clopper and Egon *Pearson in 1934 for obtaining a *confidence interval for an unknown *binomial probability, p. Suppose x out of n binomial trials are successes. Let p_L be the value of p for which $P(X \leq x) = \frac{1}{2}\alpha$ and let p_U be the value for which $P(X \geq x) = \frac{1}{2}\alpha$, then the $(1 - \alpha)$ Clopper–Pearson interval for p is (p_L, p_U).

closed question (multiple choice question) A question in a *questionnaire for which the respondent must select one from a number of given answers. By contrast an **open question** allows the respondent to answer in any fashion.

clt *Abbreviation for* CENTRAL LIMIT THEOREM.

cluster analysis A method for identifying *data items that closely resemble one another, assembling them into **clusters**. A number of characteristics are measured for each of several items (which might be, for example, people, plants, machines, etc.). The process of formation of the clusters is often represented using a *dendrogram. The most commonly used methods are the *agglomerative clustering methods.

clustered *See* INDEX OF DISPERSION.

clusters *See* CLUSTER ANALYSIS; CLUSTER SAMPLING.

cluster sampling An economical method for *sampling a scattered *population. When, as is usually the case, a geographical population is scattered, it would be uneconomic to visit the scattered individuals chosen by simple random sampling. Instead, the population is subdivided into a large number of geographically compact regions (the **clusters**) and a random sample of clusters is selected. These are referred to as the **primary sampling units (PSUs)**. In single-stage cluster sampling all members of the selected cluster are interviewed. In multi-stage cluster sampling, further subdivisions take place.

cobweb diagram A diagram that illustrates the *associations between the categories of two or more *categorical variables by means of lines whose widths indicate the strengths of the association and whose colour indicates whether the association is positive or negative.

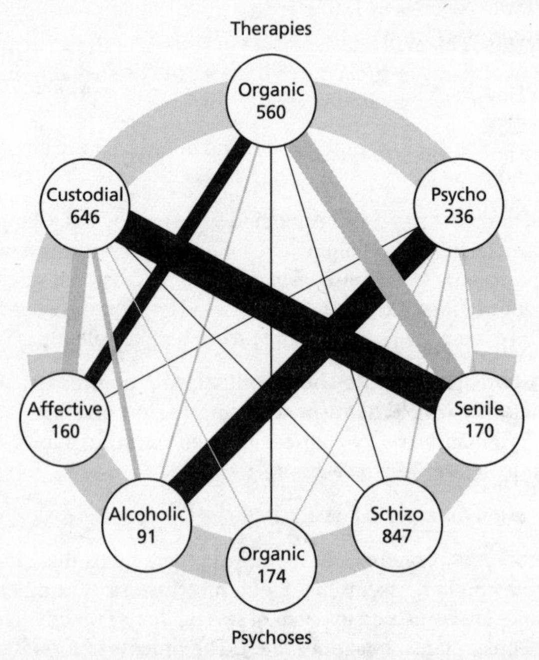

Cobweb diagram. The diagram shows the relation between types of insanity and the prescribed treatment. The thickness of the lines indicates the strength of the association between the linked categories. Black lines indicate positive association and grey lines indicate negative association. The diagram demonstrates that senile patients are particularly likely to be given custodial therapy and particularly unlikely to be given organic therapy.

Cochran, William Gemmell (1909–80; b. Rutherglen, Scotland; d. Orleans, MA) Scottish statistician who spent most of his career in the USA. Cochran studied mathematics first at Glasgow U and then at Cambridge U (supervised by *Wishart). During this time his first papers were published; the second of these (in 1934) introduced *Cochran's theorem. At the height of the depression Cochran abandoned his doctoral studies when offered a job as assistant to *Yates at *Rothamsted. In 1939 he moved to the USA, where, successively, he held chairs at Princeton U, North Carolina State U, Johns Hopkins U, and Harvard U. His co-authored books *Experimental Designs* (with Gertrude *Cox) and *Statistical Methods* (with George *Snedecor) were regarded as compulsory reading by the next generation of statisticians. He was President of the *IMS in 1946 and was its 1967 *Rietz Lecturer. He was President of the *ASA in 1953 and was its *Wilks Award winner in 1967. He was Editor of the **Journal of the American Statistical Association* from 1946 to 1950, and was the *COPSS *Fisher lecturer in 1972. He was elected to the NAS in 1974 and was elected an Honorary Life Member of the *IBS in 1976.

SEE WEB LINKS

- Biographical memoir with photograph.

Cochran-Armitage test A test (named after *Cochran and *Armitage) for trend in a $2 \times k$ *contingency table where the columns correspond to categories of an *ordinal variable. Denote the *frequency in row i and column j by f_{ij} with $i = 1$ or 2, and $j = 1, \dots, J$, denote the row totals by f_{10} and f_{20}, the column totals by $f_{01}, \dots, f_{0J}$, and the grand total by f_{00}. The test examines the *null hypothesis that, as j increases, the population proportion for a given row consistently increases (or decreases). The test statistic is T given by

$$T = \frac{\sum_{j=1}^{J} t_j (f_{1j} f_{20} - f_{2j} f_{10})}{\sqrt{f_{10} f_{20} \left(\sum_{j=1}^{J} t_j^2 f_{0j} (f_{00} - f_{0j}) - 2 \sum_{j=1}^{J-1} \sum_{k=j+1}^{J} t_j t_k f_{0j} f_{0k} \right) / f_{00}}},$$

where $t_1, \dots, t_k$ are constants chosen by the user (the optimal choice is usually $t_j = j$, for all j). In the absence of trend, T has an approximate standard *normal distribution.

The test is popularly used in the genetic context where $J = 3$, the column categories are three possible genotypes labelled *aa*, *Aa*, and *AA*, and the rows refer to two types of individual.

Cochrane–Orcutt estimation An *iterative algorithm, suggested in 1949, for the *estimation of the *parameters of a *linear regression relation when there is non-zero *autocorrelation between successive errors.

Cochran C test *See* TEST FOR EQUALITY OF VARIANCE.

Cochran Q test *See* MCNEMAR TEST.

Cochran's theorem A theorem, given by *Cochran in 1934, concerning sums of *chi-squared variables. Let **Y** be an $n \times 1$ vector of *independent standard normal variables (*see* NORMAL DISTRIBUTION) and let $\mathbf{A}_1, \mathbf{A}_2, \ldots, \mathbf{A}_k$ be non-zero symmetric *matrices such that

$$\sum_{j=1}^{k} \mathbf{Y}'\mathbf{A}_j\mathbf{Y} = \mathbf{Y}'\mathbf{Y} \text{ for all } \mathbf{Y},$$

where $\mathbf{Y}'$ is the transpose of $\mathbf{Y}$. Write $Q_j = \mathbf{Y}'\mathbf{A}_j\mathbf{Y}$. Cochran's theorem, published in 1934, states that, if any one of the following three conditions is true, then so are the other two:

(i) The ranks of $\mathbf{A}_1, \mathbf{A}_2, \ldots, \mathbf{A}_k$ sum to n.
(ii) Each of $Q_1, Q_2, \ldots, Q_k$ has a chi-squared distribution.
(iii) Each of $Q_1, Q_2, \ldots, Q_k$ is independent of all the others.

code; codebook; coding A number entered into a computer to signify each answer to a *questionnaire. The list of codes for each question is stored in the codebook. The process of translating the questionnaire responses into codes is called coding the data.

coded data *See* WORKING MEAN.

coefficient of alienation *See* ANOVA.

coefficient of concordance A *measure of agreement between m observers who *rank n items in order according to some characteristic. Let R_j be the total of the ranks assigned to the jth item and let S be given by

$$S = \sum_{j=1}^{n} \left\{ R_j - \frac{1}{2}m(n+1) \right\}^2 .$$

The coefficient of concordance, W, is given by

$$W = \frac{12S}{m^2(n^3 - n)},$$

which takes values from 0 (no general agreement) to 1 (complete agreement). The coefficient was proposed by Sir Maurice *Kendall and Bernard Babington-Smith in 1939.

coefficient of correlation *Alternative name for the* *SAMPLE CORRELATION COEFFICIENT.

coefficient of determination *See* ANOVA.

coefficient of variation The *standard deviation divided by the *mean. A term introduced by Karl *Pearson in 1896.

cofactor *See* MATRIX.

Cohen, Jacob (1923–98; b. New York City; d. New York City) American psychologist and a pioneer of research methods. Cohen enrolled in the City College (now City U) of New York at the age of fifteen. After work in intelligence in the Second World War, he gained a PhD in clinical psychology at New York U, where he spent most of his teaching career. He was a co-author of a popular introductory statistics book for the behavioural sciences, first published in 1971.

SEE WEB LINKS

- Fuller biography and photograph.

Cohen's kappa (κ) A *measure of agreement between two observers, suggested by *Cohen in 1960. Suppose that the observers are required, independently, to assign items to one of m classes. Let f_{jk} be the number of individuals assigned to class j by the first observer and to class k by the second observer. Let $f_{j0} = \sum_{k=1}^{m} f_{jk}$, $f_{0k} = \sum_{j=1}^{m} f_{jk}$, and $f_{00} = \sum_{j=1}^{m} \sum_{k=1}^{m} f_{jk}$. Define the quantities O and E by

$$O = \sum_{j=1}^{m} f_{jj}, \quad E = \sum_{j=1}^{m} \frac{f_{j0} f_{0j}}{f_{00}},$$

so that O is the total number of individuals on which the observers are in complete agreement, and E is the expected total number of agreements that would have occurred if the observers had been statistically independent. The formula for Cohen's kappa is

$$\kappa = \frac{O - E}{f_{00} - E}.$$

A value of 0 indicates statistical independence, and a value of 1 indicates perfect agreement.

coherence A term used to describe the resemblance between the fluctuations displayed by two *time series; an analogue of *correlation.

cohort study A *longitudinal study of the same group of people (the **cohort**) over time. By contrast with a *panel study, different members of the cohort may be studied at each time point. Usually the members of a cohort are of approximately the same age—for example, all those with 21st birthdays during the year 2000.

coiflet *See* WAVELET.

cokriging *See* KRIGING.

cold deck *See* IMPUTATION.

collaborative filtering A type of *recommender system used by organizations to infer characteristics of individuals, usually with the aim of encouraging those individuals to make future purchases. For an individual (B, say), the aim is to use the apparent ratings of possible purchases that are made by individuals that the organization perceives as being similar to B, to predict the ratings that B would give: the top-rated predictions are then presented as suggestions to B. Internet shopping provides organizations with both rating evidence and the opportunity to present suggestions. Techniques used include *correlation and the use of *graphs.

collapsing A term used in connection with the merging of neighbouring categories. For example, if the *random variable X takes values 0, 1, and 2 with high *probability and values 3, 4, and 5 with low probability, then it may be sensible to collapse the latter categories into a single '3 or more' category.

As a more extreme example, suppose that, in an $I \times J \times K \times L$ *contingency table, for variables A, B, C, D, it turns out that D is of no interest. All the categories of D may then be collapsed together to give an $I \times J \times K$ table—the original table has been collapsed over D.

collinearity *See* MULTIPLE REGRESSION MODEL.

column vector *See* MATRIX.

combination An unordered selection of r objects from a set of n $(\geq r)$ different objects. The number of different combinations is often denoted by ${}^nC_r$ or ${}_nC_r$. In fact,

$$ {}^nC_r = \binom{n}{r} $$

is the *binomial coefficient. Special values are ${}^nC_0 = 1$, ${}^nC_n = 1$, ${}^nC_1 = n$.

A frequently used relationship is

$$ {}^{n+1}C_r = {}^nC_r + {}^nC_{r-1}, $$

which is the defining relationship for *Pascal's triangle. For ordered selection, *see* PERMUTATION.

combinatorial optimization An optimization technique in which the values of the *variables are restricted to integers. Examples include the *knapsack and *travelling salesman problems.

combinatorics The study of the numbers of ways of selecting, or arranging, objects from a finite set. Its main applications in the theory of *probability are in calculating the probability of an event such as r of the objects chosen having a particular property, or the probability of a sequence of events with a particular property. *See also* BINOMIAL DISTRIBUTION; HYPERGEOMETRIC DISTRIBUTION; NEGATIVE BINOMIAL DISTRIBUTION.

Combinatorial results often involve *factorials and *binomial coefficients. For example, the probability that, in a random deal of 13 cards from a normal pack, Player A receives 13 spades is

$$\frac{1}{\binom{52}{13}} \approx 1.57 \times 10^{-12},$$

and the probability that in a random deal to four players, each player receives 13 cards of the same suit is

$$\frac{4!}{\binom{52}{13}\binom{39}{13}\binom{26}{13}} \approx 4.47 \times 10^{-28}.$$

combined probability of inclusion *See* RANDOM MAN NOT EXCLUDED.

Committee of Presidents of Statistical Societies *See* COPSS.

communicating states *See* MARKOV PROCESS.

competing risks *See* RISK.

complementary event The complementary event A' to an event A is the event 'A does not occur'. It satisfies $A \cup A' = S$, where S is the *sample space, and $A \cap A' = \phi$, where ϕ is the empty set. The complements of *intersections and *unions are given by de Morgan's laws:

$$(A \cap B)' = A' \cup B',$$
$$(A \cup B)' = A' \cap B'.$$

See also BOOLEAN ALGEBRA; SAMPLE SPACE.

complementary log-log link *See* GENERALIZED LINEAR MODEL.

complete linkage clustering A method of collecting *multivariate data into clusters (*see* CLUSTER ANALYSIS). For a description, *see* AGGLOMERATIVE CLUSTERING METHODS.

completely randomized design An *experimental design in which *treatments are allocated to *experimental units at random.

complexity A measure of the computer time or space required to solve a problem by means of an *algorithm of interest, expressed as a function of the dimensions of the problem. If a problem with n dimensions can be solved in at most P(n) time units, where P is a polynomial, then the algorithm is said to have **polynomial-time complexity**.

components of variance model (variance components model) A *linear model in which the explanatory variables all have random effects (*see* EXPERIMENTAL DESIGN). The analysis concentrates on the relative sizes of the variances of these effects.

composite hypothesis *See* HYPOTHESIS TEST.

compositional variable A *variable describing the breakdown of some quantity into components that sum to the whole. An example is the outcome of the vote in a general election: in each constituency the separate votes for the competing parties sum to 100% of the votes cast. The composition varies from constituency to constituency. When the composition has just three parts a ternary diagram provides an appropriate display.

compound bar chart A *bar chart used when *data are cross-categorized according to two *categorical variables. One variable is regarded as the main variable and the bars representing this variable are subdivided according to the other variable.

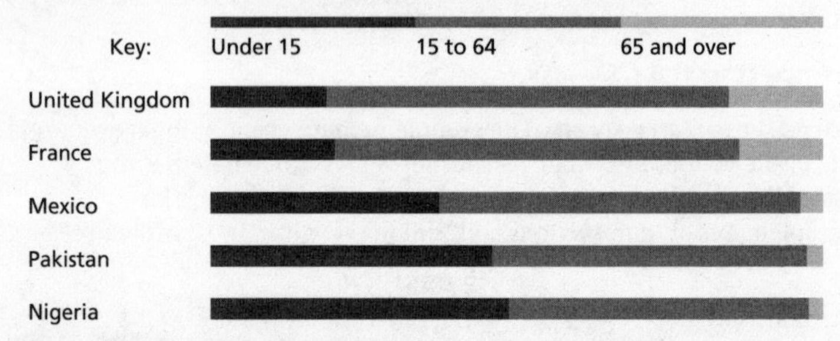

Compound bar chart. The diagram shows the similarity between the age distributions of two European countries and the difference between these and the age distributions of countries in other continents.

compound distribution A *distribution that results from allowing the *parameter of a distribution to vary. For example, if the success *probability of a *binomial distribution is not constant but has a *beta distribution, then the result is a *random variable having a *beta-binomial distribution.

compound Poisson process *See* POISSON PROCESS.

compound probability *See* MULTIPLICATION LAW FOR PROBABILITIES.

Computational Statistics & Data Analysis The official journal of the *International Association for Statistical Computing. It is published monthly.

SEE WEB LINKS

- Journal home page.

computer-intensive method A description of a method that relies on extensive computer calculations to obtain numerical results.

concentration *See* CYCLIC DATA.

concordance probability Let (X_1, Y_1) and (X_2, Y_2) represent two random observations from a *bivariate distribution. The concordance probability, K, is the *conditional probability that Y_2 is not less than Y_1, given that X_2 is not less than X_1. *See also* COEFFICIENT OF CONCORDANCE.

conditional autoregressive model A model of a spatial process in which the *variance of the value at each point in space is the same and the *expected value of the value at each point is a weighted sum of the observed values at neighbouring points.

conditional distribution; conditional expectation *See* BIVARIATE DISTRIBUTION.

conditional independence If, for each value of the *variable C, the variables A and B are *independent of one another, then they are said to exhibit conditional independence. If the variables are *categorical, with p_{jkl} denoting the *probability of an outcome in *cell (j, k, l), then A and B are conditionally independent, given the category of C, if and only if

$$p_{jkl} = \frac{p_{j0l}p_{0kl}}{p_{00l}},$$

for all j, k, and l, where $p_{j0l} = \sum_k p_{jkl}$, $p_{0kl} = \sum_j p_{jkl}$, and $p_{00l} = \sum_j \sum_k p_{jkl}$.

conditional probability If A and B are events (*see* SAMPLE SPACE), and $\mathrm{P}(B) > 0$, the conditional probability of A given B is defined by

$$\mathrm{P}(A|B) = \frac{\mathrm{P}(A \cap B)}{\mathrm{P}(B)},$$

or equivalently

$$\mathrm{P}(A \cap B) = \mathrm{P}(B) \times \mathrm{P}(A|B).$$

If $\mathrm{P}(A) > 0$ also, then $\mathrm{P}(A \cap B) = \mathrm{P}(A) \times \mathrm{P}(B|A)$, so

$$\mathrm{P}(A) \times \mathrm{P}(B|A) = \mathrm{P}(B) \times \mathrm{P}(A|B).$$

The '$A|B$' notation was introduced by *Jeffreys and popularized by *Feller. *See also* BAYES'S THEOREM; INDEPENDENT EVENTS.

Condorcet paradox A voting paradox noted by the Marquis de Condorcet in an essay published in 1785. For example, suppose there are three candidates, A, B, and C, and three voters whose preferences are as follows:

	Preference		
	First	Second	Third
Voter 1:	A	B	C
Voter 2:	B	C	A
Voter 3:	C	A	B

A is preferred to B by a majority of voters and B is preferred to C by a majority. However, it is also the case that C is preferred to A by a majority.

confidence ellipse, confidence ellipsoid The analogue of *confidence interval when a statement is to be made about the likely values of two or more unknown *parameters.

confidence interval (interval estimate) An $\alpha\%$ confidence interval for an unknown *population *parameter θ, say, is an interval, calculated from *sample values by a procedure such that if a large number of *independent samples is taken, $\alpha\%$ of the intervals obtained will contain θ. The term 'confidence interval' was introduced in 1934 by *Neyman.

A confidence interval can also be thought of as a single *observation of a random interval, calculated from a random sample by a given procedure, such that the *probability that the interval contains θ is $\alpha\%$. For example, if $X_1, X_2, \ldots, X_n$ is a random sample from a *normal distribution with unknown *mean μ and known *variance σ^2, and writing $\bar{X} = (X_1 + X_2 + \cdots + X_n)/n$,

$$\mathrm{P}\left(-1.96\frac{\sigma}{\sqrt{n}} < \bar{X} - \mu < 1.96\frac{\sigma}{\sqrt{n}}\right) = 0.95,$$

so

$$\mathrm{P}\left(\bar{X} - 1.96\frac{\sigma}{\sqrt{n}} < \mu < \bar{X} + 1.96\frac{\sigma}{\sqrt{n}}\right) = 0.95\%.$$

Hence, if $\bar{x}$ is the observed value of the sample mean, the end points of the corresponding 95% confidence interval for the mean, μ, are $\bar{x} \pm 1.96\sigma/\sqrt{n}$. This is a **symmetric confidence interval**. It is possible to have a **one-sided confidence interval**. For example $\mu > \overline{x} - 1.645\,\sigma/\sqrt{n}$ is a one-sided 95% confidence interval for the mean, μ.

If the population variance is not known then, to find a confidence interval for the mean, the *t*-distribution can be used and the end points of the 95% confidence interval for the mean are

$$\overline{x} \pm t_{n-1}(0.025)\frac{s}{\sqrt{n}},$$

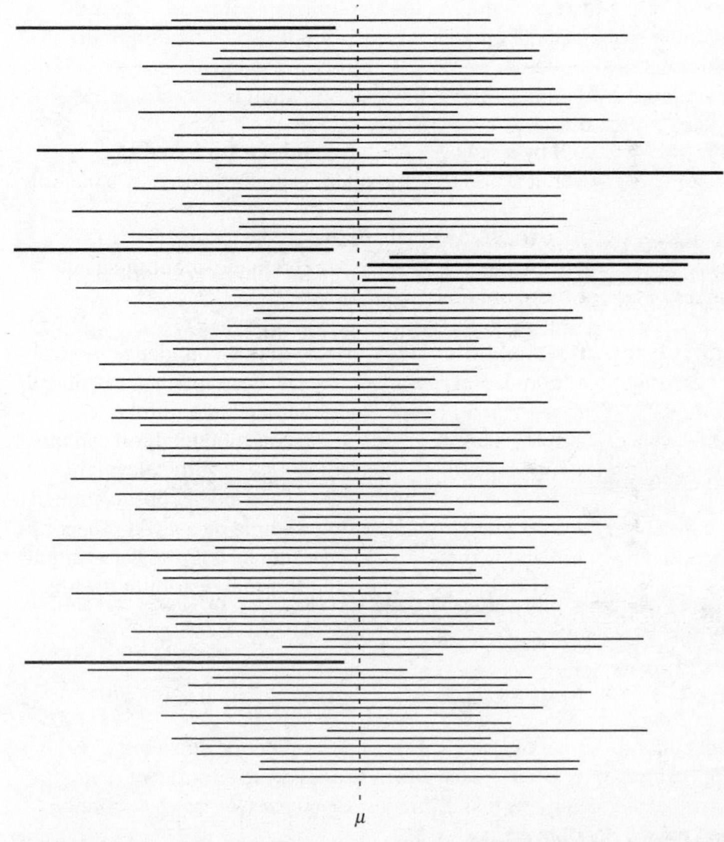

Confidence interval. The illustration shows one hundred 95% confidence intervals for the population mean. Each confidence interval is derived from a random sample from the same distribution. The intervals differ in width and location because of variations in the sample means and variances. On average 95% of such confidence intervals will include the true value of the population mean μ.

where $t_{n-1}(0.025)$ is the *critical value corresponding to an upper-tail probability of 2.5% for a t-distribution with $(n-1)$ *degrees of freedom, and s is the unbiased estimate of the population variance based on the sample values.

C

In the case when the population is not known to be normal the *central limit theorem may be used, provided n is reasonably large, to give the values $\bar{x} \pm 1.96\, s/\sqrt{n}$ as an approximate symmetric 95% confidence interval for μ.

Finding a confidence interval for a population *proportion is difficult, owing to the discrete nature of the *binomial distribution, unless the *sample size n is large enough for the normal approximation to the binomial distribution to be valid. In this case the ends of the α% symmetric confidence interval for the population proportion are the values p such that

$$p = \hat{p} \pm K\sqrt{\frac{p(1-p)}{n}},$$

where $\hat{p}$ is the sample proportion and K is the critical value corresponding to an upper-tail probability of $\frac{1}{2}(100-\alpha)$% for a standard normal distribution. *See also* CLOPPER-PEARSON METHODS.

Equivalently, the α% confidence limits are the roots of the quadratic equation

$$p^2\left(1+\frac{K^2}{n}\right) - p\left(2\hat{p}+\frac{K^2}{n}\right) + \hat{p}^2 = 0.$$

An often used, but not recommended, approximate formula is

$$p = \hat{p} \pm K\sqrt{\frac{\hat{p}(1-\hat{p})}{n}}.$$

Since $\hat{p}(1-\hat{p}) \approx \frac{1}{4}$ for values of $\hat{p}$ not too close to 0 or 1, an even simpler form is

$$p = \hat{p} \pm K\frac{1}{2\sqrt{n}}.$$

Hence for a sample of size 1000 the 90% confidence limits are approximated by $p = \hat{p} \pm 0.03$, which is possibly the source of the oft-repeated statement that estimates of percentages from an opinion poll have a possible error of $\pm 3\%$.

A confidence interval for a *population variance σ^2 can be found, under the assumption that the population is normal. If s^2 is the unbiased estimate of the population variance, based on a sample of size n, then the α% confidence for σ^2 is given by

$$\frac{(n-1)s^2}{U} < \sigma^2 < \frac{(n-1)s^2}{L},$$

where L is the critical value corresponding to a lower-tail probability of $\frac{1}{2}(100-\alpha)\%$, for a *chi-squared distribution with $(n-1)$ degrees of freedom, and U is the critical value corresponding to an upper-tail probability of the same size (*see* APPENDIX VIII).

SEE WEB LINKS

- Applet.

C

confidence limit The end point of a *confidence interval.

confidence region A confidence region is an extension of the idea of *confidence interval in the context of two or more unknown parameters.

conformal prediction theory A theory used in the context of *machine learning where a sequence of predictions is made. The accuracies of past predictions are used to derive both the future prediction and an interval within which the future value will lie with a specified *probability. No distributional assumptions are made, but exchangeability (*see* EXCHANGEABLE) is assumed.

confounded *See* FACTORIAL DESIGN.

congruential generator method A method for generating a sequence of *pseudo-random numbers. Let x_n, the nth number in the sequence, be an integer such that $0 \le x_n \le m-1$. Then x_{n+1}, the next number in the sequence, is given by the relation

$$x_{n+1} = (a + bx_n) \qquad \mod(m),$$

where a and b are constants. The right-hand side of this equation should be interpreted as an instruction to subtract a suitable integer multiple of m from $(a+bx_n)$ so as to obtain a value x_{n+1} such that $0 \le x_{n+1} \le m-1$. The values of a, b, and m have to be chosen carefully in order to get a useful sequence of x-values.

conjugate Latin squares Two *Latin squares are conjugate if the rows of one are the columns of the other in the same order.

conjugate prior *See* BAYESIAN INFERENCE.

connected graph *See* GRAPH.

consequent variable *See* ANTECEDENT VARIABLE.

consistency The property of a method that always produces a consistent *estimator. The word 'consistency' was first used in this way by Sir Ronald *Fisher in 1922.

consistent estimator *See* ESTIMATOR.

consumer's risk *See* ACCEPTANCE SAMPLING.

C

contagious distribution A term used to describe the distribution of a sum of N *random variables when N itself is a random variable.

contingency table (cross-classification; cross-tabulation) A table displaying the *frequencies for each combination of two or more *variables. The variables are either *categorical variables, or numerical variables for which the possible outcomes have been arranged in groups. The term was first used by Karl *Pearson in 1904. Each location in a table is called a **cell**, and the corresponding frequency is the **cell frequency**.

Suppose A and B are two categorical variables having J and K categories, respectively. There are therefore JK possible category combinations. The table described would be called a $J \times K$ table. One simple *model for such a table is the *independence model (*see also* CHI-SQUARED TEST). **Multidimensional contingency tables** summarize information from more than two categorical variables. A three-variable table might be called a $J \times K \times L$ table. Models used include *logit models and, most commonly, *log-linear models.

continuity correction A correction term used when the distribution of a *discrete random variable is approximated by that of a *continuous random variable. For a discrete random variable, X, taking values ..., $x-1, x, x+1, \ldots$, the *probability of X taking the value x would be approximated by

$$\mathrm{P}\left(x-\frac{1}{2}<Y<x+\frac{1}{2}\right),$$

where Y is the approximating continuous random variable. *See also* BINOMIAL DISTRIBUTION; POISSON DISTRIBUTION; TWO-BY-TWO TABLE.

continuous distribution The *probability distribution of a *continuous random variable.

continuous random variable A *continuous variable subject to random variation. The *probability distribution is defined by the *probability density function. It is usual also to require that, for each real number x, $\mathrm{P}(X=x)=0$. As a result it is immaterial whether '$<$' or '$\leq$' is used in probability statements, since e.g. $\mathrm{P}(X \leq x)=\mathrm{P}(X<x)$ and $\mathrm{P}(x_1 \leq X \leq x_2)=\mathrm{P}(x_1<X<x_2)$.

continuous-time Markov chain *See* MARKOV PROCESS.

continuous variable A *variable whose set of possible values is a continuous interval of real numbers x, such that $a < x < b$, in which a can be $-\infty$ and b can be ∞.

contour diagram By analogy with a map, a diagram that illustrates how the value of one quantity, for example the *likelihood of a set of *data, varies with changes in the values of two *parameters.

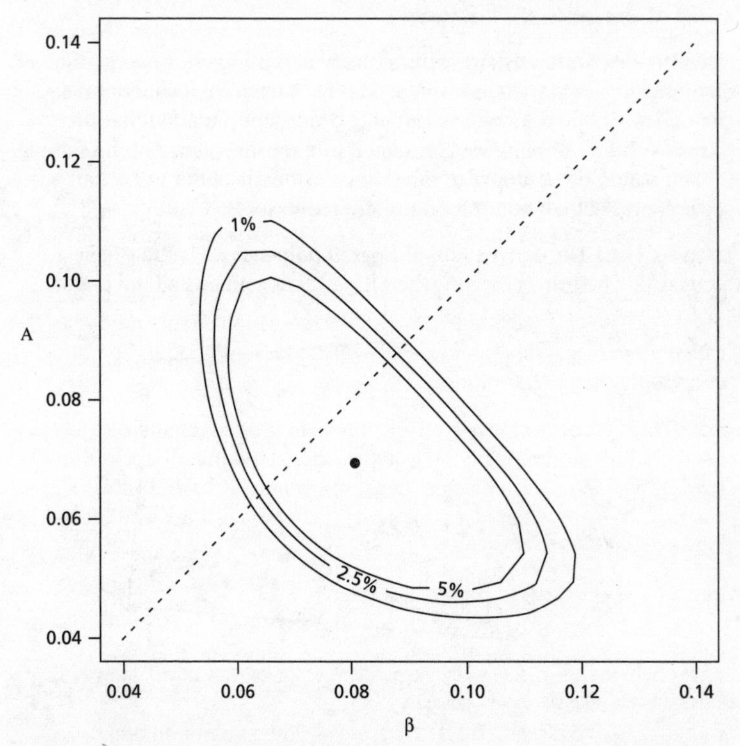

Contour diagram. The likelihood changes with the values of two parameters (A, β). The maximum is indicated by the dot. The contours show values that differ significantly from the maximum at various levels. The dotted line of equality penetrates the inner contour indicating that the hypothesis that the parameters are equal is acceptable at the 5% level.

contrasts *See* ANOVA.

control chart *See* QUALITY CONTROL.

control group The group that experiences the *control treatment.

controlled variable *See* REGRESSION.

control location *See* BACI DESIGN.

control treatment (control) The standard treatment against which new treatments are compared. *See also* RETROSPECTIVE STUDY.

control variable *See* COVARIATE.

convenience sampling (opportunity sampling) A cheap method of obtaining a *sample. An example would be interviewing supermarket customers. This would be a reasonable procedure provided that the purpose of the sampling was unrelated to the convenience of the sample (thus it would be appropriate to ask the customers about car colour preferences, but not about food preferences).

convex hull The convex hull of a set of points in $\mathbb{R}^n$ is the smallest *convex polyhedron (polygon when $n=2$) that contains all the points.

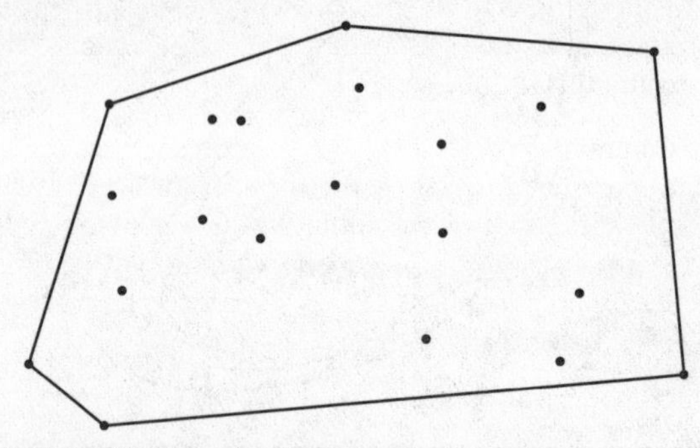

Convex hull. The hull is defined by the positions of the extreme points. In two dimensions the hull is a convex polygon.

convex polyhedron A finite region bounded by a finite number of *hyperplanes, in the sense that the interior of the region lies entirely on one side of each hyperplane.

convolution Given two *independent *random variables X and Y, the *probability distribution of their sum, Z, is called the convolution of the distributions of X and Y. If $\mathrm{f}_X(x)$, $\mathrm{f}_Y(y)$, and $\mathrm{f}_Z(z)$ denote the *probability density functions, then

$$f_Z(z) = \int_{-\infty}^{\infty} f_X(z-y)f_Y(y)dy, \text{ or, equivalently, } \int_{-\infty}^{\infty} f_Y(z-x)f_X(x)dx.$$

For *discrete random variables the equivalent result is

$$P(Z=z) = \sum_y P(X=z-y)P(Y=y) = \sum_x P(Y=z-x)P(X=x).$$

C

Cook, (Ralph) Dennis (1944– ; b. Williston, ND) American statistician. A graduate of Northern Montana College, Cook gained his PhD in 1971 from Kansas State U and joined the faculty at U Minnesota. His research interests include *experimental design, *regression diagnostics, and dimension reduction. He was the *COPSS *Fisher Lecturer in 2005.

SEE WEB LINKS

- University website.

Cook's statistic *See* REGRESSION DIAGNOSTICS.

cophenetic correlation coefficient The *sample correlation coefficient using the distances at which a pair of *observations are joined in a *dendrogram and the *dissimilarity (or similarity) values for that pair. It is a measure of how successful *cluster analysis has been in partitioning the *data.

COPSS The Committee of Presidents of Statistical Societies (COPSS) aims to promote the common interests of the societies involved, which are the *American Statistical Association, the *Institute of Mathematical Statistics, the *Statistical Society of Canada, and the two North American branches of the *International Biometrics Society. Its most prestigious prize is the Fisher Lectureship (*see* FISHER LECTURER).

SEE WEB LINKS

- Committee home page.

copula A function that relates a joint *cumulative distribution function to the distribution functions of the individual variables. If the individual distribution functions are known, but the joint distribution is unknown, then a copula can be used to suggest a suitable form for the joint distribution.

Let F be the *multivariate distribution function for the *random variables $X_1, X_2, \ldots, X_n$ and let the cumulative distribution function of X_j be F_j (for all j). Define random variables $U_1, U_2, \ldots, U_n$ by $U_j = F_j(X_j)$ for each j, so that the marginal distribution of each U_j has a continuous *uniform distribution in the interval (0, 1). Assume that for each value u_j there is a unique value $x_j = F_j^{-1}(u_j)$ and let the joint cumulative distribution function of $U_1, U_2, \ldots, U_n$ be C. Then

$$C(u_1, u_2, \ldots, u_n) = P(U_j < u_j \text{ for all } j) = F\{F_1^{-1}(u_1), F_2^{-1}(u_2), \ldots, F_n^{-1}(u_n)\},$$

for all $u_1, u_2, \ldots, u_n$ in (0, 1), since $U_j < u_j$ if and only if $X_j < F_j^{-1}(u_j)$. The function C is called the copula. An equivalent equation to the above is

$$C\{F_1(x_1), F_2(x_2), \ldots, F_n(x_n)\} = F(x_1, x_2, \ldots, x_n),$$

for all $x_1, x_2, \ldots, x_n$, where $u_j = F_j(x_j)$ for each j. **Sklar's theorem**, formulated by Abe Sklar of the Illinois Institute of Technology and published in 1959, states that, for a given F, there is a unique C such that this equation holds.

Note that it may well be that it is not possible to express the inverse functions F_j^{-1} in a simple form (an example is the *multivariate normal distribution).

Assuming that the copula and the marginal distribution functions are differentiable, the corresponding result for *probability density functions is that

$$f(x_1, x_2, \ldots, x_n) = c\{F_1(x_1), F_2(x_2), \ldots, F_n(x_n)\}f_1(x_1)f_2(x_2)\ldots f_n(x_n).$$

The trivial case where $c\,\{F_1(x_1), F(x_2), \ldots, F(x_n)\} = 1$ corresponds to the case where the n X-variables are *independent. Thus the copula encapsulates the interdependencies between the X-variables and is therefore also known as the **dependence function**. If $c(u_1, u_2, \ldots, u_n)$ is the joint probability density function of $U_1, U_2, \ldots, U_n$, then

$$c(u_1, u_2, \ldots, u_n) = f(x_1, x_2, \ldots, x_n)/\{f_1(x_1)f_2(x_2)\ldots f_n(x_n)\},$$

where $x_j = F_j^{-1}(u_j)$, for each j.

Cornish, Edmund Alfred (1909–73; b. Perth, Australia; d. Adelaide, Australia) Australian statistician and biometrician. Cornish studied agricultural biochemistry at U Melbourne, graduating in 1931. His first job was at an agricultural research institute where he was confronted by the need for statistics. His earliest work was concerned with the 23-year rainfall cycle in Adelaide (related to the well-known sunspot cycle). In 1937, at his own expense, he visited Sir Ronald *Fisher in England. This visit initiated fundamental work on approximations to *distributions. Subsequently he headed the CSIRO, Australia's largest scientific and industrial research agency. He was President of the *IBS from 1970 to 1972.

SEE WEB LINKS

- Fuller biography and photograph.

Cornish–Fisher expansion A form of the *Edgeworth expansion, introduced by *Cornish and Sir Ronald *Fisher in 1937. In its most-used inverse form, it relates the *cumulative distribution function of a *normal distribution to some distribution of interest.

Denote the 100p *percentiles of the normal distribution and of the distribution of interest by u_p and x_p, respectively, and let k_r be the rth *cumulant of the distribution of interest. Then, for all p,

$$x_p = u_p + \frac{1}{6}(u_p^2 - 1)k_3 + \frac{1}{24}(u_p^3 - 3u_p)k_4 - \frac{1}{36}(2u_p^3 - 5u_p)k_3^2$$
$$+ \frac{1}{120}(u_p^4 - 6u_p^2 + 3)k_5 - \frac{1}{24}(u_p^4 - 5u_p^2 + 2)k_3k_4$$
$$+ \frac{1}{324}(12u_p^4 - 53u_p^2 + 17)k_3^3 + \frac{1}{720}(u_p^5 - 10u_p^3 + 15u_p)k_6 - \cdots.$$

corrected moment *See* MOMENT.

correlated variables *Variables that display a non-zero *correlation. Correlated variables are not statistically independent.

correlation A general term used to describe the fact that two (or more) *variables are related. *Galton, in 1869, was probably the first to use the term in this way (as 'co-relation'). Usually the relation is not precise. For example, we would expect a tall person to weigh more than a short person of the same build, but there will be exceptions.

Although the word 'correlation' is used loosely to describe the existence of some general relationship, it has a more specific meaning in the context of linear relations between variables (*see* POPULATION CORRELATION COEFFICIENT; SAMPLE CORRELATION COEFFICIENT).

correlation coefficient Depending on context this refers either to the *population correlation coefficient, or the *sample correlation coefficient.

correlation matrix A square symmetric *matrix in which the element in row j and column k is the *population correlation coefficient between *random variables X_j and X_k. The diagonal elements are each equal to 1.

correlogram *See* AUTOCORRELATION.

correspondence analysis A technique, proposed by *Hartley in a 1935 paper, that results in the display of *data from a *contingency table in a *scatter diagram that includes points representing row categories and points representing column categories. If row points are positioned near each other in the diagram then this implies that the patterns of *counts along those rows are very similar. The same applies for groups of column points. If a row point and a column point are positioned close to one another then this implies a positive *association between the two. The calculations involved resemble those for *principal components analysis.

count A synonym for *frequency.

countable (denumerable) A set is countable if its members can be listed as a finite or infinite sequence, $x_1, x_2, \ldots$. The rational numbers are countable, but the irrational numbers, even in a finite interval, are not.

C

counting process A *stochastic process, $X_1, X_2, \ldots$, in which X_t is the number of events (for some definition of an event; *see* SAMPLE SPACE) that have occurred by time t. One example is a *Poisson process.

coupon-collecting distribution *See* ARFWEDSON DISTRIBUTION.

covariance A term used to refer to either the *population covariance or the *sample covariance, depending on context.

covariance matrix *Alternative name for* VARIANCE-COVARIANCE MATRIX.

covariate (control variable) A *variable that has an effect that is of no direct interest. The analysis of the variable of interest is made more accurate by controlling for variation in the covariate.

covariogram *See* AUTOCORRELATION.

Cox, Sir David Roxbee (1924– ; b. Birmingham, England) English statistician knighted for his services to Statistics. Cox was an undergraduate at Cambridge U and gained his doctorate at Leeds U. After employment in the Royal Aircraft Establishment and the Wool Industries Research Establishment, he took on successive academic posts at Cambridge U (1950). Birkbeck College, London (1955), and IC (1966). From 1989 to 1994 he was Warden of Nuffield College, Oxford. He was Editor of **Biometrika* from 1965 to 1991. He was the *IMS *Rietz Lecturer in 1973 and its *Wald Lecturer in 1990. In 1989 he was the *COPSS *Fisher Lecturer. In turn he was President of the *Bernoulli Society (1979), the *RSS (1980) and the International Statistical Institute (ISI) (1995). He received the *Guy Medal of the RSS in Silver in 1961, and in Gold in 1973. He was elected FRS in 1973 and in 2010 he was awarded that Society's Copley Medal 'for his seminal contributions to the theory and applications of statistics'. He was elected to membership of NAS in 1998 and to Honorary Life Membership of the *IBS in 2001. He is also an Honorary Life Member of the ISI. His 62 research students include *Aranda-Ordaz and *Gehan. He was knighted in 1985.

SEE WEB LINKS

- Fuller biography, interview, and photographs.

Cox, Gertrude Mary (1900–78; b. Dayton, IA; d. Durham, NC) American biometrician. Cox initially intended to be a deaconess in the Methodist Episcopal Church and did not start her studies at Iowa State

College until 1927. By 1933 however she was a faculty member specializing in *experimental design. In 1940 she became head of the new Department of Experimental Statistics in the School of Agriculture at North Carolina State College. She was a pioneer in the use of computer programs, with her staff developing many of the early *SAS algorithms. She was joint author, with *Cochran, of the statistical classic *Experimental Designs*. In 1956, she was President of the *ASA. She was founding Editor of the journal **Biometrics* in 1945, remaining Editor until 1955. She was President of the *IBS in 1968 having been elected an Honorary Life Member in 1964. She was elected to the NAS in 1975.

SEE WEB LINKS

- Biographical memoir with photograph.

Cox–Mantel test A *non-parametric test for comparing two survival curves (*see* HAZARD RATE), which results from the work of Sir David *Cox and Nathan *Mantel. Denote the total number of deaths in the second group by D_2, and let the ordered survival times of the combined group be $t_{(1)} < t_{(2)} < \cdots < t_{(k)}$. The test statistic, C, is given by $C = U/\sqrt{I}$, where

$$U = D_2 - \sum_{j=1}^{k} m_{(j)} p_{(j)},$$

$$I = \sum_{j=1}^{k} \frac{m_{(j)}(d_{(j)} - m_{(j)})}{d_{(j)} - 1} p_{(j)}(1 - p_{(j)}),$$

and $m_{(j)}$ is the number of survival times equal to $t_{(j)}$, $d_{(j)}$ is the total number of individuals who died (or were censored) at time $t_{(j)}$, and $p_{(j)}$ is the proportion of these who were in the second group. If the differences between the survival curves are attributable to random variation then C has a *standard normal distribution.

Cox process (doubly stochastic point process) A *Poisson process, introduced in 1955 by Sir David *Cox, in which the *mean is not constant but varies randomly in space or time.

Cox regression model *See* HAZARD RATE.

Cox–Snell R^2 *See* ANOVA.

Cox–Snell residuals Residuals introduced in 1968 by Sir David *Cox and E. Joyce Snell, for assessing the validity of a survivor function (*see* HAZARD RATE) that has been proposed for a set of survival *data. The value of the survivor function depends on the time, t, and on one or more *parameters estimated by the *vector $\hat{\boldsymbol{\theta}}$. The Cox–Snell residual r_j, corresponding to time t_j, is given by

$$r_j = -\ln\{S(t_j; \hat{\boldsymbol{\theta}})\},$$

where $S(t_j; \hat{\boldsymbol{\theta}})$ is the value of the estimated survivor function at time t_j. If the model is correct, then the residuals should have an *exponential distribution with mean 1.

C_p *See* MODEL SELECTION PROCEDURE.

CPI *See* RANDOM MAN NOT EXCLUDED.

Cramér, Carl Harald (1893–1985; b. Stockholm, Sweden; d. Stockholm, Sweden) Swedish mathematical statistician who spent his entire career at Stockholm U. He entered as a student in 1912 and retired as its President in 1961. His research centred on probability, risk theory, and the mathematical underpinnings of Statistics. His best known work is *Mathematical Methods in Statistics*, published in 1945. He was the 1953 *IMS *Rietz Lecturer. He received the *Guy Medal in Gold of the *RSS in 1972, and was elected to the NAS in 1984.

SEE WEB LINKS

- Fuller biography and portrait.

Cramér–Rao inequality; Cramér–Rao lower bound *See* FISHER'S INFORMATION.

Cramér's *V* *See* CHI-SQUARED TEST.

Cramér–von Mises test An alternative to the *Kolmogorov–Smirnov test for testing the *hypothesis that a set of *data come from a specified *continuous distribution. The *test was suggested independently by *Cramér in 1928 and *von Mises in 1931. The test statistic W (sometimes written as W^2) is formally defined by

$$W = n\int_{-\infty}^{\infty}\{F_n(x) - F_0(x)\}^2 f_0(x)dx,$$

where $F_0(x)$ is the *distribution function specified by the *null hypothesis, $F_n(x)$ is the *sample distribution function, and $f_0(x) = F_0'(x)$. In practice the statistic is calculated using

$$W = \frac{1}{12n} + \sum_{j=1}^{n}\left(t_j - \frac{2j-1}{2n}\right)^2,$$

where

$$t_j = F_0(x_{(j)}),$$

and $x_{(j)}$ is the jth ordered observation ($x_{(1)} \le x_{(2)} \le \cdots \le x_{(n)}$).

The test has been adapted for use with *discrete random variables, for cases where *parameters have to be estimated from the data, and for comparing two samples. A modification leads to the *Anderson–Darling test.

craps A game played with two dice. A roll totalling 2, 3, or 12 is a loss. A total of 7 or 11 is a win. If any other total (t, say) is obtained then the dice are rolled repeatedly until either another total of t is obtained (win), or a 7 is obtained (lose). The *probability of a win is $244/495 \approx 0.493$, very close to but (for a betting person) worryingly less than a half.

credibility theory When alternative views of the future are presented (for example, views concerning the total claims to be met by an insurance company in the next year) then there is a need to take a weighted average of these views. Credibility theory is concerned with the optimal development of the weights to be used.

critical path analysis A method of analysis aimed to schedule a set of tasks so that they are all completed in the shortest possible overall time. The difficulty is that the various tasks take different lengths of time to complete and that each task cannot be started until certain prerequisite other tasks have been completed. *See also* NETWORK FLOW PROBLEM.

critical region (rejection region) The set of values of the *statistic, in a *hypothesis test, which lead to rejection of the *null hypothesis. The phrase was introduced by *Neyman and Egon *Pearson in 1933.

critical value An end-point of a *critical region. In a *hypothesis test, comparison of the value of a test statistic with the appropriate critical value determines the result of the test. For example, 1.96 is the critical value for a two-tailed test in the case of a *normal distribution and a 5% significance level: thus if the test statistic z is such that $|z| > 1.96$ then the *alternative hypothesis is accepted in preference to the *null hypothesis.

Cronbach, Lee Joseph (1916–2001; b. Fresno, CA; d. Palo Alto, CA) American psychologist. Cronbach graduated from Fresno State College in 1934, gaining his MS from UCB in 1934 and his PhD in educational psychology from U Chicago in 1940. After posts at several universities, he joined the faculty at Stanford U in 1964, where he became Professor of Education. His classic book *Essentials of Psychological Testing* was published in 1949, with a 5th edition in 1990. He was elected to the NAS in 1974.

SEE WEB LINKS

- Biographical memoir and photograph.

Cronbach's alpha *See* RELIABILITY.

cross-classification *See* CONTINGENCY TABLE.

cross-correlation The *correlation between a selected set of *data and the corresponding set displaced in time and/or space.

crossed design An *experimental design in which every *level of one *variable occurs in combination with every level of every other variable. An example is a *randomized block design, in which each *treatment occurs once within each block. *See also* NESTED DESIGN.

Cross–Fratar procedure *See* DEMING–STEPHAN ALGORITHM.

crossover trial An *experimental design in which each *experimental unit is used with each *treatment being studied. The simplest crossover trial uses two groups of experimental units (e.g. hospital patients), 1 and 2, and two treatments (e.g. medicines), A and B. The trial uses two equal-length time periods. In the first period, group 1 is assigned treatment A and group 2 is assigned treatment B. In the second period, the assignments are reversed.

Balaam's design uses four groups of patients to compare two treatments in two periods. The assignments are AA, AB, BA, and BB. However, a complication in all crossover trials, unless there is a protracted gap between the two periods, is that the treatment allocated in the first period may continue to have an effect (the **carry-over effect**) during the second period. More complex allocations aim to estimate these effects. An example—involving two treatments, four groups, and three time periods—is designed to help with the estimation of the carry-over effects: The analysis of such a design is not simple; from the statistician's viewpoint (though not the patient's) it would be preferable to minimize the carry-over effects by allowing an interval (the **wash-out period**) between successive treatments.

	GROUP 1	GROUP 2	GROUP 3	GROUP 4
time 1	A	A	B	B
time 2	B	B	A	A
time 3	A	B	A	B

cross-sectional study A study of a human *population by means of a *sample that includes representatives of all sections of society. An alternative to a *longitudinal study.

cross-tabulation *See* CONTINGENCY TABLE.

cross-validation A method of assessing the accuracy and validity of a statistical *model. The available *data are divided into two parts. Modelling of the data uses one part only. The model selected for this part is then used to predict the values in the other part of the data. A valid model should show good predictive accuracy. *See also* MACHINE LEARNING.

cube law An empirical law relating to the outcome of two-party multiple constituency elections. It states that if the votes gained by the parties are in the ratio p to $(1-p)$, then the numbers of constituencies won by the parties will be in the ratio p^3 to $(1-p)^3$. A slight majority of votes (e.g. $p=55\%$) leads to a much larger imbalance in constituencies won, since $0.55^3/(0.55^3+0.45^3)=65\%$. In the United Kingdom the law worked well for many years, though more recently the imbalance has been less extreme.

cubic regression model *See* MULTIPLE REGRESSION MODEL.

cumulant An alternative to a *moment as part of a summary of the form of a *distribution. If the *moment-generating function of a distribution exists, then its natural logarithm exists and is called the **cumulant-generating function (cgf)**. The coefficient of $t^r/r!$ in the Taylor expansion (*see* TAYLOR SERIES) of the cgf is called the rth cumulant and is denoted by κ_r (where κ is kappa). The cumulants can be expressed in terms of the central moments, and vice versa. In particular, denoting the *mean by μ and the moments by $\mu_2, \mu_3, \ldots,$

$$\begin{aligned}\kappa_1 &= \mu,\\ \kappa_2 &= \mu_2,\\ \kappa_3 &= \mu_3,\\ \kappa_4 &= \mu_4-3\mu_2^2,\\ \kappa_5 &= \mu_5-10\mu_3\mu_2.\end{aligned}$$

For an example of the application of cumulants, *see* CORNISH–FISHER expansion. The term 'cumulant' was introduced by Sir Ronald *Fisher and *Wishart, following a suggestion by *Hotelling.

cumulant-generating function (cgf) *See* CUMULANT.

cumulative distribution function (cdf; distribution function) The function F, for a *random variable X, defined for all real values of x by

$$F(x) = P(X \le x).$$

Clearly, $F(-\infty) = 0$, and $F(\infty) = 1$, where $F(-\infty)$ and $F(\infty)$ are the limits of $F(x)$ as x tends to $-\infty$ and ∞, respectively. This function is a non-decreasing function such that if $x_2 > x_1$ then $F(x_2) \geq F(x_1)$. If X is a *continuous random variable then F is a continuous function, and conversely. If X has *probability density function f then

$$F(x) = \int_{-\infty}^{x} f(t)\,dt,$$

and $f(x) = F'(x)$, where $F'(x)$ denotes the derivative of $F(x)$.

A useful property of F is that, for any value of x, there is a corresponding value u, $0 \leq u \leq 1$, such that

$$u = F(x).$$

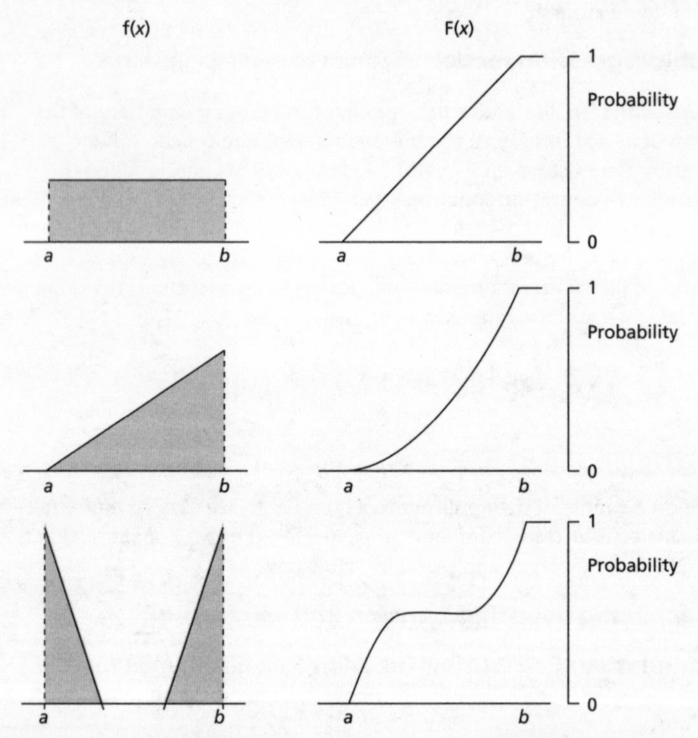

Cumulative distribution function. In each example the probability density function f defined in the interval (a, b) is shown on the left, with the corresponding cumulative distribution function F on the right. The scale from 0 to 1 refers to F.

In the case where F is a continuous and increasing function for $a \leq x \leq b$, the random variable U defined by $U = \mathrm{F}(X)$ has a continuous *uniform distribution on the interval $0 \leq u \leq 1$, and, for a given value of U, the corresponding value of X is given by $\mathrm{F}^{-1}(U)$. *See* SIMULATION.

In the case of a *discrete random variable, the distribution function is a step function, in which the step at x_j is $\mathrm{P}(X = x_j)$, and $\mathrm{F}(x) \to \mathrm{F}(x_j)$ as $x \to x_j$ from above, but $\mathrm{F}(x) \to \mathrm{F}(x_{j-1})$ as $x \to x_j$ from below.

cumulative frequency For a *sample of numerical data the cumulative frequency corresponding to a number x is the total number of *observations that are $\leq x$.

cumulative frequency polygon A diagram representing grouped numerical data in which *cumulative frequency is plotted against upper *class boundary, and the resulting points are joined by straight line segments to form a polygon. The polygon starts at the point on the x-axis corresponding to the lower class boundary of the lowest class. *See also* OGIVE; STEP DIAGRAM.

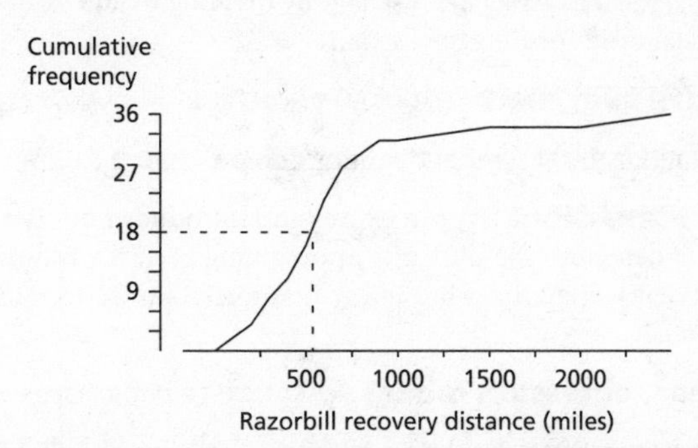

Cumulative frequency polygon. The diagram refers to the distances between the breeding colony and the point of recovery for a group of razorbills. The outline is typical, with slow increases at the left and right ends of the polygon indicating the scarcity of the corresponding values.

cumulative odds ratio A function used in models for *ordinal variables. Let j denote a possible value of the ordinal variable, Y, and let E_1 and E_2 be two possible events. The ratio R_j, given by

$$R_j = \frac{\mathrm{P}(Y \leq j|E_1)}{\mathrm{P}(Y > j|E_1)} \bigg/ \frac{\mathrm{P}(Y \leq j|E_2)}{\mathrm{P}(Y > j|E_2)},$$

is called a cumulative odds ratio (*see* TWO-BY-TWO TABLE). If E_1 and E_2 represent different values for a *vector of explanatory variables, the model $R_j = R$, where R is a constant, is a **proportional-odds model**.

C

cumulative probability For a *discrete random variable X that can take only the ordered values $x_{(1)} < x_{(2)} < \ldots$, the cumulative probability of a value less than or equal to $x_{(k)}$ is

$$\sum_{j=1}^{k} \mathrm{P}(X = x_{(j)}).$$

cumulative probability function An alternative term for the *cumulative distribution function in the case of a *discrete random variable.

cumulative relative frequency *Cumulative frequency divided by total sample size.

cumulative relative frequency diagram (cumulative relative frequency graph) A *cumulative frequency polygon or a *step diagram in which the vertical axis has been scaled by dividing by the sample size so that the maximum *ordinate value is 1.

cumulative sum chart *See* QUALITY CONTROL.

cup A name for the ∪ symbol denoting *union.

curse of dimensionality An expression introduced by *Bellman in 1961 that describes the difficulty of obtaining accurate estimates (*see* ESTIMATOR) when there are many *parameters to be estimated simultaneously.

curvilinear regression model *See* MULTIPLE REGRESSION MODEL.

curvilinear relation A relation between two *variables that is not linear but appears as a curve when the relation is graphed.

cusum chart *See* QUALITY CONTROL.

cut *See* NETWORK FLOW PROBLEM.

cut vector *See* RELIABILITY THEORY.

Cuzick trend test Suppose that subject j (belonging to a particular group) has a value that is ranked r_j amongst the n values being considered. The G groups are supposed to be ordered (for example, the classes in a school) and a score is awarded to each according to its position in that order. Typically these are the scores 1, 2, . . . , G. The test statistic is the sum, over all subjects, of the product of the rank and the group score.

cycle A repeating pattern in a *time series; examples are annual and daily patterns.

cyclic data *Data consisting of directions or times in which the measurement scale is cyclic (after 23.59 comes 00.00, after 359° comes 0°, after 31 December comes 1 January). In the case of two-dimensional directions the data may be referred to as **circular data** or **directional data** (for data in three dimensions, *see* SPHERICAL DATA).

For *observations $\theta_1, \theta_2, \ldots, \theta_n$, the variability of the data is measured by the **concentration**, $\bar{R}$, defined by

$$\bar{R} = \frac{1}{n}\sqrt{\left(\sum_{j=1}^{n} \sin\theta_j\right)^2 + \left(\sum_{j=1}^{n} \cos\theta_j\right)^2}.$$

The **circular mean**, $\bar{\theta}$, is defined only when $\bar{R} \neq 0$ and is then the angle $(0° \leq \bar{\theta} < 360°)$ such that

$$n\bar{R}\sin(\bar{\theta}) = \sum_{j=1}^{n} \sin\theta_j, \qquad n\bar{R}\cos(\bar{\theta}) = \sum_{j=1}^{n} \cos\theta_j.$$

One way of representing cyclic data is to regard each observation as a move of length 1 unit in the stated direction. The quantity R is therefore the length of the **resultant vector**. The complete sequence of such moves, taken in any order, will end at a finishing point that is a distance $R = n\bar{R}$ from the start. The direction of this finishing point from the start will be the angle $\bar{\theta}$. *See also* CIRCULAR DISTRIBUTION.

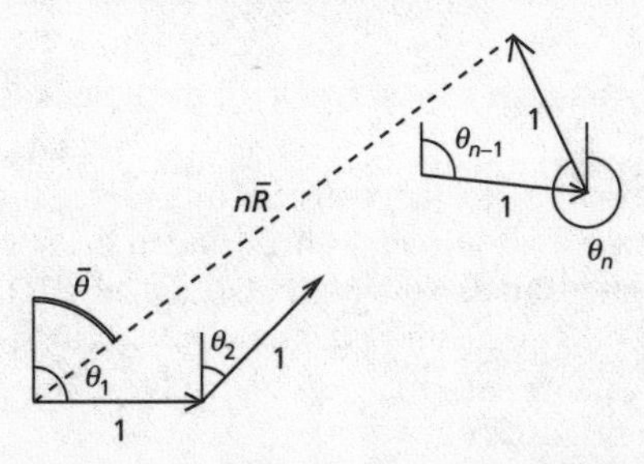

Cyclic data. Each data item is represented by a unit vector. The diagram shows the first two such vectors, and also the last two. The resultant vector, with length $n\bar{R}$, connects the start of the first unit vector to the end of the last unit vector. The direction of the resultant vector is $\bar{\theta}$.

For cyclic data the *histogram is replaced by the **circular histogram** (apparently first used by Florence *Nightingale as a means of representing the numbers of deaths in the Crimean War) or the *rose diagram.

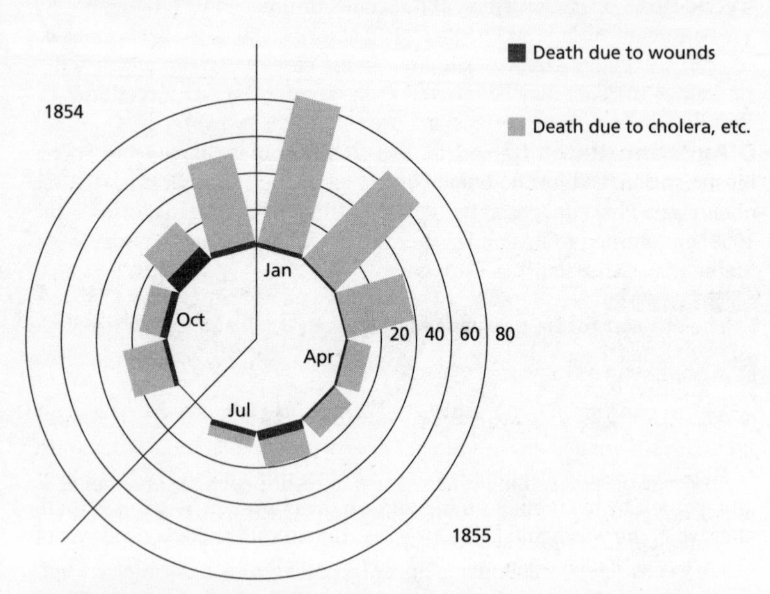

Circular histogram. Florence Nightingale invented this diagram to illustrate that the main cause of death in the Crimean War was disease rather than the enemy. The reduction in deaths after March 1855 illustrates the efficacy of Nightingale's measures to reduce disease.

cyclic permutation A rearrangement of an ordered list in which items from the end of the list are successively moved to the start. For example, the cyclic permutations of the letters UPTON are NUPTO, ONUPT, TONUP, and PTONU, but neither PUTON nor NOTUP.

D'Agostino, Ralph Benedict (1940– ; b. Somerville, MA) American biostatistician. D'Agostino obtained his first degrees from Boston U. After obtaining a PhD supervised by *Mosteller and *Cochran at Harvard U in 1968, he returned to Boston U, where *inter alia* he is Chairman of the Mathematics and Statistics Department.

SEE WEB LINKS

- University news item with photograph.

D'Agostino test *See* TEST FOR NORMALITY.

d'Alembert, Jean le Rond (1717–83; b. Paris, France; d. Paris, France) French mathematician who was a pioneer of partial differential equations. D'Alembert was the illegitimate son of a former nun and an artillery officer and was left by his mother on the steps of the church of St Jean Le Rond, after which he was named. Amongst his students at the Paris Academy of Science was *Laplace. He was elected FRS in 1748.

DALYs *See* DISABILITY-ADJUSTED LIFE YEARS.

damned lies In his autobiography Mark Twain attributed to Disraeli the statement 'There are three sorts of lies: lies, damned lies and statistics'. This might be interpreted as a comment on the difficulty of interpreting data.

Daniell weights *See* MOVING AVERAGE.

Daniell window *See* PERIODOGRAM.

Danish Society for Theoretical Statistics One of the four organizations that jointly publish the **Scandinavian Journal of Statistics*. It also publishes its own Danish language journal, *Meddelelser*. The society has nearly 300 members.

SEE WEB LINKS

- Society home page.

Dantzig, George Bernard (1914–2005; b. Portland, OR; d. Palo Alto, CA) American specialist in *operational research who was the originator of the *simplex method and has been described as the father of *linear programming. Dantzig, the son of a mathematician, was named in honour of George Bernard Shaw. He obtained his AB in mathematics and physics at U Maryland and his MA at U Michigan. During the Second World War he was head of the Combat Analysis Branch of the United States Air Force statistics section. His work in the air force involved 'programming', which in those days meant the production of efficient scheduling for the training and deployment of men. He was awarded his PhD (supervised by *Neyman) at UCB in 1946. He devised the simplex method in August 1947 while working at the Pentagon. Subsequent appointments included chairs at UCB and at Stanford U. He was awarded the National Medal of Science in 1975.

SEE WEB LINKS

- Obituary with photograph.

Darling test *See* SHAPIRO–WILK TEST.

data Information, usually *numerical or *categorical.

database A record, kept in a suitably accessible form on a computer, in which the values of several variables, which may be *categorical or *numerical, are separately recorded for each sampling unit (*see* SAMPLE).

data depth A measure that provides a partial ordering of points (*observations) in multidimensional space. One such measure is **Tukey depth** (also called **half-space depth**). Multidimensional space is divided in two by a *hyperplane passing through the point of interest. The hyperplane is rotated, and the Tukey depth is the minimum number of points recorded in a half-space (one half of the multidimensional space that is separated by a hyperplane from the other half). A value for which the Tukey depth is a maximum is the **Tukey median**.

data dredging Allowing the data to suggest hypotheses to be tested. The results of the ensuing tests must be treated with caution. Sometimes called a **fishing expedition**.

data fusion The process of synthesizing information from different sources so as to obtain a more accurate understanding of the situation under study.

data mining The exploration of a large set of *data with the aim of uncovering relationships between *variables.

data reduction The process of reducing an unmanageably large amount of *data to produce a few *summary statistics (and graphical displays such as *histograms and *scatter diagrams) that encapsulate the useful content of the original data.

datum A single item of *data.

Daubechies, Ingrid (1954– ; b. Houthalen, Belgium) Belgian mathematician now working in the USA. Daubechies was educated at the Free U, Brussels, where she studied physics, gaining her doctorate in 1980 and joining the faculty in 1984. She has developed an international reputation for her work on *wavelets. In 1987 she moved to AT&T Bell Laboratories in the USA. She has subsequently held posts at Rutgers U (1991), Princeton U (1993), and Duke U (2011). She was elected a Fellow of the AAAS in 1993 and of the NAS in 1998.

SEE WEB LINKS

- University website.

daughter wavelet *See* WAVELET.

death rate *See* MORTALITY RATE.

decile An approximate value for the *r*th decile of a *data set can be read from a *cumulative frequency graph as the value of the *variable corresponding to a *cumulative relative frequency of $10r$%. So the 5th decile is the *median and the 2nd decile is the 20th *percentile. The term decile was introduced by *Galton in 1882.

decision rule *See* GAME THEORY.

decision theory A method of deciding between various alternative experiments and actions. The decision process is informed by assigning *probabilities to the various alternatives and assigning costs (or benefits or *utilities) to the alternative outcomes. The decision process is governed by the wish to minimize the expected cost or to maximize the expected benefit or utility.

decision tree A graphical representation of the alternatives in a decision-making problem. As an example (*see diagram overleaf*), suppose that we are choosing between two machines. One costs 100 units and has a 20% probability of breaking down within a year. The other costs 120 units and has a 5% breakdown probability. A breakdown costs 60 units. Ignoring the possibility of multiple breakdowns, which machine should we buy? Reading the decision tree from right to left, we see that it is cheaper to buy the 100-unit machine.

In *machine learning the 'decisions' are, in effect, the answers to questions, with the final values at the ends of the tree being predicted values or classes. **Pruning** is a technique for simplifying such trees by removing sections that have minimal impact on the final decision or value.

decomposable model *See* DIRECT MODEL.

defective An item that does not meet the required standard of a production process.

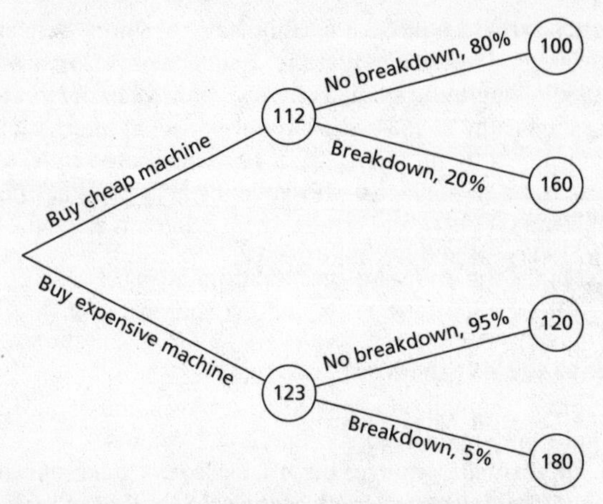

Decision tree. A circle at the right contains the total cost if this path occurs. An intermediate circle gives the expected costs at this stage. Thus, for example, $112 = 100 \times 80\% + 160 \times 20\%$. If cost is the only consideration, then in the case illustrated the cheaper machine is the better buy.

defender's fallacy A mis-statement of a *probability as a result of a misunderstanding of *conditional probability. *See also* PROSECUTOR'S FALLACY. As an example, suppose that a blood type possessed by only 1% of the population is found at a crime scene. The accused has blood of this type. The defender argues that, since the population size is one million, there are 10 000 with this blood type and therefore the probability that it is the accused who is guilty is 1 in 10 000.

For a randomly chosen person, let A be the event (*see* SAMPLE SPACE) 'person chosen has blood of this type' and let B be the event 'person chosen is guilty'. The figure quoted by the defender is $P(B|A)$, whereas the probability of interest is P(accused is guilty).

de Finetti, Bruno (1906–85; b. Innsbruck, Switzerland; d. Rome, Italy) Italian probabilist with interests in financial and actuarial mathematics. De Finetti studied mathematics first at Milan Polytechnic, where he wrote his first paper (on population genetics), and then at Milan U, where he graduated in 1927. In 1939 he was appointed Professor of Financial Mathematics at Trieste U, moving to U Rome in 1954. He is known for his work on *exchangeability and subjective probability (*see* BAYESIAN INFERENCE).

- Photographs.

defining contrast *See* FACTORIAL DESIGN.

degree *See* POLYNOMIAL.

degrees of freedom (df) A *parameter that appears in some *probability distributions used in *statistical inference, particularly the *t-distribution, the *chi-squared distribution, and the *F-distribution. The phrase 'degrees of freedom' was introduced by Sir Ronald *Fisher in 1922.

In the case of the t-distribution, the term usually reflects the fact that the *population variance has been estimated. The number of degrees of freedom is equal to the number of independent pieces of information concerning the variance. In the most familiar case of n *observations, $x_1, x_2, \ldots, x_n$, from a *population with unknown *mean and variance, there are $(n-1)$ independent deviations from the mean, since $x_n - \bar{x} = -\sum_{j=1}^{n-1}(x_j - \bar{x})$, where $\bar{x} = (\sum_{j=1}^{n} x_j)/n$. In this case, therefore, there are $(n-1)$ degrees of freedom.

In the case of a *random variable with a chi-squared distribution, if it can be expressed as the sum of squares of m *independent standard normal variables (*see* NORMAL DISTRIBUTION), then the distribution has m degrees of freedom.

In the case where the chi-squared distribution is used as a *goodness-of-fit test, each independent parameter estimated from the data represents another constraint, so the number of degrees of freedom is $(c-1-p)$, where c is the number of cells, p is the number of parameters estimated from the sample data, and there is the constraint that the sum of the observed frequencies is the sum of the expected frequencies.

In the case where the chi-squared distribution is used to test the *null hypothesis of *independence in a $J \times K$ *contingency table, there are $(J-1)(K-1)$ degrees of freedom.

Delaunay, Charles Eugene (1816–72; b. Lusigny-sur-Barse, France; d. Cherbourg, France) French mathematician. Delaunay studied at the Sorbonne and taught at the École Polytechnique. His theory of lunar motion advanced the development of theories of planetary motion. He drowned in a boating accident. A street in Paris and a lunar crater are named after him.

SEE WEB LINKS

- Photograph.

Delaunay triangles Triangles, named after *Delaunay, that provide a tessellation of the plane that is related to the *Dirichlet *tessellation. If two measurement points lie on either side of a Dirichlet tile, then the line joining them forms a side of a Delaunay triangle (*see diagram overleaf*).

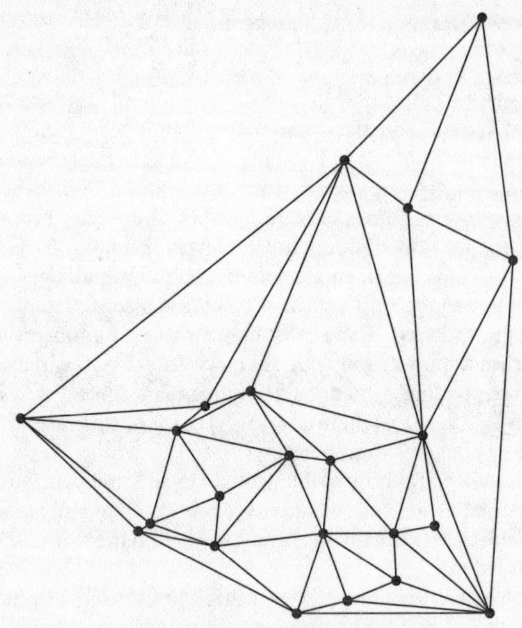

Delaunay triangles. The diagram shows the tessellating triangles corresponding to the locations of 22 rain-gauges in the Bolton area, England.

deletion residual A *statistic that provides an effective way of detecting an *outlier observation. The deletion residual corresponding to the *j*th of *n* *observations is calculated by comparing its value with the corresponding *fitted value based on the *model with the *parameters estimated from the remaining $(n-1)$ observations. *See* REGRESSION DIAGNOSTICS.

de Méré's problem French gambler Antoine Gomband, the Chevalier de Méré, conjectured that the chance of getting one '6' with four rolls of a fair die was greater than that of getting one 'double-six' with twenty-four rolls of two dice. Unable to prove this, in 1654 he posed this as a problem to *Pascal who demonstrated that de Méré was correct.

Deming, William Edwards ('Ed') (1900–93; b. Sioux City, IA; d. Washington, DC.) American engineer and statistician. Deming was a pioneer of *quality control. He was voted by business staff of the *Los Angeles Times* as being one of the 50 most influential business people of the century, though he described himself as 'Consultant in Statistical Studies'. He studied electrical engineering at U Wyoming, graduating in

1921. As a summer job he worked for the Western Electric Company in Chicago where he encountered *Shewhart's work on quality control. He obtained his MS in mathematics and mathematical physics from U Colorado in 1925 and his PhD from Yale U in 1928. He began working first for the US Department of Agriculture and then for the US Bureau of the Census. In 1947 he spent three months in Japan helping with the Japanese census. On his return to Japan in 1950 he gave an extended course in quality control; the course was so successful and influential that he was invited back on many occasions, being received by Emperor Hirohito and awarded the Second Order of the Sacred Treasure. He was President of the *IMS in 1945. In 1955 he was awarded the *Shewhart Medal of the *ASQ, in 1983 the *Wilks Award of the *ASA, and in 1987 the National Medal for Technology. He once wrote 'The only useful function of a statistician is to make predictions and thus to provide a basis for action'.

SEE WEB LINKS

- Fuller biography and photograph.

Deming–Stephan algorithm (iterative proportional fitting) An algorithm, given by *Deming and *Stephan in 1940, that can be used to fit *log-linear models to *contingency tables. The algorithm matches marginal totals by using iterative scaling of a table whose entries are initially all equal to unity. Essentially the same algorithm allows one to revise sample margins to preserve known population values: depending on context this is known as the **Cross–Fratar procedure**, **raking**, or **structure-preserving estimation** (**SPREE**).

demography The study of human populations, particularly with respect to births, marriages, deaths, employment, migration, health, etc.

de Moivre, Abraham (1667–1754; b. Vitry, France; d. London, England) French mathematical prodigy. De Moivre studied at U Sedan when aged eleven, and later at U Saumur and the Sorbonne. In 1688 he emigrated to London to avoid persecution (he was a Protestant). In London he initially made a living by giving advice to gamblers and stock-exchange workers in the coffee-houses. In 1697 he was elected FRS. In his 1711 paper *de Mensura Sortis* he derived both the distribution now called the *Poisson distribution and the *hypergeometric distribution. He is also famed for his work on *probability, encapsulated in his book *The Doctrine of Chances*, and he is credited with the first derivation of the *probability density function of the *normal distribution.

SEE WEB LINKS

- Fuller biography and portrait.

de Moivre–Laplace theorem *See* BINOMIAL DISTRIBUTION.

de Morgan, Augustus (1806–71; b. Madura, India; d. London, England) English mathematician. De Morgan established the logical footing of mathematical induction, and his interest in algebra led to the formulation of the de Morgan Laws (*see* BOOLEAN ALGEBRA). He did not excel at school, having lost the sight of an eye shortly after birth, but nevertheless went to Cambridge U when aged sixteen. He refused to take the theological test required for an MA and, as a result, was not eligible for a fellowship. In 1828, after studying law at Lincoln's Inn for a year, he successfully applied for the chair of mathematics at the newly founded UCL despite his lack of mathematical publications. He was a major contributor to the popular *Penny Cyclopedia*, contributing over 700 articles. He has been described as a 'dry, dogmatic pedant' and during his career he twice resigned his chair on matters of principle—being re-appointed each time—and he refused to allow his name to be put forward for election to the Royal Society. He was the first President of the LMS. A lunar crater is named after him.

SEE WEB LINKS

- Fuller biography and photograph.

de Morgan laws *See* BOOLEAN ALGEBRA.

Dempster, Arthur Pentland (1929- ; b. Toronto, Canada) Canadian statistician specializing in *statistical inference whose career has been in the United States. Dempster obtained his BA and MA from U Toronto and his PhD (supervised by *Tukey) from Princeton U in 1956. From 1958 to 2005 he was a faculty member at Harvard. In 1968 he introduced a generalization of *Bayesian inference. In 1977 he was the senior author of the paper introducing the *EM algorithm that dealt with the analysis of incomplete *multivariate data. He was the *IMS *Neyman Lecturer in 1986 and the *COPSS *Fisher Lecturer in 1998.

SEE WEB LINKS

- University website.

Dempster-Shafer theory of evidence *See* BAYESIAN INFERENCE.

dendrogram A diagram used in the context of *cluster analysis to trace the stages in the aggregation (or disaggregation) of clusters. Thus, in the *diagram opposite*, reading from left to right, items 1 and 2 are the first to join together. Then 3 and 4 join together and are subsequently joined by 5. Finally, all five items join in a single cluster. The horizontal axis is used to indicate the distances at which the joins occur; in the example items 1 and 2 are much closer together than items 3 and 4.

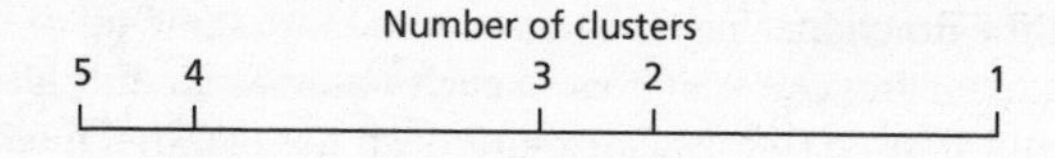

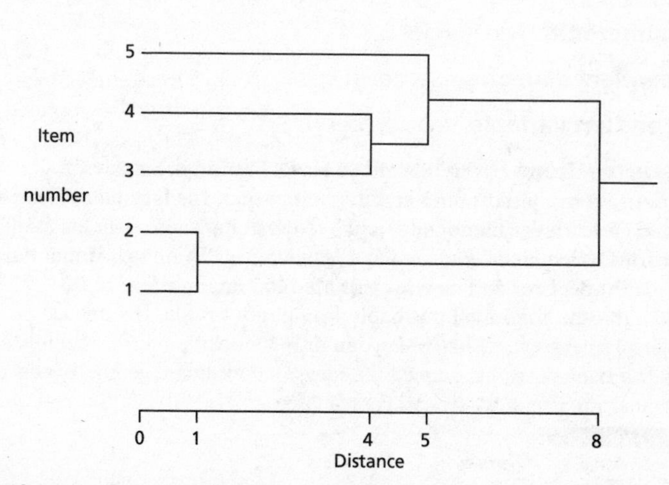

Dendrogram. The horizontal distances at which joins occur indicate the distances between the items.

An alternative is the **icicle plot**, which is read from bottom to top. For the situation illustrated in the dendrogram the icicle plot is:

		Item number								
		1		2		3		4		5
Number of clusters	1	X	X	X	X	X	X	X	X	X
	2	X	X	X		X	X	X	X	X
	3	X	X	X		X	X	X		X
	4	X	X	X		X		X		X
	5	X		X		X		X		X

Reading from the bottom we see that the icicles for items 1 and 2 join at the four-cluster step, with 3 and 4 joining at the three-cluster step, and so on.

density estimation The process of estimating the *probability density function from a *sample of *observations. The question is how best to estimate the function. One standard method is to build up the estimate by assigning a rectangular unit area to each observation: this leads to the *histogram. A more sophisticated approach uses *kernel methods.

density function *See* PROBABILITY DENSITY FUNCTION.

denumerable *See* COUNTABLE.

dependence function *See* COPULA.

dependent variable *See* REGRESSION.

Descartes, René (1596–1650; b. La Haye, France; d. Stockholm, Sweden) French philosopher and mathematician. His birthplace is now named Descartes (a lunar crater and a street in Paris also bear his name). Descartes was initially educated at a Jesuit college in Anjou. At that time his health was poor and he was permitted to remain in bed until 11.00 a.m.—he continued this habit throughout his life. His treatise on universal science published at Leyden in 1637 included three appendices. One, '*La Géométrie*', introduced the ideas of coordinate geometry and in particular the use of *Cartesian coordinates.

SEE WEB LINKS

- Fuller biography and portrait.

deseasonalized data If a *time series exhibits regular seasonal fluctuations then for the purposes of analysis (for example, to estimate an underlying trend (*see* MOVING AVERAGE)) it is often necessary to remove the *seasonality to leave deseasonalized data.

design matrix *See* MULTIPLE REGRESSION MODEL.

detection error tradeoff graph *See* TWO-BY-TWO TABLE.

determinant *See* MATRIX.

deterministic An outcome that can be predicted exactly (or a method with a perfectly predictable outcome).

detrending The removal of a trend (*see* MOVING AVERAGE) from a *time series.

Deutsche Statistische Gesellschaft The German Statistical Society, founded in 1911. The society has about 800 members.

SEE WEB LINKS

- Society home page.

deviance A measure for judging the extent to which a *model explains the variation in a set of *data when the *parameter estimation is carried out using the method of *maximum likelihood. The deviance, D, is given by

$$D = -2(\ln L_c - \ln L_s) \qquad \text{or, equivalently,} \qquad D = -2\ln\left(\frac{L_c}{L_s}\right),$$

where L_c is the *likelihood of the current model, and L_s is the likelihood of the *saturated model. The deviance has an approximate *chi-squared distribution; the number of *degrees of freedom is given by the difference in the numbers of parameters in the two models.

A summary of how well a model fits the data is provided by **McFadden's R^2** (which is also known as the **likelihood ratio index**) defined by

$$R^2 = 1 - \frac{\ln(L_c)}{\ln(L_0)},$$

where L_0 is the likelihood corresponding to the model that describes the response variable as being a constant unaffected by any explanatory variables. The term 'deviance' was introduced by *Nelder and *Wedderburn in 1972.

deviance information criterion (DIC) A modification of the AIC (*see* MODEL SELECTION PROCEDURE) for use in cases where the *model parameters are not fixed but are subject to *random variation.

deviance residual An alternative to an *Anscombe residual or a Pearson residual (*see* CHI-SQUARED TEST). In the case of a *generalized linear model with *random errors having a *Poisson distribution, a deviance residual is

$$r = \frac{y - \hat{y}}{|y - \hat{y}|}\sqrt{2y\ln\left(\frac{y}{\hat{y}}\right) - y + \hat{y}},$$

where y and $\hat{y}$ are the corresponding *observed and *fitted values.

df The number of *degrees of freedom, often denoted by ν (nu) or d.

DFBETA; DFFITS *See* REGRESSION DIAGNOSTICS.

diagnostic test In a medical context, a test as to whether or not a patient has a disease. *See* TWO-BY-TWO TABLE.

diagonal matrix *See* MATRIX.

diagonals model *See* SQUARE TABLE.

DIC *Abbreviation for* DEVIANCE INFORMATION CRITERION.

dice problem A *probability problem concerning the outcomes of rolling dice. One such problem considered at length by *Fermat and

*Pascal concerned the number of times that one must throw a pair of dice before obtaining a double six. Their correspondence laid the foundation for the modern study of probability. *See* GEOMETRIC DISTRIBUTION.

dichotomous *See* CATEGORICAL VARIABLE.

d

Dickey, David Alan (1945– ; b. Painesville, OH) American statistician specializing in *time series analysis. Dickey obtained his AB (1967) and MA (1969) in mathematics at Miami U and his PhD in 1976 under the supervision of *Fuller at ISU. Since 1976 he has been on the faculty at North Carolina State U.

SEE WEB LINKS

- University website.

Dickey–Fuller test A test for *stationarity in a *time series. The original test (suggested by *Dickey and *Fuller in 1979) was designed to determine whether an AR(1) model (*see* AUTOREGRESSIVE MODEL) was appropriate:

$$X_j = \alpha X_{j-1} + \varepsilon_j.$$

Here X_j is the value at time j, α is an unknown *parameter and ε_j is a random error. The test statistic is the product of the number of time points and $(\hat{\alpha} - 1)$, where $\hat{\alpha}$ is the ordinary least squares estimate (*see* METHOD OF LEAST SQUARES) of α. The test was extended in 1981 for use with other autoregressive models; the resulting test is called the **augmented Dickey–Fuller test** or the **ADF test**. The **Phillips–Perron tests** are related alternative tests suggested in 1988.

die The singular form of 'dice'.

differencing The process of subtracting successive terms in a series of equi-spaced observations. Used in *numerical analysis and as a simple method for removing trends (*see* MOVING AVERAGE) in *time series. Suppose, for example, that y_t, the value at time t, is given (for $t = 1, 2, 3, \ldots$) by $y_t = a + bt$. Then, for each value of $t > 1$, the **first difference**, Δy_t, is given by

$$\Delta y_t = y_t - y_{t-1} = (a + bt) - \{a + b(t-1)\} = b,$$

which is no longer dependent on t. In the same way, further differences will remove higher polynomial dependencies on time. For example, the **second difference**, $\Delta^2 y_t$ $(= \Delta y_t - \Delta y_{t-1})$, will remove any quadratic dependence on time.

diffusion process *See* STOCHASTIC PROCESS.

digraph *See* GRAPH.

directed acyclic graph *See* GRAPHICAL MODEL.

directed arc *See* NETWORK FLOW PROBLEM.

directed divergence *See* KULLBACK–LEIBLER INFORMATION.

directed graphical model *See* GRAPHICAL MODEL.

directional data *See* CYCLIC DATA.

direct model (decomposable model) A *log-linear model for which simple formulae exist for the *expected frequencies. The best known is the *independence model. A direct model that can be interpreted in terms of *conditional independence is a *graphical model.

Dirichlet, Johann Peter Gustav Lejeune (1805–59; b. Düren, Germany; d. Göttingen, Germany) German mathematician. It is claimed that as a twelve-year-old Dirichlet spent his pocket money on mathematics books. He graduated from U Cologne aged sixteen and went to Paris to act as a tutor. While there he taught *Poisson and published a paper on *Fermat's last theorem which brought him fame. Returning to Germany, he joined the faculty at U Cologne before moving to U Berlin in 1828. In 1855 he succeeded *Gauss as Professor at U Göttingen and, in the same year, he was elected FRS. He married a sister of the composer Mendelssohn and has a lunar crater named after him.

SEE WEB LINKS

- Portrait.

Dirichlet distribution The *multivariate distribution corresponding to the *beta distribution. In *Bayesian inference this distribution is used as the conjugate prior for the parameters of a *multinomial distribution.

Dirichlet tessellation A *tessellation that divides a region into n subregions, one for each measurement point, with the jth subregion consisting of all points in the region that are nearer to the jth measurement point than to any other (*see diagram overleaf*).

The Dirichlet tessellation has been rediscovered many times. Other names include **Voronoi polygons** and **Thiessen polygons**. Every edge of a Dirichlet subregion separates two of the original measurement points. Joining each such pair produces a new tessellation in which the subregions are *Delaunay triangles.

dirty data *Data with unusual, missing, or incorrect values. Most real sets of data are dirty.

disability-adjusted life years (DALYs) A measure of the impact of a disease. The sum of the estimated reduction in life span and the estimated number of years during which an individual lives with a crippling

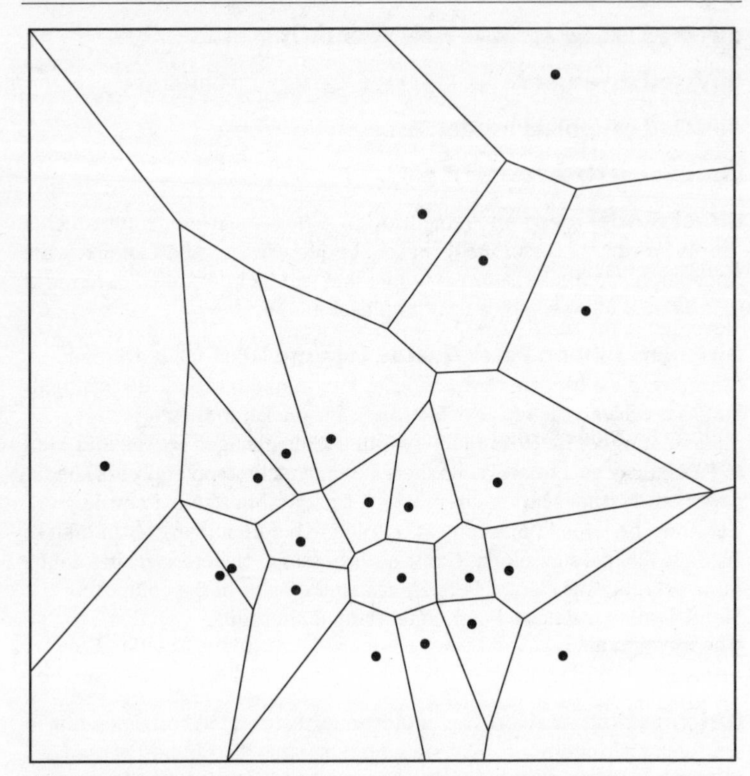

Dirichlet tessellation. The tessellation is defined by the locations of 22 rain-gauges in the Bolton area, England.

disability. A measure that brings out the impact of psychiatric and other chronic diseases.

discrete distribution *See* DISCRETE RANDOM VARIABLE.

discrete random variable A *random variable whose set of possible values is a finite or infinite sequence of numbers $x_1, x_2, \ldots$. The set of values is often a subset of the non-negative integers 0, 1, ... The *probability distribution of a discrete random variable is referred to as a **discrete distribution**.

discrete-time Markov chain *See* MARKOV PROCESS.

discriminant analysis A procedure for the determination of the group to which an individual belongs, based on the characteristics of that individual. Suppose we have measurements on p characteristics for each of a *sample of individuals. We know that each individual belongs to one of g groups, but we do not know which. Discriminant analysis attempts to maximize the *probability of correct allocation. It differs from *cluster analysis in that we have an initial data set, the **training set**, whose group allocations are known.

For the case $g=2$, suppose that the $p \times 1$ *vectors of sample means are $\bar{\mathbf{x}}_1$ and $\bar{\mathbf{x}}_2$. Let n_1 and n_2 be the numbers of members of the training set falling in the two groups and let $\mathbf{S}_1$ and $\mathbf{S}_2$ be the *variance–covariance matrices for the two parts of the training set. If we define the matrix $\mathbf{S}$ by

$$\mathbf{S} = \frac{(n_1\mathbf{S}_1 + n_2\mathbf{S}_2)}{(n_1 + n_2)},$$

a future individual with measurement vector $\mathbf{x}$ will be assigned to group 1 if and only if

$$\mathbf{a}'\mathbf{x} > \frac{1}{2}\mathbf{a}'(\bar{\mathbf{x}}_1 + \bar{\mathbf{x}}_2),$$

where $\mathbf{a}'$ is the transpose of the vector $\mathbf{a}$ given by

$$\mathbf{a} = \mathbf{S}^{-1}(\bar{\mathbf{x}}_1 - \bar{\mathbf{x}}_2).$$

The function $\mathbf{a}'\mathbf{x}$ is called **Fisher's linear discriminant function**. The terms 'discriminant analysis' and 'discriminant function' were coined by Sir Ronald *Fisher in his articles in 1936 and 1938 that introduced the procedure. *See also* MACHINE LEARNING; NEURAL NET.

discrimination information *See* KULLBACK–LEIBLER INFORMATION.

disjunctive kriging *See* KRIGING.

dispersion *See* INDEX OF DISPERSION.

dispersion matrix *See* VARIANCE–COVARIANCE MATRIX.

dispersion test *See* INDEX OF DISPERSION.

dissimilarity A measure of the difference between two observations in a set of *multivariate data.

In the case of *multivariate presence/absence observations, the **Hamming distance** is the number of variables in which the observations differ. For example, the Hamming distance between the two binary strings 00101 and 11110 is four, since the strings differ in four positions.

By contrast, **similarity** is a measure of the resemblance between observations of multivariate data. Suppose two individuals are each assessed with respect to N characteristics that are either present or absent.

For each characteristic there are three possible outcomes: 'Absent for both individuals', 'Present for just one of the individuals', and 'Present for both individuals'. With the corresponding *counts denoted by n_0, n_1, and n_2, respectively, the **matching coefficient** is the proportion of variables in which the two classifications of the two variables agree:

$$\frac{n_0 + n_2}{N}.$$

An alternative, that ignores characteristics that are absent for both individuals, is the **Jaccard coefficient**

$$\frac{n_2}{n_1 + n_2}.$$

See also DISTANCE MEASURE.

distance measure A measure of the distance between points in multidimensional space (also called a **metric**). Distance measures are used with techniques such as *cluster analysis and *multidimensional scaling. The two measures most commonly used are Euclidean distance and the city-block metric.

Euclidean distance is the straight-line distance between two points. In d dimensions, if the positions of P and Q are given by the coordinates $(p_1, p_2, \ldots, p_d)$ and $(q_1, q_2, \ldots, q_d)$ then the Euclidean distance between P and Q is given by

$$\sqrt{\sum_{j=1}^{d} (p_j - q_j)^2}.$$

The **city-block metric** in two dimensions measures the distance between two points in a city if, for example, the only directions in which one could travel were north-south and east-west. It is also called the **Manhattan distance**. In d dimensions the city-block distance between P and Q is given by

$$\sum_{j=1}^{d} |p_j - q_j|.$$

A generalization of the previous measures is the **Minkowski distance**

$$\left\{\sum_{j=1}^{d} |p_j - q_j|^k\right\}^{\frac{1}{k}},$$

where k is a positive integer. Euclidean distance is the Minkowski distance of order 2, and Manhattan distance is the Minkowski distance of order 1.

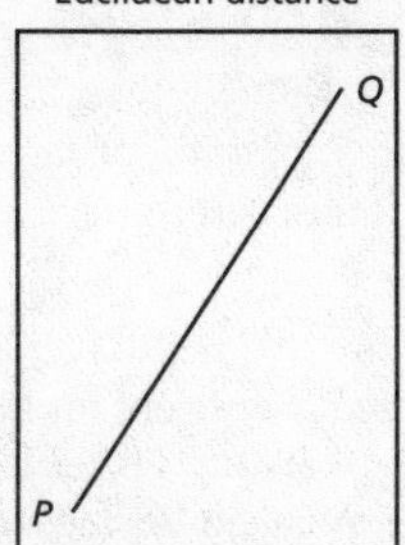

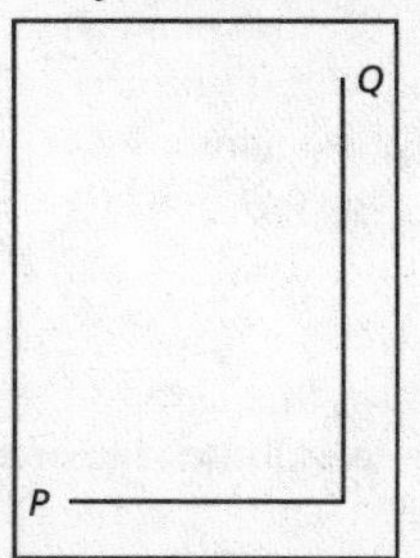

Distance measure. These are the two simplest of an infinite number of possible measures of distance.

The **Chebyshev distance** is the largest difference over the d dimensions

$$\max_{j=1}^{d} |p_j - q_j|.$$

Two other measures that have been proposed are the **Canberra distance**, given by

$$\sum_{j=1}^{d} \left\{ \frac{|p_j - q_j|}{|p_j| + |q_j|} \right\},$$

and, in the context of counts of organisms, the **Bray-Curtis distance** (also called the **Sorensen distance**):

$$\frac{\sum_{j=1}^{d} |p_j - q_j|}{\sum_{j=1}^{d} (p_j + q_j)}.$$

distance model *See* SQUARE TABLE.

distributed lags model A model that relates y_t, the value of a dependent variable evaluated at time t, to the values taken by an explanatory variable x at times t, $t-1, \ldots$. The general model is

$$y_t = \alpha + \sum_{s=0}^{\infty} \beta_s x_{t-s} + \varepsilon_t,$$

for suitable values of the constants $\alpha, \beta_0, \beta_1, \ldots$. Here, x_{t-s} is the value of x at lag s, and $\varepsilon_t, \varepsilon_{t-1}, \ldots$, are the (possibly *correlated) *random errors.

One simple model of this type is the three-parameter **Koyck model** for which $\beta_s = \gamma\beta^s$, where β and γ are positive constants. Another is the $(k+1)$-parameter **Almon model** in which

$$y_t = \sum_{s=0}^{n} \beta_s x_{t-s} + \varepsilon_t,$$

where n is known and

$$\beta_s = \sum_{j=0}^{k} \gamma_j s^j,$$

for constant $\gamma_0, \gamma_1, \ldots, \gamma_k$.

The term 'distributed lag' was introduced by Irving *Fisher in a 1925 paper entitled 'Our Unstable Dollar and the so-called Business Cycle'.

distribution The set of values of a set of *data, possibly grouped into classes, together with their *frequencies or relative frequencies. In the case of *random variables the distribution is the set of possible values together with their probabilities in the *discrete case and the *probability density function in the case of a *continuous variable.

distribution-free test *See* NON-PARAMETRIC TEST.

distribution function Shortened name for *cumulative distribution function.

diversity index A measure of the extent to which the different types in a *population are unequally common. A value of 0 is attained if there is only one type in the population. One index is the **Shannon index**:

$$-\sum_{j=1}^{k} p_j \log(p_j),$$

where there are k types and p_j is the proportion of type j. The base of the logarithm is arbitrary, but the usual choice is 2 (so that $\log_2\left(\frac{1}{2}\right) = -1$, $\log_2\left(\frac{1}{4}\right) = -2$, etc.) in which case the index may alternatively be referred to as the **entropy** of the classifying *variable. An alternative is the **Simpson index**:

$$1 - \sum_{j=1}^{k} p_j^2.$$

Dixon test *See* OUTLIER.

DK Short for 'Don't know' as a *questionnaire answer.

Doll, Sir (William) Richard Shaboe (1912–2005; b. Hampton, England; d. Oxford, England) English physician who, together with *Hill, provided in 1952 the first statistical evidence linking smoking and lung cancer. Doll studied medicine at St Thomas Hospital in London, qualifying in 1937.

After a spell in the Royal Army Medical Corps, he joined the faculty at LSHTM in 1948. Appointed Professor of Medicine at Oxford U in 1969, he was Warden of Green College from 1979 to 1983. Appointed FRS in 1966, he was awarded the Society's Royal Medal in 1986. He was knighted in 1971 and made a Companion of Honour in 1996.

SEE WEB LINKS

- Fuller biography and photograph.

dominance A term used in *game theory and *decision theory to describe the situation in which one action is never worse than an alternative action.

Donoho, David Leigh (1957– ; b. Los Angeles, CA) American mathematical statistician known for his application of *wavelets to image processing. He obtained his AB at Princeton U (where he was supervised by *Tukey) and his PhD at Harvard U (supervised by *Huber). After joining the UCB faculty in 1984 he joined the faculty at Stanford U in 1990. He was the 1997 *IMS *Wald Lecturer and the 2003 IMS *Le Cam Lecturer. He has been elected to the NAS and AAAS and was awarded the 2013 Shaw Prize for Mathematics (worth $1 million) 'for his profound contributions to modern mathematical statistics'.

SEE WEB LINKS

- University webpage.

Doob, Joseph Leo (1910–2004; b. Cincinnati, OH; d. Urbana, IL) American mathematician and probabilist known for his work on *stochastic processes. Doob entered Harvard U intending to study physics because of his passion for radio. Switching to mathematics, he obtained his AB in 1930, his AM in 1931 and his PhD in 1932, specializing in analytic functions. After a short spell at Columbia U with *Hotelling, he joined the faculty of U Illinois in 1935. One of his first research students was *Blackwell. He was President of the *IMS in 1950 and President of the AMS in 1963. He was the 1965 *COPSS *Rietz Lecturer. He was a member of the NAS and received the National Medal of Science in 1979. He was elected an Honorary Fellow of the *RSS in 1987.

SEE WEB LINKS

- Fuller biography, interview, and photographs.

D-optimality *See* EXPERIMENTAL DESIGN.

dot plot An alternative to a *bar chart or line graph when there are very few *data values. Each value is recorded as a dot, so that the *frequencies for each value can easily be counted. *See diagram overleaf.*

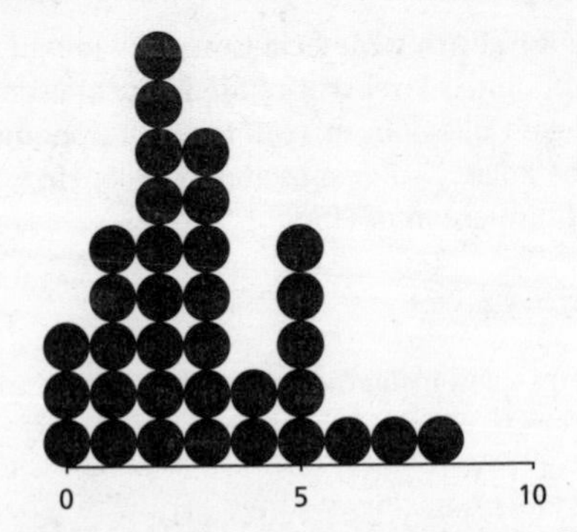

Dot plot. The diagram illustrates the numbers of goals scored in football games on one particular day. The modal number of goals was 2.

double-blind trial *See* BLINDING.

double bootstrap *See* BOOTSTRAP.

double exponential distribution *Alternative name for the* LAPLACE DISTRIBUTION.

double sampling *Acceptance sampling in which the initial *sample leads either to a decision to immediately accept or reject the *batch being sampled, or to a decision to postpone the decision until a second sample has been taken.

doubly stochastic point process *Alternative name for the* COX PROCESS.

draftsman's plot (pairs plot) A method for looking at the interrelations between variables in *multivariate data. The plot consists of a simple array of two-variable *scatter diagrams.

drunkard's walk *See* RANDOM WALK.

duality A property of an *optimization problem. Duality relates any linear maximization problem to an equivalent minimization problem. For example, with non-negative x-variables and y-variables, the maximum value for A, where

$$A = 3x_1 + 6x_2 + 2x_3,$$

subject to

$$\begin{aligned} 3x_1 + 4x_2 + x_3 &\le 2, \\ x_1 + 3x_2 + 2x_3 &\le 1, \end{aligned}$$

is equal to the minimum value of B, where

$$B = 2y_1 + y_2,$$

subject to

$$3y_1 + y_2 \geq 3,$$
$$4y_1 + 3y_2 \geq 6,$$
$$y_1 + 2y_2 \geq 2.$$

Duckworth-Lewis method A procedure used in cricket to decide the target scores in limited-overs matches that are rain-affected. It relies on the use of historical data to determine the *expected value of the number of runs scored by a team having w wickets left with b balls remaining.

dummy variable A *variable, taking only the values 0 and 1, derived from a polytomous *categorical variable. If the categorical variable has k categories then $(k-1)$ dummy variables are required. For example, with four categories the three dummy variables (x_1, x_2, x_3) could be assigned the values (1, 0, 0) for category 1, (0, 1, 0) for category 2, (0, 0, 1) for category 3, and (0, 0, 0) for category 4. Dummy variables enable the inclusion of categorical information in *multiple regression models.

Duncan test *See* MULTIPLE COMPARISON TEST.

Dunnett, Charles William (1921–2007; b. Windsor, Ontario; d. Hamilton, Ontario) Canadian statistician. Dunnett obtained his BS in mathematics and physics from McMaster U in 1942. He was then seconded to the English Royal Navy to work on the radar detection of submarines. After the war he returned to Canada, obtaining his MA from U Toronto in 1946. From 1953 to 1974 he worked as a statistician for the American Cyanamid Co, obtaining his PhD from U Aberdeen in 1960. In 1974 he returned to McMaster U, first as Professor of Biostatistics, then as Professor of Applied Mathematics. He was President of *SSC in 1982 and was awarded the Society's Gold Medal in 1986.

SEE WEB LINKS

- Photographs and conversation.

Dunnett test *See* MULTIPLE COMPARISON TEST.

Dunn test *See* MULTIPLE COMPARISON TEST.

Durbin, James (1923–2012; b. Wigan, England; d. London, England) English statistician noted for his co-authorship of the 1950 paper introducing the *Durbin–Watson statistic. Durbin's mathematical studies at Cambridge U were interrupted by the Second World War, during which he did simple statistical work. On his return to Cambridge U he concentrated on statistics, researching in *econometric models. His subsequent career was spent on the faculty at LSE. He was one of three individuals (with *Armitage and *Plackett) to have been awarded all three of the *Guy Medals of the *RSS (Bronze, 1966; Silver, 1976; Gold, 2008). He was President of that Society in 1986, having been President of the *International Statistical Institute (ISI) in 1983. He was elected an Honorary Life Member of the ISI in 1999 and a Fellow of the British Academy in 2001.

SEE WEB LINKS

- Obituary and photograph.

Durbin–Watson statistic A statistic proposed by *Durbin and Geoffrey *Watson in 1950. Suppose the sequence of *observations y_1, $y_2, \ldots, y_t$ forms a *time series. Suppose also that there are k explanatory variables (*see* REGRESSION) taking values $x_{1t}, x_{2t}, \ldots, x_{kt}$ at time point t. It is proposed that the variation over time in the value of y_t, at time t, may be explained by the *multiple regression model,

$$y_t = \beta_0 + \sum_{j=1}^{k} \beta_j x_{jt} + \varepsilon_t,$$

where ε_t is the *random error and $\beta_0, \beta_1, \ldots, \beta_k$ are unknown *parameters.

The Durbin–Watson statistic, d, given by

$$d = \frac{\sum_{m=2}^{t} (\hat{\varepsilon}_m - \hat{\varepsilon}_{m-1})^2}{\sum_{m=1}^{t} \hat{\varepsilon}_m^2},$$

where $\hat{\varepsilon}_m$ is the ordinary least squares (*see* METHOD OF LEAST SQUARES) estimate of ε_m, tests whether the sequence $\varepsilon_1, \varepsilon_2, \ldots, \varepsilon_t$ consists of *independent values (in which case d has expected value 2), or forms a *Markov process in which each random error is related to its predecessor.

Dutch book *See* ARBITRAGE.

dynamic programming The problem of optimizing a sequence of decisions in which each decision must be made after the outcome of the previous decision becomes known.

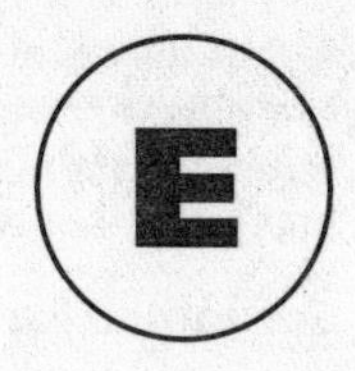

e *See* EXPONENTIAL.

ecological fallacy An error of reasoning brought about by arguing that the relationship between properties of groups of people must apply to the people themselves (or vice versa). The analogous situation with categorical data is described as *Simpson's paradox.

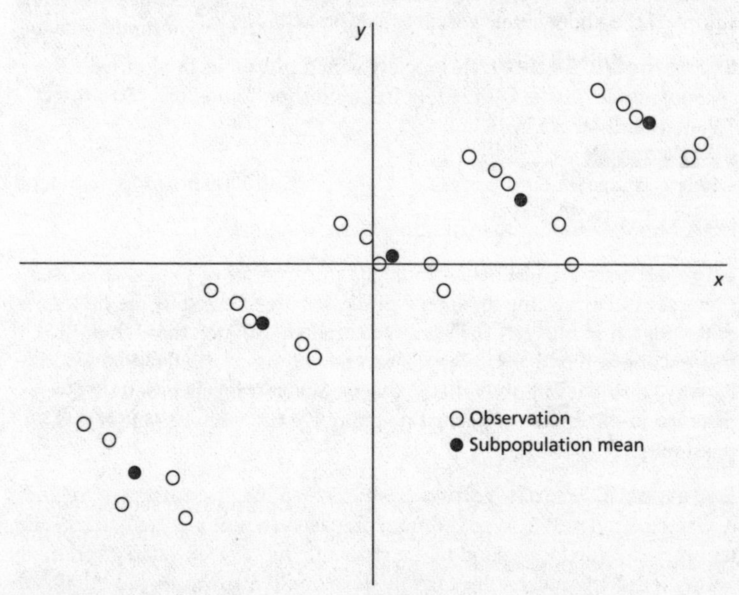

Ecological fallacy. Considering only the subpopulation means (black dots), we would conclude that the correlation between the random variables *X* and *Y* was positive. However, the data for each subpopulation (the groups of open circles) show that *X* and *Y* are in fact negatively correlated.

ecological inference A method for estimating probabilities in a *transition matrix. The method has mainly been used in the context of political data. It is assumed that for every pair of political parties the *probability of a supporter changing from one party to the other is the same in every constituency. If that is the case then, by studying the pairs of vote totals for two elections, it will be possible to identify the *transition matrix.

Econometrica The bi-monthly journal of the *Econometric Society. The first issue appeared in 1933.

SEE WEB LINKS

- Journal home page.

econometric models Models that describe the processes in an economic system. Typically an economic system is described by a set of simultaneous *multiple regression models that describe the links between a number of dependent (*endogenous) variables and a number of predetermined (*exogenous) variables (which may include previous values of the endogenous variables).

Econometric Society The Society which publishes the journal **Econometrica*. It was founded by Irving *Fisher, *Hotelling, *Shewhart, *Wiener, and others in 1930.

SEE WEB LINKS

- Society home page.

EDA Abbreviation for EXPLORATORY DATA ANALYSIS.

edge effects Edge effects complicate the analysis of spatial processes. Many formulae relating to *spatial processes are affected by the finiteness of the region of interest. For example, suppose we look through a microscope at blood cells. The nearest neighbour to a cell of interest may be very close but out of the field of view. Statistical tests based on the distance to the apparent nearest neighbour must take account of this possibility.

Edgeworth, Francis Ysidro (1845–1926; b. Edgeworthstown, Ireland; d. Oxford, England) Irish econometrician, lawyer, and mathematician. His family had settled in Ireland in the sixteenth century, and had given their name to the town that subsequently developed. Edgeworth was educated at Trinity College, Dublin (studying modern languages) and Oxford U (studying classics). He then studied law, being called to the bar in 1877, and also mathematics, so that his first academic post, in 1880, was as Lecturer in Logic at KCL. In 1891 he became Professor of Political Economy at Oxford U, where he inaugurated the *Economics Journal*. Though much of his time was devoted to economics, he developed the early theory of *correlation, *estimation, and the application of the *normal distribution. Edgeworth lived extremely simply, with few personal possessions: an exaggerated claim was that these were limited to red tape and gum. He was President of the *RSS from 1912 to 1914 and was awarded its *Guy Medal in Gold in 1907.

SEE WEB LINKS

- Information and photograph.

Edgeworth expansion An expansion, derived by *Edgeworth in 1905, that relates the *probability density function, f, of a *random variable, X, having *expected value 0 and *variance 1, to ϕ, the probability density function of a standard *normal distribution, using the *Chebyshev–Hermite polynomials. The first terms of the expansion are

$$\mathrm{f}(x) = \phi(x)\left\{1 + \sum_{r=3}^{6} \frac{\kappa_r}{r!}\mathrm{H}_r(x) + \frac{\kappa_3^2}{72}\mathrm{H}_6(x)\right\},$$

where κ_r is the rth *cumulant of X, and $\mathrm{H}_r(x)$ is the rth Chebyshev–Hermite polynomial. *See also* CORNISH–FISHER EXPANSION.

efficiency; efficient estimator *See* ESTIMATOR.

Efron, Bradley (1938– ; b. St Paul, MN) American statistician with a special interest in *Bayesian inference. Efron graduated in mathematics from Caltech in 1960. In 1964 he obtained his PhD at Stanford U, where he joined the faculty. He was elected a member of the AAAS in 1983 and of the NAS in 1985. He was President of the *IMS in 1988. He was awarded the *Wilks Award of the *ASA in 1990, and was its President in 2004. He was the *IMS *Rietz Lecturer in 1977 and the *Wald Lecturer in 1981. In 1996 he was *COPSS *Fisher Lecturer. He was awarded the *Parzen Prize 1998 and the *Rao Prize in 2003. He was awarded the National Medal of Science in 2007.

SEE WEB LINKS

- Fuller biography, interview, and photographs.

EFSPI The European Federation of Statisticians in the Pharmaceutical Industry. It comprises over 2000 pharmaceutical statisticians from eleven member groups across Europe, including over 1000 members from the United Kingdom (*see* PSI).

SEE WEB LINKS

- Federation home page.

EGRET A statistical package specializing in techniques suitable for the analysis of epidemiological data.

SEE WEB LINKS

- Computer package home page.

eigenvalue; eigenvector *See* MATRIX.

eighty/twenty rule A rule much quoted in business that states that eighty per cent of a firm's business is derived from twenty per cent of its customers; also that eighty per cent of its business is obtained by twenty per cent of its sales force. The rule derives from the work of *Pareto.

element *See* MATRIX.

EM algorithm An *algorithm for computing *maximum likelihood estimates of *parameters when some of the *data are missing. It is an *iterative algorithm that alternates two steps until convergence is attained to sufficient accuracy. Given some values assumed for the unknown parameters, the E step evaluates the joint *likelihood of the complete data set, suitably averaged over all values of the missing data. This is therefore an expectation (*see* EXPECTED VALUE) of the likelihood that is *conditional on the observed data. The M step maximizes this expectation over the unknown parameter values. The values providing this maximization are used for the next E step. The algorithm was introduced by *Dempster, Laird, and *Rubin in 1977.

empirical A statement based on a study of *data without the use of any mathematical *model.

empirical Bayes method *See* BAYESIAN INFERENCE.

empty set *See* SAMPLE SPACE.

endogenous variable A dependent variable in the context of an *econometric model.

Engle, Robert Fry (1942– ; b. Syracuse, NY) American econometrician specializing in *time series. Engle studied physics at Williams College (BS, 1964) and Cornell U (MS, 1966). At Cornell he switched to economics (PhD, 1969). In 1969 he joined the faculty at MIT moving to UCSD in 1977 and to NYU in 2000. He was elected to the AAAS in 1995 and to the NAS in 2005. Together with *Granger, he was awarded the Nobel Prize for Economics in 2003.

SEE WEB LINKS

- University website.

ensemble classifier A classifier (*see* CLASSIFICATION TREE) that makes its final classification on the basis of the separate classifications of a group of classifiers. Examples include *bagging, *boosting, and the *random forest. *See also* MACHINE LEARNING.

ensemble learning *See* MACHINE LEARNING.

entropy *See* DIVERSITY INDEX.

entry *See* MATRIX.

Environmetrics The journal of *The International Environmetrics Society. The first edition appeared in 1990. There are eight issues per year.

SEE WEB LINKS

- Journal home page.

E-optimality *See* EXPERIMENTAL DESIGN.

Epanechikov kernel *See* KERNEL METHOD.

equilibrium A *stochastic process is said to be in equilibrium if, for every state of the process and for each time point, the *probability of being in that state at the next time point is equal to the probability for the current time point.

equiprobable If there are several possible outcomes to a statistical experiment, with each having the same probability of occurrence, then the outcomes are described as being 'equiprobable'.

erf *See* NORMAL DISTRIBUTION.

ergodic state *See* MARKOV PROCESS.

Erlang, Agner Krarup (1878–1929; b. Lonborg, Denmark; d. Copenhagen, Denmark) Danish engineer and probabilist. Erlang, the son of the local schoolmaster, learnt to read upside down, sitting opposite his elder brother. He graduated from U Copenhagen in 1901 and joined the Copenhagen Telephone Company in 1908, with whom he remained for the rest of his career. His first publication was entitled *The Theory of Probabilities and Telephone Conversations* and his subsequent research involved many trips down the manholes of Copenhagen.

- Photograph.

Erlang distribution *See* GAMMA DISTRIBUTION.

Erlang formula In the context of a queue, suppose that there are s servers and that the mean number of arrivals whilst one demand is serviced is μ. In 1917 *Erlang showed that the proportion of arrivals that are obliged to wait for service is

$$\frac{\mu^s}{s!}\bigg/\left(\sum_{j=0}^{s}\frac{\mu^j}{j!}\right).$$

ERNIE Acronym for Electronic Random Number Indicator Equipment, the apparatus used in the United Kingdom to select winning Premium Bond numbers. The system does not use *pseudo-random numbers.

error function *See* NORMAL DISTRIBUTION.

error of the first kind; error of the second kind *See* HYPOTHESIS TEST.

error sum of squares *See* ANOVA.

estimable Adjective used to describe a quantity that can be estimated.

estimate *See* ESTIMATOR.

estimation The process of using a mathematical formula, or formulae, to calculate from observed *data a value, or values, for an unknown *parameter, or parameters. Three common methods are the *method of least squares, the *method of maximum likelihood, and the *method of moments. *See also* BAYESIAN INFERENCE; RESAMPLING.

estimator A *statistic used to estimate a *parameter. The realized value of an estimator for a particular *sample of *data is called the **estimate** (or **point estimate**).

If the *expected value of the statistic is equal to the parameter then it is described as being an **unbiased estimator** and the realized value is referred to as an **unbiased estimate**. If T is an estimator of the parameter θ and the expected value of T is $\theta+b$, where $b \neq 0$, then b is called the **bias** and the estimator is a **biased estimator**. If the bias tends to 0 as the *sample size increases, then the estimator is described as being an **asymptotically unbiased estimator**.

The **efficiency** of an unbiased estimator is the ratio of its *variance to the Cramér–Rao lower bound (*see* FISHER INFORMATION). For an **efficient estimator** the ratio is 1. The **relative efficiency** of two unbiased estimators T and T' is given by the inverse ratio of their variances.

Comparisons involving biased estimators are often based on the **mean squared error** (**MSE**) defined to be

$$E[(T-\theta)^2]=\mathrm{Var}(T)+\{\mathrm{E}(T)-\theta\}^2=\mathrm{Var}(T)+b^2,$$

where $\mathrm{E}(T)$ and $\mathrm{Var}(T)$ are, respectively, the expected value and variance of T. The **root mean square error (RMSE)** is the square root of the mean squared error and has the same units as the original data.

An estimator is said to be a **consistent estimator** if, for all positive c,

$$\lim_{n\to\infty} \mathrm{P}(|T-\theta| \geq c)=0,$$

where n is the sample size.

A **sufficient estimator** or **sufficient statistic** is a statistic that encapsulates all the information provided by the data that concerns the value of the unknown parameter.

The terms 'biased' and 'unbiased' appear in an 1897 text by *Bowley. The terms 'efficiency', 'estimate', 'estimation', and 'sufficiency' were introduced by Sir Ronald *Fisher in 1922. The term 'estimator' was introduced in a specialized sense by *Pitman in 1939.

eta-squared *See* ANOVA.

Euclid (330?–275? BC; Athens?) Author of *Elements*, a rigorous mathematical treatise setting out, *inter alia*, the foundations of elementary geometry.

Euclidean distance The straight-line distance between two points. *See* DISTANCE MEASURE.

Euclidean space A term used to describe ordinary two- or three-dimensional space. Also used in connection with the n-dimensional space R^n, in which a point is a real column vector $\mathbf{x}=(x_1\ x_2 \ldots x_n)'$, the distance between two points $\mathbf{x}$ and $\mathbf{y}$ is

$$\sqrt{\sum_{k=1}^{n}(x_k - y_k)^2} = \sqrt{(\mathbf{x}-\mathbf{y})'(\mathbf{x}-\mathbf{y})},$$

and the angle θ ($0 \le \theta \le \pi$) between the vectors $\mathbf{x}$ and $\mathbf{y}$ is given by

$$\cos\theta = \frac{\mathbf{x}'\mathbf{y}}{\sqrt{\mathbf{x}'\mathbf{x}}\sqrt{\mathbf{y}'\mathbf{y}}}.$$

Euler, Leonhard (1707–83; b. Basel, Switzerland; d. St Petersburg, Russia) Swiss mathematician. Euler attended Basel U when aged thirteen and gained a master's degree at sixteen. He was the author of more than 900 papers and introduced much of the notation in modern use (e.g. π, e, i, Σ). A street in Paris and a lunar crater are named after him.

SEE WEB LINKS

- Fuller biography and portrait.

Euler's constant The limit, as n approaches infinity, of

$$1 + \frac{1}{2} + \cdots + \frac{1}{n} - \ln n.$$

It is usually denoted by γ and its numerical value is 0.5772156649 . . . .

event *See* SAMPLE SPACE.

event history analysis Analysis of the *recall data provided by individuals about their life. The life pattern is usually characterized as being in one of a number of states (e.g. education) and the aim is to study the pattern of change between these states using *longitudinal data.

EWMA *See* EXPONENTIAL SMOOTHING.

exact test Although there are many exact tests, this term is often intended to refer to the Fisher exact test. (*See* TWO-BY-TWO TABLE.)

exchangeable; exchangeability Let $X_1, X_2, X_3, \ldots$ be an infinite sequence of *random variables. If it is the case that, for all values of *m*, any two sets of *m* variables have the same joint *distribution as one another, then the sequence is said to be exchangeable and to display exchangeability. The term was introduced by *de Finetti in the early 1930s.

exclusion process A *stochastic process governing the movement of individuals between locations in which an individual may not move to a location that already contains another individual.

exclusive *See* MUTUALLY EXCLUSIVE EVENTS.

exclusive union *See* UNION.

exhaustive Events (*see* SAMPLE SPACE) $A_1, A_2, \ldots,$ are exhaustive if $A_1 \cup A_2 \cup \cdots \cup = S$, where *S* is the sample space. *See also* UNION.

exogenous variable An independent variable (*see* REGRESSION) in the context of an *econometric model.

exp *See* EXPONENTIAL.

expectation *See* EXPECTED VALUE.

expectation algebra The mathematical rules that govern *expected values. For *random variables X, Y, X_j, with expected values E(X), E(Y), and E(X_j), and for any constants a, b, a_j, the following formulae hold, without any assumptions of *independence:

$$\begin{aligned} \mathrm{E}(X+Y) &= \mathrm{E}(X)+\mathrm{E}(Y), \\ \mathrm{E}(X-Y) &= \mathrm{E}(X)-\mathrm{E}(Y), \\ \mathrm{E}(aX+b) &= a\mathrm{E}(X)+b, \\ \mathrm{E}(aX+bY) &= a\mathrm{E}(X)+b\mathrm{E}(Y), \\ \mathrm{E}\left(\sum_{j=1}^{n} a_j X_j\right) &= \sum_{j=1}^{n} a_j \mathrm{E}(X_j). \end{aligned}$$

The following law also holds, provided *X* and *Y* are independent:

$$\mathrm{E}(XY) = \mathrm{E}(X) \times \mathrm{E}(Y).$$

expected frequency The *expected value for a cell frequency (*see* CONTINGENCY TABLE) for a *model of interest. Also called the *fitted frequency. In the context of a *chi-squared test the expected frequency for a particular outcome is the product of the *sample size and the *probability of the outcome under the *null hypothesis.

expected value The expected value of a *random variable X, is denoted by $\mathrm{E}(X)$ and may be interpreted as the long-term average value of X. In the case of a *discrete random variable, taking values $x_1, x_2, \ldots$

$$\mathrm{E}(X) = \sum_j x_j \mathrm{P}(X = x_j).$$

For a *continuous random variable with *probability density function f,

$$\mathrm{E}(X) = \int_{-\infty}^{\infty} x\mathrm{f}(x)\mathrm{d}x.$$

$\mathrm{E}(X)$ is often referred to as the **expectation** of X or as the *mean of X. The word 'expectation' has been used in this context since the use of 'expectatio' by *Huygens in his 1657 treatise on the results of playing games of chance: *De Ratiociniis in Ludo Aleae.*

If g is any function, the expected value of $\mathrm{g}(X)$, $\mathrm{E}[\mathrm{g}(X)]$, is defined by

$$\mathrm{E}[\mathrm{g}(X)] = \sum_j \mathrm{g}(x_j)\mathrm{P}(X = x_j),$$

in the case of a discrete random variable, and by

$$\mathrm{E}[\mathrm{g}(X)] = \int_{-\infty}^{\infty} \mathrm{g}(x)\mathrm{f}(x)\mathrm{d}x,$$

in the case of a continuous random variable. Thus

$$\mathrm{E}(X^2) = \sum_j x_j^2 \mathrm{P}(X = x_j) \quad \text{or} \quad \int_{-\infty}^{\infty} x^2\mathrm{f}(x)\mathrm{d}x.$$

The expected value of a random variable may be thought of as the limiting value of the arithmetic mean of the observed values as the *sample size increases. It should be noted that in the case of a discrete random variable the expected value is generally not a possible value. For example, when an unbiased coin is tossed once, the expected value of the number of heads is $\frac{1}{2}$.

Some random variables (for example, those with a *Cauchy distribution) do not have an expected value. There are also discrete random variables without expected values, though this situation can only arise when the set of possible values is infinite and the sum involved is the sum of a series that does not converge. For example, if $\mathrm{P}(X = r) = K/r^2$, for $r=1, 2, \ldots$, where $1/K = \sum 1/r^2 (= \pi^2/6)$, then the expression for $\mathrm{E}(X)$ is $\sum K/r$, and this series diverges to ∞. For the expected values of sums and products of random variables, *see* EXPECTATION ALGEBRA.

experimental design The branch of Statistics concerned with the efficient *estimation of the unknown *parameters in a *linear model. Common experimental designs include *balanced incomplete blocks, *crossover trial, factorial experiments (*see* FACTORIAL DESIGN), *Latin squares, paired comparisons, *randomized blocks, *repeated measures, and *split plots. In each case a *linear model relates the response variable (*see* REGRESSION) to one or more explanatory variables. The results are usually summarized in an *ANOVA table.

The analysis of the ANOVA table is affected by the nature of the explanatory variables. For example, if a design compares treatment A with treatment B because these specific treatments are of interest, then they are said to have **fixed effects**. On the other hand, if treatments A and B have been chosen at random from a population of possible treatments with the intention of attempting to answer the general question 'Do treatments differ?', then they are said to have **random effects**. A design including both random and fixed effects is a **mixed effects design**.

An example of a random effects model is

$$Y = \mu + \tau + \varepsilon,$$

where μ is the fixed overall mean, τ is a random effect (variance σ_t^2), and ε is a random error (variance σ^2). The overall variance for Y is therefore $\sigma_t^2 + \sigma^2$ with the proportion due to random effects being $\sigma_t^2/(\sigma_t^2 + \sigma^2)$. This is variously known as the **intraclass correlation coefficient** (see separate entry for alternative definition) or the **intracluster correlation coefficient** (a cluster being the set of observations on a particular treatment). This is not a *correlation coefficient of the usual form since the bounds are 0 and 1.

If a design is repeated, so that there are two or more observations being made under each experimental condition, then this is called **replication** and the separate sets of results are called **replicates**.

In most of the designs mentioned in the first paragraph, the design *matrix, $\mathbf{X}$ (*See* MULTIPLE REGRESSION MODEL) is such that the matrix product $\mathbf{X}'\mathbf{X}$ is diagonal and the *estimators of the model *parameters are *uncorrelated variables. There are three other general classes of models having desirable properties. Models with **A-optimality** are such that the trace (*see* MATRIX) of $(\mathbf{X}'\mathbf{X})^{-1}$ is minimized. This corresponds to a minimization of the average *variance of the parameter estimators. Models with **D-optimality** are obtained by minimizing the determinant (*see* MATRIX) of $(\mathbf{X}'\mathbf{X})^{-1}$, which implies that in this case the *covariances of the estimators are also taken into account. Models with **E-optimality** are obtained by minimizing the largest eigenvalue (*see* MATRIX) of $(\mathbf{X}'\mathbf{X})^{-1}$.

experimental error The difference between an *observation and the value predicted by a model.

experimental unit A person, a plot of land, a machine, etc. that forms the 'material' on which an experiment is performed. In an agricultural context the term **plot** is usually preferred. *See* EXPERIMENTAL DESIGN.

expert system A computer application that can carry out the same problem-solving functions as a human expert in a particular area, for example, financial forecasts or medical diagnostics. Human expert knowledge in the area is collected and used as a base, with rules for solving relevant problems. The system is designed by studying how human experts make decisions and translating their methods into the software. The application may be interactive, such that the user is asked to clarify the problem.

explanatory variable *See* REGRESSION.

exploratory data analysis (EDA) The search for relationships between variables (as in *data mining) by the use of graphics such as *boxplots and techniques such as *classification trees. Writing about his work over the previous decade, *Tukey used the phrase as the title of his 1977 book.

exponent *See* INDEX.

exponential A function of a real variable defined for all real x by the sum to infinity of the (convergent) series

$$e^x = 1 + x + \frac{x^2}{2!} + \frac{x^3}{3!} + \cdots.$$

The notation reflects the **index property** that $e^x \times e^y = e^{x+y}$. The notation 'exp(x)' is sometimes used for e^x, particularly when x is a complicated algebraic expression. The number e is given by the above series with $x=1$, so

$$e = 1 + 1 + \frac{1}{2!} + \frac{1}{3!} + \cdots \approx 2.718\ 28.$$

The exponential and *natural logarithm functions are related by $\exp(\ln x) = x$ for $x > 0$ and $\ln(e^x) = x$ for all values of x.

exponential distribution (negative exponential distribution) A *random variable X with an exponential distribution has *probability density function f given by

$$f(x) = \lambda e^{-\lambda x}, \qquad x \geq 0,$$

where λ is a positive constant. The *cumulative distribution function F is given by

$$F(x) = 1 - e^{-\lambda x}, \qquad x \geq 0.$$

The distribution has *expected value $1/\lambda$ and *variance $1/\lambda^2$. The distribution has the *forgetfulness property. *See also* POISSON PROCESS.

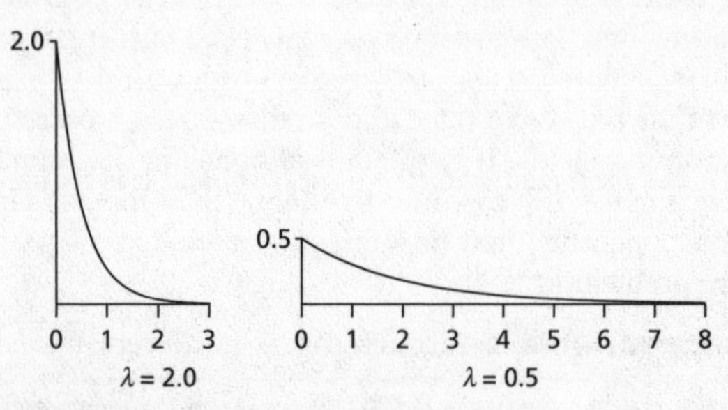

Exponential distribution. All exponential distributions have a range from 0 to ∞, a mode at 0, and a mean equal to the standard deviation.

exponential family If a *random variable *X* has a *probability distribution (*discrete case) or a *probability density function (*continuous case) that can be written in the form

$$\exp\{a(x)b(\theta) + c(\theta) + d(x)\},$$

where θ is a parameter and a, b, c, and d are known functions, then it is a member of the exponential family. The *Poisson distribution, the *binomial distribution, and the *normal distribution are members of this family. If $a(x)=x$, then the distribution is in **canonical form** and $b(\theta)$ is a **natural parameter** of the distribution.

exponentially weighted moving average *See* EXPONENTIAL SMOOTHING.

exponential smoothing A method of smoothing a *time series. Let $x_1, x_2, \ldots, x_n$ denote the observed values, and let $\hat{x}_1, \hat{x}_2, \ldots, \hat{x}_n$ be a corresponding set of smoothed values computed using the formulae

$$\hat{x}_1 = x_1,$$
$$\hat{x}_t = \alpha x_t + (1 - \alpha)\hat{x}_{t-1}, \quad t = 2, 3, \ldots, n,$$

where α $(0 < \alpha < 1)$ is a constant. An equivalent formula is

$$\hat{x}_t = \alpha \sum_{k=1}^{t} (1-\alpha)^{t-k} x_k + (1-\alpha)^t x_1.$$

The value of α is usually chosen to minimize $\sum_{t=1}^{n}(x_t - \hat{x}_t)^2$. With this value of α, $\hat{x}_t$ is called the **exponentially weighted moving average (EWMA)**.

With this value of α, the formula may be adapted for *forecasting a time series having neither trend (*see* MOVING AVERAGE) nor *seasonality by writing

$$\hat{x}_{t+1} = \alpha x_t + (1-\alpha)\hat{x}_t, \qquad t = n, n+1, \ldots,$$

where $\hat{x}_{t+1}$ is the predicted value for time $(t+1)$ and it is assumed that a long series of values exists.

extinction *See* BRANCHING PROCESS.

extra-binomial variation Greater variability in repeat estimates of a population proportion than would be expected if the population had a *binomial distribution. For example, suppose that *n* *observations are taken on *independent *Bernoulli variables that take the value 1 with *probability *p*, and the value 0 with probability $1-p$. The expected value of the total of the observations will be np and the *variance will be $np(1-p)$. However, if the probability varies from variable to variable, with overall mean *p* as before, then the expected value of the variance of the total of the observations will now be $> np(1-p)$. In the context of plant and animal populations, extra-binomial variation may be termed **overdispersion**. *See also* INDEX OF DISPERSION.

extrapolation The use of a formula, deduced from an initial set of *data, for data values outside the range of the initial data set. Compare INTERPOLATION. Excessive extrapolation may give unreliable values and is not recommended.

For example, if we discover that 1 kg of fertilizer per ha results in a yield of 50 kg of produce per ha, that 1.5 kg gives 60 kg, and that 2 kg gives 70 kg, then we might reasonably extrapolate that 2.5 kg would give 80 kg. However, it would be excessive extrapolation to expect 10 kg to give 230 kg.

extreme-value distribution A *distribution that accounts for the presence of extremely large (or small) values. The best-known distributions of this type are the *Fréchet distribution, the *Gumbel distribution, and the *Weibull distribution.

factor Depending on the context, *see* FACTOR ANALYSIS; FACTORIAL DESIGN.

factor analysis A *multivariate method in which the basic *data consist of the observed values of the p characteristics of each of n individuals. The method, introduced in 1931 by *Thurstone, aims to explain the $p \times p$ *variance-covariance matrix for these data in terms of the relationships between a much smaller number (q) of unobserved variables (*latent variables) called **factors**. The underlying proposition is that the jth observed value, x_j, is the sum of a linear combination of the factors $f_1, f_2, \ldots, f_q$ and an error term, ε_j:

$$x_j = \lambda_{j1} f_1 + \lambda_{j2} f_2 + \ldots + \lambda_{jq} f_q + \varepsilon_j.$$

The multipliers $\lambda_{j1}, \lambda_{j2}, \ldots, \lambda_{jq}$ are called **factor loadings**.

Unlike *principal components analysis, which it resembles, the methodology makes many assumptions (for example, that factors exist and that there are exactly q of them). Its use has therefore attracted considerable criticism. Its popularity might be thought to be based on the ability of the method to produce the result desired by the researcher!

factorial If n is a positive integer, then factorial n, written as $n!$, is defined by

$$n! = 1 \times 2 \times \cdots \times n,$$

with 0! defined to be 1. The notation was introduced in about 1810 by the French mathematician Christian Kramp (1760–1826). Factorial n is the number of different orderings of n individuals. The numerical value of $n!$ increases rapidly with n:

$$1! = 1, 2! = 2, 3! = 6, 4! = 24, 5! = 120, \ldots, 10! = 3\,628\,800, \ldots.$$

For $n > 1$, successive values of $n!$ can be calculated using

$$n! = n \times (n-1)!$$

If n is not an integer, factorial n can be defined via the *gamma function Γ as

$$n! = \Gamma(n+1).$$

For large values of n an approximation is provided by *Stirling's formula.

factorial design; factorial experiment An *experimental design to investigate the effects of several explanatory variables, in this context called **factors**, on a single response variable. Each factor takes only a small number of different values (typically two), which may not be quantitative and are called **levels**. For example, in an agricultural context the two levels might be two different varieties of a plant. Sir Ronald *Fisher introduced the term 'factor' in 1929 and used the description 'factorial design' as a chapter heading in his seminal work *The Design of Experiments*, published in 1935.

An experiment with k two-level factors (the upper and lower levels) is called a 2^k-factorial. The standard notation denotes factors by capital letters $A, B, \ldots$. Each **experimental unit** is subject to one or other of the levels of each factor. Lower-case letters are used to indicate factor combinations, so ac indicates that A and C occur at the higher level for that experimental unit, and other factors occur at the lower level. The combination with all factors at the lower level is indicated as (1), so the eight treatment combinations for a 2^3-factorial are

$$(1), a, b, ab, c, ac, bc, abc.$$

A 2^3-factorial enables one to estimate the three main effects (see below) as well as the three *interactions involving two variables and the three-variable interaction. The **main effect** of A compares the mean value of the observations at the higher level (a, ab, ac, abc), for example, with those at the lower level ((1), b, c, bc).

When 2^k is large it may not be feasible to maintain constant testing conditions. One solution is to test fractions at a time using *randomized blocks:

Block 1	Block 2
(1) b bc c	a ab abc ac

Suppose that the values in the second block of experiments are much larger than those in the first block. It is impossible to tell whether this is because of a difference between the two blocks or a difference between the two levels of A: the effects of A and of the blocks are said to be **confounded**. A better treatment allocation, in which it is the three-variable interaction that is confounded, is

Block 3	Block 4
(1) *ab* *bc* *ac*	*c* *a* *b* *abc*

It will be seen that each of the treatment combinations in block 3 has an even number of letters in common with *abc* while each of those in block 4 has an odd number of letters in common with *abc*. In the previous example the reference combination was *a*. The reference combination, which determines the composition of the blocks, is called the **defining contrast**. An experiment using all four of these blocks provides an example of **partial confounding**, since two quantities of interest (the main effect of *A* and the *ABC* interaction) can only be estimated from part of the available data.

An alternative to performing a complete factorial using blocks is to use a **fractional replicate**, also called a **fractional factorial design**. As an example, a **half-replicate** with four factors would consist of eight rather than sixteen treatment combinations; these might be as follows:

(1), *ab*, *ac*, *ad*, *bc*, *bd*, *cd*, *abcd*.

Since only a fraction of the full number of treatment combinations is tested, only a fraction of the full number of main effects and interactions can be estimated. In effect, what happens is that main effects and interactions are confounded with one another. Two such confounded effects are said to be **aliases** of each other. In the example no main effects are aliases of each other; the main effect of *A*, for example, is aliased with the *BCD* interaction. Fractional replicates are used in the exploratory stages of investigations with factors that are *continuous variables, where the aim is to find the global maximum (or minimum) of a *response surface.

If it can be assumed that there are no interactions then a **Plackett and Burman design** is appropriate. This design, introduced in a 1946 paper by *Plackett and *Burman, is very economical in terms of the number of observations that need to be performed. For example, with seven factors (*A*–*G*) each having two levels, the seven main effects can be estimated with the following set of treatment combinations:

dfg, *aeg*, *abf*, *bcg*, *acd*, *bde*, *cef*, *abcdefg*.

A design that has more factors than observations may be described as a **supersaturated design**.

factor loadings *See* FACTOR ANALYSIS.

failure rate *Alternative name for* HAZARD RATE.

false discovery rate (FDR) In a situation where there are many hypotheses of interest (for example, in the testing of genetic material), some hypotheses may be true and others false. A *hypothesis test may give an apparently significant result either by incorrectly rejecting a true hypothesis or by correctly rejecting a false hypothesis. The FDR is the proportion of the total number of rejected hypotheses that have been incorrectly rejected. The **Benjamini-Hochberg test** aims to control that proportion. *See also* MULTIPLE COMPARISON TEST.

false-positive rate *See* TWO-BY-TWO TABLE.

far out *See* QUARTILE.

Farr, William (1807–83; b. Kenley, England; d. Bromley, England) English physician and statistician. The son of poor parents, Farr was adopted at the age of two by a local squire. On the death of his benefactor in 1829, Farr inherited £500 which he used to fund his study of hygiene and medical statistics at UCL. In 1837 he was appointed to the General Register Office where he instigated a proper statistical record of causes of death. In 1864 he published a *life table for England based on his analysis of more than six million deaths (for which he used a new-fangled calculating machine designed by *Babbage). He was elected FRS in 1855.

SEE WEB LINKS

- Fuller biography and photograph.

fast Fourier transform (FFT) A numerical method for rapid computation of the coefficients in a finite *Fourier transform. The method was introduced in 1965 by *Tukey and the American mathematician James W. Cooley. If the (possibly complex) values of a function $f(x)$ are known at N equally spaced points, $x_0, x_1, \ldots, x_{N-1}$, where $x_k = 2\pi k/N$ for $k = 0, 1, \ldots, N-1$, the coefficients c_n are such that

$$f(x_k) = \sum_{n=0}^{N-1} c_n e^{ink}, \qquad k = 0, 1, \ldots, N-1,$$

where $i = \sqrt{-1}$.

father wavelet *See* WAVELET.

FDA *Abbreviation for* FUNCTIONAL DATA ANALYSIS.

FDR *Abbreviation for* FALSE DISCOVERY RATE.

***F*-distribution** If Y_1 and Y_2 are *independent random variables each having a *chi-squared distribution with ν_1 and ν_2 *degrees of freedom, respectively, then the ratio X, given by

$$X = \frac{Y_1}{\nu_1} \bigg/ \frac{Y_2}{\nu_2},$$

is said to have an *F*-distribution with ν_1 and ν_2 degrees of freedom. This may be written as the F_{ν_1,ν_2}-distribution. Evidently $1/X$ will have a F_{ν_2,ν_1}-distribution.

The form of the distribution was first given in 1922 by Sir Ronald *Fisher, and it is sometimes still referred to as **Fisher's *F*-distribution**. In 1934 the distribution was tabulated (*see* APPENDIX VII) by *Snedecor, who used the letter *F* in Fisher's honour. The distribution is therefore also referred to as the **Snedecor *F*-distribution**. The *probability density function f of the F_{ν_1,ν_2}-distribution is given by

$$f(x) = \frac{\nu_1^{\frac{1}{2}\nu_1} \nu_2^{\frac{1}{2}\nu_2} x^{\frac{1}{2}\nu_1 - 1}}{B(\frac{1}{2}\nu_1, \frac{1}{2}\nu_2)(\nu_2 + \nu_1 x)^{\frac{1}{2}(\nu_1+\nu_2)}}, \qquad x > 0,$$

where B is the *beta function.

The distribution has *mean $\nu_2/(\nu_2 - 2)$ provided that $\nu_2 > 2$. The distribution has *variance

$$\frac{2\nu_2^2(\nu_1 + \nu_2 - 2)}{\nu_1(\nu_2 - 2)^2(\nu_2 - 4)},$$

provided that $\nu_2 > 4$. If $\nu_1 \leq 2$ there is a mode at 0, otherwise the mode is at

$$\frac{\nu_2(\nu_1 - 2)}{\nu_1(\nu_2 + 2)}.$$

If *X* has a *t-distribution with ν degrees of freedom, then X^2 has an $F_{1,\nu}$ distribution.

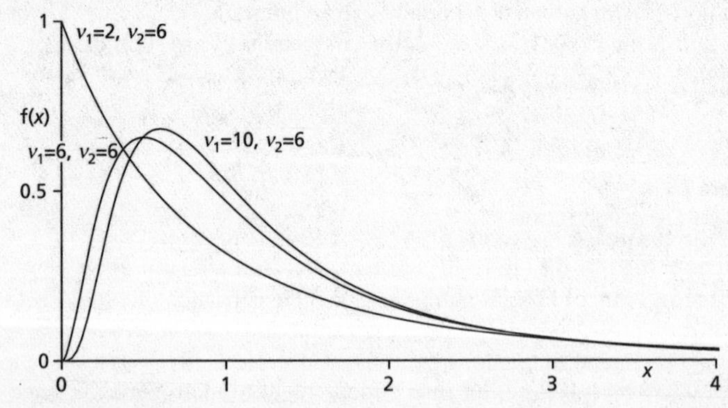

***F*-distribution.** All *F*-distributions take values from 0 to ∞ and, for ν_1, $\nu_2 > 4$, have mean and mode near 1.

feasible region In the context of *linear programming, the region in which all the constraints are satisfied. All feasible solutions must lie in this feasible region.

feasible solution *See* LINEAR PROGRAMMING.

Feller, Vilibald Srećko 'William' (1906–70; b. Zagreb, Croatia; d. New York City) Croatian probabilist who spent most of his career in the USA. Feller was author of the influential two-volume *An Introduction to Probability Theory and its Applications*. Volume 1 (1950) remained the standard introductory text for *stochastic processes for several decades. He obtained his MSc at U Zagreb in 1925, and his PhD at U Göttingen in 1926 (aged 20). Following posts at U Kiel, U Copenhagen, U Stockholm, and U Lund, Feller emigrated to the USA where he held successive posts at Brown U (1939), Cornell U (1945), and Princeton U (1950). He was President of the *IMS in 1947 and its *Rietz Lecturer in 1955. Feller was elected to membership of the NAS in 1960. In 1970 he was posthumously awarded the National Medal of Science. An asteroid is named after him.

SEE WEB LINKS

- Fuller biography, bibliography, and photograph.

Feller's coin-tossing constants Constants related to $w(n)$, the *probability that no *run of k consecutive heads is obtained when a fair coin is tossed n times. In a 1949 paper, *Feller showed that $w(n)$ obeyed the condition

$$\lim_{n\to\infty} w(n)\alpha^{n+1} = \beta,$$

where α is the smaller $(1 < \alpha < 2)$ positive root of the equation

$$1 - x + \left(\frac{1}{2}x\right)^{k+1} = 0,$$

and

$$\beta = \frac{2-\alpha}{k+1-k\alpha}.$$

fence *See* QUARTILE.

Fermat, Pierre de (1601–65; b. Beaumont-de-Lomagne, France; d. Castres, France) French mathematician. He is particularly famous for his 'last theorem', which he discovered in about 1637, and of which he claimed he had a 'marvellous demonstration'. He died without revealing his proof and it was not until 1994 that the English mathematician Andrew Wiles gave a full proof. The son of a wealthy leather merchant, Fermat was educated at a Franciscan monastery and studied law at U Toulouse. He became a judge, but he had a passion for mathematics and obtained many mathematical theorems, which he communicated to fellow mathematicians, always remaining very secretive about his proofs. The correspondence between Fermat and *Pascal laid the foundations of the modern theory of *probability. A lunar crater and a street in Paris are named after him.

SEE WEB LINKS

- Fuller biography and portrait.

FFT *Abbreviation for* FAST FOURIER TRANSFORM.

fiducial inference A method of *statistical inference proposed by Sir Ronald *Fisher in 1930. Fisher's aim was to prescribe an entirely objective procedure that avoided prior assumptions or hypotheses. Fisher's approach is not easy to understand and continues to be the subject of discussion.

Fieller's theorem A theorem first stated by E. C. Fieller in 1940. Suppose that the *random variables X and Y have *expected values μ_x and μ_y, respectively, and a bivariate normal distribution (*see* MULTIVARIATE NORMAL DISTRIBUTION). The quantity of interest is γ, the ratio of the two expected values:

$$\gamma = \frac{\mu_x}{\mu_y}.$$

Let $\overline{x}$ and $\overline{y}$ be estimates (*see* ESTIMATOR) of μ_x and μ_y, respectively, and let the respective estimates of the *variances and *covariance be denoted by s_x^2, s_y^2 and c_{xy}, with v *degrees of freedom. Fieller's theorem states that

$$\frac{\overline{x} - \gamma\overline{y}}{s_x^2 - 2\gamma c_{xy} + \gamma^2 s_y^2}$$

has a *t-distribution with v degrees of freedom.

filter A procedure that converts one *time series into another. A simple example is a *moving average, which, because it is a linear combination of the terms in the time series, is described as a **linear filter**. A filter that removes short-term random fluctuations is called a **low-pass filter**, and one that removes long-term fluctuations is called a **high-pass filter**.

filter coefficient *See* WAVELET.

finite mixture model In a finite mixture model the *distribution of the *random variable X has a known form (e.g. *normal distribution), but the values of the *parameters of the distribution are not known. Instead, it is known that the *vector of the parameters takes one of the values $\boldsymbol{\theta}_1, \boldsymbol{\theta}_2, \ldots, \boldsymbol{\theta}_m$ with associated probabilities $w_1, w_2, \ldots, w_m$. If, for a *continuous random variable, the *probability density function of the kth of the possible distributions is denoted by $f(x, \boldsymbol{\theta}_k)$, then the finite mixture distribution of X is given by

$$\sum_{k=1}^{m} w_k f(x, \boldsymbol{\theta}_k).$$

An equivalent result holds for a *discrete random variable.

finite population correction *See* SAMPLING FRACTION.

Finnish Statistical Society This society was founded in 1920 and has over 400 members. The Society is one of the four societies that jointly publish the **Scandinavian Journal of Statistics*.

SEE WEB LINKS

- Society home page.

first difference *See* DIFFERENCING.

Fisher, Irving (1867–1947; b. Saugerties, NY; d. New York City) America's first mathematical economist. Fisher was co-founder and first President of the *Econometrics Society. His academic life was spent at Yale U, where he studied mathematics and then economics, gaining his PhD (the first in economics at Yale) in 1891. He then joined the mathematics faculty before moving into the social sciences, becoming Professor of Political Economy from 1898 until his retirement in 1935.

SEE WEB LINKS

- Fuller biography and photograph.

Fisher, Sir Ronald Aylmer (1890–1962; b. East Finchley, England; d. Adelaide, Australia) English statistician; arguably the most influential statistician of the twentieth century. Fisher was educated at Harrow and at Cambridge U, where he studied mathematics. His initial interest in statistics developed because of his interest in genetics, and he pursued both subjects for the rest of his life. Fisher's first paper introduced the method of *maximum likelihood, his second the mathematical derivation of the *t-distribution, and his third the distribution of the *correlation coefficient. Following the First World War (which saw Fisher employed as a teacher because his dreadful eyesight prevented him from fighting), he joined the agricultural research centre at *Rothamsted. During his time there he virtually invented the subjects of *experimental design and *ANOVA, which motivated his derivation of the *F-distribution. In 1925 the first edition of his *Statistical Methods for Research Workers* appeared. The extent of the influence of this work can be gauged from the fact that there was a new edition about every three years until 1958, and a posthumous fourteenth edition in 1970. At a conference in India he said 'To call in the statistician after the experiment is done may be no more than asking him to perform a postmortem examination: he may be able to say what the experiment died of'. In 1933 he became Professor of Eugenics at UCL and in 1943 Professor of Genetics at Cambridge U, where *Rao was one of his research students. He was elected FRS in 1929 and was awarded the Society's Copley Medal in 1955. He was elected to membership of the NAS in 1948 and was knighted in 1952. He was President of the *RSS in 1952 and was awarded its *Guy Medal in Gold in 1946.

f

SEE WEB LINKS

- Web archive with photograph and links.

Fisher–Behrens problem *See* BEHRENS–FISHER PROBLEM.

Fisher distribution *See* *F*-DISTRIBUTION; LANGEVIN DISTRIBUTION.

Fisher exact test *See* TWO-BY-TWO TABLE.

Fisher index *See* PRICE INDEX.

Fisher information The amount of information that a *sample provides about the value of an unknown *parameter. Writing L as the *likelihood for n *observations from a *distribution with parameter θ, Sir Ronald *Fisher in 1922 defined the information, $\mathrm{I}(\theta)$, as being given by

$$\mathrm{I}(\theta) = \mathrm{E}\left\{\left(\frac{\partial \ln L}{\partial \theta}\right)^2\right\}.$$

If T is an unbiased estimator of θ then a lower bound to the variance of T is

$$\frac{1}{\mathrm{I}(\theta)}.$$

Since 1948 this has generally been referred to as the **Cramér–Rao lower bound** in recognition of the independent proofs by *Cramér and *Rao. The inequality $\mathrm{Var}(T) \geq 1/I(\theta)$ (where 'Var' denotes *variance) is the **Cramér–Rao inequality.**

If the distribution involves several parameters, $\theta_1, \theta_2, \ldots, \theta_p$, then the **Fisher information matrix** has elements given by

$$I_{jk} = \mathrm{E}\left\{\left(\frac{\partial \ln L}{\partial \theta_j}\right)\left(\frac{\partial \ln L}{\partial \theta_k}\right)\right\}.$$

Fisher information matrix *See* FISHER INFORMATION.

Fisher Lecturer; Fisher Lectureship This lectureship is awarded annually by *COPSS in honour of Sir Ronald *Fisher as a 'very high recognition of meritorious achievement and scholarship in statistical science'.

Fisher linear discriminant function *See* DISCRIMINANT ANALYSIS.

Fisher's *F*-distribution *See* *F*-DISTRIBUTION.

Fisher's *z*-transformation A transformation of the sample *correlation coefficient, r, suggested by Sir Ronald *Fisher in 1921. The statistic z is given by

$$z = \frac{1}{2}\ln\left(\frac{1+r}{1-r}\right).$$

For samples from a bivariate normal distribution (*see* MULTIVARIATE NORMAL DISTRIBUTION) with sample sizes of 10 or more, the *distribution of z is approximately a *normal distribution with *mean and *variance

$$\frac{1}{2}\ln\left(\frac{1+\rho}{1-\rho}\right) + \frac{\rho}{2(n-1)} \qquad \text{and} \qquad \frac{1}{n-3},$$

respectively, where n is the sample size and ρ is the *population correlation coefficient. The transformation is remarkably *robust.

fishing expedition *See* DATA DREDGING.

fitted distribution *See* FITTED FREQUENCY.

fitted frequency The frequency of a value or category that results from some *model. When a distribution of a specified type is matched to a set of observed frequencies, the value corresponding to a particular observed frequency is the fitted frequency and the set of fitted frequencies is the **fitted distribution**. Usually the observed and fitted frequencies will be compared using a *goodness-of-fit test. The *parameters of the fitted distribution will often have been estimated using the *method of maximum likelihood. *See also* EXPECTED FREQUENCY.

fitted values The values predicted by a *model fitted to a set of *data.

five-barred gate *See* TALLY CHART.

five-number summary For a set of numerical *data, the least value, the lower *quartile, the *median, the upper quartile, and the greatest value, in that order. *See* BOXPLOT.

five-sigma rule The rule apparently used by particle physicists to decide that an event has not occurred by chance. The description refers to an *observation that is more than five *standard deviations from a hypothesized mean. For a *normal distribution this corresponds to a *one-tail probability of about 1 in a million.

fixed effects *See* EXPERIMENTAL DESIGN.

Fligner–Killeen test *See* TEST FOR EQUALITY OF SCALE.

flowchart A diagram showing the structure of a computer program, and the different pathways that can be taken.

forecasting The prediction of future values in a *time series. Methods include *exponential smoothing, *Holt-Winters forecasting, and the *Box-Jenkins procedure.

f

forgetfulness property (lack of memory) The property possessed by the *Poisson process, and by its discrete analogue (a sequence of Bernoulli trials (*see* BERNOULLI DISTRIBUTION)), that *probabilities of future occurrences are uninfluenced by past events. This is just a restatement of *independence. The corresponding *distributions of the time to the first observation (either *exponential or *geometric distributions) and their extensions to the time to the *n*th observation (either *gamma or *negative binomial distributions) all have the forgetfulness property.

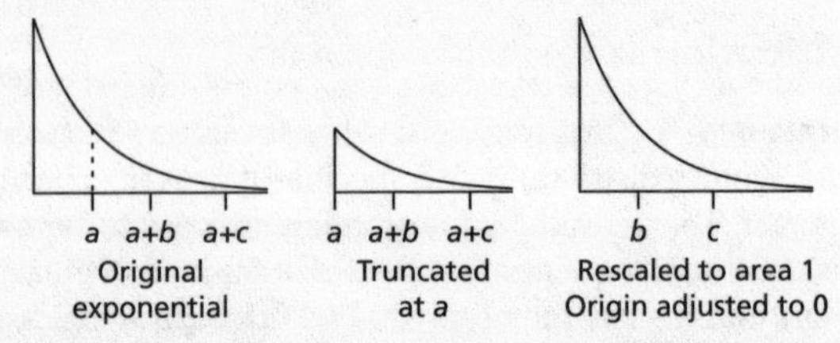

Forgetfulness property. For the exponential distribution, if we know that the observed value is greater than *a*, then the resulting conditional distribution is a facsimile of the original distribution.

FORTRAN Abbreviation for Formula Translator—one of the earliest computer programming languages.

forward selection *See* STEPWISE PROCEDURE.

fourfold table *See* TWO-BY-TWO TABLE.

Fourier series If f is an integrable function on $(-\pi, \pi)$ then the Fourier series for f is

$$\frac{1}{2}a_0 + \sum_{j=1}^{\infty} a_j\cos(jx) + b_j\sin(jx),$$

where

$$a_j = \frac{1}{\pi}\int_{-\pi}^{\pi} f(t)\cos(jt)dt \quad \text{and} \quad b_j = \frac{1}{\pi}\int_{-\pi}^{\pi} f(t)\sin(jt)dt, \quad j = 0, 1, \ldots.$$

Fourier transform If f is a function of the real variable x, its Fourier transform F is defined by

$$F(t) = \frac{1}{\sqrt{2\pi}}\int_{-\infty}^{\infty} e^{itx} f(x)dx,$$

where $i = \sqrt{-1}$ and t is a real variable. The **inverse Fourier transform** is given by

$$f(x) = \frac{1}{\sqrt{2\pi}}\int_{-\infty}^{\infty} e^{-itx}\, F(t)dt.$$

fractal An infinite set of points, usually in a plane, whose *fractal dimension is not an integer. Fractals are usually self-similar, i.e. any part of the fractal contains scaled-down versions of the whole fractal. The word 'fractal' was coined by *Mandelbrot in 1975. For a well-known example of a fractal, *see* SNOWFLAKE CURVE.

fractal dimension A generalization of the usual idea of dimension that permits non-integer values. For a plane set of points, the fractal dimension is estimated as follows. A lattice of square boxes of side s is superimposed over the set. The number, $N(s)$, of these boxes that contain a portion of the set is determined. This is repeated for a range of values of s. When $\ln\{N(s)\}$ is plotted against $-\ln(s)$ the result will be an approximate straight line. The fractal dimension is the slope of this line. An ordinary curve has a fractal dimension equal to 1. Any set of points in a plane has a fractal dimension not greater than 2.

fractional factorial design *See* FACTORIAL DESIGN.

fractional replicate *See* FACTORIAL DESIGN.

frailty model A *model of survival that takes account of the fact that some people (or components) are more vulnerable (or delicate) than others. *See also* COX REGRESSION MODEL.

frame Short for sampling frame. *See* SAMPLE.

Fréchet distribution A type of *extreme-value distribution. An example of the *probability density function of a *random variable, X, having a Fréchet distribution is:

$$f(x) = ax^{-(a+1)}e^{-(x^{-a})},$$

where a is a positive constant and $x > 0$.

Freeman–Tukey deviate *See* FREEMAN–TUKEY TRANSFORMATION.

Freeman–Tukey test A *goodness-of-fit test introduced in 1950. If there are n observed frequencies, $f_1, f_2, \ldots, f_n$ with, according to a p-parameter model, corresponding *expected frequencies $e_1, e_2, \ldots, e_n$, then the Freeman–Tukey statistic, T^2, is given by

$$T^2 = \sum_{k=1}^{n}\left(\sqrt{f_k} + \sqrt{f_k + 1} - \sqrt{4e_k + 1}\right)^2.$$

f

If the model is correct then the distribution of T^2 is approximately *chi-squared with $(n - p)$ *degrees of freedom.

Freeman–Tukey transformation If X is a binomial variable (*see* BINOMIAL DISTRIBUTION) with *parameters n and p, then the transformed variable Y, given in radians by

$$Y = \frac{1}{2}\left\{\sin^{-1}\left(\sqrt{\frac{X}{n+1}}\right) + \sin^{-1}\left(\sqrt{\frac{X+1}{n+1}}\right)\right\},$$

has an approximate *normal distribution with *mean $\sin^{-1}(\sqrt{p})$ and *variance $1/(4n + 2)$. The approximation improves as np increases and should not be used if $np < 1$.

If X is a Poisson variable (*see* POISSON DISTRIBUTION) with expected value μ, then, for $\mu > 1$, the transformed variable Z, given by

$$Z = \sqrt{X} + \sqrt{X+1} - \sqrt{4\mu + 1},$$

has an approximate standard *normal distribution. Observed values of Z are referred to as **Freeman–Tukey deviates**. Both transformations were proposed in a 1950 paper by M. F. Freeman and *Tukey.

frequency

1. The number of times that a particular data value is obtained in a *sample. For example, the frequency of 5 in the sample 4, 6, 5, 7, 4, 5, 2, 5 is 3. The sum of the frequencies is the *sample size. The term is also used in connection with a set of values. For example, the number of people aged between 20 and 30, or the number of people with blue or green eyes. The term **relative frequency** is used for the ratio (frequency) ÷ (sample size). In the context of a *goodness-of-fit test, a frequency is referred to as an **observed frequency** and usually denoted by O, or f_O.

2. In the context of a *time series, the number of *cycles per unit time.

frequency distribution A summary of the values obtained and the frequencies with which these values have occurred. The term was first used by Karl *Pearson in 1895.

frequency domain *See* PERIODOGRAM.

frequency polygon A graphical method for displaying *data that is an alternative to the *bar chart for displaying data on a *discrete random variable, and an alternative to the *histogram as a method of approximating the graph of a *probability density function.

Suppose a set of discrete numerical *data consists of the observed value x_1 with *frequency f_1, the observed value x_2 with frequency f_2, and so on. The frequency polygon consists of the straight line segments joining the points with coordinates (x_1, f_1), (x_2, f_2), ... It is conventional to start and finish the polygon on the x-axis at appropriate points.

In the case of grouped numerical data with classes of equal width d, say, the frequency polygon consists of the straight-line segments joining the points with coordinates (x_1, f_1), (x_2, f_2), ..., (x_n, f_n), where $x_1, x_2, \ldots, x_n$ are now the midpoints of the class intervals and $f_1, f_2, \ldots, f_n$ are the corresponding class frequencies. The polygon starts at $(x_1 - d, 0)$ and finishes at $(x_n + d, 0)$. A suitable change of scale gives the **relative frequency polygon**. As n increases and d decreases, the relative frequency polygon approaches the graph of the corresponding probability density function.

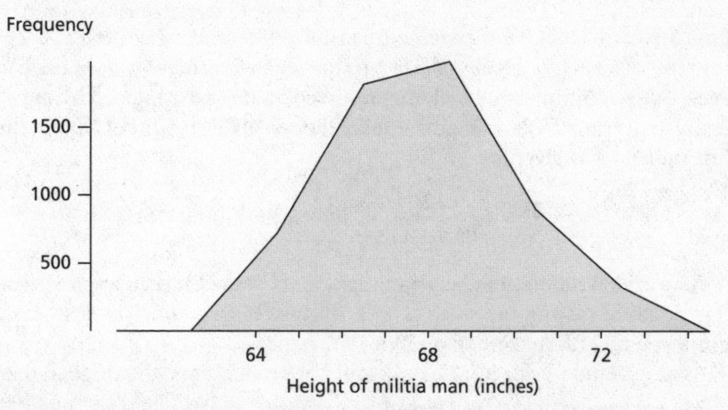

Frequency polygon. This polygon illustrates the frequencies of the heights of 5732 Scottish militiamen in 1817.

frequentist One who views *probability as being equal to the limiting relative *frequency as the sample size increases. The term was coined by Sir Maurice *Kendall who apparently had in mind *Venn and *von Mises amongst others.

Friedman, Jerome Harold (1939–; b. Siskiyou, CA) American statistician who has developed new methods of *exploratory data analysis and *machine learning. A physics graduate at UCB, he began developing statistical *algorithms when he joined Stanford U in the 1970s. He was the *IMS *Rietz Lecturer in 1999 and the *Wald Lecturer in 2009. He was awarded the *Parzen Prize in 2004, elected a fellow of the AAAS in 2005 and of the NAS in 2010.

SEE WEB LINKS

- University website.

Friedman, Milton (1912–2006; b. Brooklyn, NY; d. San Francisco, CA) American Nobel Prize winner for Economics in 1976. He was the son of an immigrant merchant who died when Milton was fifteen. A graduate of Rutgers U in 1932, Friedman obtained his PhD from Columbia U in 1946, by which time he was Professor of Economics at U Chicago. From 1937 to 1981 he was associated with the National Bureau of Economics Research. During this time he wrote many books on aspects of economics. In 1988 he was awarded the Presidential Medal of Freedom and the National Medal of Science.

SEE WEB LINKS

- Obituary with photographs.

Friedman test A *non-parametric test for differences between $t\ (>2)$ *treatments using b *blocks of size t, so that each treatment is used once in each block. Within each block the observed values are replaced by the equivalent *ranks. Denoting the total of the ranks for treatment i by R_i, the test statistic T is given by

$$T = \left\{\frac{12}{bt(t+1)}\sum_{i=1}^{t} R_i^2\right\} - 3b(t+1).$$

If there are no differences between treatments, then T has an approximate *chi-squared distribution with $(t-1)$ *degrees of freedom. The test was proposed by Milton *Friedman in a 1937 paper.

As an example, with $b=7$ blocks and $t=3$ treatments, the original data

A	52	63	45	53	47	62	49
B	45	79	57	51	50	72	52
C	38	50	39	43	56	49	40

are replaced by

A	1	2	2	1	3	2	2	$R_A = 13$
B	2	1	1	2	2	1	1	$R_B = 10$
C	3	3	3	3	1	3	3	$R_C = 19$

giving

$$T = \frac{12}{7 \times 3 \times 4}(13^2 + 10^2 + 19^2) - 3 \times 7 \times 4 = 90 - 84 = 6.$$

Since 6 just exceeds the upper 5% point of a χ_2^2 distribution it suggests that there are significant differences between the treatments.

The **Quade test** is a variant in which the ranks are multiplied by block scores that reflect the variability of the values within a block. The resulting test pays more attention to those blocks that provide the clearest evidence of differences between the treatments.

***F*-test** *See* TEST FOR EQUALITY OF VARIANCE.

***F* to enter; *F* to remove** *See* STEPWISE PROCEDURE.

Fuller, Wayne Arthur (1931-   ; b. Corning IA) American statistician specializing in *econometrics, survey sampling, and *time series. Fuller spent his entire career at ISU. As a student there he obtained his BS (1955), MS (1957), and PhD (1959) in agricultural economics. Joining the faculty he was appointed Professor in 1966 and Distinguished Professor in 1983. He is best known for his 1979 paper with *Dickey that introduced the eponymous *Dickey-Fuller test for *stationarity in a time series. Dickey was one of Fuller's 67 PhD students.

full rank *See* MATRIX.

functional data analysis (FDA) The study of the variability of a smooth function, $y = \mathrm{f}(x)$, say, based on data from several examples. For each example several data pairs (x,y) are available. The function f is taken to be the same for each example and the question of interest is the estimated functional form and the precision of its estimation. An example could be readings of the wave pattern corresponding to an individual repeatedly saying the word 'functional'.

funnel plot A plot used when collating findings from independent studies of the same topic (for example, the outcome of a medical treatment). The plot is a *scatter diagram of the estimated treatment effects in the individual studies against the study size. Small studies will give more variable estimates and hence greater scatter. Thus the diagram should give the appearance of a funnel. The plot is used to check for *publication bias (since, in small samples, the extreme estimates may have been published, whereas the less extreme estimates may not have been reported). *See diagram overleaf.*

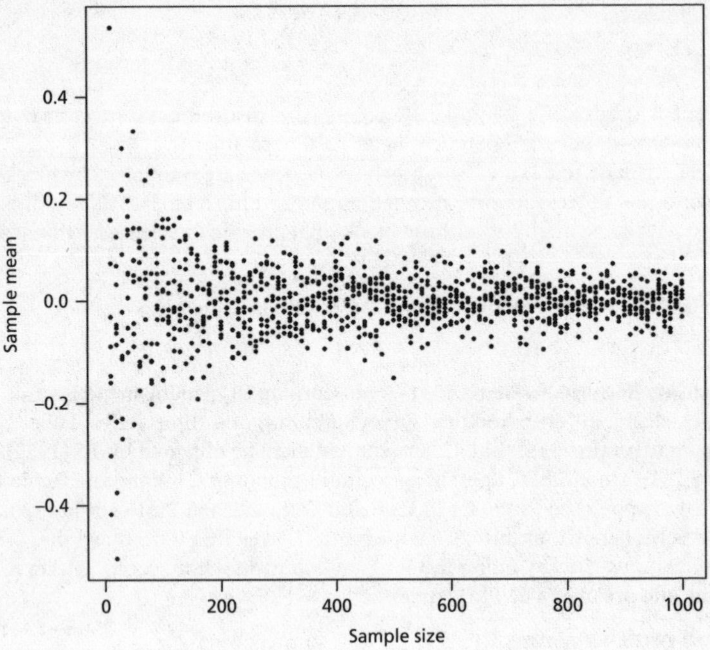

Sample mean
0.4
0.2
0.0
−0.2
−0.4
0
200
400
600
800
1000
Sample size

Funnel plot. Each point indicates the mean of a random sample taken from a standard normal distribution. Apparently unusual results are more likely when the sample size is small.

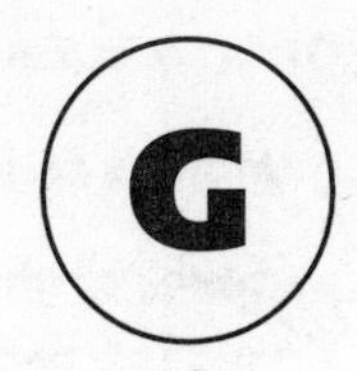

Gabriel, Kuno Ruben (1929–2003; b. Berlin, Germany; d. Rochester, NY) German-Jewish statistician known for his introduction of the *biplot. Gabriel spent his youth first in France and then in Palestine, where he studied through correspondence courses. He gained his BA at the LSE and his PhD (in *demography) in 1957 from Hebrew U in Jerusalem where he joined the faculty specializing in applying Statistics to meteorology. He moved to U Rochester in the USA in 1975.

SEE WEB LINKS

- Obituary.

Gabriel biplot *See* BIPLOT.

Gabriel test *See* MULTIPLE COMPARISON TEST.

Galton, Sir Francis (1822–1911; b. Birmingham, England; d. Haslemere, England) English doctor, explorer, meteorologist, biometrician, and statistician. Galton, who was a first cousin of Charles Darwin, the author of *The Origin of Species*, studied medicine at Cambridge U. On coming into money, he abandoned this career and spent the period 1850–2 exploring Africa; he received the Gold Medal of the Royal Geographical Society in recognition of his achievements. In the 1860s he turned to meteorology and devised an early form of the weather maps used by modern meteorologists. He coined the term 'anticyclone'. Subsequently, perhaps inspired by Darwin's work, he turned to inheritance and the relationships between the characteristics of successive generations. In his 1869 book *Hereditary Genius* he used the term *correlation in its statistical sense. His best-known work, published in 1889, was entitled *Natural Inheritance*. He made great use of the *normal distribution and illustrated it in a lecture to the Royal Institution in 1874 using a *quincunx. He is quoted as saying 'Whenever you can, count'. He was elected FRS in 1860.

SEE WEB LINKS

- Website with links and photographs.

Galton–Watson model *See* BRANCHING PROCESS.

GAM *See* GENERALIZED ADDITIVE MODEL.

gambler's fallacy *See* LAW OF AVERAGES.

gambler's ruin A classic *random walk problem that can be solved using the theory of *Markov processes. At each play a gambler is supposed to have *probability p of winning one unit and probability q $(= 1 - p)$ of losing one unit. The gambler starts with j units. Betting continues until either the gambler's fortune reaches N (in which case the gambler retires) or the gambler is ruined. If $p = \frac{1}{2}$ then the probability of ruin is $1 - (j/N)$, and otherwise it is

$$\left\{\left(\frac{q}{p}\right)^j - \left(\frac{q}{p}\right)^N\right\} \Big/ \left\{1 - \left(\frac{q}{p}\right)^N\right\}.$$

Games–Howell test *See* MULTIPLE COMPARISON TEST.

game theory A theory that deals with the determination of the optimal strategies for each of the players in a game with well-defined rules. In a typical scenario, two players, A and B, are playing a game. Player A is required to make a decision in ignorance of a simultaneous decision made by player B. The outcome is a consequence of the two decisions. In a **zero-sum game** A wins what B loses. Many real-life situations can be modelled in game-theoretic terms. The rules used by the players to determine their strategies are called **decision rules**.

gamma (γ) A measure of association (*see* ASSOCIATION), of the type known as *proportional reduction in error, suitable for examining the association between two *ordinal variables.

gamma distribution The general form of the gamma distribution has *probability density function f given by

$$\mathrm{f}(x) = \frac{(x-\gamma)^{\alpha-1}\exp\left[\frac{-(x-\gamma)}{\beta}\right]}{\beta^{\alpha}\Gamma(\alpha)}, \qquad x > \gamma,$$

where α (> 0), β (> 0) and γ are *parameters, and Γ is the *gamma function. The distribution has *mean $\alpha\beta + \gamma$ and *variance $\alpha\beta^2$. If $\alpha > 1$ then the distribution has *mode at $x = \gamma + \beta\,(\alpha - 1)$; otherwise the mode is at $x = \gamma$.

The case $(\alpha = \frac{1}{2}, \beta = 2, \gamma = 0)$ corresponds to the *chi-squared distribution with ν *degrees of freedom. The case $(\beta = 1, \gamma = 0)$ gives the standard form of the distribution:

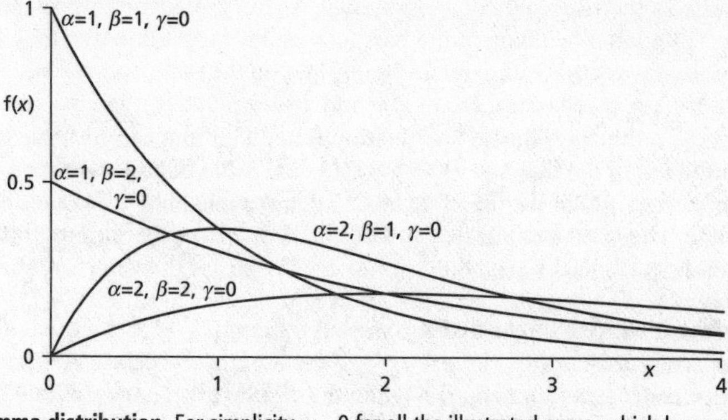

Gamma distribution. For simplicity $\gamma = 0$ for all the illustrated cases, which have mean $\alpha\beta$. When $\alpha \geq 1$ the mode is at $x = \beta(\alpha - 1)$, otherwise it is at $\alpha = 0$.

$$f(x) = \frac{x^{\alpha-1}e^{-x}}{\Gamma(\alpha)}, \qquad x > 0.$$

The case ($\alpha = 1, \beta = 1, \gamma = 0$) gives the *exponential distribution, and the case ($\alpha = k, \beta = 1, \gamma = 0$), where k is a positive integer, gives the **Erlang distribution**, which is the distribution of the time to the kth event in a *Poisson process. If Z is a standard normal variable (*see* NORMAL DISTRIBUTION) then $\frac{1}{2}Z^2$ has a gamma distribution with parameters $\alpha = \frac{1}{2}$, $\beta = 1$, $\gamma = 0$.

gamma function The gamma function, Γ, is defined by

$$\Gamma(x) = \int_0^\infty s^{x-1}e^{-s}ds.$$

In general $\Gamma(x+1) = x\Gamma(x)$. If x is a positive integer then $\Gamma(x) = (x-1)!$ (*see* FACTORIAL). If x is positive and an odd multiple of $\frac{1}{2}$ then the value of $\Gamma(x)$ is given by

$$\Gamma\left(\frac{1}{2}\right) = \sqrt{\pi},$$

$$\Gamma\left(\frac{2k+1}{2}\right) = \frac{2k-1}{2} \times \frac{2k-3}{2} \times \cdots \times \frac{1}{2}\sqrt{\pi}, \qquad k = 1, 2, \ldots.$$

gamma sampling *See* POISSON SAMPLING.

Gantt chart A project-planning tool that can be used to represent the scheduling of tasks required to complete a project. Each task corresponds to a row. Dates run along the top of the chart in suitable increments of time. The expected time for each task is represented by a horizontal bar whose left end marks the expected beginning of the task and whose right end marks the expected completion date. As the project progresses, the chart is updated by filling in the bars to a length proportional to the fraction of work that has been accomplished on the task. Completed tasks lie to the left of the current date-line and are completely filled in. Current tasks cross the date-line and are behind schedule if their filled-in section is to the left of the line and ahead of schedule if the filled-in section stops to the right of the line. Future tasks lie completely to the right of the line. The chart was devised in 1917 by Henry L. Gantt, an American engineer and social scientist.

garbage in, garbage out A phrase that reminds us that if the basic *data are untrustworthy (for example, if most of the observed variation results from *experimental error) then the results of any analysis will be valueless.

GARCH (Generalized Autoregressive Conditional Heteroscedasticity) GARCH *models are used to predict the *volatility of financial *random variables.

Gauss, Johann Carl Friedrich (1777–1855; b. Brunswick, Germany; d. Göttingen, Germany) German mathematician and astronomer. Gauss was responsible, in a paper published in 1809, for developing the statistical theory underlying the *method of least squares. He was educated at U Göttingen and U Helmstedt, obtaining his doctorate from the latter in 1799. His work on least squares, which was a consequence of his appointment as director of the observatory at Göttingen in 1807, also entailed his deriving an appropriate error distribution—the distribution now called the *normal distribution or Gaussian distribution. He was elected FRS in 1804 and awarded the Society's Copley Medal in 1838. He was elected FRSE in 1820. A lunar crater is named after him.

SEE WEB LINKS

- Information and portraits.

Gaussian distribution *See* NORMAL DISTRIBUTION.

Gaussian kernel *See* KERNEL METHODS.

Gauss inequality Let X be a *continuous random variable with a *distribution having a single *mode m. Let τ^2 be the *expected value of $(X-m)^2$. The inequality, published in 1823, is

$$P(|X - m| \geq k\tau) \leq \begin{cases} \left(\frac{2}{3k}\right)^2, & k \geq \frac{2}{\sqrt{3}}, \\ 1 - \frac{k}{\sqrt{3}}, & 0 \leq k < \frac{2}{\sqrt{3}}. \end{cases}$$

Gauss–Markov theorem *See* MULTIPLE REGRESSION MODEL.

Geary test *See* TEST FOR NORMALITY.

GEE model *Abbreviation for* GENERALIZED ESTIMATING EQUATIONS MODEL.

g

Gehan, Edmund Alpheus (1931-   ; b. Brooklyn, NY) American biostatistician specializing in applications to cancer research. A mathematics graduate of Manhattan College, New York, Gehan obtained his PhD in experimental statistics and public health at North Carolina State U in 1957 (partly supervised by Sir David *Cox). In 1994 he was appointed Director of Biostatistics at the Georgetown U Medical Center in Washington. In 1987 he was President of the *ISCB.

SEE WEB LINKS

- Photograph.

Gehan test A test, introduced by Gehan in 1965, as an alternative to the *Mantel–Haenszel test for comparing two survivor functions (*see* HAZARD RATE) using *censored data (because, for example, some patients are still living). With two samples, of sizes m and n, the test statistic is W, given by

$$W = \frac{1}{mn}(N_{A>B} - N_{B>A}).$$

The calculation of $N_{A>B}$ and $N_{B>A}$ requires examination of the mn pairs of observations (with one member of each pair from each sample). The quantity $N_{A>B}$ is the number of pairs in which the member chosen from sample A is certain to have a longer lifetime than the member chosen from sample B.

generalized additive model (GAM) Model analogous to a *generalized linear model in which the linear combination of explanatory variables (*see* REGRESSION) is replaced by a linear combination of *scatterplot smoothers.

generalized estimating equations model Linear model for *discrete data when the *observations cannot be regarded as independent. Typically, the data consist of *repeated measures on individuals.

generalized inverse *See* MATRIX.

generalized least squares (GLS) The extension of the *method of least squares procedure to the case where the *observations have been taken on *random variables that are not all independent of one another. The GLS estimate of the $(p+1)\times 1$ *parameter *vector β in the *multiple regression model

$$\mathrm{E}(\mathbf{Y}) = \mathbf{X}\beta$$

is given by

$$\hat{\beta} = (\mathbf{X}'\boldsymbol{\Sigma}^{-1}\mathbf{X})^{-1}\mathbf{X}'\boldsymbol{\Sigma}^{-1}\mathbf{y},$$

where $\mathbf{y}$ is an $n\times 1$ column vector of observations, $\boldsymbol{\Sigma}^{-1}$ is the inverse of the *variance-covariance matrix, and, with p explanatory variables and a constant term in the model, $\mathbf{X}$ is the $n\times(p+1)$ design matrix (*see* MULTIPLE REGRESSION MODEL), and $\mathbf{X}'$ is the transpose of $\mathbf{X}$. *See also* WEIGHTED LEAST SQUARES.

g

generalized linear model (GLM) A phrase introduced by *Nelder and *Wedderburn in 1972 to describe any model that relates μ, the *expected value of the response variable (*see* REGRESSION) Y, to a linear combination of the explanatory variables $x_1, x_2, \ldots, x_p$ using the model

$$\mathrm{g}(\mu) = \beta_0 + \beta_1 x_1 + \cdots + \beta_p x_p,$$

where $\beta_1, \beta_2, \ldots, \beta_p$ are unknown *parameters and g is the **link function**. Examples of link functions are $\mathrm{g}(\mu) = \mu$ (the **identity link**), and $\mathrm{g}(\mu) = \ln(\mu)$ (the **logarithmic link**). These correspond to random variables having a *normal distribution and a *Poisson distribution, respectively.

For different families of models that deal with variation in the parameter p of a binomial distribution:

$\mathrm{g}(p) = \log(p) - \log(1-p)$ (the **logistic link**, also called the **logit link**)
$\mathrm{g}(p) = \log\{-\log(1-p)\}$ (the **complementary log-log link**)
$\mathrm{g}(p) = -\log\{-\log(p)\}$ (the **negative log-log link**)
$\mathrm{g}(p) = \tan\{\pi(p-\frac{1}{2})\}$ (the **cauchit link**, also called the **inverse Cauchy link**)
$\mathrm{g}(p) = \Phi^{-1}(p)$ (the **probit link**),

where Φ denotes the *cumulative distribution function of the standard *normal distribution and here the logarithms are to base e.

generalized logistic equation *See* GROWTH CURVE.

generating function A function of an arbitrary variable (usually t) which, when expanded as a power series (in t), yields coefficients of interest to statisticians and others. Examples are the *moment-generating function and the *probability-generating function.

genetic algorithm An *algorithm used in particular for *combinatorial optimization and *machine learning. Later search locations are derived by reproduction, mutation, or crossover, which represent three different rules for adjusting or combining the coordinates of earlier search locations.

GENSTAT (Generalized Statistical Package) A statistical computer package particularly used in analysing natural science data.

SEE WEB LINKS

- Computer package home page.

geometric distribution A particular discrete distribution. An experiment or trial with two possible results, usually classified as 'success' or 'failure', is repeated independently until the first success is obtained. If the *probability of a success, p ($\neq 0, 1$), is the same for each trial, then the *distribution of the total number of trials, up to and including the one in which the first success is obtained, is a geometric distribution with *parameter p. The *probability function is given by

$$P(X = r) = p(1 - p)^{r-1}, \qquad r = 1, 2, \ldots.$$

The reason for the term 'geometric' is that the sequence of probabilities $P(X=r)$, for $r=0, 1, 2, \ldots$, forms a geometric progression with first term p and common ratio $(1 - p)$. The *mean is $1/p$ and the *variance is $(1 - p)/p^2$. The *mode is 1 for all values of p.

The fact that the mode is always 1 means that, even for a very rare event (e.g. a lottery win), the most probable number of trials necessary to obtain the first success is 1. This is sometimes regarded as a paradox.

If we write $q=1-p$, cumulative probabilities are given by $P(X\geq r)=q^{r-1}$ and $P(X\leq r)=1-q^r$. The geometric distribution is a discrete analogue of the *exponential distribution and shares the 'non-ageing' or *forgetfulness property:

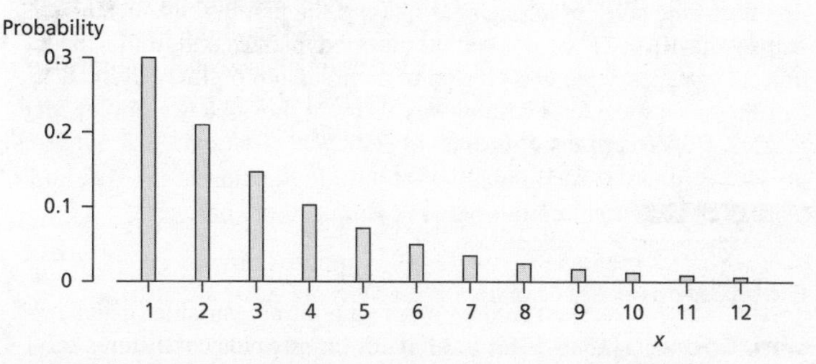

Geometric distribution. All geometric distributions have their mode at the lowest possible value, with successive probabilities having the common ratio p.

$$P(X = r + s | X \geq s) = P(X = r), \qquad s \geq 0.$$

If $X_1, X_2, \ldots, X_n$ are independent random variables each having a geometric distribution with parameter p, then $\sum_{k=1}^{n} X_k$, which is the number of trials up to and including the nth success, has a *negative binomial distribution with parameters n and p.

Some authors take the definition of a geometric distribution to be the number of trials before the first successful trial. This leads to $P(X = r) = p(1-p)^r$, for $r = 0, 1, \ldots$, for which the mean is $(1-p)/p$ and the mode is 0, though the variance is unchanged.

geometric mean For a set of positive observations $x_1, x_2, \ldots, x_n$, the geometric mean, g, is given by

$$g = (x_1 \times x_2 \times \cdots \times x_n)^{\frac{1}{n}}.$$

geometric variable A *random variable having a geometric distribution.

George W. Snedecor Award This bi-annual award by *COPSS, named in honour of *Snedecor, honours an individual who has made a noteworthy recent publication in *biometry.

geostatistics A subject involving ideas from geology, mathematics, mining engineering, and statistics. It is principally concerned with methods of analysis for *spatial processes, the motivation being the accurate detection of mineral deposits. Its scope was well defined by the work of *Krige and *Matheron.

Gibbs, Josiah Willard (1839–1903; b. New Haven, CT; d. New Haven, CT) American pioneer of statistical mechanics. Gibbs entered Yale U in 1854 to study mathematics and Latin. He remained at Yale, becoming its first doctorate student in engineering (and the first in the USA) in 1863. During the 1870s Gibbs worked on thermodynamics and in the 1880s on vector analysis. He was elected to membership of the NAS in 1879. In 1897 he was elected FRS and was awarded the Society's Copley Medal in 1901. His *Elementary Principles of Statistical Mechanics* was published in 1902. A lunar crater is named after him.

SEE WEB LINKS

- Biographical memoir and photograph.

Gibbs sampler *See* MARKOV CHAIN MONTE CARLO METHODS.

Gini, Corrado (1884–1965; b. Motta di Livenza, Italy; d. Rome, Italy) Italian statistician. Gini studied Statistics as a compulsory part of his study of law at U Bologna. He realized that Statistics provided a means of measuring social phenomena and this became his specialism. He was the founder of the first Department of Statistics in Italy (at U Rome), in 1936.

SEE WEB LINKS

- Information and photograph.

Gini coefficient; Gini index *See* LORENZ CURVE.

Gini statistic A measure of the spread or dispersion of a *probability distribution. Let X_1 and X_2 be *independent random variables with this distribution; then the Gini statistic is the *expected value of $|X_1 - X_2|$.

g

Gittins, John Charles (1938-   ; b. Durham, England) English statistician. Gittins studied mathematics at Cambridge U and obtained his PhD (supervised by *Lindley, his research concerned optimal resource allocation) from University College, Aberystwyth. After working for the Steel Company of Wales, he joined the faculty at Oxford U in 1974, becoming Professor of Statistics in 1996. He was awarded the *Guy Medal in Silver of the *RSS in 1984.

SEE WEB LINKS

- University website.

Gittins index *See* BANDIT PROBLEMS.

GLIM (Generalized Linear Interactive Modelling) A statistical computer package especially suitable for fitting *generalized linear models. The package was commissioned by the *RSS in 1972 with the guidance of *Nelder.

GLM *See* GENERALIZED LINEAR MODEL.

global maximum *See* LOCAL MAXIMUM.

GLS *See* GENERALIZED LEAST SQUARES.

Gnanadesikan, Ramanathan (1932-   ; b. Madras (Chennai), India) Indian mathematical statistician, specializing in the analysis of *multivariate data, whose career was in the USA. A graduate of U Madras, Gnanadesikan obtained his PhD at UNC in 1957. In 1959 he joined AT&T Labs working alongside *Wilk and *Mallows. In 1991 he joined the faculty at Rutgers U. He was President of the *IASC in 1981 and President of the *IMS in 1989.

SEE WEB LINKS

- Fuller biography, interview, and photographs.

Gompertz, Benjamin (1779–1865; b. London, England; d. London, England) English actuary. Gompertz's parents were Jewish Dutch merchants. Because of his faith he was denied entry to university, but he nevertheless became a member of the LMS when aged eighteen. He joined the Stock Exchange in 1810 and was elected FRS in 1819. As an actuary he was concerned with the expectation of life. Statisticians remember Gompertz for his 1825 'Law of Mortality' (*see* GOMPERTZ DISTRIBUTION), while animal rights campaigners recall that his youngest brother, Lewis, struck by 'the similitude between man and other animals' was one of the founders, in 1824, of the Society for Prevention of Cruelty to Animals (now the RSPCA). In 1834 Benjamin was a founder member of the *RSS.

SEE WEB LINKS
- Fuller biography.

Gompertz distribution A probability distribution now better known as the *Gumbel distribution. In 1825 *Gompertz proposed a probability model for a 'Law of Mortality', assuming the 'average exhaustion of a man's power to avoid death to be such that at the end of equal infinitely small intervals of time he lost equal portions of his remaining power to oppose destruction which he had at the commencement of these intervals'.

Gompertz equation *See* GROWTH CURVE.

Goodman, Leonard Albert (1928– ; b. New York City) American statistician and sociologist. A graduate of mathematics and sociology at Syracuse U, Goodman obtained his PhD at Princeton U (supervised by *Tukey and *Wilks) in 1950. His academic career was spent at U Chicago and UCB. He was the *COPSS *Fisher Lecturer in 1968. He was elected to membership of the AAAS in 1973 and the NAS in 1974. He was the *IMS *Rietz Lecturer in 1983 and was presented with the *ASA *Wilks Award in 1985.

SEE WEB LINKS
- Fuller biography, interview, and photographs.

Goodman and Kruskal's gamma (γ) *See* ASSOCIATION.

Goodman and Kruskal's lambda (λ) *See* ASSOCIATION.

goodness-of-fit test A test of the fit of some *model to a set of *data. The most commonly used tests are the *chi-squared test and the *likelihood-ratio test. *See also* INDEX OF DISPERSION; KOLMOGOROV–SMIRNOV TEST.

gooseberry bushes An example of *nonsense correlation. If the number of gooseberry bushes in London gardens is plotted against the number of births in London in successive decades of the twentieth century, then the resulting graph is likely to show a strong positive nonsense correlation, since both are a reflection of increasing population size. In the same way, increases in births can be linked to the numbers of storks nesting on the rooftops of European cities.

Gosset, William Sealy (1876–1937; b. Canterbury, England; d. Beaconsfield, England) English statistician. Gosset was a pioneer of the statistical theory associated with small *sample sizes. He was educated at Winchester College and Oxford U, where he studied chemistry. In 1899 he joined the staff of Arthur Guinness Son & Co. Ltd as a 'brewer'. An early task was to investigate the relationship between the quality of the final product and that of the raw materials (such as barley and hops). The difficulty with this task was the expense and time involved in obtaining an observation, so large samples were not available. Gosset correctly mistrusted the existing theory and, in a paper published in 1908, entitled *The Probable Error of a Mean*, he conjectured the form of the *t-distribution relevant for small samples. Guinness company policy at the time meant that Gosset was obliged to publish under a pseudonym: being naturally modest he chose the pen-name 'Student', and the t-distribution is still sometimes referred to as 'Student's t-distribution'.

SEE WEB LINKS

- Fuller biography and photograph.

Graeco-Latin square *See* LATIN SQUARE.

Gram–Charlier expansion An expansion relating a *probability density function, f, to ϕ, the probability density function of a standard *normal distribution, using *Chebyshev–Hermite polynomials. The alternative *Edgeworth expansion is usually preferred.

grand mean When the *data come from different groups (e.g. 'males' and 'females') then the grand mean is the mean of all the values, irrespective of their group.

Granger, Sir Clive William John (1934–2009; b. Swansea, Wales; d. San Diego, CA) Welsh econometrician, specializing in *time series, who spent most of his career in the USA. Granger studied at U Nottingham, obtaining his BA in mathematics in 1955. He promptly joined the faculty as lecturer in statistics, obtaining his PhD in 1959. After a period in the United States he returned to Nottingham as Professor of Economics and Statistics. In 1974 he moved to UCSD. He was elected to the AAAS in 1994.

Together with *Engle, he was awarded the Nobel Prize for Economics in 2003. He was knighted in 2004.

SEE WEB LINKS

- Obituary with photograph.

graph A graph consists of a set of **nodes** usually represented by small circles, and a set of **arcs** usually represented by lines. If a path can be found that connects all the nodes, then a graph is said to be a **connected graph**. If no pair of nodes is connected by more than one arc then the graph is said to be a **simple graph**. A graph in which each arc has an associated direction is a **digraph**.

A summary of the connections between the nodes of a graph is provided by an **adjacency matrix**. For a simple graph with n nodes, the entry in cell (i, j) of the $n \times n$ *matrix will be 1 if nodes i and j are **adjacent** (i.e. directly connected), and 0 otherwise.

graphical model A *graph in which nodes represent *random variables. Pairs of nodes connected by arcs correspond to variables that are not *independent of one another. If two sets of nodes A and B are not connected, except via paths that pass through a node in a third set, C, then the variables represented by the nodes in the set A are *conditionally independent of the variables represented by the nodes in the set B, given the values of the variables represented by the nodes in the set C.

If an arc has no direction attached, then this implies that there is *association but not causation. An **undirected graphical model** (also called a *Markov random field) is a model in which no arcs have directions attached. A model in which all arcs have directions attached is called a **directed graphical model** (or **Bayes network** or **Bayes net**). If we consider a directed path as referring to 'generations' of variables, then a simple example of conditional independence (not the only possibility) occurs when a 'child variable' is conditionally independent of earlier 'ancestor variables', given the values of the 'parent variables'. A special case is the *hidden Markov model. A directed graphical model having no path that leads from a node back to that same node is termed a **directed acyclic graph**. Such graphs occur in *expert systems. A graph containing a mixture of directed and undirected arcs is a **chain graph**.

Graunt, John (1620–74; b. London, England; d. London, England) Prosperous London haberdasher and Freeman of the Drapers' Company. Graunt became interested in the information implicit in the weekly 'Bills of Mortality' for London and, in 1662, he published *Natural and Political Observations Mentioned in a following Index and Made Upon the Bills of Mortality*. This was well received by the Royal Society, which duly elected him to be a Fellow. At one time he had a fine collection of paintings, described by the diarist Samuel Pepys as 'the best collection of anything almost I saw'. However in 1666 he lost much property in the Great Fire of London and he died in poverty. His book contains the foundations of *demography, showing understanding of the concepts of *sample and *population and introducing the *life table.

SEE WEB LINKS

- Fuller biography and his original data.

gravity model A *model of spatial mobility. If locations i and j, at a distance d_{ij} apart, have populations p_i and p_j respectively, then the gravity model states that the number of individuals moving from i to j is proportional to $p_i^a\, p_j^b\, d_{ij}^c$, where a, b, and c are constants.

Graybill, Franklin Arno (1927–2012; b. Carso, 1A; d. Mesa, AZ) American statistician. Supervised by *Kempthorne, Graybill obtained his PhD from Iowa State U in 1952. In 1961, at Colorado State U, he founded the statistical laboratory that bears his name. He was co-author with *Mood of the 1963 textbook *Introduction to the Theory of Statistics*, which for many years was the recommended introductory textbook worldwide. In 1976 Graybill became President of the *ASA after a contested election (an unusual occurrence—the loser was *Kruskal).

SEE WEB LINKS

- Fuller biography and photograph.

Green, Peter James (1950– ; b. Solihull, England) English mathematical statistician specializing in *Markov Chain Monte Carlo methods. A graduate of Oxford U, Green gained his PhD at Sheffield U in 1976. His academic career has taken him from U Bath (1974) to Durham U (1978), and to U Bristol (1989). He was awarded the *Guy Medal in Bronze of the *RSS in 1987, and in Silver in 1999. He was President of the RSS in 2001 and was elected FRS in 2003.

SEE WEB LINKS

- University website.

Greenhouse, Samuel William (1918–2000; b. New York City; d. Rockville, MD) American biometrician. Greenhouse was educated at City College (now City U) of New York, gaining his BS in mathematics in 1938. On graduating he joined the Bureau of Census, working with *Deming. After serving in the army during the war, Greenhouse was recruited, in 1948, to be an inaugural member (with *Mantel) of the first *biometry group in the National Cancer Institute. His subsequent career took him through the leadership of several sections of the National Institute of Health (NIH). At the same time, from 1946 he had been a part-time faculty member at GWU, where he also studied under *Kullback for his PhD. On his retirement from NIH in 1974 (as an Associate Director), he took a full-time post at GWU, retiring in 1988.

SEE WEB LINKS

- Fuller biography and photograph.

Greenhouse–Geisser correction An adjustment made to the numbers of *degrees of freedom, in analysis of variance (*see* ANOVA) when it is known that the *observations do not obey the usual assumptions of being *uncorrelated with constant *variance.

g

Greenwood, Major (1880–1949; b. London, England; d. London, England) English medical statistician. Greenwood qualified in medicine from UCL in 1904. His work as a medical statistician began in 1910. After a spell in the Royal Army Medical Corps (just failing to become Major Major Greenwood!) he joined the Ministry of Health. He was appointed Professor of Vital Statistics and Epidemiology in 1928 at LSHTM. He was elected FRS in 1928. He was awarded the *Guy Medal in Gold of the *RSS in 1945, having been its President in 1934.

SEE WEB LINKS

- Photograph.

Greenwood formula *See* KAPLAN–MEIER ESTIMATE.

group-average clustering A method for collecting *multivariate data into clusters. *See* AGGLOMERATIVE CLUSTERING METHODS; CLUSTER ANALYSIS; CLUSTER SAMPLING.

grouped data The *frequencies of occurrence of values in specified intervals (*classes). *Histograms and *frequency polygons are used to illustrate grouped data.

grouped mean *See* SAMPLE MEAN.

growth curve A curve describing the growth of a *population. Let y denote the size of the population at time t, with c denoting the ultimate size and b denoting the **growth rate** (the parameter that governs the rate at which the population approaches its maximum value). Simple two-parameter possibilities include

$$y = ce^{-e^{-bt}} \quad \text{(the \textbf{Gompertz equation})}$$
$$y = c(1 - e^{-bt}) \quad \text{(the \textbf{Mitscherlich equation})}$$
$$y = ct/(b+t) \quad \text{(the \textbf{Michaelis-Menten equation}).}$$

Taking reciprocals in the last of these gives the **Lineweaver–Burk equation**

$$\frac{1}{y} = \frac{1}{c} + \frac{b}{c} \times \frac{1}{t},$$

which is a simple *linear model connecting $1/y$ and $1/t$.

A more general growth equation is the **generalized logistic equation**, also called the **Richards equation**:

$$y = a + (c - a)\left\{1 + de^{-b(t-t_m)}\right\}^{-1/d},$$

which describes growth from a low of a to a high of c, with t_m being the time of maximum growth and d controlling whether that time occurs when the value of y is nearer to a or to c.

Grubbs, Francis Ephraim (1913–2000; d. Five Points, AL) American operational researcher. With degrees in electrical engineering and mathematical statistics, Grubbs began his military career in the United States Army reserve in 1941. During the Second World War he served in the Ordnance Department and the Ballistics Research Laboratory (BRL). In 1944 he supervised the stockpiling of ammunition in England prior to the D-Day landings, dividing the ammunition into categories based on their ballistic characteristics. After the war, whilst continuing to work for BRL, he obtained his PhD (on the detection of *outliers) at U Michigan in 1949. He is credited with initiating *Box's career in the USA. In 1964 he was the first recipient of the *Wilks Award of the *ASA. In 1971 he was awarded the *Shewhart Medal of the *ASQ. He was the author of the interestingly entitled *Statistical measures of accuracy for riflemen and missile engineers*.

Grubbs test *See* OUTLIER.

G^2 *See* LIKELIHOOD-RATIO GOODNESS-OF-FIT STATISTIC.

Gumbel, Emil Julius (1891–1966; b. Munich, Germany; d. New York City) German-Jewish statistician who spent most of his career in exile. He was educated at Munich U, obtaining a PhD in population statistics in 1914. In 1923 he joined the faculty of U Heidelberg but he was an outspoken pacifist and his political publications led to exile, first in France in 1932 where he worked at U Lyon, and then in the USA in 1940. It was while in France that he published the definitive study of the *extreme-value distribution that bears his name.

SEE WEB LINKS

- Archive and photograph.

Gumbel distribution An *extreme-value distribution, introduced by *Gumbel in 1935, that has *probability density function f given by

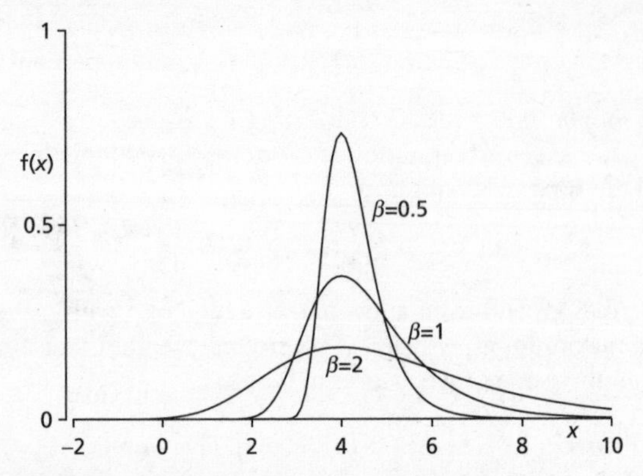

Gumbel distribution. All the distributions illustrated have $\alpha = 4$.

$$f(x) = \frac{1}{\beta}\exp\left\{-\frac{(x-\alpha)}{\beta} - \exp\left(-\frac{x-\alpha}{\beta}\right)\right\}, \qquad -\infty < x < \infty.$$

The distribution has *mode α, *mean $\alpha + \gamma\beta$ (where $\gamma = 0.5772156649\ldots$ is *Euler's constant), and *variance $\frac{1}{6}\beta^2\pi^2$.

Guttman, Louis Eliahu (1916–87; b. Brooklyn, NY; d. Jerusalem, Israel) American-born psychometrician who spent most of his working life in Israel. Guttman's PhD, at U Minnesota, was in *factor analysis and led to the development of the *Guttman scale. He was the founder of the Israel Institute of Applied Social Research.

SEE WEB LINKS

- Fuller biography.

Guttman scale A method of quantifying the strength of a person's opinion on a topic of interest. The idea is that a series of k related yes/no questions in a *questionnaire may be ordered in the sense that when the questions have been arranged in the appropriate order, an individual who answers 'Yes' to question j ($<k$) will probably answer 'Yes' to questions $j+1, j+2, \ldots, k$. For example, a Guttman scale is formed by the series of questions asking whether an individual deserves the death penalty for killing another when the killing was

1. by accident;
2. in self-defence;
3. in the heat of the moment;
4. premeditated.

Usually the ordering of the questions will be less obvious.

Guy, William Augustus (1810–85; b. Chichester, England; d. London, England) English physician and statistician. Guy was appointed Professor of Medical Jurisprudence at KCL in 1838. In 1844 he published a paper entitled *On the Value of Numerical Methods as applied to Science, but especially to Physiology and Medicine*. An advocate of sanitary reform, he was elected FRS in 1866 and was President of the *RSS from 1873 to 1875.

- Obituary in *New York Times*.

Guy Medal Award founded by the *RSS in honour of their past President, William *Guy. Currently, the Guy Medal in Gold is awarded once every three years to individuals judged to have made innovative contributions to the theory or application of Statistics. Guy Medals in Silver and Bronze are awarded annually to Fellows of the Society for papers published in the Society's journals or given at a meeting organized by the Society.

- Photograph.

H_0; H_1 *See* HYPOTHESIS TEST.

Haar, Alfréd (1885–1933; b. Budapest, Hungary; d. Szeged, Hungary) Hungarian mathematician. Haar obtained his PhD in mathematics from U Göttingen in 1909. In 1912 he returned to Hungary and after the First World War he founded the Mathematical Centre at Szeged U.

Haar wavelet *See* WAVELET.

Hadamard, Jacques Salomon (1865–1963; b. Versailles, France; d. Paris, France) French mathematician. Hadamard was educated in Paris where, after graduating from the École Normale Supérieure in 1888, he obtained his doctorate in 1892. His wife was a relative of Alfred Dreyfus, who was convicted of selling military secrets to the Germans. It was subsequently discovered that the evidence against Dreyfus had been manufactured and Hadamard was instrumental in helping to clear his name. Hadamard's research spanned many areas of mathematics. He was elected a member of the NAS in 1926, FRS in 1932, and FRSE in 1946.

SEE WEB LINKS

- Photograph.

Hadamard matrix A square *matrix $\mathbf{H}_n$ with n rows and columns, in which every entry is either 1 or – 1, arranged so that $\mathbf{H}'_n\,\mathbf{H}_n = n\mathbf{I}$, where $\mathbf{I}$ is an $n \times n$ identity matrix and $\mathbf{H}'_n$ is the transpose of $\mathbf{H}_n$. These matrices are useful in the construction of *experimental designs. The possible values of n are 1, 2, and multiples of 4.

Haenszel, William Manning (1910–98; b. New York City; d. Wheaton, IL) American epidemiologist. Haenszel was a graduate of U Buffalo. He joined the New York State Department of Health in 1934. Much of his time there was spent in the National Cancer Institute, where he worked with *Mantel. He was co-author (with Mantel) of the 1959 paper which introduced the eponymous *Mantel–Haenszel test of *independence. From 1976 to 1995 he was Professor of Epidemiology at U Illinois.

SEE WEB LINKS

- Fuller biography, interview, and photograph.

Haldane, John Burdon Sanderson (1892–1964; b. Oxford, England; d. Bhubaneswar, India) Pioneering English geneticist who applied Statistics to human genetics. His career included posts at UCL, U Cambridge and U Oxford. His *A Mathematical Theory of Natural and Artificial Selection* was published in 10 parts between 1924 and 1934. He was elected FRS in 1932 and was awarded the Society's Darwin Medal in 1952.

SEE WEB LINKS

- Biography, photograph, and links.

Haldane estimator *See* TWO-BY-TWO TABLE.

half-normal plot *See* NORMAL PROBABILITY PAPER.

half-replicate *See* FACTORIAL DESIGN.

h

half-space depth *See* DATA DEPTH.

Hall, Peter Gavin (1951– ; b. Sydney, Australia) Australian statistician specializing in non-parametric statistics (*see* NON-PARAMETRIC TEST). In 1974 Hall graduated in mathematical statistics from U Sydney, and in 1976 he obtained both an MSc from ANU and a PhD (supervised by *Kingman) from U Oxford. He has held posts at ANU and (currently) U Melbourne. He was elected a Fellow of the AAS in 1987, and FRS in 2001. He was awarded the Pitman Medal (*see* PITMAN) of the *SSAI in 1990, and the *Hannan Medal of the AAS in 1994. He was President of the *Bernoulli Society in 2001, the *IMS *Wald Lecturer in 2006, and the *ASA *Wilks Award winner in 2012.

SEE WEB LINKS

- University website.

halo effect A bias affecting judgements of performance. If an individual does well in one aspect then this is likely to trigger favourable reports on other aspects.

Hammersley, John Michael (1920–2004; b. Helensburgh, Scotland; d. Oxford, England) English mathematician and statistician. Hammersley's education at Cambridge U was interrupted by war service spent in the Royal Artillery. Graduating in 1948, Hammersley joined the statistics faculty at Oxford U. In 1964 he was co-author of one of the first books on *Monte Carlo methods. He was elected FRS in 1976.

SEE WEB LINKS

- Fuller biography and portrait.

Hammersley–Clifford theorem A theorem, formulated by *Hammersley and Peter Clifford, in 1971, that is concerned with the values at a set of interconnected locations. The theorem defines the class of *probability distributions that are consistent with the Markov property that the value at a location is dependent only on the values of its immediate neighbours. *See* MARKOV RANDOM FIELD.

Hamming, Richard Wesley (1915–98; b. Chicago, IL; d. Monterey, CA) American computing expert. Hamming studied at U Chicago and U Nebraska before obtaining his PhD from U Illinois at Urbana-Champaign. From 1946 to 1976 he worked at Bell Telephones. He ended his career as Professor of Computer Science at the Naval Postgraduate School at Monterey, California. Hamming once wrote 'The purpose of computing is insight, not numbers'.

SEE WEB LINKS

- Fuller biography and photographs.

Hamming distance *See* DISSIMILARITY.

Hamming window *See* PERIODOGRAM.

hanging rootogram A diagram suggested by *Tukey in 1971, for comparing an observed *bar chart or *histogram (with equal-width

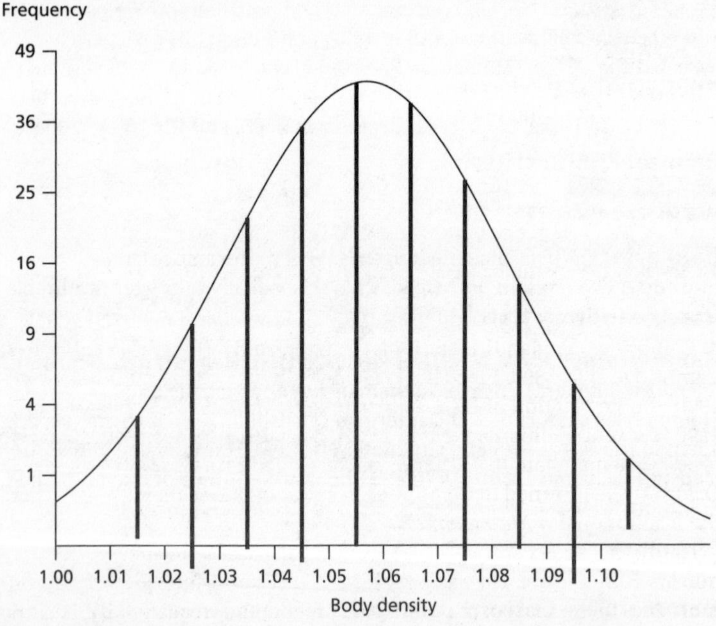

Hanging rootogram. The data are the body densities (in gm/cm^3) of 200 men. The diagram suggests that a *normal distribution provides an acceptable description.

categories) with a theoretical *probability distribution. The comparison is made easier by 'hanging' the observed results from the theoretical curve, so that the discrepancies are seen by comparison with the horizontal axis rather than a sloping curve. As in the *rootogram, the vertical axis is scaled to the square-root of the frequencies so as to draw attention to discrepancies in the tails of the distribution.

Hannan, Edward James (1921–94; b. Melbourne, Australia; d. Canberra, Australia) Australian statistician specializing in *time series. Hannan left school aged fifteen to work in the Commercial Bank of Australia, before enlisting in the army through the Second World War. After the war, he took a BCom degree at U Melbourne and then resumed work in the bank where he calculated interest rates. During this period he taught himself mathematical statistics. A chance encounter with *Moran led to an invitation to Hannan to study with Moran at ANU. After a period at Canberra U, Hannan returned to ANU in 1970 as Professor of Statistics. In 1979 he was awarded the Lyle Medal of the AAS. He was President of the *SSAI in 1981 and was awarded its Pitman Medal (*see* PITMAN) in 1986.

SEE WEB LINKS

- Fuller biography and portrait.

Hannan Medal A medal awarded every two years by the AAS. It honours the contribution of *Hannan to *time series analysis.

Hannan–Quinn criterion *See* MODEL SELECTION PROCEDURE.

hardcore process A *point process which is such that no two points can be within some minimum distance of each other. Appropriate for modelling, for example, the locations of mature oak trees.

Hardy, Godfrey Harold (1877–1947; b. Cranleigh, England; d. Cambridge, England) English mathematician. Hardy graduated from Cambridge U in 1898. Elected a Fellow of Trinity College, Cambridge in 1900, the first edition of his *A Course of Pure Mathematics* appeared in 1908 (the tenth edition appeared in 1993). In 1914 he hosted the Indian prodigy, Ramanujan. In 1919, he was appointed Professor of Geometry at Oxford U. He returned to Cambridge in 1931 and it was there that he wrote the inspirational *A Mathematician's Apology*.

Hardy was an eccentric—on entering hotel rooms, he covered all mirrors with a towel. He was a keen fan of cricket, which he watched on most days in the season; in order to ensure that the weather stayed fine he would arrive at the cricket ground with thick sweaters and an umbrella, which he referred to as his 'anti-God battery'.

Hardy made significant advances in the theory of numbers, particularly the prime number theorem. Amongst his sayings is the statistical observation that 'It is not worth an intelligent man's time to be in the majority. By definition, there are already enough people to do that'.

Hardy was elected FRS in 1910 and was awarded the Society's Copley Medal in 1947. He was elected to membership of the NAS in 1927. He was President of the LMS in 1926 and again in 1939 and was awarded that Society's de Morgan Medal in 1929.

SEE WEB LINKS

- Fuller biography and photograph.

Hardy–Weinberg law A law that states that under random mating in a *population the *proportions with particular genetic characteristics remain constant over the generations. Suppose that a large population contains individuals with gene pairs *AA*, *Aa*, and *aa*, and that individuals mate with randomly chosen partners. Their offspring receive one gene from each parent. If, from each gene pair, the gene is chosen at random, then the Hardy–Weinberg law, first published independently by *Hardy and by Wilhelm Weinberg, both in 1908, states that the proportions of the three types of gene pair will remain constant. The population is in *equilibrium.

harmonic mean The *reciprocal of the *mean of the reciprocals of a set of non-zero values. For values $x_1, x_2, \ldots, x_n$ the harmonic mean, h, is given by

$$h = n \Big/ \sum_{j=1}^{n} \frac{1}{x_j}.$$

Hartley, Herman Otto (1912–80; b. Berlin, Germany; d. Durham, NH) German statistician who spent the majority of his career in the USA. He was born with the surname Hirschfeld and was known by his colleagues as HOH. Hartley gained his PhD in mathematics at U Berlin in 1933. In the following year, he emigrated to England to work with Egon *Pearson, gaining in 1940 a PhD in statistics at Cambridge U, where he was supervised by *Wishart. In 1946 he joined Pearson on the faculty at UCL where he supervised *Box. He was co-author, with Pearson, of the widely used *Biometrika Tables for Statisticians* In 1953 he moved to the USA, with posts successively at Iowa State College, Texas A & M U (1963), and Duke U (1979). He was President of the *ASA in 1979 and the Association's *Wilks Award winner in 1973.

SEE WEB LINKS

- Photograph and obituary.

Hartley test *See* TEST FOR EQUALITY OF VARIANCE.

Hasse diagram A type of diagram used by the German mathematician Helmut Hasse (1898–1979) to illustrate relations in a finite partially ordered set, such as the set of possible *multiple regression models.

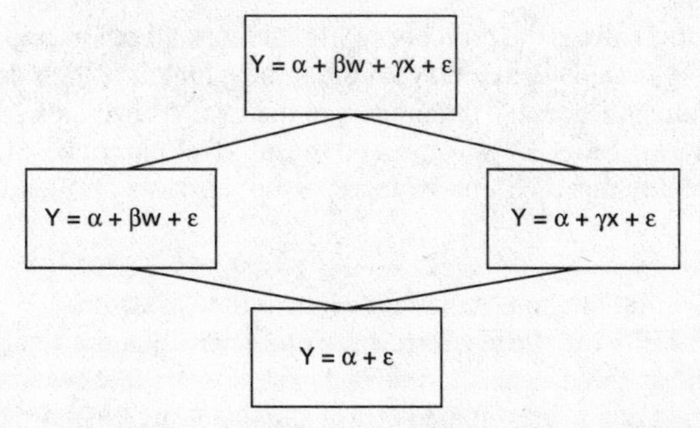

Hasse diagram. The diagram shows the relations between four possible regression models.

Hastings, W. Keith (1930– ; b. Toronto, Canada) Canadian statistician. Hastings was a student at U Toronto where his PhD (1962) was supervised by Geoffrey *Watson. After two years at U Canterbury in New Zealand, and two at Bell Labs, he returned in 1966 to U Toronto where he wrote the paper (in **Biometrika*) that placed an algorithm due to *Metropolis in a statistical context (*see* MARKOV CHAIN MONTE CARLO METHODS). In 1971 he moved to U Victoria, BC.

SEE WEB LINKS
- Biographical memoir.

hat matrix *See* REGRESSION DIAGNOSTICS.

Hawthorne effect *See* BLINDING.

hazard rate (age-specific failure rate; failure rate) A measure, $h(t)$, of the chance that a component (or person), still working (or alive) at age t, is about to fail (or die). Formally, the hazard rate is defined by

$$\mathrm{h}(t) = \lim_{\delta t \to 0}\left\{\frac{\mathrm{P}[\text{a component, working at } t, \text{fails in the interval } (t, t+\delta t)]}{\delta t}\right\}.$$

The **proportional-hazards model** (also termed the **Cox regression model**) was proposed by Sir David *Cox in 1972. It is a *linear model in which the logarithm of the hazard rate, h(t), is related to one or more explanatory variables x_1, x_2, ... as follows:

$$\ln\{h(t)\} = g(t) + \beta_1 x_1 + \beta_2 x_2 + \cdots,$$

where g(t) is an unspecified function of time and β_1, β_2, ... are *parameters.

The **hazard ratio** is the ratio of two hazard rates. If both hazard rates are described by the above model, with the same g function, then the hazard ratio is a function of the β-parameters of the two hazard rates.

The **survivor function** (also termed the **survival function**), S(t), is the *probability that a component (or person) survives until time t. Thus

$$S(t) = 1 - F(t),$$

where F(t) is the lifetime *cumulative distribution function.

The **reliability function**, which is the probability that the component is still working at time t, is an alternative description of the survivor function.

The hazard rate is related to the *probability density function, f, and the survivor function by the equation

$$f(t) = h(t)S(t).$$

For many situations the graph of a hazard rate is a **bathtub curve**: initially the rate is high as the component beds in, there is then a constant hazard rate, and finally the component starts to wear out.

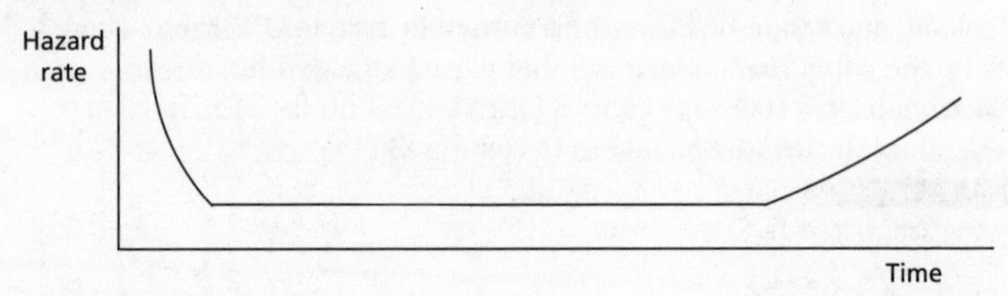

Hazard rate. The typical bathtub curve that results from a plot of the hazard rate against time. A high initial rate falls to a 'normal level' and then slowly increases as ageing sets in.

A plot of the survivor function against time is termed the **survival curve**. The survival time is the time until the occurrence of a particular event such as death or the failure of a component.

hazard ratio *See* HAZARD RATE.

heavy-tailed distribution A *distribution in which the probability of an extreme value is greater than it would be for a *normal distribution having the same *quartiles. For the *random variable X, one definition is that the distribution has a heavy right tail if

$$\lim_{x\to\infty} e^{\lambda x} P(X > x) = \infty, \text{for all } \lambda > 0.$$

If, in addition the distribution satisfies the requirement that, for all $t > 0$,

$$\lim_{x\to\infty} P(X > x + t | X > x) = 1$$

then the distribution is said to have a long right tail and may be described as a **long-tailed distribution.**

Examples of heavy-tailed distributions are the *Cauchy distribution, the *lognormal distribution, the *Pareto distribution, and the *t-distribution.

h

Heckman, James Joseph (1944– ; b. Chicago, IL) American econometrician. Heckman studied mathematics at Colorado College (BA, 1965) followed by economics at Princeton U (MA, 1968; PhD, 1971). He joined the faculty of U Chicago in 1973. He was elected to the AAAS in 1985 and the NAS in 1992. Together with *McFadden, he was awarded the Nobel Prize for Economics in 2000.

SEE WEB LINKS

- University website.

Hellinger, Ernst David (1883–1950; b. Strzegom, Poland; d. Chicago, IL) Polish mathematician. Hellinger was a Polish Jew, educated at several German universities, obtaining his doctorate from U Göttingen in 1907. In 1914 he was appointed Professor of Mathematics at U Frankfurt. In 1938 he was arrested (because he was a Jew) and taken to Dachau concentration camp. The following year he was allowed to emigrate to the USA, where he joined the faculty at Northwestern U.

SEE WEB LINKS

- Web archive.

Hellinger distance A measure of the distance between populations with *multivariate distributions having *probability density functions f and g. The distance is given by $\sqrt{2(1-\rho)}$, where

$$\rho = \int_{-\infty}^{\infty} \cdots \int_{-\infty}^{\infty} \sqrt{f(x_1, x_2, \ldots, x_n) g(x_1, x_2, \ldots, x_n)} dx_1 dx_2 \ldots dx_n.$$

See also BHATTACHARYA DISTANCE; KULLBACK–LEIBLER INFORMATION.

Helmert, Friedrich Robert (1843–1917; b. Freiburg, Germany; d. Potsdam, Germany) German mathematical physicist. Helmert's interest in Statistics resulted from his research in geodesy (the study of distances and angles between widely separated locations on the Earth's surface). He became Professor of Geodesy at U Aachen in 1872, and at U Berlin in 1887, where he became director of the Prussian Geodetic Institute. His research included fundamental work on the *chi-squared distribution.

SEE WEB LINKS

- Photograph.

Helmert contrast *See* ANOVA.

Herfindahl index (Herfindahl–Hirschman index) A measure, H, of variability. Denoting the frequency of the jth distinct value in a sample of size N by n_j,

$$H = \sum \left(\frac{n_j}{N}\right)^2.$$

h

Often used in the context of market shares, when it is usually multiplied by 10000.

heritability (H^2) The proportion of the *variance of some population characteristic that is attributable to genetic variation. Often estimated using either the *path analysis of *Wright, or via *ANOVA using a *components of variance model.

hessian matrix The *matrix of second-order (partial) derivatives of a multivariate function. If $y = \mathrm{f}(x_1, x_2, \ldots, x_n)$ the element in row j and column k is

$$\frac{\partial^2 y}{\partial x_j \partial x_k}.$$

The matrix is $n \times n$ and symmetric (provided that its elements are continuous).

heteroskedastic A term used of a set of *random variables that have different *variances. By contrast, a set of variables all having the same variance is described as **homoskedastic**.

heuristic A general recommendation based on statistical evidence. For example, ‘Smoking may severely damage your health’.

Heyde, Christopher Charles (1939–2008; b. Sydney, Australia; d. Canberra, Australia) Australian statistician, who specialized in applications of *quasi-likelihood. A graduate of U Sydney, Heyde obtained his PhD at ANU (in 1965, supervised by *Moran), joining the ANU faculty in 1968. After spells at the CSIRO and U Melbourne, Heyde rejoined ANU, dividing his time between ANU and Columbia U, where he was Director of the Center for Applied Probability. He was President of the *SSAI from 1979 to 1981, and was awarded the Society’s Pitman Medal (*see* PITMAN) in

1988, and the *Hannan Medal of the AAS in 1994. He was President of the *BSMSP for 1985–7.

SEE WEB LINKS

- Fuller biography, interview, and photographs.

HGLM *Abbreviation for* HIERARCHICAL GENERALIZED LINEAR MODEL.

hidden Markov model A special type of *graphical model. The observed *variable, Y, has a *distribution that depends on the value of the corresponding X-variable. The idea is that the successive values of X obey the Markov property, with the value of each depending only on the value of its predecessor.

hierarchical Bayes An *estimation procedure using *Bayesian inference. Suppose that the *random variables corresponding to n *sample means are denoted by $\bar{Y}_1, \bar{Y}_2, \ldots, \bar{Y}_n$. The belief is that each sample is obtained from a *population having the same *variance, σ^2, but different means, $\mu_1, \mu_2, \ldots, \mu_n$, with the means themselves being regarded as observations from a distribution with mean μ and variance ω^2. The *parameters μ and ω^2 are called **hyperparameters** and are assigned prior distributions referred to as **hyperpriors**.

hierarchical generalized linear model (HGLM) A *hierarchical model extending the *generalized linear model. It was proposed in 1996 by the Korean statistician Youngjo Lee and *Nelder, and allows for additional random terms in the linear predictor (*see* REGRESSION).

hierarchical models Two or more *models that are nested in the sense that the more complex model includes all the *parameters of the simpler model together with at least one other parameter. *See also* LOG-LINEAR MODEL.

high-order interaction An *interaction involving more than two *variables.

high-pass filter *See* FILTER.

highspread The difference between the greatest value in a *sample and the sample *median. *See also* LOWSPREAD; RANGE.

Hill, Sir Austin 'Tony' Bradford (1897–1991; b. London, England; d. Cumbria, England) English epidemiologist and statistician. Having contracted tuberculosis as a pilot in the First World War, Hill collected a war-related disability pension for the next 75 years! Hill's first degree was in economics. In 1923 he began work for the National Institute for Medical Research as a statistician. In 1933 he moved to the LSHTM, becoming

Professor of Medical Statistics in 1947. Together with *Doll, he provided in 1952 the first statistical evidence linking smoking and lung cancer. He was President of the *RSS in 1950, and was awarded its *Guy Medal in Gold in 1953. The Society now awards a medal every three years in his honour. He was elected FRS in 1954 and was knighted in 1961.

SEE WEB LINKS

- Fuller biography and photograph.

hill-climbing *See* STEEPEST ASCENT.

hinge *See* QUARTILE.

histogram A diagram representing a *sample of numerical *data, in which rectangles are used to represent *frequency. It differs from the *bar chart in that the rectangles may have differing widths, but the key feature is that, for each rectangle, the area is proportional to the frequency represented. The term 'histogram' was introduced by Karl *Pearson in his lectures in 1891.

SEE WEB LINKS

- Applet.

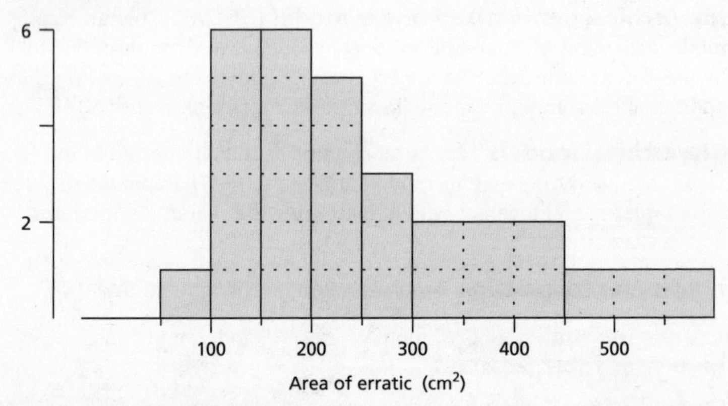

Histogram. This histogram represents data on the cross-sectional area of 30 erratics (boulders left behind by retreating glaciers). Note the use of wider intervals for the classes corresponding to the scarcer larger boulders. In a histogram, area is proportional to frequency.

Hochberg test *See* MULTIPLE COMPARISON TEST.

Hodges, Joseph Lawson, Jnr (1922–2000; b. Shreveport, LA; d. Berkeley, CA) American mathematical statistician. Hodges entered UCB aged sixteen and graduated in 1942. After the Second World War, during which he served with *Lehmann in the operations analysis group in Guam, he obtained his doctorate at UCB (under the supervision of *Neyman) and promptly joined the faculty, retiring in 1991. He was Editor of the *Annals of Mathematical Statistics* during 1961–4.

SEE WEB LINKS

- University website.

Hodges–Lehman estimators *Robust estimators suggested by *Hodges and *Lehmann in 1963. The estimate of location for a *sample of n observations is the *median of the averages of the $\frac{1}{2}n(n-1)$ possible pairs of observations. For example, with the four observations 22, 24, 26, and 60, the six possible averages are 23, 24, 41, 25, 42 and 43, so that the Hodges–Lehmann location estimate is 33 (equal to the *mean in this case).

In the case of samples (of sizes m and n) from two populations, where an estimate is required of the difference in their locations, the Hodges–Lehmann estimate is the median of the mn possible differences resulting from taking one observation from each sample.

Hoeffding, Wassily (1914–91; b. Mustamäki, Finland; d. Chapel Hill, NC) Finnish mathematical statistician who spent most of his career in the USA. After study in Denmark and Germany, Hoeffding gained his PhD from U Berlin in 1940. In 1946 he moved to the United States and joined the faculty at UNC, retiring in 1979. He was President of the *IMS in 1969 and its *Wald Lecturer in 1967. He was elected to membership of the NAS in 1976.

SEE WEB LINKS

- Fuller biography and photographs.

Hoeffding's *D* A rather complicated function of the *ranks of *observations. It was suggested by *Hoeffding in 1948 as the basis of a *non-parametric test of the *independence of two *random variables.

Hölder inequality For *random variables X and Y, and constants p and q, both > 1 and satisfying $p^{-1}+q^{-1}=1$,

$$\mathrm{E}(XY) \le \{\mathrm{E}(|X|^p)\}^{\frac{1}{p}} \times \{\mathrm{E}(|Y|^q)\}^{\frac{1}{q}}.$$

See also BERNSTEIN INEQUALITY; CHEBYSHEV INEQUALITY; KOLMOGOROV INEQUALITY; MARKOV INEQUALITY; MINKOWSKI INEQUALITY.

Holt–Winters forecasting An application of *exponential smoothing to a *time series that displays a trend (*see* MOVING AVERAGE) and *seasonality.

homoskedastic *See* HETEROSKEDASTIC.

honestly significant difference test (HSD test) *See* MULTIPLE COMPARISON TEST.

Hong Kong Statistical Society The Society, which was founded in 1977, has about 300 members.

SEE WEB LINKS

- Society home page.

Hooke–Jeeves pattern search A search procedure used in an attempt to identify the maximum (or minimum) of a function of several variables. An original base point is chosen and the function evaluated. This is followed by exploratory moves, changing the values of one variable at a time, and resulting, in general, in a new base point, with a higher value for the function than the original base point. A pattern move is now made, the new point being dictated by the values of the function at the preceding base points. The alternating sequence of exploratory and pattern moves continues. If the process converges the value obtained may only be a *local maximum.

Hopfield net A *network, introduced by American scientist John Hopfield in 1982, in which every node is connected to every other node (but not to itself) and the connection lengths are symmetrical in that the length from node i to node j is the same as that from node j to node i. Hopfield networks are used in the context of supervised learning (*see* MACHINE LEARNING).

hot deck *See* IMPUTATION.

Hotelling, Harold (1895–1973; b. Fulda, MN; d. Chapel Hill, NC) American statistician and economist. Hotelling's first degree, a BA in journalism at U Washington in 1919, was interrupted by the First World War. Scheduled to fight in the trenches, his life was saved by a mule named Dynamite that broke his leg and caused him to be invalided out of service. After an MA in mathematics at U Washington, he obtained his PhD (in topology) at Princeton U in 1924. His interest in Statistics resulted from his first job, which was in the Food Research Institute at Stanford U. In 1931 he was appointed Professor of Economics at Columbia U. In 1946 he was invited to found the Department of Statistics at UNC. He was President of the *Econometrics Society in 1936 and of the *IMS in 1941. He was the IMS *Rietz Lecturer in 1951. He was elected to membership of the NAS in 1970.

SEE WEB LINKS

- Biographical memoir and photograph.

Hotelling–Lawley trace *See* MULTIVARIATE ANALYSIS OF VARIANCE.

Hotelling T^2 A *statistic having a *multivariate distribution which is the analogue of the univariate *t-distribution. The statistic was introduced in 1931 by *Hotelling. A sample of size n is drawn from a *multivariate normal distribution with mean *vector $\boldsymbol{\mu}$ and *variance-covariance matrix estimated as $\mathbf{S}$. If the column vector of *sample means is denoted by $\bar{\mathbf{x}}$ the statistic T^2 is given by

$$T^2 = n(\bar{\mathbf{x}} - \boldsymbol{\mu})'\mathbf{S}^{-1}(\bar{\mathbf{x}} - \boldsymbol{\mu}).$$

In the case of p variables, $\{(n-p)T^2\}/\{(n-1)p\}$ has an *F-distribution with p and $(n-p)$ *degrees of freedom. *See also* MAHALANOBIS D^2.

Hovmöller diagram A two-dimensional plot showing how the value of some quantity varies in space-time; one axis refers to time and the other to spatial location. It is particularly used in meteorology.

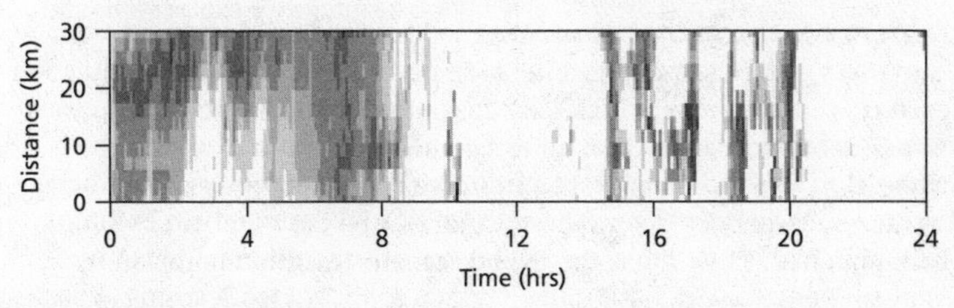

Hovmöller diagram. The diagram shows the recorded rain intensity along a 30km path during a 24-hour period.

HSD test *See* MULTIPLE COMPARISON TEST.

H-spread *See* QUARTILE.

H^2 *See* HERITABILITY.

Hsu MCB test *See* MULTIPLE COMPARISON TEST.

Huber, Peter Jost (1934– ; b. Wohlen, Switzerland) Swiss statistician known for his work on *robust statistics. In 1961 he obtained his PhD from ETH Zurich. His subsequent career alternated between universities in Switzerland and the USA where he supervised *Donoho at Harvard U. He was the 1972 *IMS *Wald Lecturer.

SEE WEB LINKS

- Fuller biography, interview, and photographs.

Huber function *See* *M*-ESTIMATE.

Huygens, Christiaan (1629–95; b. The Hague, Netherlands; d. The Hague, Netherlands) Dutch mathematician, astronomer, and horologist.

The son of a diplomat, Huygens studied law and mathematics, first at U Leiden and then at the College of Orange at Breda. His principal interest was astronomy, but this required precision instrumentation and led to his devising new methods for polishing lenses and keeping accurate time—in 1656 he patented the first pendulum clock. The previous year he had discovered the first of Saturn's moons. In 1657 he wrote the first printed work on probability, *De Ratiociniis in Ludo Aleae* (*Calculation in Games of Chance*).

SEE WEB LINKS

- Fuller biography.

h

Huynh–Feldt correction A correction applied in *ANOVA calculations for *longitudinal data when the usual assumption of independent experimental errors, of constant *variance, is untrue.

hypergeometric distribution The *distribution of the number of 'successes' when *sampling without replacement from a finite *population each of whose members is classified as either a 'success' or a 'failure'. As an example, suppose that an urn contains N balls of which w are white. Suppose n balls are taken from the urn at random and without replacement, and let the *random variable X be the number of white balls obtained. Then X has the hypergeometric distribution given by

$$\mathrm{P}(X=r)=\frac{\binom{w}{r}\binom{N-w}{n-r}}{\binom{N}{n}}, \qquad 0\leq r\leq w, 0\leq(n-r)\leq(N-w).$$

The initial derivation of the distribution was published by *de Moivre in 1711. The *mean of the distribution is nw/N and the *variance is

$$\frac{nw(N-n)(N-w)}{N^2(N-1)}.$$

For large values of N the hypergeometric distribution may be approximated by the *binomial distribution $\mathrm{B}(n, p)$, where $p=w/N$, since in this case sampling without replacement is approximately equivalent to sampling with replacement.

hyperparameter; hyperprior *See* HIERARCHICAL BAYES.

hyperplane A hyperplane in $\mathbb{R}^n$ is the set of points $(x_1, x_2, \ldots, x_n)$ such that $a_1x_1+a_2x_2+\cdots+a_nx_n=c$, where $a_1, a_2, \ldots, a_n$ are constants, not all zero, and c is a constant. In the case $n=2$ a hyperplane is a line, and in the case $n=3$ it is a plane.

hypothesis test A procedure (based on the *Neyman–Pearson lemma) for deciding between two hypotheses on the basis of the value of a *statistic called the **test statistic**, which is a function of the *observations in a random *sample. In a test concerning the value of an unknown *parameter, the **null hypothesis** specifies a particular value for the parameter, whereas the **alternative hypothesis** specifies either an alternative value or, more usually, a range of alternative values. The null hypothesis is often denoted by H_0 or NH, and the alternative hypothesis by H_1 or AH.

A typical null hypothesis might state that the *population mean $\mu=20$. The alternative hypothesis $\mu<20$ is described as **one-sided** and the test procedure is described as **one-tailed**. By contrast, the alternative hypothesis $\mu\neq 20$ is **two-sided** and the test is **two-tailed**.

If necessary, assumptions are made about the *distribution type of the population, so that the *probability distribution of the statistic can be determined assuming H_0. The probability of obtaining a value as extreme as the observed value or a more extreme value (taking account of the alternative hypothesis), is called the ***p*-value**. If the actual value of the statistic is too far from its *expected value the test is deemed to be **significant** and the decision is to reject H_0 in favour of H_1. If the actual value of the statistic is close to its expected value the test is deemed to be **not significant** and the decision is not to reject H_0. The set of values of the statistic that lead to rejection of H_0 is called the **critical region** or **rejection region**, and the set of values that do not lead to rejection of H_0 is called the **acceptance region**.

There are two cases when the test leads to a correct result. These occur when H_0 is true and the test leads to its acceptance and when H_1 is true and the test leads to rejection of H_0. On the other hand there are two cases when the test leads to an incorrect result. These occur when H_0 is true but the test leads to its rejection (a **Type I error**, or **error of the first kind**) and when H_1 is true but the test leads to acceptance of H_0 (a **Type II error**, or **error of the second kind**).

The size of the critical region is determined by the desired **significance level** of the test, often denoted by α (alpha), which is the probability of making a Type I error. The smaller the significance level, the smaller the critical region. The significance level is usually expressed as a percentage. The word 'significance' seems to have been introduced into Statistics by *Edgeworth in 1885. The ideas concerning the two types of error were introduced by *Neyman and Egon *Pearson in 1928.

As an example, suppose X has a *normal distribution $N(\mu, 9)$ and it is desired to test H_0: $\mu=20$ against H_1: $\mu>20$, using a sample of size 25.

An appropriate statistic is the sample mean $\overline{X}$, which has distribution N(20, 9/25) under H_0, or its standardized value

$$Z = \frac{\overline{X} - 20}{\sqrt{9/25}},$$

which has distribution N(0, 1). If the desired significance level is 5%, the critical region, from tables of **critical values** for the standard normal distribution (*see* APPENDIX IV) is $Z > 1.645$. In terms of $\overline{X}$, the critical region is $\overline{X} > 20 + 1.645\sqrt{9/25}$ or, equivalently, $\overline{X} > 20.99$.

The probability of making a Type II error is often denoted by β (beta). The **power** of the test, which is the probability of accepting the alternative hypothesis when it is in fact true, is $1 - \beta$ and its value depends on the value of the parameter under test. A plot of the relation between power and the parameter value is called the **power curve**. If there is a choice of test, with a predetermined significance level, it is usual to choose the test (if one exists) that maximizes the power.

However, the variance may not be known. With the same assumption of a normal distribution, as before, a *t*-test is appropriate. Suppose (with the same hypotheses as previously) that the sample of 25 observations has sample mean and variance (using the $(n-1)$ divisor) given, respectively, by $\overline{x} = 21.4$ and $s^2 = 12.25$. The test statistic *t* is given by

$$t = \frac{21.4 - 20}{\sqrt{12.25/25}} = 2.0$$

The upper 5% point (*see* PERCENTAGE POINT) of a *t*-distribution with $24(=25-1)$ *degrees of freedom is 1.711 (*see* APPENDIX VI). Since $2.0 > 1.711$, the null hypothesis would be rejected in favour of the alternative hypothesis.

Frequently, hypothesis tests involve the comparison of two populations. The case where unrelated random samples are taken from the populations gives rise to **two-sample tests**. A two-sample test of the equality of two population means in which the populations have the same (unknown) variance requires the use of a pooled estimate of the common variance (*see* POOLED ESTIMATE OF COMMON MEAN). The case where the population variances are unknown and cannot be assumed to be equal is the *Behrens–Fisher problem.

If the two random samples have the same size and matched pairs of individuals are obtained, one from each population, then, for a test of equality of population means, a **paired-sample test** is appropriate. Here the differences within each pair of values constitute the observations. When the data have a normal distribution (or the samples are large) a *t*-test is performed, the null hypothesis being that the mean difference is 0, and the sample variance being the variance of the differences.

A hypothesis that specifies several simultaneous conditions is described as a **composite hypothesis**. For example, with samples from k populations, the null hypothesis might specify that all k population means are equal.

IAOS *Abbreviation for* INTERNATIONAL ASSOCIATION FOR OFFICIAL STATISTICS.

IASC *Abbreviation for* INTERNATIONAL ASSOCIATION FOR STATISTICAL COMPUTING.

IASE *Abbreviation for* INTERNATIONAL ASSOCIATION FOR STATISTICAL EDUCATION.

IASS *Abbreviation for* INTERNATIONAL ASSOCIATION OF SURVEY STATISTICIANS.

IBS *Abbreviation for* INTERNATIONAL BIOMETRIC SOCIETY.

icicle plot *See* DENDROGRAM.

icon plot A diagram for displaying multidimensional observations. The diagram consists of an array of individual diagrams (icons), one for each data item. In the individual diagrams a separate feature corresponds to each of the variables measured. One example of an icon plot is the use of *Chernoff faces.

ICSA *Abbreviation for* INTERNATIONAL CHINESE STATISTICAL ASSOCIATION.

ideal index *See* PRICE INDEX.

idempotent matrix *See* MATRIX.

identifiability A term introduced in about 1945 by *Koopmans in a discussion of whether a *parameter may be *estimable from a set of *data.

identity link *See* GENERALIZED LINEAR MODEL.

identity matrix *See* MATRIX.

ignorable non-response *See* NON-RESPONSE.

i.i.d. Abbreviation for 'independent and identically distributed'. Thus i.i.d. *random variables are independent variables (*see* REGRESSION) all having the same *distribution. The most common situation involving i.i.d. random variables arises when a random *sample of *observations is taken from a single *population.

ill-conditioned *See* MULTIPLE REGRESSION MODEL.

impact location *See* BACI DESIGN.

importance sampling A method of reducing the *variance of an estimate obtained via *simulation. Suppose, for example, that we wish to estimate the quantity I, given by

$$I = \int_0^1 \mathrm{g}(x)\mathrm{d}x,$$

where g is a given function. An obvious method is to generate *pseudo-random numbers $u_1, u_2, \ldots, u_n$, in the interval (0, 1). Writing $U_1, U_2, \ldots,$ U_n for the corresponding random variables, the *estimator, I_1, is given by

$$I_1 = \frac{1}{n}\sum_{j=1}^{n} \mathrm{g}(U_j).$$

Importance sampling makes a more representative choice of values: suppose that f is a *probability density function that resembles g in its general shape, and let F be the corresponding *distribution function. Instead of working with $U_1, U_2, \ldots, U_n$, we work with $V_1, V_2, \ldots, V_n$, where $\mathrm{F}(V_j) = U_j$. The resulting estimator, I_2, given by

$$I_2 = \frac{1}{n}\sum_{j=1}^{n} \frac{\mathrm{g}(V_j)}{\mathrm{f}(V_j)},$$

will have a smaller variance than I_1.

improper prior *See* BAYESIAN INFERENCE.

imputation The process of replacing missing values in (usually) a large-scale social survey. Suppose, for example, that the salary information is missing for an individual who is known to be a doctor aged 55. One approach would be to determine the *average salary of all 55-year-old doctors and to replace the missing value with this average value (or some estimate obtained, for example, by *multiple regression of salary on other *variables). There are two possible objections to this approach: (i) the imputed value might not be a salary actually obtained by any 55-year-old doctor, and (ii), if there are many doctors of this age with missing salary information and if each were given the same imputed salary, this would give a very misleading idea of salary variability.

The first objection can be met by insisting that the value imputed must be a real salary. Both objections can be met if the imputed value is taken to be the most recently encountered actual salary of a doctor of that age. This is called **hot deck** imputation, which refers to a time when the records on each individual were on a separate card. Imagine examining the deck of cards a card at a time, imputing missing values. The salary chosen for the doctor is that most recently encountered in the cards containing information on 55-year-old doctors. The alternative, in which all imputation takes place after all the cards have been examined, is called **cold deck** imputation—the most recently encountered value is used to replace all those missing (so that only the first objection is met).

IMS *Abbreviation for* INSTITUTE OF MATHEMATICAL STATISTICS.

inadmissible decision rule A decision rule (*see* GAME THEORY) that is never better than a given alternative rule. Suppose that δ_1 and δ_2 are two decision rules and that the outcome of each rule is dependent on the *parameter *vector $\boldsymbol{\theta}$. If, for all $\boldsymbol{\theta}$, the expected loss from using rule δ_2 is at least as large as the loss from using rule δ_1 and if it is greater for at least one $\boldsymbol{\theta}$, then the rule δ_2 is described as being inadmissible.

incidence rate The number of new cases of a disease occurring during a given period as a proportion of the number of people in the *population. It is usually expressed as cases per 1000, or per 100000 per annum. *See* AGE-SPECIFIC RATE; MORBIDITY RATE; MORTALITY RATE; SEX-SPECIFIC RATE.

inclusion-exclusion principle Let $E_1, E_2, \ldots, E_n$ be n distinct events (*see* SAMPLE SPACE). The *probability that at least one occurs is $\mathrm{P}(E_1 \cup E_2 \cup \cdots \cup E_n)$. The principle states that this can be expressed in terms of the probabilities of *intersections as

$$\sum \mathrm{P}(E_i) - \sum \mathrm{P}(E_i \cap E_j) + \cdots \pm \sum \mathrm{P}(E_i \cap E_j \cap \cdots \cap E_r) + \cdots + (-1)^{(n+1)} \mathrm{P}(E_1 \cap E_2 \cap \cdots \cap E_n),$$

where the sums are over all combinations of different events.

increment A small change in the value of a *variable.

independence For independence of events, *see* INDEPENDENT EVENTS; for independence of random variables, *see* INDEPENDENT RANDOM VARIABLES.

independence model A *model in which two (or more) *random variables are *independent of one another: the value taken by one variable is completely unaffected by the value taken by the other variable. Suppose, for example, that A, B, and C are three *categorical random variables, with p_{jkl} denoting the *probability of an outcome in cell (j, k, l); then the independence model states that

$$p_{jkl} = p_{j00}\, p_{0k0}\, p_{00l},$$

where

$$p_{j00} = \sum_k \sum_l p_{jkl}, \qquad p_{0k0} = \sum_j \sum_l p_{jkl}, \qquad p_{00l} = \sum_j \sum_k p_{jkl}.$$

With just two variables, A and B, the corresponding result is

$$p_{jk} = p_{j0}\, p_{0k},$$

where

$$p_{j0} = \sum_k p_{jk}, \qquad p_{0k} = \sum_j p_{jk}.$$

i

independent events Two events (*see* SAMPLE SPACE) A and B are independent events if

$$\mathrm{P}(A \cap B) = \mathrm{P}(A) \times \mathrm{P}(B),$$

or, equivalently, if $\mathrm{P}(A|B) = \mathrm{P}(A)$, or if $\mathrm{P}(B|A) = \mathrm{P}(B)$. Three events A, B, and C are said to be independent (or **mutually independent**) if each pair is independent and if, in addition,

$$\mathrm{P}(A \cap B \cap C) = \mathrm{P}(A) \times \mathrm{P}(B) \times \mathrm{P}(C).$$

For a set of more than three events to be independent the *multiplication law for probabilities must hold for all possible subsets. *See also* CONDITIONAL PROBABILITY; INTERSECTION; PAIRWISE INDEPENDENT.

independent random variables Two *random variables are said to be independent if the value taken by one has no effect on the value taken by the other.

independent variable *See* REGRESSION.

index (exponent) In the term a^x, the index is x. The plural is **indices**.

index number A measure of the value of a variable relative to its value at some base date or state (the **base period**). The index is often scaled so that its base value is 100. Such an index may be described as a **base-weighted index**. *See* PRICE INDEX; RETAIL PRICE INDEX.

index of dispersion A measure of the extent to which a set of observed *frequencies (for example the numbers of plants in randomly distributed *quadrats) follow a *Poisson distribution. For a sample of n observations, let $\bar{x}$ and s^2 denote, respectively, the *sample mean and the *sample variance (using the $(n-1)$ divisor). Under the *null hypothesis of a Poisson distribution the quantity I (the index of dispersion), given by

$$I = \frac{(n-1)s^2}{\bar{x}},$$

has an approximate *chi-squared distribution with $(n-1)$ *degrees of freedom. According to the null hypothesis the value of I should be near $(n-1)$. This test is called the **dispersion test**.

If I is unusually large then the data are described as being **over-dispersed**, and, in the case of plants (or stars, or other point objects), the data are described as **clustered**. If I is unusually small then the data are described as displaying **regularity**. *See* EXTRA-BINOMIAL VARIATION.

index of diversity *See* LOGARITHMIC DISTRIBUTION.

i

index property *See* EXPONENTIAL.

Indian Statistical Institute (ISI) The Institute, which is located at Kolkata, was founded by *Mahalanobis in 1931. It publishes the journal **Sankhya*.

SEE WEB LINKS

- Institute home page.

indicator kriging *See* KRIGING.

indicator variable *Alternative name for a* DUMMY VARIABLE.

indices *See* INDEX.

inference The process of deducing properties of the underlying *distribution by analysis of *data.

inflation The percentage increase in the *retail price index relative to its value one year earlier.

influence *See* REGRESSION DIAGNOSTICS.

information *See* FISHER INFORMATION.

informative prior *See* BAYESIAN INFERENCE.

inner fence *See* QUARTILE.

inspection by attributes; inspection by variables *See* ACCEPTANCE SAMPLING.

Institute of Mathematical Statistics (IMS) The Institute is one of the major American societies for statisticians. Founded in 1935 at Ann Arbor in Michigan, it currently has about 4500 members worldwide. The IMS publishes the **Annals of Applied Probability*, **Annals of Probability*,

*Annals of Statistics, *Statistical Science* and **Annals of Applied Statistics*. The first three of these titles superseded its initial publication, the **Annals of Mathematical Statistics*.

SEE WEB LINKS

- Institute home page.

Institute of Statistical Mathematics (ISM) Japanese research institute based in Tokyo. Its publications include the **Annals of the Institute of Statistical Mathematics*.

SEE WEB LINKS

- Institute home page.

Institute of Statisticians (IoS) A British society founded in 1948 for non-academic statisticians. It published a journal originally called *The Incorporated Statistician,* though the title was soon changed to *The Statistician* (*see* JOURNAL OF THE ROYAL STATISTICAL SOCIETY). The Institute merged with the *Royal Statistical Society in 1993, with *The Statistician* becoming the fourth of the Society's journals.

instrumental variable *See* MULTIPLE REGRESSION MODEL.

integer programming *See* LINEAR PROGRAMMING.

intelligence quotient (IQ) In 1905, the French psychologists Alfred Binet and Théodore Simon proposed the concept of an intelligence test that measured mental age. In 1912, the German psychologist William Stern proposed the division of mental age by chronological age, to give the intelligence quotient. In 1916 the test was introduced into the USA by Lewis Terman of Stanford U. Terman proposed scaling the quotient by 100, to remove awkward decimal places, and as a result the test became known as the Stanford–Binet test. The idea of a mental age to chronological age ratio only works well for the young, however. As a consequence the original ratio concept has been abandoned. In 1939 the Romanian-born US psychologist David Wechsler introduced a statistical definition of IQ as a *random variable having a *normal distribution with *mean 100 and *standard deviation 15, so that most people then have IQs in the range 70 to 130, with IQs over 150 being extremely rare.

interaction A term particularly used in the contexts of factorial experiments (*see* FACTORIAL DESIGN) and *log-linear models to describe cases where the combined effects of two *variables is not a simple sum of their separate effects. For example, we might write

$$\mathrm{E}(Y) = \beta_0 + \beta_1 x_1 + \beta_2 x_2 + \beta_{1,2} x_1 x_2,$$

the final term quantifying the interaction between the two x-variables.

International Association for Official Statistics (IAOS) Founded in 1985 as an autonomous section of the *International Statistical Institute. It has nearly 400 members.

SEE WEB LINKS
- Association home page.

International Association for Statistical Computing (IASC) An autonomous section of the *International Statistical Institute. Having about 700 members, it publishes the journal **Computational Statistics & Data Analysis*.

SEE WEB LINKS
- Association home page.

International Association for Statistical Education (IASE) The education section of the *International Statistical Institute. It publishes the **Statistics Education Research Journal* and has over 300 members.

SEE WEB LINKS
- Association home page.

International Association of Survey Statisticians (IASS) Founded in 1973 as an autonomous section of the *International Statistical Institute. With over 500 members it publishes the journal **Survey Statistician* twice a year.

SEE WEB LINKS
- Association home page.

International Biometric Society (IBS) A society, founded in 1947, devoted to the mathematical and statistical aspects of biology. Through its regional organizations the Society sponsors regional and local meetings and it has more than 6000 members worldwide. Its principal publications are **Biometrics* and (jointly with the *American Statistical Association) the **Journal of Agricultural, Biological, and Environmental Statistics*.

SEE WEB LINKS
- Society home page.

International Chinese Statistical Association (ICSA) An association, founded in 1987, with a membership primarily consisting of Chinese statisticians in the USA, Hong Kong, and Taiwan. The ICSA has published **Statistica Sinica* since 1991.

SEE WEB LINKS
- Association home page.

International Society for Bayesian Analysis (ISBA) A society founded in 1992 to promote the application of *Bayesian inference. It has about 600 members. The first edition of its journal **Bayesian Analysis* appeared in 2005.

SEE WEB LINKS

- Society home page.

International Society for Clinical Biostatistics (ISCB) A society founded in 1980 to stimulate research on the biostatistical principles and methodology used in clinical research. It has about 800 members.

SEE WEB LINKS

- Society home page.

International Statistical Institute (ISI) An autonomous society, founded in 1885, that seeks to develop and improve statistical methods and their application, through the promotion of international activity and co-operation. It has about 2000 members worldwide. Its principal publications are the **Bulletin of the International Statistical Institute*, **International Statistical Review*, and **Statistical Theory and Methods Abstracts*.

SEE WEB LINKS

- Institute home page.

International Statistical Review A journal of the *International Statistical Institute. Founded in 1933, the journal contains a high proportion of historical and review papers. There are three issues per year.

SEE WEB LINKS

- Journal home page.

interpolation The use of a formula to estimate an intermediate data value. *Compare* EXTRAPOLATION. For example, if we discover that 1 kg of fertilizer per hectare results in a yield of 50 kg of produce per hectare, and that 2 kg gives 70 kg, then we might reasonably interpolate between these values to deduce that 1.5 kg would give a yield of about 60 kg ha^{-1}.

In general, if $x_1 < x_0 < x_2$ and the values of y corresponding to x_1 and x_2 are y_1 and y_2, respectively, then the estimate of y_0, the value corresponding to x_0, is given by **linear interpolation** as

$$y_0 = \frac{y_2(x_0 - x_1) - y_1(x_0 - x_2)}{x_2 - x_1} = y_1 + \frac{x_0 - x_1}{x_2 - x_1}(y_2 - y_1).$$

interquartile range *See* QUARTILE.

inter-rater reliability A measure of the extent to which raters (for example, judges of an ice-skating contest) agree.

intersection The intersection of two events (*see* SAMPLE SPACE) A and B is the event 'both A and B occur'. It is denoted by $A \cap B$. Clearly, for any events A, B, C,

$$A \cap A = A, \qquad A \cap B = B \cap A,$$
$$A \cap S = A, \qquad A \cap \phi = \phi,$$
$$A \cap (B \cap C) = (A \cap B) \cap C,$$

where S is the sample space and ϕ is the empty set. *See also* BOOLEAN ALGEBRA; VENN DIAGRAM.

The intersection of the n events $A_1, A_2, \ldots, A_n$ is the event 'all of $A_1, A_2, \ldots, A_n$ occur'. It is denoted by $A_1 \cap A_2 \cap \cdots \cap A_n$.

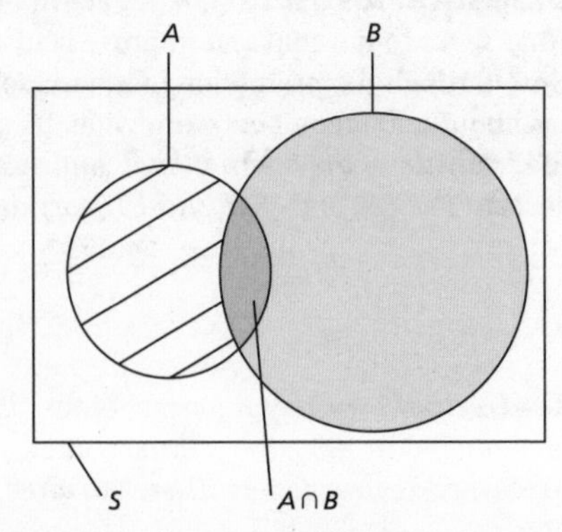

Intersection. This Venn diagram shows the intersection of two events as the overlapping part of the circles that represent the individual events.

interval estimate *See* CONFIDENCE INTERVAL.

interval scale A scale of measurement which can be used to measure the difference, or distance, between two general states or points. For example, use of a ruler to measure length, use of a stop-watch to measure a time interval, or measurement of a musical interval (octave, fifth, etc.).

In a **ratio scale** the difference, or distance, is measured between a state or point of interest and a standard state or point. For example, height above sea-level, distance from London, frequency of a musical note (in cycles per second), temperature in °Kelvin (above absolute zero), clock-time (13:05 on 5 November 2001 AD). Confusingly, comparison of ratio scale measurements gives an interval scale, and in music an interval is measured by the ratio of the frequencies.

intraclass correlation coefficient An extension of the *sample correlation coefficient to the case where the n pairs of values $(x_{1,1}, x_{1,2})$,

. . . , $(x_{n,1}, x_{n,2})$ are drawn from a bivariate *population. The *population mean and *population variance are therefore estimated using *pooled estimates (*see* POOLED ESTIMATE OF COMMON MEAN):

$$\overline{x} = \frac{1}{2n}\sum_{i=1}^{n}(x_{i,1} + x_{i,2}) \quad \text{and} \quad s^2 = \frac{1}{2n-1}\sum_{i=1}^{n}\left\{(x_{i,1} - \overline{x})^2 + (x_{i,2} - \overline{x})^2\right\}.$$

The intraclass correlation coefficient, r, is given by

$$r = \frac{1}{ns^2}\sum_{i=1}^{n}(x_{i,1} - \overline{x})(x_{i,2} - \overline{x}).$$

For an alternative definition, *see* EXPERIMENTAL DESIGN.

intracluster correlation coefficient *See* EXPERIMENTAL DESIGN.

inverse Cauchy link *See* GENERALIZED LINEAR MODEL.

inverse Fourier transform *See* FOURIER TRANSFORM.

inverse Gaussian distribution *See* INVERSE NORMAL DISTRIBUTION.

inverse matrix *See* MATRIX.

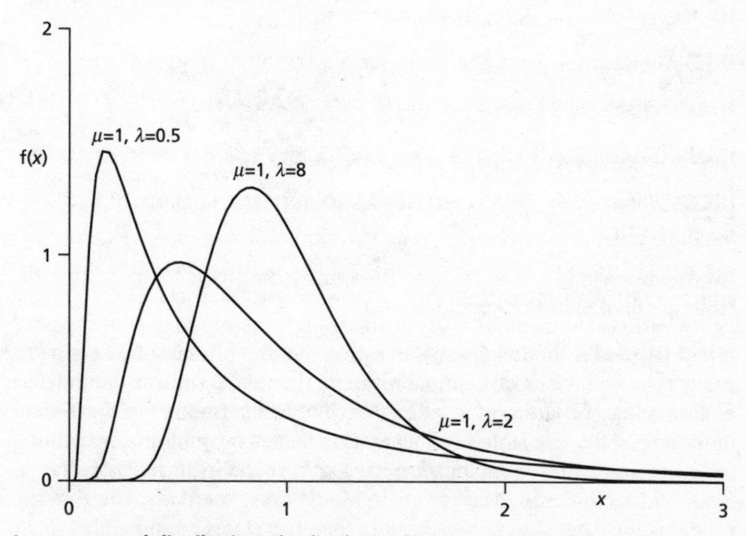

Inverse normal distribution. The distribution has two parameters: the mean μ and a parameter λ related to the variance.

inverse normal distribution (inverse Gaussian distribution; Wald distribution) A *continuous distribution with *probability density function f is given by

$$f(x) = \left(\frac{\lambda}{2\pi x^3}\right)^{\frac{1}{2}} \exp\left\{-\frac{\lambda}{2\mu^2 x}(x-\mu)^2\right\}, \qquad x > 0,$$

where $\mu > 0$, $\lambda > 0$ (*see diagram on preceding page*). The distribution has *mean μ, *variance μ^3/λ, and *mode

$$\frac{\mu}{2\lambda}\left(\sqrt{4\lambda^2 + 9\mu^2} - 3\mu\right).$$

inverse transformation method A general method for the *simulation of *observations of a *continuous random variable X.

Let u be a *pseudo-random number in the interval (0, 1) and let F be the *distribution function of X. Then x, given by

$$x = F^{-1}(u),$$

is a random observation of X. For example, an *exponential random variable X, with mean μ, has distribution function $(1 - e^{-x/\mu})$; hence $-\mu\ln(1-u)$ is an observation of X.

invertible *See* MATRIX.

IQ *Abbreviation for* INTELLIGENCE QUOTIENT.

IRLS *Abbreviation for* ITERATIVELY REWEIGHTED LEAST SQUARES.

irreducible *See* MARKOV PROCESS.

ISBA *Abbreviation for* INTERNATIONAL SOCIETY FOR BAYESIAN ANALYSIS.

ISCB *Abbreviation for* INTERNATIONAL SOCIETY FOR CLINICAL BIOSTATISTICS.

ISI An abbreviation used for both the *Indian Statistical Institute and the *International Statistical Institute.

Ising model A *model originated by the German physicist Ernst Ising to predict the behaviour of a simple magnet. The model regards the magnet as consisting of a lattice of locations, each of which is either in the + state or in the – state. The state of a location is affected by random fluctuations in the energy levels of the location itself and, crucially, its neighbours. Similar ideas underlie *Markov chain Monte Carlo methods. The **Potts model** is the extension in which more than two states are possible.

ISM *Abbreviation for* INSTITUTE OF STATISTICAL MATHEMATICS.

isotropy; isotropic If a *time series would have the same basic properties were time to run backwards, then it displays isotropy and is said to be isotropic. Similarly, if the properties of a *spatial process are the same in all directions then it is isotropic. However, if, for example, the spatial *autocorrelation is different in the north–south direction and the east–west direction, then it displays **anisotropy** and is said to be **anisotropic**.

item non-response *See* NON-RESPONSE.

item reliability *See* RELIABILITY.

item response function In the context of an examination, the *probability that an individual will respond correctly to a question (or 'item') is assumed to be a function of the individual's ability.

A typical *model has the form

$$P(\theta) = \frac{1 + ab\exp(-c\theta)}{1 + b\exp(-c\theta)},$$

where θ is the *parameter that represents the individual's ability and a, b, and c are item parameters. The parameter a, with a range from 0 to 1, represents the probability of a guess being correct, the parameter b reflects the difficulty of the question, and the parameter c measures the extent to which the question discriminates between individuals with low and high ability. A related simpler model is the *Rasch model.

iteration *See* ITERATIVE ALGORITHM.

iterative algorithm A numerical method usually used when explicit formulae are unavailable. The idea is that a repetition of (usually simple) calculations will result in a sequence of approximate values for the quantity of interest. The differences between successive values will usually diminish rapidly and a usual termination rule is based on the difference having reached an acceptably small value. Each repetition is called an **iteration**. Examples include the *Deming–Stephan algorithm, the *EM algorithm, and *iteratively reweighted least squares.

iteratively reweighted least squares (IRLS) An *iterative algorithm for fitting a *linear model in the case where the *data may contain *outliers that would distort the *parameter estimates if other *estimation procedures were used. The procedure uses *weighted least squares, the influence of an outlier being reduced by giving that observation a small *weight. The weights chosen in one iteration are related to the magnitudes of the residuals in the previous iteration—with a large residual being given a small weight. This is one of a number of methods for *robust regression, the weights being related to **M*-estimates.

iterative proportional fitting *See* DEMING–STEPHAN ALGORITHM.

Ito, Kiyoshi (1915–2008; b. Hokusei, Japan; d. Kyoto, Japan) Japanese mathematician known for his introduction of the eponymous *Ito calculus. A graduate of Tokyo U, his career was spent on the faculty of Kyoto U. He was awarded the Japanese Order of Culture in 2008.

SEE WEB LINKS

- Fuller biography and photograph.

Ito calculus A framework proposed by *Ito in the 1940s for dealing with models of random behaviour such as *Brownian motion. The calculus is used in the study of stochastic differential equations that are used in modern financial models such as the *Black-Scholes model.

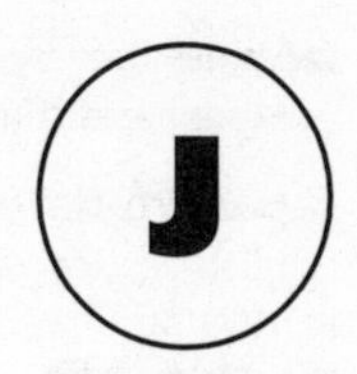

JABES *Abbreviation for* JOURNAL OF AGRICULTURAL, BIOLOGICAL, AND ENVIRONMENTAL STATISTICS.

Jaccard coefficient *See* DISSIMILARITY.

jackknife A *computer-intensive *resampling method for the *estimation of an unknown *parameter of a *distribution while making minimal assumptions. In this respect it resembles the *bootstrap. Denote the parameter by θ and its usual estimate, based on a *sample of n *observations, by $\hat{\theta}$. For example, if the parameter were the *mean of a distribution then the usual estimate would be the sample mean.

The jackknife procedure produces an alternative estimate $\tilde{\theta}$, together with an estimate of the bias (*see* ESTIMATOR) of the usual estimate. Let $\hat{\theta}_{-j}$ be the usual estimate of θ calculated from the same sample but with the jth observation omitted. Now define the **pseudovalue** $\tilde{\theta}_j$ by

$$\tilde{\theta}_j = n\,\hat{\theta} - (n-1)\,\hat{\theta}_{-j}.$$

The jackknife mean $\tilde{\theta}$ and variance s^2 are given by

$$\tilde{\theta} = \frac{1}{n}\sum_j \tilde{\theta}_j, \qquad s^2 = \frac{1}{n-1}\sum_j (\tilde{\theta}_j - \tilde{\theta})^2.$$

The estimated bias is $\hat{\theta}-\tilde{\theta}$. The ratio

$$\frac{\hat{\theta} - \theta}{s/\sqrt{n}}$$

has an approximate standard *normal distribution. The method also applies to the estimation of more complex characteristics, such as the *correlation in a set of *bivariate observations. The term jackknife was coined by *Tukey in the early 1960s.

James–Stein estimator *See* STEIN EFFECT.

Japanese Federation of Statistical Science An association, formed in 2005, that encompasses the *Japan Statistical Society, the Japan Classification Society, the Japanese Society of Applied Statistics, the

Japanese Society of Computational Statistics, the Behaviormetric Society of Japan, and the Biometric Society of Japan.

Japan Statistical Society The Society, which publishes a biannual English version of its journal, is based in Tokyo. The Society was founded in 1931 and has about 1500 members.

SEE WEB LINKS

- Society home page.

JASA *Abbreviation for* JOURNAL OF THE AMERICAN STATISTICAL ASSOCIATION.

JBES *Abbreviation for* JOURNAL OF BUSINESS & ECONOMIC STATISTICS.

JCGS *Abbreviation for* JOURNAL OF COMPUTATIONAL AND GRAPHICAL STATISTICS.

Jeffreys, Sir Harold (1891–1989; b. Durham, England; d. Cambridge, England) English astronomer, geophysicist, and mathematician. Jeffreys studied at Cambridge U, being made a Fellow of St John's College in 1914 and holding this title for a record 75 years. From 1946 to 1958 he was Plumian Professor of Astronomy and Experimental Philosophy. He was interested in scientific method and developed *Bayesian inference. He was elected FRS in 1925 and was awarded the Royal Society's Royal Medal in 1948 and its Copley Medal in 1960. He was elected to membership of the NAS in 1945 and was elected FRSE in 1953. In 1953 he was knighted. He was awarded the *Guy Medal in Gold of the *RSS in 1962.

SEE WEB LINKS

- Interview and photograph.

Jeffreys–Lindley paradox A situation studied by *Jeffreys in 1939 and termed a paradox by *Lindley in 1957. The paradox concerns the possibility that a *hypothesis test may lead to rejection of the *null hypothesis in favour of an alternative that is judged (using *Bayesian inference) to be less plausible. The possible interpretations of this paradox have given rise to much debate.

Jeffreys prior *See* BAYESIAN INFERENCE.

Jenkins, Gwilym Meirion (1933–82; b. Gowerton, Wales; d. Lancaster, England) Welsh statistician. Jenkins was both an undergraduate and a post-graduate of UCL, gaining his PhD (supervised by *Johnson) in 1956. His knowledge of the theoretical side of the analysis of *time series was augmented by practical insights gained from two years at the Royal Aircraft Establishment. In 1957 he joined the Statistics faculty at IC. From 1965 to 1974 he was Professor of Systems Engineering at

Lancaster U. He was co-author with *Box of the influential 1970 book *Time Series Analysis: Forecasting and Control*, which set out what is now referred to as the *Box–Jenkins procedure.

SEE WEB LINKS

- Memorial lecture.

Jensen, Johan Ludwig William Valdemar (1859–1925; b. Nakskov, Denmark; d. Copenhagen, Denmark) Danish mathematician. Jensen had his first mathematics papers published whilst still a student at Copenhagen College of Technology. He spent his entire career working for the Copenhagen division of the International Bell Telephone Company, treating mathematics as a hobby.

SEE WEB LINKS

- Fuller biography.

Jensen inequality For a *random variable, X, the inequality (given by *Jensen in 1906) states that $\mathrm{E}[g(X)] \geq g[\mathrm{E}(X)]$, where E denotes *expected value and g denotes any convex function.

Jewell estimator *See* TWO-BY-TWO TABLE.

jitter A procedure for improving the display of *bivariate data. If the variables are *discrete, or if they are *continuous, but the values have

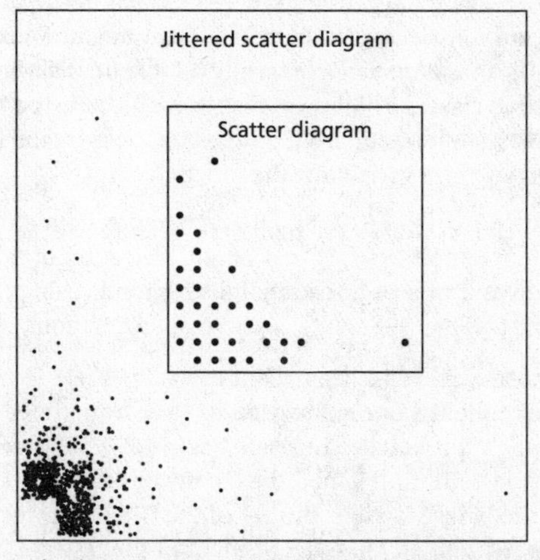

Jitter. The axes in the diagrams represent the numbers of tips reported by two tipping-bucket rain-gauges in September 2000. The jittered scatter diagram illustrates the amounts of rain for the 813 two-minute periods for which at least one gauge reported a tip. The inner diagram shows the corresponding 34-point scatter diagram.

been rounded, then some value pairs may occur very often, but would be represented by a single point on a normal *scatter diagram. A better impression of the data can be obtained by adding small random quantities to all values before plotting.

Johnson, Norman Lloyd (1917–2004; b. Ilford, England; d. Chapel Hill, NC) English statistician whose career was divided between England and the USA. Johnson was a graduate of UCL, joining its faculty on either side of the Second World War, during which he was seconded to the London Ordnance Board. In 1948 he obtained his PhD (supervised by Egon *Pearson). Whilst at UCL he supervised *Jenkins and *Mallows. In 1962 he moved to UNC. He was presented with the *Shewhart Medal of the *ASQ in 1984 and the *Wilks Award of the *ASA in 1993.

SEE WEB LINKS

- Fuller biography, interview, and photographs.

j

Johnson distributions *Probability distributions, of widely differing shapes, that can be obtained by simple transformations of the standard *normal distribution. The relationship to the normal distribution makes it easy to calculate, for example, a tail probability (*see* TAIL AREA). The entire family is defined by

$$Y = \mathrm{g}\left(\frac{Z-\alpha}{\beta}\right),$$

where α and β are constants, g is a function, Z is a standard normal *random variable and Y is a random variable having a Johnson distribution.

There are three classes of Johnson distributions, referred to as the S_U, S_B, and S_L classes, which result from different choices for the function $\mathrm{g}(x)$. In the order given, for a variable x, the functions are

$$\sinh x, \qquad \frac{1}{2}(1+\tanh x), \qquad \text{and} \qquad \mathrm{e}^x.$$

The last case gives a type of *lognormal distribution.

Johnstone, Iain (1956– ; b. Melbourne, Australia) Australian statistician whose career has been spent in the USA. He is known for his work on *wavelets. Johnstone's first degrees were from the ANU, where he obtained his MSc in probability and statistics in 1978. He obtained his PhD from Cornell U in 1981, and joined the faculty of Stanford U. In 1995 he was awarded the *Guy Medal in Bronze of the *RSS and was the *COPSS President's Award winner. He was President of the *IMS in 2001 and its *Wald Lecturer in 2004.

SEE WEB LINKS

- University website.

joint distribution *See* BIVARIATE DISTRIBUTION.

joint independence The *random variables *U* and *V* are jointly independent of *W* if their *joint probability distribution is unaffected by the value of *W*. For *categorical variables, with p_{jkl} denoting the probability of the combination (j, k, l), the condition for *U* and *V* to be jointly independent of *W*, is that

$$p_{jkl} = p_{jk0}p_{00l},$$

for all *j*, *k*, and *l*, where

$$p_{jk0} = \sum_l p_{jkl}, \qquad \text{and} \qquad p_{00l} = \sum_j \sum_k p_{jkl}.$$

joint probability The joint probability of a set of events (*see* SAMPLE SPACE) is the *probability that all occur simultaneously.

j

joint probability density function *See* BIVARIATE DISTRIBUTION.

Jonckheere–Terpstra test A *k*-sample analogue of the *Mann–Whitney test. The test was suggested independently by Jonckheere (in 1954) and Terpstra (in 1952). The assumption is that all *k* populations have the same continuous distribution, but that the population means may change (monotonically) during the ordered sequence of samples.

Journal de la SFdS The quarterly journal of the *Société Française de Statistique; first published (under a different title) in 1859.

SEE WEB LINKS

- Journal home page.

***Journal of Agricultural, Biological, and Environmental Statistics* (*JABES*)** A quarterly journal, first published in 1996. It is a joint publication of the *American Statistical Association and the *International Biometric Society.

SEE WEB LINKS

- Journal home page.

***Journal of Business & Economic Statistics* (*JBES*)** A quarterly jounal of the *American Statistical Association, first published in 1982.

SEE WEB LINKS

- Journal home page.

***Journal of Computational and Graphical Statistics* (*JCGS*)** A quarterly journal first published in 1992. It is a joint publication of the

*American Statistical Association, the *Institute of Mathematical Statistics, and the Interface Foundation of North America.

SEE WEB LINKS

- Journal home page

Journal of the American Statistical Association **(*JASA*)** The principal journal published by the *American Statistical Association. It was first published under a slightly different name in 1888 and under its current name since 1922. Its emphasis is on developments in statistical theory. It appears quarterly.

SEE WEB LINKS

- Journal home page.

Journal of the Royal Statistical Society **(*JRSS*)** The first regularly published Statistics journal in the world. Currently the journal is published as three series. Series A, now quarterly and entitled *Statistics in Society*, was first published in 1839. Series B, *Statistical Methodology*, was first published in 1934. Series C, *Applied Statistics*, was first published in 1954. These last two journals now appear five times a year. A fourth series (D), *The Statistician*, formerly the journal of the *Institute of Statisticians, ceased publication in 2003.

SEE WEB LINKS

- Journals home page.

JRSS *Abbreviation for* JOURNAL OF THE ROYAL STATISTICAL SOCIETY.

j

Kaiser rule *See* PRINCIPAL COMPONENTS ANALYSIS.

Kalman, Rudolf Emil (1930- ; b. Budapest, Hungary) Hungarian electrical engineer. Kalman, a graduate of MIT (BSc, 1953; MSc, 1954) gained his DSci from Columbia U (1957). In papers published in 1960 and 1961 he introduced what is now termed the *Kalman filter. At that time he was at the Research Institute for Advanced Study in Baltimore. He was appointed Professor of Engineering Mathematics at Stanford U in 1964, moving to U Florida in 1971 as Professor of System Theory. He is an elected member of NAS, NAE, and AAAS. He was awarded the 2008 National Medal of Science.

SEE WEB LINKS

- Fuller biography and photograph.

Kalman filter A computationally efficient method of updating the estimates of the time-dependent *parameters of a *multiple regression model as successive values in the *time series of values of the dependent variable (*see* REGRESSION) become available. *Exponential smoothing provides an extremely simple example of the recursive calculations involved. The procedure was introduced by *Kalman in 1960.

Kaplan, Edward Lynn (1920–2006; b. Philadelphia, PA; d. Corvallis, OR) American mathematician. Kaplan obtained his BS in Mathematics at Carnegie Institute of Technology in 1941 and then worked at the US Naval Ordnance Lab in Maryland. In 1951 he obtained his PhD at Princeton U, where he worked with *Feller and *Wilks. From 1950 to 1957 he worked at Bell Labs, NJ, where he studied the estimation of the survivor function (*see* HAZARD RATE). His classic paper with *Meier was published in 1958. In 1961 he was appointed Professor of Mathematics at Oregon State U, where he specialized in *linear programming.

SEE WEB LINKS

- Obituary.

Kaplan–Meier estimate (product-limit estimate) A procedure for estimating the survivor function (*see* HAZARD RATE) from *observations of lifetimes (of people, machine components, etc.) when some observations are *censored (e.g. people move away from the observation site, or functioning components are removed before the end of the experimental period). The estimate is calculated as follows. First, order the *data by length of lifetime. Suppose there are m distinct lifetimes (where, for example, people alive at the time of calculation are given their current lifetime). Let $t_{(j)}$ be the jth of these ordered lifetimes, let n_j individuals have lifetimes of $t_{(j)}$ or more, and let d_j individuals have lifetimes of exactly $t_{(j)}$. The Kaplan–Meier estimate of the survivor function is given by

$$\hat{S}(t_{(j)}) = \prod_{k=1}^{j}\left(1 - \frac{d_k}{n_k}\right).$$

The corresponding *estimator has *variance estimated by the **Greenwood formula**

$$[\hat{S}(t_{(j)})]^2 \sum_{k=1}^{j} \frac{d_k}{n_k(n_k - d_k)}.$$

k

Closely related to the Kaplan–Meier estimate is the **Nelson–Aalen** estimate of the cumulative hazard rate, given by

$$\hat{\Lambda}(t_{(j)}) = \sum_{k=1}^{j} \frac{d_k}{n_k}.$$

SEE WEB LINKS

- Applet.

Kempthorne, Oscar (1919–2000; b. St Tudy, England; d. Annapolis, MD) English statistician, specializing in *experimental design, whose career was largely in the USA. After graduation from Cambridge U, Kempthorne joined Sir Ronald *Fisher and *Yates at *Rothamsted in 1941. In 1947 he joined the fledgling Statistics Department at ISU where his 44 research students included *Graybill and *Wilk. He was the *COPSS *Fisher Lecturer in 1965, the President of the *IMS in 1985, and was awarded Honorary Fellowship of the *RSS in 1988.

SEE WEB LINKS

- Fuller biography, interview, and photographs.

Kendall, David George (1918–2007; b. Ripon, England; d. Cambridge, England) English mathematical statistician and probabilist. A graduate of Oxford U, where he was supervised by *Bartlett, Kendall worked during the Second World War for the Ministry of Supply, receiving his statistics training in evening discussions with *Anscombe. In 1946 he was elected a Fellow of Magdalen College, Oxford. In 1962 he was appointed Professor of Mathematical Statistics at Cambridge U, where he supervised *Kingman and *Silverman. He was elected FRS in 1964 and was awarded the Royal Society's Sylvester Medal in 1976. He was the President of the LMS in 1972 and President of the *BSMSP in 1975. He was awarded the *Guy Medal of the *RSS in Silver in 1955 and in Gold in 1981.

SEE WEB LINKS

• Fuller biography, interview, and photographs.

Kendall, Sir Maurice George (1907–83; b. Kettering, England; d. Redhill, England) English statistician. The son of a publican, Kendall failed to get a place in his local grammar school, but nevertheless won a scholarship to study mathematics at Cambridge U. On graduating he initially worked at the Ministry of Agriculture. His work there underlies the modern approach to the analysis of *time series. He also became interested in methods for measuring *correlation; his initial paper on rank correlation (*see* RANK CORRELATION COEFFICIENT) was published in 1938, though *Kendall's tau became widely used only after the publication in 1948 of his book *Rank Correlation Methods*. He may be best known as the author of the two-volume *The Advanced Theory of Statistics* (the volumes appearing in 1943 and 1946). For many years these volumes were the first reference used whenever there was a statistical inquiry. Further editions, with other authors, have appeared regularly ever since (currently with Kendall's name transferred to the title of the book). He became Professor of Statistics at the LSE in 1949, where he remained until 1961. From 1960 to 1962 he was President of the *RSS and was awarded its *Guy Medal in Silver in 1945 and in Gold in 1968. From 1972 to 1980 he was Director of the World Fertility Study. He was knighted in 1974.

SEE WEB LINKS

• Obituary.

Kendall rank correlation coefficient *Alternative name for* KENDALL'S TAU.

Kendall's tau (τ) A *correlation coefficient that can be used as an alternative to *Spearman's rho for *data in the form of *ranks. It is a simple function of the minimum number of neighbour swaps needed to produce one ordering from another. Its properties were analysed by Sir Maurice *Kendall in a paper published in 1938.

As an example, suppose that we have four objects (A–D) and the two orderings (D,B,A,C) and (A,B,D,C). To convert the first ordering into the second using neighbour swaps we could begin by swapping A and B to get (D,A,B,C). Now the swap of A and D gives (A,D,B,C) and then the swap of

B and D gives the desired (A,B,D,C). The reordering cannot take fewer than the three swaps used. In a general case with n items to order and a minimum of Q swaps required, τ (which takes values in the interval -1 to 1, inclusive) is given by

$$\tau = 1 - \frac{4Q}{n(n-1)}.$$

An easy way of counting the minimum number of neighbour swaps is as follows. First, write down the items in the order specified by the first ordering. Next, write down the list of ranks assigned to these items by the second ordering. For each number in this list in turn, count how many subsequent numbers are smaller than it. The sum of these *counts is Q.

kernel density estimation; kernel function *See* KERNEL METHOD.

kernel method (kernel density estimation) A method for the *estimation of a *probability density function. Suppose X is a *continuous random variable with unknown probability density function f. A random *sample of *observations of X is taken. If the sample values are denoted by $x_1, x_2, \ldots, x_n$, the estimate (*see* ESTIMATOR) of f is given by

$$\tilde{\mathrm{f}}(x) = A\sum_{j=1}^{n} \mathrm{K}(x, x_j),$$

where K is a **kernel function** and the constant A is chosen so that $\int_{-\infty}^{\infty} \tilde{\mathrm{f}}(x)\mathrm{d}x = 1$. The observation x_j may be regarded as being spread out between $x_j - a$ and $x_j + b$ (usually with $a = b$). The result is that the naive estimate of f as being a function capable of taking values only at $x_1, x_2, \ldots,$ x_n, is replaced by a continuous function having peaks where the data are densest. Examples of kernel functions are the **Gaussian kernel,**

$$\mathrm{K}(x, t) = \exp\left\{-\frac{(x-t)^2}{2h^2}\right\}, \qquad -\infty < x < \infty,$$

and the **Epanechikov kernel,**

$$\mathrm{K}(x, t) = \begin{cases} 1 - \dfrac{(x-t)^2}{5h^2}, & -\sqrt{5}h < x - t < \sqrt{5}h, \\ 0, & \text{otherwise.} \end{cases}$$

The constant h is the **window width** or **bandwidth**. The choice of h is critical: small values may retain the spikes of the naive estimate, and large values may oversmooth so that important aspects of f are lost (*see diagram opposite*).

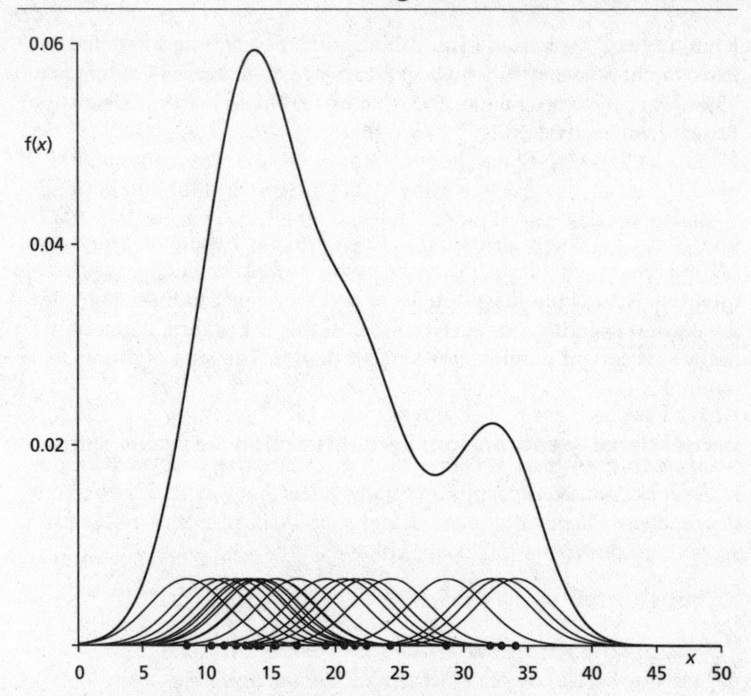

Kernel method. In this case a sample of twenty observations have been generated randomly from a chi-squared distribution with twenty degrees of freedom and a Gaussian kernel with $h = 3$ has been used to generate the kernel density estimate.

kernel regression *See* NON-PARAMETRIC REGRESSION.

Khinchin, Aleksandr Yakovlevich (1894–1959; b. Kondrovo, Russia; d. Moscow, USSR) Russian probabilist. A graduate of Moscow U in 1916, he soon joined the faculty. He was elected to the USSR Academy of Sciences in 1939. His 1943 book on the mathematical principles of statistical mechanics provided the first coherent mathematical treatment of the subject. His statement of the *law of the iterated logarithm appeared in 1924.

SEE WEB LINKS

- Photograph.

Kingman, Sir John Frank Charles (1939– ; b. Beckenham, England) English probabilist, specializing in queueing theory (*see* QUEUE) and *stochastic processes. A graduate of Cambridge U, where his research advisors were *Whittle and David *Kendall, Kingman was appointed Professor at Sussex U at the age of 27, moving to a chair at Oxford U in 1969. He was elected FRS in 1971, and was awarded the Royal Society's Royal Medal in 1983. In 1977 he was the *IMS *Wald Lecturer. From 1985 to 2001 he was Vice-Chancellor of U Bristol. He was President of the *RSS in 1987, and was awarded its *Guy Medal in Silver in 1981 and in Gold in 2013. He was made an Honorary Fellow of the Society in 1993. In 1990 he was President of the LMS. He has been Director of the Isaac Newton Institute (2001–6) and Chairman of the UK's Statistics Commission (2001–3). He was knighted in 1985.

SEE WEB LINKS

- Website.

Klotz test *See* TEST FOR EQUALITY OF SCALE.

knapsack problem A *combinatorial optimization problem. Given a knapsack of limited capacity and various items that vary in value and size, the problem is to determine which items should be placed in the knapsack so as to maximize the value of its contents.

Kohonen map *See* SELF-ORGANIZING FEATURE MAP.

Kolmogorov, Andrei Nikolaevich (1903–87; b. Tambov, Russia; d. Moscow, Russia) Russian mathematician and probabilist. Kolmogorov's mother died in childbirth and he was brought up by his aunt. Before enrolling at Moscow U in 1920 he spent some time as a railway conductor. A brilliant student, he obtained fundamental results in Fourier series (1922), logic (1925), and probability (1929). In 1929 he joined the faculty of Moscow U, becoming Professor in 1931. He was successively appointed Professor of Probability in 1938, of Statistical Methods in 1966, of Mathematical Statistics in 1976, and of Logic in 1980. Kolmogorov also worked on such practical projects as turbulence, the motion of the planets, the theory of fire, telecommunications scheduling, and the landing of planes on aircraft carriers. His work is now especially remembered by statisticians in the context of the *Chapman–Kolmogorov equations for *Markov processes and the *non-parametric *Kolmogorov–Smirnov test for a specified distribution. He was devoted to the development of talent in others, particularly the young, and worked to improve the teaching of mathematics in secondary schools. He was elected FRS in 1964 and a member of the NAS in 1967.

SEE WEB LINKS

- Website with links.

Kolmogorov inequality A stronger form of the *Chebyshev inequality. Let $X_1, X_2, \ldots, X_n$ be *independent random variables with finite *expected values and *variances. Define the partial sum S_k by

$$S_k = \sum_{j=1}^{k} X_j, \qquad k = 1, 2, \ldots, n.$$

The inequality states that, for all positive values of ϵ,

$$\mathrm{P}\left(\max_k \left\{\frac{|S_k - \mu_k|}{\sigma_k}\right\} < \varepsilon\right) \geq 1 - \frac{1}{\varepsilon^2},$$

where, for each k, μ_k and σ_k^2 are, respectively, the expected value and variance of S_k. *See also* BERNSTEIN INEQUALITY; HÖLDER INEQUALITY; MARKOV INEQUALITY; MINKOWSKI INEQUALITY.

Kolmogorov–Smirnov test A *non-parametric test for the *null hypothesis that a random *sample has been drawn from a specified *distribution (either *discrete or *continuous). There are several similar tests, each involving a comparison of the sample *distribution function with that hypothesized. For example, let the sample values, in increasing order, be $x_{(1)}, x_{(2)}, \ldots, x_{(n)}$. Let the hypothesized *probability of a value less than or equal to $x_{(j)}$ be p_j. Let u_j and v_j be defined by

$$u_j = \frac{j}{n} - p_j, \qquad v_j = p_j - \frac{j-1}{n} = \frac{1}{n} - u_j.$$

The test statistic is the largest of the absolute magnitudes of these $2n$ differences. A two-sample version of the test compares the two sample distribution functions. In the single-sample case, approximate *critical values are $1.36/\sqrt{n+1}$ at the 5% level and $1.63/\sqrt{n+1}$ at the 1% level. The test was introduced by *Kolmogorov in 1933, and further developed by *Smirnov in 1939.

As described, the test refers to a fully prescribed distribution. However, by using special tables of critical values and estimating unknown parameters from the sample data, its use has been extended to testing for *exponential, *extreme-value, *logistic, *normal (the *Lilliefors test), and *Weibull distributions with unspecified parameters.

As an example, to test the hypothesis that the values 0.273, −1.184, 1.456, −0.655, −0.323, −0.733, −1.600, 0.819, 0.081, 0.971 have been drawn from a standard normal distribution the results given in the table (*see overleaf*) are obtained. The value of the test statistic is 0.144, which is much less than $1.36/\sqrt{10+1} = 0.41$: so the null hypothesis is acceptable.

Koopmans, Tjalling Charles (1910–85; b. s'Graveland, Netherlands; d. New Haven, CT) Dutch econometrician. Having initially studied

value	j/n	p_j	$(j-1)/n$	u_j	v_j
−1.600	0.1	0.055	0.0	0.045	0.055
−1.184	0.2	0.118	0.1	0.082	0.018
−0.733	0.3	0.232	0.2	0.068	0.032
−0.655	0.4	0.256	0.3	0.144	−0.044
−0.323	0.5	0.373	0.4	0.127	−0.027
0.081	0.6	0.532	0.5	0.068	0.032
0.273	0.7	0.608	0.6	0.092	0.008
0.819	0.8	0.793	0.7	0.007	0.093
0.971	0.9	0.834	0.8	0.066	0.034
1.456	1.0	0.927	0.9	0.073	0.027

mathematics, Koopmans obtained his MSc in theoretical physics from U Utrecht in 1933 and his PhD in the emerging area of *econometrics from U Leiden in 1936. In 1940 Koopmans moved to Washington, working as a statistician. He joined U Chicago in 1944 and then, in 1955, Yale U. He was awarded the Nobel Prize in Economics in 1975.

SEE WEB LINKS

- Autobiography and photograph.

Koyck model *See* DISTRIBUTED LAGS MODEL.

Krige, Danie Gerhardus (1919–2013; b. Bothaville, RSA; d. Johannesburg, RSA) South African mining engineer and pioneer of *geostatistics. Krige developed statistical methods for assessing the size of buried mineral deposits. Aged 19, he graduated in mining engineering from Witwatersrand U. In 1981, having retired from a career in the mining industry, he returned to the University to spend ten years as Professor of Mineral Economics.

SEE WEB LINKS

- Obituary with photograph.

kriging A method of spatial prediction, named by *Matheron in honour of *Krige. The values of some quantity, s, are known for n points in space. Estimates for intervening locations are required—for example, we might wish to derive contour lines from a series of spot heights. This is also called **point kriging**. If, instead, the aim is to estimate the total amount (of e.g. oil, or gold) in a region, rather than at a point, then that is called **block kriging**. The predictions are to be calculated as the weighted sum of the observed values. The question is what weights (*see* WEIGHTED AVERAGE) to use so as to minimize the average squared difference between the predicted value and the true value.

If, at each spatial location, *data are collected on several *variables, then the *interpolation of values for the variable of prime interest may be improved by noting the variation in the other variables—this is called **cokriging**.

Other variants include **disjunctive kriging**, which uses linear combinations of functions of the data, **indicator kriging**, which uses *binary indicators (e.g. presence/absence) in place of the observed data, and **universal kriging**, in which the surface being estimated is an unknown linear combination of known spatial surfaces.

Kruskal, William Henry (1919–2005; b. New York City; d. Chicago, IL) American statistician. Kruskal was co-author (with *Wallis) of the 1952 paper that introduced one of the best-known *non-parametric tests. He was educated at Harvard U and Columbia U (where his PhD was supervised by *Scheffé and *Levene). He joined the Statistics Department at U Chicago in 1950, where he worked with *Goodman and Wallis. A nice quote is 'Each of us has been doing statistics all his life, in the sense that each of us has been busily reaching conclusions based on empirical observations ever since birth'. He was Editor of the **Annals of Mathematical Statistics* during 1958–61. He was President of the *IMS in 1971 and was its *Rietz Lecturer in 1979. He was the 1978 *COPSS *Fisher Lecturer. He was President of the *ASA in 1982, having been presented with its *Wilks Award in 1978. He was made an Honorary Fellow of the *RSS in 1989.

SEE WEB LINKS

- Fuller biography, interview, and photographs.

Kruskal stress *See* MULTIDIMENSIONAL SCALING.

Kruskal–Wallis test A *non-parametric test, introduced in 1952 by *Kruskal and *Wallis. It is a k-sample extension of the two-sample *Mann–Whitney test. It tests the *null hypothesis that the k sampled *populations have the same *distribution function.

With n_j *observations in the jth *sample, denote the total number of observations, $\sum_{j=1}^{k} n_j$, by N. Arrange these N observations in order of size and replace their values with the corresponding *ranks. Let the total of the ranks in sample j be R_j. The test statistic is H, given by

$$H = \frac{12}{N(N+1)} \sum_{j=1}^{k} \frac{R_j^2}{n_j} - 3(N+1).$$

Under the null hypothesis, for large N, with no n_j small, H has an approximate *chi-squared distribution with $(k-1)$ *degrees of freedom.

Kuder–Richardson formulae *See* RELIABILITY.

Kullback, Solomon (1903–94; b. Brooklyn, NY; d. Boynton Beach, FL) American statistician and cryptographer. A graduate of City College (now City U) New York, Kullback obtained his PhD from GWU in 1934. By this time he had been a member of the United States Army's Signals Intelligence Service for four years. The war years were busy ones for code-breakers and he was part of the three-man team that broke the Japanese codes. He did not leave the National Security Agency until 1962, when he joined the faculty at GWU where *Greenhouse was one of his research students. He received the *Wilks Award of the *ASA in 1976. Among his other honours was the Legion of Merit from the United States Army.

SEE WEB LINKS

- Fuller biography and photograph.

k

Kullback–Leibler information (directed divergence; discrimination information) A measure of the difference between two *probability density functions (f and g, say) taking non-zero values on the same interval (a, b). Usually denoted by I, where

$$I = \int_a^b \ln\left\{\frac{\mathrm{f}(x)}{\mathrm{g}(x)}\right\}\mathrm{f}(x)\mathrm{d}x.$$

It is also called the 'information in favour of f'. An equivalent expression holds for the comparison of two *discrete distributions. The measure was introduced in 1951 by *Kullback and *Leibler. *See also* BHATTACHARYA DISTANCE; HELLINGER DISTANCE.

Kulldorff scan statistic A *statistic that evaluates reported spatial or space-time disease clusters to determine their significance (*see* HYPOTHESIS TEST). Conversely, it can be used to search for clusters of events in space, time, or space-time.

kurtosis A measure of the peakedness of a *distribution. The measure (usually denoted by β_2 or κ), was introduced by Karl *Pearson before 1905. The measure is given by

$$\beta_2 = \frac{\mu_4}{\mu_2^2},$$

where μ_4 is the fourth central *moment of the distribution and μ_2 is the *variance. It is invariant under a change of scale or origin.

An alternative definition uses $(\beta_2 - 3)$ to give a value of 0 for the *normal distribution. With this definition, a distribution having negative kurtosis is described as being **platykurtic** (flatter), and a distribution having positive kurtosis is described as being **leptokurtic** (more peaked). A distribution having kurtosis zero is described as **mesokurtic**.

lack of memory *See* FORGETFULNESS PROPERTY.

lag *See* AUTOCORRELATION.

lag window *See* PERIODOGRAM.

lambda *See* ASSOCIATION.

Lancaster, Oliver Henry (1913–2001; b. Sydney, Australia; d. Sydney, Australia) Australian medical statistician. Lancaster trained as a doctor, graduating in medicine from U Sydney in 1937. During the Second World War he served as a pathologist. He continued medical work whilst researching the application of the *chi-squared distribution to discrete distributions (*see* DISCRETE RANDOM VARIABLE), gaining his PhD in 1953. In 1947 he was a founder member of the organization that became the *SSAI. In 1959 he was appointed to the Chair of Mathematical Statistics at U Sydney. For twelve years he was Editor of the Society's journal and served as its President in 1965. He was awarded the Society's Pitman Medal (*see* PITMAN) in 1980.

SEE WEB LINKS

- Obituary and photograph.

Langevin, Paul (1872–1946; b. Paris, France; d. Paris, France) French physicist. Langevin derived the *Langevin distribution during a study in 1905 of the magnetic properties of molecules.

SEE WEB LINKS

- Biography and photograph.

Langevin distribution (Fisher distribution) A *distribution used to model *spherical data. The *probability density function f in the direction (θ, ϕ) is given by

$$f(\theta,\phi) = c\exp[\kappa\{\cos\theta_0\cos\theta + \sin\theta_0\sin\theta\cos(\phi-\phi_0)\}],\quad 0 \le \theta \le \pi, -\pi \le \phi < \pi,$$

where c is a positive constant and the non-negative *parameter κ is a measure of the concentration of the distribution, which is symmetrical about the mode. The case $\kappa = 0$ corresponds to a uniform distribution over the entire sphere. If $\kappa \neq 0$ the distribution has *mode at (θ_0, ϕ_0).

Laplace, Marquis Pierre-Simon (1749–1827; b. Beaumont-en-Auge, France; d. Paris, France) French mathematician. Laplace was born into the French middle class but died a Marquis having prospered as well during the French Revolution as during the monarchy. His name is well known to mathematicians both for his work on transformations and for his work on planetary motion. He had a deterministic view of the universe. Reputedly, his reply, when asked by Napoleon where God fitted in, was 'I have no need of that hypothesis'. Napoleon appointed him as Minister of the Interior—but removed him six weeks later for trying 'to carry the spirit of the infinitesimal into administration'. He studied under *d'Alembert in Paris, where he later taught *Poisson at the École Polytechnique. In Statistics he worked on many probability problems, including the *Laplace distribution, and he is credited with independently discovering *Bayes's Theorem. He was elected FRS in 1789 and FRSE in 1813. A street in Paris is named after him, as is the Promontorium Laplace on the Moon.

SEE WEB LINKS

- Fuller biography.

Laplace distribution A distribution, first given by *Laplace in 1774, that has *probability density function f given by

$$f(x) = \frac{1}{2\phi}\exp\left(-\frac{|x|}{\phi}\right), \qquad -\infty < x < \infty,$$

where ϕ is a positive *parameter. The distribution is also called the **double exponential distribution** and is the distribution of the difference of two *independent random variables, each with the same *exponential distribution. The distribution is symmetrical about 0, which is therefore both its mean and its *mode. The distribution has *variance $2\phi^2$.

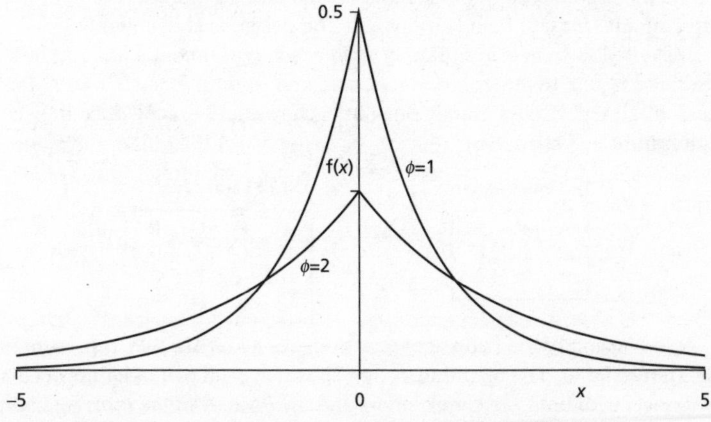

Laplace distribution. Each distribution has variance equal to $2\phi^2$.

LARS *See* STEPWISE PROCEDURE.

Laspeyres, Ernst Louis Étienne (1834–1913; b. Halle/Saale, Germany; d. Giessen, Germany) German economist. Laspeyres was the son of a law professor. He studied law and public finance at a succession of universities, gaining his PhD in political science and public finance from U Heidelberg. From 1874 to 1900 he was Professor of Political Science at Giessen U. He was an active member of the *International Statistical Institute. He is remembered for the Laspeyres price index (*see* PRICE INDEX), though he was not the first to suggest its use and apparently did not use it himself.

SEE WEB LINKS

- Fuller biography.

Laspeyres price index *See* PRICE INDEX.

lasso Acronym derived from 'least absolute shrinkage and selection operator'. *See* MULTIPLE REGRESSION MODEL.

latent class model A model for cross-tabulated *data. The idea is that the apparently complex relationship between the observed *variables can be explained by a simple relationship that holds for unobserved *latent variables. *See also* CONTINGENCY TABLE.

latent variable An unobserved *variable that may account for variation in the data and/or for apparent relations between observed variables. *Compare* MANIFEST VARIABLE.

Latin square An *experimental design that extends the *randomized blocks design. In a Latin square two types of *blocks are used. The numbers of both type of block must be equal to the numbers of *treatments (m, say) being compared. The design is conveniently represented as an $m \times m$ square in which each experimental unit has been assigned to one treatment, one row, and one column in such a way that each treatment occurs exactly once in each row and exactly once in each column:

3×3 Latin square

A	C	B
B	A	C
C	B	A

4×4 Latin square

A	D	C	B
B	C	A	D
D	A	B	C
C	B	D	A

An extension of the Latin square introduces a fourth effect, represented by a Greek letter. The resulting design, in which each pair of letters occurs just once, with both the Greek letters and the Roman letters forming Latin squares, is called a **Graeco-Latin square**. Removal of a row or column from a Latin square results in a so-called **Youden square** (actually a rectangle). Youden squares can be used to generate *balanced incomplete block designs.

Graeco-Latin square

Cδ	Bα	Dγ	Aβ
Aα	Dδ	Bβ	Cγ
Dβ	Aγ	Cα	Bδ
Bγ	Cβ	Aδ	Dα

Youden square

A	D	C	B
B	C	A	D
C	B	D	A

Latin squares were introduced into Statistics by Sir Ronald *Fisher in 1925. Graeco-Latin squares were introduced by Fisher and *Yates in 1934. In 1938 Fisher and Yates introduced Youden squares which they named after *Youden.

lattice model A model describing the variation in the values at the nodes of a *network.

lattice square An *experimental design related to *Latin square and *balanced incomplete block designs. As an example, suppose there are nine treatments A–I. The design consists of two squares:

A	B	C
D	E	F
G	H	I

and

A	F	H
I	B	D
E	G	C

Considering the squares separately, we find that each pair of treatments occurs exactly once—either in a row or in a column. In general, with n^2 treatments, the design requires $(n-1)$ separate $n \times n$ squares.

Lauritzen, Steffen Lilholt (1947- ; b. Frederiksberg, Denmark) Danish statistician specializing in *graphical models. Having been both a student and a faculty member at Copenhagen U, Lauritzen joined the faculty at Aalborg U in 1981, moving to Oxford U in 2004. Elected an Honorary Member of the *RSS in 1992, he was awarded its *Guy Medal in Silver in 1996. He was appointed a Knight of the Order of the Dannebrog in 1999 and was elected FRS in 2011. He was the 2012 *Wald Lecturer of the *IMS.

law of averages As a *sample size increases so the *data will increasingly resemble the *population being sampled and will take on its characteristics (*see* LAWS OF LARGE NUMBERS). However, there is nothing *deterministic about this process. It is easy to be deluded by the **gambler's fallacy**, which advocates betting on numbers that 'should have come up'. As an example, suppose that a fair die is rolled 24 times and no six is obtained. According to the gambler's fallacy a six is long overdue and must be a good bet for the 25th roll—the truth is that its *probability will be the same as for every other roll, namely, 1/6.

law of compound probability *See* MULTIPLICATION LAW FOR PROBABILITIES.

law of the iterated logarithm The sequence $X_1, X_2, \ldots, X_n$ of independent random variables, each with mean 0 and variance σ^2, obeys the law of the iterated logarithm if

$$\lim_{n \to \infty} \sup \left\{ \sum_{j=1}^{n} X_j \Big/ \sigma\sqrt{2n\ln\{\ln(n)\}} \right\} = 1.$$

The law was first given by *Khinchin in 1924.

law of total probability *See* ADDITION LAW FOR PROBABILITIES.

laws of large numbers Laws that describe the way in which the *sample mean approaches the *population mean as the sample size increases. The phrase is due to *Poisson, who, in 1835, referred to 'La loi des grands nombres'. The **weak law of large numbers** states that if $X_1, X_2, \ldots, X_n$ are a set of *independent identically distributed *random variables, each with *expected value μ, and if $\bar{X} = \frac{1}{n}\sum_{j=1}^{n} X_J$ then, for every positive ε,

$$\lim_{n\to\infty} \mathrm{P}(|\bar{X} - \mu| \geq \varepsilon) = 0.$$

An equivalent statement is that

$$\bar{X} \to \mu \qquad \text{'almost surely'}, \quad \text{as} \quad n \to \infty.$$

The **strong law of large numbers** states that, for every positive δ, it is always possible to find positive values ε and N such that

$$\mathrm{P}(|\bar{X} - \mu| \geq \varepsilon) \leq \delta, \qquad n = N, N+1, \ldots.$$

An equivalent statement is that

$$\bar{X} \to \mu \qquad \text{'in probability'}, \quad \text{as} \quad n \to \infty.$$

SEE WEB LINKS

- Animation.

leading diagonal *See* MATRIX.

least angle regression selection *See* MODEL SELECTION PROCEDURE.

least significant difference (LSD) *See* MULTIPLE COMPARISON TEST.

least squares *See* METHOD OF LEAST SQUARES.

Le Cam, Lucien Marie (1924–2000; b. Croze, France; d. San Pablo, CA) French mathematical statistician, specializing in *statistical inference, whose career was spent in the USA. In 1942, intending to study chemistry,

Le Cam arrived at U Clermont-Ferrand to be told that term had started and there was no space in the laboratories—so he studied mathematics instead, graduating in 1944. The next year he studied at U Paris and was then employed on hydrological work determining flood probabilities. In 1950 he met *Neyman and moved to UCB, obtaining his PhD (completed in six months) in 1952 and joining the UCB faculty. He was the *IMS *Wald Lecturer in 1963 and the Institute's President in 1973. He was elected to AAAS in 1976.

SEE WEB LINKS

- Fuller biography, interview, and photographs.

Le Cam Lecturer; Le Cam Lectureship A lectureship, established in 2003 by the *IMS in honour of *Le Cam. It is awarded every three years to an individual who is held to have made a fundamental contribution to the development of mathematical statistics or probability.

Legendre, Adrien-Marie (1752–1833; b. Paris, France; d. Paris, France) French mathematician. Legendre was educated at the Collège Mazarin in Paris. In 1775 he taught with *Laplace at the École Militaire. Legendre's career covered many branches of mathematics and, in 1805, he published the first account of the *method of least squares.

SEE WEB LINKS

- Fuller biography.

Lehmann, Erich Leo (1917–2009; b. Strasbourg, France; d. Berkeley, CA) German mathematical statistician who spent most of his career in the USA. Lehmann's family left Germany for Zurich in 1933, when Hitler came to power. After study at Cambridge U, Lehmann moved to UCB, where he obtained his PhD (supervised by *Neyman) in 1946. He promptly joined the UCB faculty, where his 42 research students included *Bickel and *Puri. He was Editor of the **Annals of Mathematical Statistics* during 1953–5. He was President of the *IMS in 1961 and was its *Wald Lecturer in 1964. He was elected a member of the AAAS in 1975 and of the NAS in 1978. He was the *COPSS *Fisher Lecturer in 1988. He was created an Honorary Fellow of the *RSS in 1986 and was presented with the *Wilks Award of the *ASA in 1996.

SEE WEB LINKS

- Fuller biography, interview, and photographs.

Lehmann–Scheffé theorem If T is a sufficient statistic (*see* ESTIMATOR) for the *parameter θ, then the minimum variance unbiased estimator of θ is given by $\mathrm{E}(\hat{\theta}|T)$, where $\hat{\theta}$ is any unbiased estimator of θ. The theorem, published in 1950, is an extension of the *Rao–Blackwell theorem.

Leibler, Richard Arthur (1914–2003; b. Chicago, IL; d. Reston, VA) American mathematician and cryptoanalyst. Leibler, who obtained his PhD at U Illinois, worked in the National Security Agency with *Kullback. Their 1951 paper introduced the eponymous *Kullback-Leibler information measure. In 1958 he was appointed to the Institute for Defense Analysis, becoming its Director in 1963. In 1977 he was appointed Chief of the Office of Research at the National Security Agency.

SEE WEB LINKS

- Obituary and photograph.

Leibniz, Gottfried Wilhelm (1646–1716; b. Leipzig, Germany; d. Hannover, Germany) German philosopher and mathematician. Leibniz graduated from U Leipzig at the age of seventeen. In 1667 he was awarded a doctorate in law by U Altdorf. In 1673 he was elected FRS. In 1675, in Paris, he developed the dx notation for differentials. In 1679 he developed the binary system. Leibniz was an assiduous letter writer with more than 600 correspondents including all the leading mathematicians of the day. He studied all areas of mathematics and is credited with devising the diagrams popularized by *Venn. His later years were much taken up with disputes on priority concerning the work on differentiation and integration: he was accused of plagiarism by a supporter of Sir Isaac Newton (who had derived equivalent results independently of Leibniz).

SEE WEB LINKS

- Fuller biography and portrait.

leptokurtic *See* KURTOSIS.

Leslie matrix; Leslie model A *matrix that captures the evolution of the age distribution of the females in a *population. The female population is divided into g age-groups, each of width k years. The *probability that a female, in group j at time t, will be alive in group $(j+1)$ at time $(t+k)$ is denoted by S_j. The mean number of female offspring born to a mother in group j between times t and $(t+k)$, and alive at time $(t+k)$, is denoted by M_j. The Leslie matrix, $\mathbf{L}$, is given by

$$\mathbf{L} = \begin{pmatrix} M_0 & M_1 & M_2 & M_3 & \dots \\ S_0 & 0 & 0 & 0 & \dots \\ 0 & S_1 & 0 & 0 & \dots \\ 0 & 0 & S_2 & 0 & \dots \\ \vdots & \vdots & \vdots & \vdots & \ddots \end{pmatrix}.$$

The entries in this matrix are the mean numbers at time $(t+k)$ resulting from single individuals in each group at time t. If $\mathbf{N}(t)$ is the $g \times 1$ column *vector with entries being the mean numbers of females in the groups at time t then the **Leslie model** states that

$$\mathbf{N}(t+k) = \mathbf{LN}(t).$$

***L*-estimate** A *robust *measure of location that is a *linear combination of *order statistics. Examples are the *median and the *trimean.

level

1. In significance testing (*see* HYPOTHESIS TEST), the designated *probability for rejection of the *null hypothesis—the significance level (e.g. 5%).

2. In *experimental design, the category of a *treatment (typically, in a 2^n factorial experiment (*see* FACTORIAL DESIGN), reference is made to upper and lower levels of a treatment).

Levene, Howard (1914–2003; b. New York City; d. New York City) American biostatistician and geneticist. A graduate of New York U, in 1941 Levene moved to Columbia U, working on *quality control for the war effort. Supervised by *Wolfowitz, he completed his PhD in 1947 and joined the Columbia faculty as a biostatistician, retiring in 1982. *Kruskal was one of his research students.

SEE WEB LINKS

- Obituary and photograph.

Levene test *See* TEST FOR EQUALITY OF SCALE.

leverage *See* REGRESSION DIAGNOSTICS.

Lévy, Paul (1886–1971; b. Paris, France; d. Paris, France) French mathematician regarded as the founder of modern probability theory. A graduate of the École Polytechnique, his first paper was published in 1905. After study under *Hadamard, he obtained his doctorate in 1912. After a period as Professor at the École Nationale des Mines (where he supervised *Matheron), he returned in 1920 to the École Polytechnique (where he supervised *Mandelbrot).

SEE WEB LINKS

- Fuller biography and portrait.

Lévy process A continuous-time *stochastic process in which the value at any time point is equal to the value at the previous time point altered by an amount which is an observation from some specified distribution. Examples are *Poisson processes and Wiener processes (*see* WIENER).

Lévy stable distribution A *probability distribution, with *parameters α, β, c, and m (the *mode). The parameter β $(-1 \leq \beta \leq 1)$ determines the skewness (*see* SKEWED), with $\beta = 0$ corresponding to a distribution symmetric about m. The spread of the distribution depends on the value of c. The shape of the distribution depends on α $(0 < \alpha \leq 2)$, with $\alpha = 2$ corresponding to the *normal distribution. The case $\alpha = 1$ and $\beta = 0$ corresponds to the *Cauchy distribution. In general, the distribution has an infinite *variance. The epithet 'stable' is a consequence of the property that (with $m = 0$) any *linear combination of *random variables having this distribution will also have the distribution (with the same values for α, β, and c). Although the *characteristic function for a general Lévy stable distribution is known, there is no general formula for the corresponding *probability density function.

Lexis, Wilhelm (1837–1914; b. Eschweiler, Germany; d. Göttingen, Germany) German social scientist. Lexis graduated in 1859 from U Bonn, where he had studied mathematics and physics. His subsequent career focused on the social sciences: at various times he held chairs in economics, geography, and political science. In 1879 he published *On the Theory of the Stability of Statistical Series* which began the study of time series. Lexis was a pioneer of the analysis of demographic *time series.

SEE WEB LINKS

- Photograph.

Lexis diagram A plot of age against time, consisting of lines inclined at 45° representing the lifetimes of individuals. Each line starts with birth (or

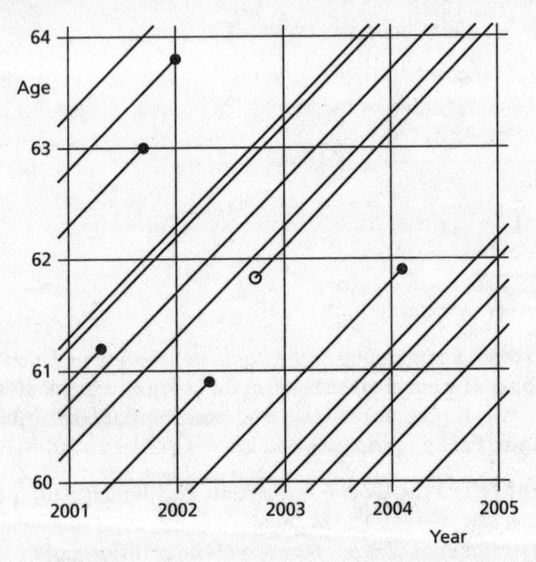

Lexis diagram. Diagonal lines show lifetimes of individuals; vertical lines show points in time. There were three people aged 60 at the beginning of 2001. Of these, one died later that year aged 61. In 2002 a 61-year-old immigrant entered the population.

immigration into the population) and ends with death (or emigration from the population).

life table A table, first developed by *Graunt, that presents the *probability of dying in the next time interval, as a function of current age. The table provides a means of calculating life expectancy, conditional on current age.

	Males		Females	
Age at start of decade	Probability of death in next decade	Number alive at start of decade	Probability of death in next decade	Number alive at start of decade
0	0.00926	100 000	0.00763	100 000
10	0.00427	99 074	0.00216	99 237
20	0.01393	98 651	0.00505	99 023
30	0.01581	97 276	0.00821	98 523
40	0.03164	95 738	0.01929	97 714
50	0.07060	92 709	0.04187	95 829
60	0.14302	86 164	0.09292	91 817
70	0.31342	73 841	0.22397	83 285
80	0.63525	50 698	0.52771	64 632
90	0.92343	18 492	0.87984	30 525
100	0.99763	1 416	0.94980	3 668

Life table. The table records the probabilities for males and females in the USA, using 2007 information. As an example, the probability that, in the USA, a 50-year-old male lives to be at least 70 is estimated as $(1-0.07060)\,(1-0.14302) = 0.79648$.

likelihood For a random sample $x_1, x_2, \ldots$ of *observations on a *discrete random variable X, the likelihood (a term coined by Sir Ronald *Fisher in 1921) is proportional to the product of the *probabilities of the individual values:

$$\prod_j \mathrm{P}(X = x_j).$$

When X is *continuous, a reported value x_j has to be regarded as meaning that the observed value is (in general) in the interval $(x_j - \delta, x_j + \delta)$, where δ represents the accuracy of measurement. The likelihood is then proportional to

$$\prod_j \mathrm{P}(x_j - \delta < X < x_j + \delta).$$

If δ is very small then this is approximately proportional to

$$\prod_j \mathrm{f}(x_j),$$

where f is the *probability density function of X.

likelihood principle A principle introduced by *Barnard in 1949. It asserts that all the information provided by a *sample about an unknown *parameter θ is provided by the *likelihood.

likelihood-ratio goodness-of-fit statistic (G^2) A *statistic used in an alternative to the *chi-squared test of goodness-of-fit. The two test statistics usually have similar values and both have approximate *chi-squared distributions when the *model under test is a correct description of the *data. The G^2 test is usually preferred when comparing *hierarchical models. If the *observed cell frequencies are denoted by $O_1, O_2, \ldots, O_k$ and the corresponding expected cell frequencies (*see* EXPECTED FREQUENCY) by $E_1, E_2, \ldots, E_k$, respectively, G^2 is given by

$$G^2 = 2\sum_{j=1}^{k} O_j \ \ln\left(\frac{O_j}{E_j}\right),$$

where ln is the *natural logarithm. The test results from the work of *Neyman and Egon *Pearson, who first used the term 'likelihood ratio' in a 1931 paper.

likelihood-ratio index *See* DEVIANCE.

likelihood residuals *Alternative name for* DELETION RESIDUALS.

Likert, Rensis (1903–81; b. Cheyenne, WY; d. Ann Arbor, MI) American psychologist. Likert graduated from U Michigan with an AB in economics and sociology in 1922. He received his doctorate in psychology from Columbia U in 1932. During the next decade his work involved the interpretation of surveys and the handling of questionnaires. He joined the faculty of U Michigan in 1946, becoming the founding Director of the Institute for Social Research in 1949.

SEE WEB LINKS

- Fuller biography and photograph.

Likert scale A scale, usually of approval or agreement, used in *questionnaires. The *respondent is asked to say whether, for example, they ‘Strongly agree’, ‘Agree’, ‘Disagree’, or ‘Strongly disagree’ with some statement.

Lilliefors test A modification of the *Kolmogorov–Smirnov test for testing whether a *population has a *normal distribution.

Lindley, Dennis Victor (1923–2013; b. London, England; d. Minehead, England) English statistician, specializing in *Bayesian inference. Lindley’s mathematical studies at Cambridge U were interrupted by the Second World War, during which he worked with *Barnard. He joined the Cambridge faculty in 1948, moving in 1960 to establish the Statistics Department at U Wales, Aberystwyth (where he supervised *Gittins). In 1967 he moved to UCL (where he supervised *Smith). He was awarded the *Guy Medal of the *RSS in Silver in 1968, and in Gold in 2002. He was the *IMS *Wald Lecturer in 1988. The *ISBA awards a biennial prize in his honour.

SEE WEB LINKS

- Fuller biography, interview, and photographs.

linear combination A *variable y is a linear combination of the variables $x_1, x_2, \dots, x_m$ if

$$y = \sum_{j=1}^{m} a_j x_j,$$

where $a_1, a_2, \dots, a_m$ are constants. *See also* MATRIX.

linear constraints *See* LINEAR PROGRAMMING.

linear dependence The *variables y and x display linear dependence (*see also* MATRIX) if, for some constants a and b ($\neq 0$),

$$y = a + bx.$$

linear filter *See* FILTER.

linear interpolation *See* INTERPOLATION.

linearizing The process of turning a non-linear relation into a linear relation. For example, after taking natural logarithms, the non-linear relation $y = ae^{bx}$ becomes $\ln y = \ln a + bx$.

linear-logistic model *See* LOGISTIC REGRESSION MODEL.

linearly dependent *See* MATRIX.

linearly independent *See* MATRIX.

linear model A *regression model that represents the *expected value of a *random variable, Y, as a *linear combination of functions of the explanatory variables $x_1, x_2, \ldots$. An example is the model

$$E(Y) = \alpha + \beta x_1 + \gamma x_2 + \delta x_2^2 + \eta \sin(x_3),$$

where α, β, γ, δ and η are unknown *parameters. *See also* GENERALIZED LINEAR MODEL; LINEAR REGRESSION; MULTIPLE REGRESSION MODEL.

linear piecewise regression model (break-point model) A *model that describes the situation where the graph relating the *expected value of the *random variable Y and the variable x is linear for $x \leq x_0$ and is also linear for $x \geq x_0$ but with a different slope. For example:

$$\begin{aligned} E(Y) &= \alpha + \beta_1 x, & x \leq x_0, \\ E(Y) &= \{\alpha + (\beta_1 - \beta_2)x_0\} + \beta_2 x, & x \geq x_0. \end{aligned}$$

linear programming A method for determining the optimal use of limited resources. The phrase 'linear programming' was used by *Dantzig in 1949. A typical problem requires the maximization (or minimization) of a linear combination of the non-negative variables $x_1, x_2, \ldots, x_n$:

$$c_1x_1 + c_2x_2 + \cdots + c_nx_n,$$

where the coefficients $c_1, c_2, \ldots, c_n$ have known values. The function is known as the **objective function**. The maximization is subject to a set of **linear constraints** such as

$$\begin{aligned} a_{11}x_1 + a_{12}x_2 + \cdots + a_{1n}x_n &\leq a_{10}, \\ a_{21}x_1 + a_{22}x_2 + \cdots + a_{2n}x_n &\geq a_{20}, \\ &\vdots \\ a_{m1}x_1 + a_{m2}x_2 + \cdots + a_{mn}x_n &= a_{m0}, \end{aligned}$$

where the coefficients $\{a_{ij}\}$ have known values.

In order to solve the problem, inequalities are turned into equalities by introducing non-negative **slack variables** so that, for example, the first inequality would become

$$a_{11}x_1 + a_{12}x_2 + \cdots + a_{1n}x_n + x_{n+1} = a_{10}.$$

Any solution satisfying all the constraints (with the addition of slack variables as required) is called a **feasible solution**. A feasible solution involving exactly m non-zero x-variables (including any slack variables) is called a **basic feasible solution**.

One standard approach to the general problem is to use the *simplex method (introduced by Dantzig in 1947). Special algorithms are required for **integer programming** (where the x-values are required to be integers) and for *assignment problems, *network flow problems, and *transportation problems.

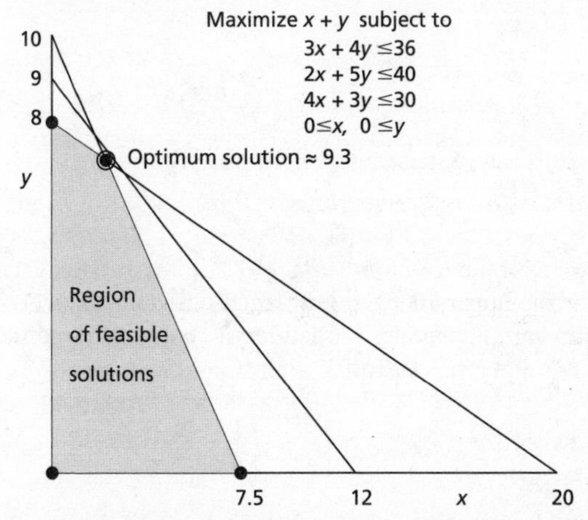

Linear programming. The solution must lie at one of the inner vertices marked with dots: these are the basic feasible solutions. The values of $x + y$ at the other three vertices are 0, 7.5, and 8.

linear regression The simplest and most used of all statistical regression *models. The model states that the *random variable Y is related to the variable x by

$$Y = \alpha + \beta x + \varepsilon,$$

where the *parameters α and β correspond to the intercept and the slope of the line, respectively, and ε denotes a *random error. With observations $(x_1, y_1), (x_2, y_2), \ldots, (x_n, y_n)$ the usual assumption is that the random errors are *independent observations from a *normal distribution with *mean 0 and *variance σ^2. In this case the parameters are usually estimated using ordinary least squares (*see* METHOD OF LEAST SQUARES). The estimates, denoted by $\hat{\alpha}$ and $\hat{\beta}$, are given by

$$\hat{\alpha} = \overline{y} - \hat{\beta}\overline{x}, \qquad \text{and} \qquad \hat{\beta} = \frac{S_{xy}}{S_{xx}},$$

where $\overline{x}$ and $\overline{y}$ are the means of $x_1, x_2, \ldots, x_n$ and $y_1, y_2, \ldots, y_n$, respectively, and where

$$S_{xy} = \sum_{j=1}^{n} x_j y_j - n\overline{x}\,\overline{y}, \qquad \text{and} \qquad S_{xx} = \sum_{j=1}^{n} x_j^2 - n\overline{x}^2.$$

The variance σ^2 is estimated by

$$\hat{\sigma}^2 = \frac{1}{n-2}\sum_{j=1}^{n}\left\{y_j - \left(\hat{\alpha} + \hat{\beta}x_j\right)\right\}^2$$
$$= \frac{S_{xx}S_{yy} - S_{xy}^2}{(n-2)S_{xx}},$$

where $S_{yy} = \sum_{j=1}^{n} y_j^2 - n\bar{y}^2$.

A 100(1−2θ)% *confidence interval for β is provided by

$$\hat{\beta} \pm t_{n-2}(\theta)\sqrt{\frac{\hat{\sigma}^2}{S_{xx}}},$$

where $t_\nu(\theta)$ is the upper 100θ% point (*see* PERCENTAGE POINT) of a *t-distribution with ν *degrees of freedom. A 100(1−2θ)% confidence interval for the expected value of Y when $x = x_0$ is

$$(\hat{\alpha} + \hat{\beta}x_0) \pm t_{n-2}(\theta)\sqrt{\hat{\sigma}^2\left\{\frac{1}{n} + \frac{(x_0 - \bar{x})^2}{S_{xx}}\right\}}.$$

A 100(1−2θ)% *prediction interval for the value y_0 of Y when $x = x_0$ is

$$(\hat{\alpha} + \hat{\beta}x_0) \pm t_{n-2}(\theta)\sqrt{\hat{\sigma}^2\left\{1 + \frac{1}{n} + \frac{(x_0 - \bar{x})^2}{S_{xx}}\right\}}.$$

See also MULTIPLE REGRESSION MODEL; REGRESSION DIAGNOSTICS; REGRESSION THROUGH THE ORIGIN.

SEE WEB LINKS

- Applet.

line graph *See* BAR CHART.

line transect A path, usually straight, across a region of interest, that will be used for sampling. If the path has constant width then the sampling procedure may be to count the organisms in predefined regions of the path (for example, in a sequence of *quadrats). Alternatively, the sampler may walk along the line transect either counting visible organisms or recording their positions.

Lineweaver–Burk equation *See* GROWTH CURVE.

linkage Linkage is concerned with the tendency of genes that are situated close to one another within the same chromosome to be inherited from the same parent. For a pair of genes, the **recombination probability**, p, is defined as the *probability that the genes have been inherited from different parents. If p is much less than $\frac{1}{2}$ then the genes are said to display linkage. Low values of p occur when the genes are situated close to one another within the same chromosome: this fact helps in deciding where genes are located within a chromosome.

A measure of linkage is the **lod score**, Z. In the simple case where the *data consist of n independent *observations on a particular pair of genes, of which r are recombinants, the lod score is

$$Z = r\log_{10}r + (n-r)\log_{10}(n-r) + n\log_{10}(2/n).$$

A lod score of 3 or more is generally held to provide evidence of linkage.

link function *See* GENERALIZED LINEAR MODEL.

LISREL (Linear Structural Relations) A computer program for analysing structural equation models (*see* SIMULTANEOUS EQUATION MODEL).

Little's formula *See* QUEUE.

Ljung–Box test *See* BOX–LJUNG TEST.

***L*-moments** *Measures of location, spread, and other properties of *distributions that are computed from *linear combinations of the ordered *data values $x_{(1)} \le x_{(2)} \le \cdots \le x_{(n)}$ (with *mean $\bar{x}$, which is the first L-moment). Defining the quantities $b_1, b_2, \ldots, b_{n-1}$ by

$$b_r = \frac{1}{n}\sum_{j=r+1}^{n} \frac{(j-1)(j-2)\cdots(j-r)}{(n-1)(n-2)\cdots(n-r)} x_{(j)},$$

the next three L-moments are $2b_1 - \bar{x}$, $6b_2 - 6b_1 + \bar{x}$, and $20b_3 - 30b_2 + 12b_1 - \bar{x}$. The second L-moment is related to the *Gini statistic. An advantage of L-moments is that they can be calculated even when distributions have infinite variances.

ln *See* NATURAL LOGARITHM.

local maximum A high value surrounded only by lower values, but not necessarily the highest value possible (which is the **global maximum**). Thus the peak of Ben Nevis is a local maximum of height above sea level, whereas that of Mount Everest is both a local maximum and a global maximum.

lod score *See* LINKAGE.

loess (lowess) A computationally intensive method for fitting smooth curves or surfaces to a set of data. The procedure fits *polynomials (usually linear or quadratic) to local subsets of the data, using *weighted least squares so as to pay less attention to distant points. Loess requires no predetermined model for the entire data set, but therefore provides no explicit formula for the smoothed values.

logarithm The logarithm to base a (>0) of a positive number b is denoted by $\log_a b = x$, where $a^x = b$. Logarithms have the fundamental property $\log_a(bc) = \log_a b + \log_a c$. A logarithm to base 10 is usually denoted by log and a *natural logarithm, i.e. a logarithm to base e (*see* EXPONENTIAL), by ln. Before the advent of computers and calculators, logarithms were used in evaluating products.

logarithmic distribution (log-series distribution) A discrete distribution (*see* DISCRETE RANDOM VARIABLE) with one *parameter θ $(0<\theta<1)$, for which

$$P(X=x) = -\frac{\theta^x}{x\ln(1-\theta)}, \qquad x = 1, 2, \ldots,$$

where ln is the *natural logarithm. The distribution has *mean and *variance equal to

$$-\frac{\theta}{(1-\theta)\ln(1-\theta)} \quad \text{and} \quad -\frac{\theta}{(1-\theta)^2\ln(1-\theta)}\left(1+\frac{\theta}{\ln(1-\theta)}\right),$$

respectively. A simple *recurrence relation links successive probabilities

$$P(X=x) = \frac{(x-1)\theta}{x}P[X=(x-1)], \qquad x = 2, 3, \ldots.$$

The distribution is much in use in the context of species diversity. If, when sampling a mixed-species population, n species occur just once, then it is commonly found that, for some θ, the numbers of species occurring twice, thrice, ... are approximately $\frac{1}{2}n\theta$, $\frac{1}{3}n\theta^2$, ... The ratio n/θ is the **index of diversity**. The distribution is also suitable for modelling the numbers of items of a product bought by a buyer in a given time period.

logarithmic link *See* GENERALIZED LINEAR MODEL.

logistic curve *See* GROWTH CURVE.

logistic distribution A continuous *distribution with *probability density function f given by

$$f(x) = \frac{\exp\{-(x-\alpha)/\beta\}}{\beta[1+\exp\{-(x-\alpha)/\beta\}]^2}, \qquad -\infty < x < \infty,$$

where α is the *mean of the distribution and β is a positive *parameter. The distribution has *variance $\frac{1}{3}\beta^2\pi^2$. *See diagram opposite.*

logistic link *See* GENERALIZED LINEAR MODEL.

logistic regression model A *linear model in which the dependent variable (*see* REGRESSION) is a *logit and at least one explanatory variable is

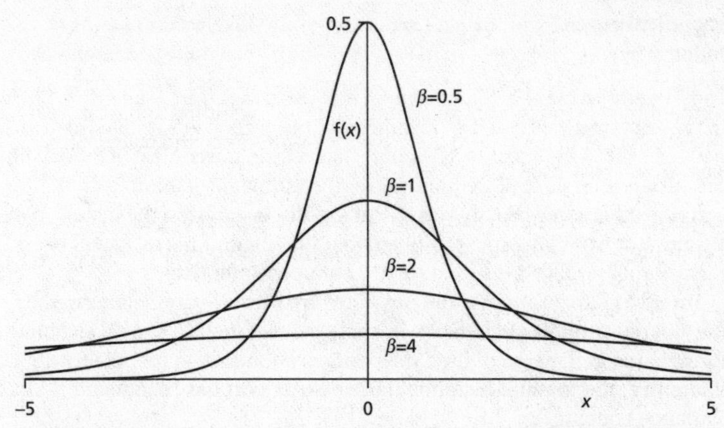

Logistic distribution. The distributions shown are for $\alpha = 0$. The variance of these distributions is $\frac{1}{3}\beta^2\pi^2$.

---

*continuous. The case where all the explanatory variables are *categorical is called a *logit model.

logistic transformation The use of the *logit, which takes values in the interval $(-\infty, \infty)$, in place of a *probability with values in the finite interval (0, 1).

logit (log-odds) The quantity

$$\ln\left(\frac{p}{1-p}\right),$$

where p is a *proportion or *probability. The term 'logit' was introduced by *Berkson in 1944. Modelling variations in proportions directly is hampered by the need to ensure that estimated probabilities lie in the interval (0, 1). Since corresponding values for the logit lie in the unrestricted interval $(-\infty, \infty)$, models for proportions are usually constructed in terms of logits. *See also* PROBIT.

logit link *See* GENERALIZED LINEAR MODEL.

logit model A *linear model in which the dependent variable is a *logit and the explanatory variables are *categorical. A logit model can often be re-expressed as a *log-linear model. If one or more of the explanatory variables is not categorical then the model is called a *logistic regression model.

log-likelihood The *natural logarithm of the *likelihood of a set of *data.

log-linear model A *linear model that describes the variation in the expected value of the *logarithm of the response variable (*see* REGRESSION). Log-linear models arise naturally in the context of modelling the variation in the cell frequencies of a *contingency table.

In the commonly used **hierarchical log-linear models**, the inclusion of a term corresponding to a multi-variable *interaction necessitates the inclusion of terms corresponding to all simpler interactions.

For all log-linear models the *expected frequencies are determined by the *marginal totals of the table. In some cases there is a simple algebraic formula (as in the case of the *independence model)—a model for which this is the case is called a *direct model. *See also* LOGIT MODEL.

lognormal distribution The *distribution of a *random variable X such that $\ln(X-\theta)$ has a *normal distribution, where θ is a *parameter. There are two other parameters, δ (>0) and γ. The *probability density function f is given by

$$\mathrm{f}(x)=\frac{\delta}{(x-\theta)\sqrt{2\pi}}\exp\left[-\frac{1}{2}\{\gamma+\delta\ln(x-\theta)\}^2\right],\qquad x>\theta.$$

If we write

$$\alpha=\exp\left(-\frac{\gamma}{\delta}\right)\qquad\text{and}\qquad\beta=\exp\left(\frac{1}{\delta^2}\right),$$

the distribution has *mean $(\theta+\alpha\sqrt{\beta})$ and *variance $\alpha^2\beta(\beta-1)$. The distribution has *mode at $\theta+\alpha/\beta$. *See* JOHNSON DISTRIBUTIONS.

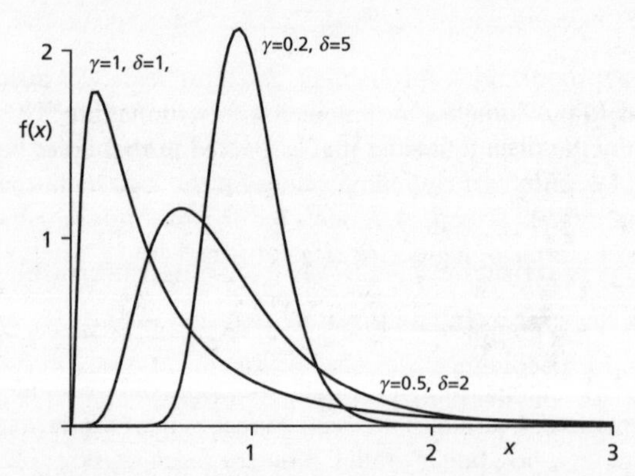

Lognormal distribution. The distributions illustrated are for $\theta=0$. They have mode at $\exp(-(1+\gamma\delta)/\delta^2)$.

log-odds *Alternative name for* LOGIT.

log-rank test A test for comparing two (or more) sets of survival times (*see* HAZARD RATE). The data consist of the completed lifetimes of individuals who have died and the current ages of those still living. The *null hypothesis is that the sets come from the same *population. Suppose there are m distinct survival times, $t_{(1)} < t_{(2)} < \cdots < t_{(m)}$. For the two-sample case the test statistic is g, given by

$$g = \sum_{j=1}^{m} \frac{n_{2j}d_{1j} - n_{1j}d_{2j}}{N_j},$$

where n_{kj} is the number of individuals in set k who are alive just before time $t_{(j)}$, d_{kj} is the number of these who die at that time, and $N_j = n_{1j} + n_{2j}$.

The distribution of g can be approximated by a *normal distribution with, under the null hypothesis, *mean 0 and *variance V, where

$$V = \sum_{j=1}^{m} \frac{n_{1j}n_{2j}D_j(N_j - D_j)}{N_j(N_j - 1)},$$

and $D_j = d_{1j} + d_{2j}$.

log-series distribution *See* LOGARITHMIC DISTRIBUTION.

longitudinal data *Data collected at a sequence of time points for each of a *sample of individuals. Because each individual contributes several *observations, these observations are usually not *independent and this has to be taken into account in the analysis.

longitudinal study A study, usually in the context of medicine or the social sciences, conducted over time. The *data that result are called *longitudinal data.

long memory A *time series is said to have a long memory if values from the distant past have an appreciable effect on present values. Financial and communications networks data provide examples.

long-tailed distribution *See* HEAVY-TAILED DISTRIBUTION.

Lorenz curve A graphical representation of social inequality introduced by the American economist Max Otto Lorenz (1876–1959) in 1905 in connection with the distribution of wealth in a population. Let w be the total income of those members of a population whose income is at most v, and let W be the total income of the population. In the Lorenz curve, w/W

is plotted against the (cumulative) proportion of the population that has income at most v. So, if f is the probability density function of income then the curve is the plot of

$$\frac{\int_0^v u\mathrm{f}(u)\mathrm{d}u}{\int_0^\infty u\mathrm{f}(u)\mathrm{d}u} \quad \text{against} \quad \int_0^v \mathrm{f}(u)\mathrm{d}u.$$

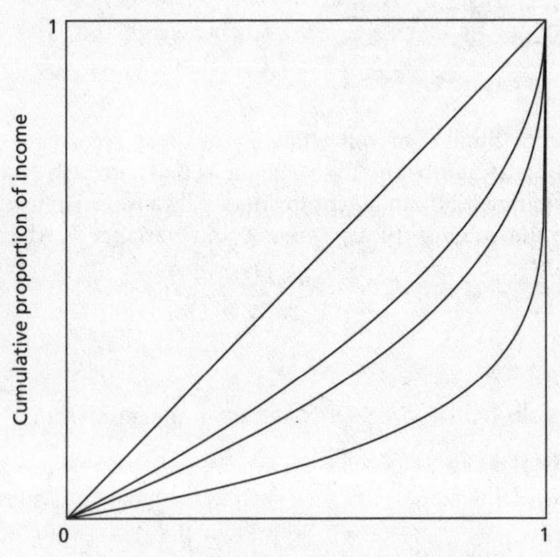

Lorenz curve. The curves shown result from *Pareto distributions of income with parameter k equal to one and parameter a taking the values 1, 1.2, 1.6, and 2.

The ratio of the area between the Lorenz curve and the 45° line to the total area below that line is the **Gini index** (also called the **Gini coefficient**), G.

Suppose a sample of n individuals have incomes $x_{(1)} \leq x_{(2)} \leq \cdots \leq x_{(n)}$. The Lorenz curve is approximated by the polygon joining the origin to the successive points with coordinates

$$\left(\frac{r}{n}, \frac{\sum_{j=1}^{r} x_{(j)}}{\sum_{j=1}^{n} x_{(j)}}\right) \qquad r = 1, 2, \ldots, n.$$

If the corresponding income proportions are $y_{(1)} \leq y_{(2)} \leq \cdots \leq y_{(n)}$ then G can be calculated using

$$G = 1 - \frac{1}{n+1} - \frac{2}{n+1}\sum_{j=1}^{n} y_{(j)}.$$

loss function The function that is minimized in the process of fitting a *model. An example is the residual sum of squares (*see* ANOVA).

lot *Alternative term for* BATCH.

lot tolerance percent defective (LTPD) *See* ACCEPTANCE SAMPLING.

lower percentage point *See* PERCENTAGE POINT.

lower quartile *See* QUARTILE.

lower-tail probability The probability of a *random variable X taking the value x or less, where, usually, x will be appreciably less than the *mean of X.

lower triangular matrix *See* MATRIX.

lowess *See* LOESS.

low-pass filter *See* FILTER.

lowspread The difference between the sample *median and the minimum value in a set of *data. *See also* HIGHSPREAD; RANGE.

loyalty model *See* SQUARE TABLE.

LSD Abbreviation for least significant difference. *See* MULTIPLE COMPARISON TEST.

LTPD Abbreviation for lot tolerance percent defective. *See* ACCEPTANCE SAMPLING.

machine learning An interdisciplinary field concerned with the interpretation of *data. For example, the sounds of breaking glass and running feet near a greenhouse might be interpreted as vandalism. With **supervised learning** the interpretation is informed by a training sample (*see* NEURAL NET) and generally takes the form of **ensemble learning** wherein the final outcome is reached algorithmically using information from many different models of the data. Examples are *bagging, *boosting, and the *random forest.

In **unsupervised learning** the aim is to find underlying patterns in the data using techniques such as *cluster analysis, the *EM algorithm, *factor analysis, *graphical models, *hidden Markov models, *multiple regression models, and *principal components analysis. *See also* CLASSIFICATION TREE.

A key feature is the intention that, whichever approach is used, the accuracy of the computer algorithm will improve over time (the 'machine' will 'learn') as a result of feedback concerning previous accuracy.

Maclaurin, Colin (1698–1746; b. Kilmodan, Scotland; d. Edinburgh, Scotland) Scottish mathematician. Orphaned by the age of nine, Maclaurin was a student at Glasgow U when aged eleven. At the age of nineteen he was appointed Professor of Mathematics at Aberdeen U. In 1719 he was elected FRS. In 1725 he was appointed Professor at Edinburgh U, where he remained for the rest of his life. A lunar crater is named after him.

SEE WEB LINKS

- Fuller biography and portrait.

Maclaurin expansion *See* MACLAURIN SERIES.

Maclaurin series (Maclaurin expansion) A series representation for a function f having continuous derivatives of all orders. The series is

$$f(0) + f'(0)\frac{x}{1!} + f''(0)\frac{x^2}{2!} + \cdots = \sum_{r=0}^{\infty} f^{(r)}(0)\frac{x^r}{r!},$$

where $f^{(r)}(0)$ means $d^r f(x)/dx^r$ evaluated at $x = 0$. The series may converge for all values of x, or for $|x| < R$, for some positive R, or it may only converge when $x = 0$. If $f(x)$ is a *polynomial then the series is finite and the sum is f(x). For most practical cases, if the series converges then its sum is f(x)—this is true for e^x and sin x, when the series is convergent for all x, and for $(1+x)^{1/2}$ and $\ln(1+x)$, when the series is convergent for $|x|<1$. In such cases the first $(n+1)$ terms give a polynomial approximation g(x) to f(x), which has the property that at $x=0$, the first n derivatives of g(x) and f(x) are equal. There are however non-zero functions f such that f(x) and all its derivatives are zero at $x=0$, and in this case the Maclaurin series vanishes—for example, $f(x)=\exp(-1/x^2)$ for $x\neq 0$, with $f(0)=0$.

The Maclaurin series is of use in the theoretical development of *moment-generating functions and *probability-generating functions. It is the particular case of a *Taylor series when $a=0$.

MAD *See* MEAN ABSOLUTE DEVIATION; MEDIAN ABSOLUTE DEVIATION.

MAE *Abbreviation for* MEAN ABSOLUTE ERROR.

Mahalanobis, Prasanta Chandra (1893–1972; b. Calcutta, India; d. Calcutta, India) Indian statistician who was the founder (in 1931) of the *Indian Statistical Institute. Mahalanobis studied physics at Calcutta U and Cambridge U. In the library of King's College, Cambridge, Mahalanobis was asked by a fellow student to express his opinion about an article in **Biometrika* It was this encounter that stimulated his subsequent research in Statistics. Practical problems led to theoretical advances in many areas of Statistics—it was an anthropometric problem that led to his construction of the eponymous *Mahalanobis distance, D^2. As statistical adviser to the Indian government, Mahalanobis established, in 1950, the first national survey of India's population. He was elected FRS in 1945.

SEE WEB LINKS

- Website and links.

Mahalanobis distance (D^2) The two-sample version of *Hotelling T^2. The quantity D^2, introduced by *Mahalanobis in a 1930 paper, may be considered to be a measure of the distance between two groups of p-dimensional observations. Let $\overline{\mathbf{x}}_j$ be the $p\times 1$ vector of means for group j ($j=1,2$), with $\mathbf{S}_j$ the corresponding *variance-covariance matrix. D^2 is given by

$$D^2=(\overline{\mathbf{x}}_1-\overline{\mathbf{x}}_2)'\mathbf{S}^{-1}(\overline{\mathbf{x}}_1-\overline{\mathbf{x}}_2),$$

where $\mathbf{S}$ is given by

$$\mathbf{S}=\frac{n_1\mathbf{S}_1+n_2\mathbf{S}_2}{n_1+n_2},$$

and n_1 and n_2 are the sample sizes.

main diagonal *See* MATRIX.

main effect *See* FACTORIAL DESIGN.

m

Mallows, Colin Lingwood (1930- ; b. Great Sampford, England) English statistician who has spent most of his career in the USA. Mallows was educated at UCL, obtaining his PhD (supervised by *Johnson) in 1953, and joining the UCL faculty in 1955. In 1960 he moved to the USA, joining the faculty at Princeton U at the invitation of *Tukey. His later research took place at AT&T Labs (1987) and Avaya Labs (2000). He was the *COPSS *Fisher Lecturer in 1997 and the *Wilks Award winner of the *ASA in 2007.

SEE WEB LINKS

- Personal website.

Mallows C_p *See* MODEL SELECTION PROCEDURE.

MA model *See* MOVING AVERAGE MODEL.

Mandelbrot, Benoît B. (1924–2010; b. Warsaw, Poland; d. Cambridge, MA) Mandelbrot was a Lithuanian Jew whose family emigrated to France. He studied first at the École Polytechnique in Paris and then at the Faculté des Sciences de Paris, where, in 1952, he obtained his PhD (supervised by *Lévy). In 1958 he joined the IBM research centre in New York State. In 1975 he coined the term *fractal. He was an advocate of the use of *heavy-tailed distributions. From 1987 to 2005 he was on the faculty of Yale U. He was made a fellow of the AAAS in 1982. He was made a Chevalier, L'Ordre National de la Légion d'Honneur in 1989. His second initial, B, was his own addition and was not an abbreviation.

SEE WEB LINKS

- University website with links.

Manhattan distance *See* DISTANCE MEASURE.

manifest variable A *variable that is directly measurable. *Compare* LATENT VARIABLE.

Mann, Henry Berthold (1905–2000; b. Vienna, Austria; d. Tucson, AZ) Austrian mathematician who spent most of his career in the USA. Mann obtained his PhD in number theory at Vienna U in 1935. He emigrated to New York in 1938 and, after a variety of temporary positions, arrived at Ohio State U in 1946, where he remained until 1964. One of his first research students was *Whitney with whom his name is now linked as joint originator of the *Mann–Whitney test. In 1964 he moved to U Wisconsin, and subsequently to U Arizona. His many publications show clearly his love of analytic number theory and its applications to *experimental design.

SEE WEB LINKS

- Appreciation and photograph.

Mann–Whitney test A two-sample *non-parametric test, equivalent to the **Wilcoxon rank-sum test**, introduced in 1947 by *Mann and *Whitney. It is assumed that the *samples are random and come from *populations (*random variables X and Y) that have the same *distribution after a translation of size k:

$$\mathrm{P}(X<x)=\mathrm{P}(Y<x+k), \qquad \text{for all values of } x.$$

The *null hypothesis is that the random variables have the same distribution (i.e. $k=0$). *See also* TEST FOR EQUALITY OF LOCATION.

With samples of sizes m and n ($m \le n$), the first stage is the replacement of the $(m+n)$ observed values by their *ranks in the combined sample. For the (smaller) sample of size m, denote the sum of the ranks by R. The distribution of R is approximately

$$\mathrm{N}\left(\frac{1}{2}m(m+n+1),\ \frac{1}{12}mn(m+n+1)\right),$$

so that the test statistic, z, is given by

$$z=\frac{R-\frac{1}{2}m(m+n+1)\pm\frac{1}{2}}{\sqrt{\frac{1}{12}mn(m+n+1)}},$$

where the $\pm\frac{1}{2}$ is a *continuity correction with sign chosen so as to reduce the absolute magnitude of the numerator.

For example, suppose that the marks obtained by a small random sample of statistics students were as follows:

Boys	10, 22, 42, 59, 61, 63, 65, 83, 85, 90, 93
Girls	36, 53, 54, 56, 69, 84, 88.

The question of interest is whether the data support the null hypothesis of a common mark distribution. The ranks are

Boys	1, 2, 4, 8, 9, 10, 11, 13, 15, 17, 18
Girls	3, 5, 6, 7, 12, 14, 16.

Working with the (smaller) set of girls, m is 7 and R is $3+5+\cdots+16=63$. Using $n=11$, the test statistic, z, is given by

$$z=\frac{(63-66.5+0.5)}{\sqrt{121.9}}=-0.272.$$

Since $|z|<1.96$, we accept, at the 5% significance level, the hypothesis that the two sets of marks have come from the same distribution.

MANOVA *Abbreviation for* MULTIVARIATE ANALYSIS OF VARIANCE.

Mantel, Nathan (1919–2002; b. New York City; d. Potomac, MD) American statistician. Mantel graduated from the City College (now City U) of New York in 1939. From 1947 to 1974 he worked at the National Cancer Institute, where *Haenszel was a colleague. Mantel was co-author (with Haenszel) of the 1959 paper which introduced the eponymous *Mantel–Haenszel test of *independence. He held research professorships at GWU (1974–82) and the American U in Washington (1982–2002). He was made an Honorary Fellow of the *RSS in 1992.

SEE WEB LINKS

- Biography, interview, and photographs.

Mantel–Haenszel statistic *See* MANTEL–HAENSZEL TEST.

Mantel–Haenszel test A test suitable for testing the *null hypothesis of *independence between two dichotomous variables (*see* CATEGORICAL VARIABLE) using *data from a *population subdivided into L classes: it is, therefore, a test for use with a $2 \times 2 \times L$ *contingency table. The test, which was introduced in 1959 by *Mantel and *Haenszel, is most used in medical contexts where, for example, one variable is outcome ('success' or 'failure'), one variable is treatment ('control' or 'new'), and the L classes correspond to different patient categories. The test assumes that any *association between the dichotomous variables is unaffected by the third variable.

m

The test statistic, M, is computed as follows. Denote by f_{jkl} the number of patients in class l $(=1, 2, \ldots, L)$ who experience outcome j $(=1$ or $2)$ when given treatment k $(=1$ or $2)$. Write $f_{0kl} = f_{1kl} + f_{2kl}$, $f_{j0l} = f_{j1l} + f_{j2l}$, and $f_{00l} = f_{10l} + f_{20l}$. Then

$$M = \left\{ \left| \sum_{l=1}^{L} \left(f_{11l} - \frac{f_{10l}\, f_{01l}}{f_{00l}} \right) \right| - \frac{1}{2} \right\}^2 \Bigg/ \sum_{l=1}^{L} \frac{f_{10l}\, f_{20l}\, f_{01l}\, f_{02l}}{f_{00l}^2 (f_{00l} - 1)}.$$

Under the null hypothesis, the distribution of M is approximated by a *chi-squared distribution with one *degree of freedom. When $L = 1$ the test is equivalent to the Yates-corrected chi-squared test (*see* TWO-BY-TWO TABLE). The $\frac{1}{2}$ term is a *continuity correction.

The test combines information from each of the L classes. In a similar way, the **Mantel–Haenszel statistic**, ψ, combines information about the strength of the relationship between the dichotomous variables. This statistic, given by

$$\psi = \sum_{l=1}^{L} \frac{f_{11l}\, f_{22l}}{f_{00l}} \Bigg/ \sum_{l=1}^{L} \frac{f_{12l}\, f_{21l}}{f_{00l}},$$

is an aggregate estimate of the odds ratio (*see* TWO-BY-TWO TABLE) for the two variables.

MAPE *See* MEAN ABSOLUTE ERROR.

MAR *See* MISSING VALUES.

marginal distribution *See* BIVARIATE DISTRIBUTION.

marginal homogeneity The property possessed by a square *contingency table, in which, for all *j*, the total for row *j* is equal to the total for column *j*. For example, the rows might refer to a set of possible birthplaces and the columns to a set of current locations, using the same location categories. Marginal homogeneity would imply that, for all *j*, the *proportion born in location *j* is the same as the proportion living in location *j*.

marginal totals If the cell frequencies of a (multidimensional) *contingency table are totalled over one or more of the categorizing variables the result is a set of marginal totals. For a two-dimensional table, the marginal totals are the row and column totals.

margin of error A 'margin of error' is often reported with opinion polls using a phrase such as 'the proportion intending to vote for the Red party is 35%, with a margin of error of 3%.' This actually means that on an average of 19 times out of 20, the sample proportion obtained will be within 3 percentage points of the true value of the population proportion; it is therefore defining a 95% *confidence interval for the population proportion, with the margin of error amounting to half the width of that interval.

Markov, Andrei Andreevich (1856–1922; b. Ryazan, Russia; d. St Petersburg, Russia) Russian mathematician. Markov was educated at St Petersburg U (where one teacher was *Chebyshev). On graduation, he joined the faculty and specialized in *probability. Markov's outstanding contribution was the idea underlying *Markov processes. He is also remembered for his contribution to the mathematics of linear models through the Gauss–Markov theorem (*see* MULTIPLE REGRESSION MODEL). A lunar crater is named after him.

SEE WEB LINKS

- Website and links.

Markov chain *See* MARKOV PROCESS.

Markov chain Monte Carlo (MCMC) methods Methods, making heavy use of computers, that are very useful in the analysis of large arrays of *observations of *correlated variables, such as satellite images of the Earth. Suppose we wish to calculate θ, the *expected value of a function h of r correlated *random variables. Let $\mathbf{X}$ be the $r \times 1$ *vector of random variables. A standard *Monte Carlo method would generate a sequence of numerical values, $\mathbf{X}_1, \mathbf{X}_2, \ldots, \mathbf{X}_n$, from which θ could be estimated by $\hat{\theta}$, given by

$$\hat{\theta} = \frac{1}{n}\sum_{j=1}^{n} \mathrm{h}(\mathbf{X}_j).$$

However, it can be difficult to generate a vector $\mathbf{X}_j$ when the joint distribution of the r variables involves a complex correlation structure—for example, when the variables refer to the values displayed in an array of neighbouring pixels in a satellite image of the Earth's surface.

Let X_{jk} be the kth random variable in $\mathbf{X}_j$. MCMC methods aim to generate r Markov chains (*see* MARKOV PROCESS), where the kth chain is the sequence of values $x_{1k}, x_{2k}, \ldots$. An arbitrary set of values is chosen for the first chain, subsequent chains being generated from their predecessors by Monte Carlo methods. In calculating $\hat{\theta}$ it is necessary to ignore the early chains in the sequence, since these will be affected by the initial choice of values. This is called the **burn-in period**.

A popular method for obtaining the required Markov chains, which need to be stationary (*see* MARKOV PROCESS), is the **Metropolis–Hastings algorithm**. Let the stationary probability for the random variable X_j being in state m be p_m and let $\mathbf{Q}$ be a known matrix with non-negative elements. The transition probability matrix governing the sequence $X_1, X_2, \ldots$ is defined by

$$\begin{aligned} p_{jk} &= c_j q_{jk} a_{jk}, \qquad \text{if } j \neq k, \\ p_{jj} &= c_j q_{jj} + c_j \sum_{m \neq j} q_{jm}(1 - a_{jm}), \end{aligned}$$

where a_{jk} is given by

$$a_{jk} = \min\left\{\frac{p_k q_{kj}}{p_j q_{jk}},\ 1\right\},$$

and each c_j is chosen so that $\sum_k p_{jk} = 1$.

The most used version of the Metropolis–Hastings algorithm incorporates the **Gibbs sampler**. The aim is to generate a random vector $\mathbf{X}$ with elements satisfying a specified relationship. Denote the initial values in $\mathbf{X}$ by $x_1^{(0)}, x_2^{(0)}, \ldots$. A single element of the vector (element j, say) is chosen at random, and a potential new value, x, is generated by random selection from the *conditional distribution of X_j given the values of the remaining variables, $\{X_k, k \neq j\}$. If the new value is in accord with the specified relationship, then it replaces the previous value so that $x_j^{(1)} = x$; otherwise the previous value is retained: $x_j^{(1)} = x_j^{(0)}$. This process is repeated until approximate *equilibrium has been reached.

The procedure was originally proposed in the context of statistical mechanics by *Metropolis and others in 1953, and was introduced into Statistics in 1970 by *Hastings.

Markov inequality Let X be a *random variable for which $|X|$ has *expected value v. Then, for any positive constant a,

$$\mathrm{P}(|X|) \geq a) \leq \frac{v}{a}.$$

See also BERNSTEIN INEQUALITY; CHEBYSHEV INEQUALITY; HÖLDER INEQUALITY; KOLMOGOROV INEQUALITY; MINKOWSKI INEQUALITY.

Markov process A *stochastic process, $X_1, X_2, \ldots$, for which the value taken by the *random variable X_m is, for all $m > 2$, *independent of X_1, $X_2, \ldots, X_{m-2}$ but may depend on X_{m-1}. The variable X_m may be described as exhibiting the **Markov property**. If there is a common finite (or *countable) number of possible outcomes for all the X-values then the process is a **Markov chain**. The term 'chain' was used by *Markov in his original 1906 description.

In a **discrete-time Markov chain** the random variables $X_1, X_2, \ldots$ are ordered in time. If $X_m = r$ then the process is said to be in **state** r at time point m. The Markov property implies that the state occupied at time point m is dependent on the state occupied at time point $(m-1)$ but not on the state occupied at any previous time point.

The probability, p_{jk}, that a chain in state j at any time point will be in state k at the next time point is called the **transition probability** and is assumed to be unchanging over time. The square *matrix, $\mathbf{P}$, in which p_{jk} is the element in the jth row and kth column, is the **transition matrix**. The elements on each row of $\mathbf{P}$ sum to 1.

If there is a state that, once entered, can never be left, then this is called an **absorbing state** (or **absorbing barrier** or **trapping state**). Such a state will be indicated by a 1 in the leading diagonal of $\mathbf{P}$.

Denote by $p_{jk}^{(n)}$ the probability that a chain in state j at any time point will be in state k at n time points later (so that $p_{jk} = p_{jk}^{(1)}$). If $p_{jk}^{(n)} > 0$, for some n, j, and k, then state k is **accessible** from state j and the two states are **communicating states**. A Markov chain is **irreducible** if all states communicate with each other.

A state is **transient** if, starting from that state, the probability of ever returning to it is < 1; if the probability is equal to 1 then the state is a **recurrent state** (or **persistent state**). A recurrent state with a finite expected time until return is called an **ergodic state**.

In a **stationary chain** $\mathrm{P}(X_m = k)$ is independent of m for all k. Denoting by p_k the probability of being in state k in a stationary chain, this implies that

$$p_k = \sum_j p_j p_{jk},$$

for all k.

In a **time-reversible stationary chain**, the probability of observing a move from state j at time m to state k at time $(m+1)$ is equal to the probability of observing a move from state k at time m to state j at time $(m+1)$:

$$p_j p_{jk} = p_k p_{kj},$$

for all j and k.

If s refers to a past time, t to the present time, and u to a future time, then, in a **continuous-time Markov chain**, the probability of being in state l at time u, given that the process is in state k at time t, is independent of the state occupied at time s. If the transition probabilities depend only on the time lag $(u - t)$ rather than on u or t, then the process is a **time-homogeneous Markov chain**. Examples of continuous-time Markov chains include *birth-and-death processes, *Brownian motion, *queues, *random walks, and *renewal processes.

Let $P_{jk}(t)$ denote the probability that a time-homogeneous Markov chain in state j at time 0 will be in state k at time t. The **Chapman-Kolmogorov equation** states that

$$P_{jk}(t) = \sum_l P_{jl}(s) P_{lk}(t - s),$$

m

for all j and k and for any s in the interval $(0, t)$.

Markov property *See* MARKOV PROCESS.

Markov random field A characterization of a *spatial process in which the value observed at a location depends only on the values at the immediately neighbouring locations. The form of the dependence is the same for all locations. The *probability distributions for which a Markov random field can exist are specified by the *Hammersley–Clifford theorem. A Markov random field is an undirected **graphical model**.

Marshall–Edgeworth index *See* PRICE INDEX.

martingale The sequence of *random variables $X_1, X_2, \ldots$ (with finite expected values) is a martingale if the conditional expected value of X_{n+1}, given the values of $X_1, X_2, \ldots, X_n$, is given by

$$\mathrm{E}(X_{n+1} | X_1 = x_1, X_2 = x_2, \ldots, X_n = x_n) = x_n.$$

An example is provided by letting X_n be the amount of a gambler's winnings after the nth of a sequence of fair games.

matched pairs Pairs of experimental units chosen to resemble each other as closely as possible. Alternatively, the phrase may refer to single experimental units measured at two points in time, or in two different situations. *See also* EXPERIMENTAL DESIGN; HYPOTHESIS TEST; WILCOXON SIGNED-RANK TESTS.

matching coefficient *See* DISSIMILARITY.

matching problem A typical matching problem is as follows: 'There are n stockbrokers, each wearing his own bowler hat. On arrival at work, each stockbroker leaves his hat in the cloakroom. On leaving work, however, each stockbroker is given a randomly chosen hat by the cloakroom attendant. What is the *probability that at least one stockbroker is given his own hat?' Using the *inclusion-exclusion principle, the answer is

$$1-\frac{1}{2!}+\frac{1}{3!}-\frac{1}{4!}+\cdots+\frac{(-1)^n}{n!}\approx 1-\frac{1}{\mathrm{e}}=0.632 \text{ (to 3 dp)}.$$

Convergence is so quick ($n=6$ gives 0.632 to 3 dp) that the answer is almost unaffected by the number of stockbrokers. *See also* SECRETARY PROBLEM.

Matheron, Georges Francois Paul Marie (1930–2000) French geologist and statistician. Matheron was a graduate of the École Polytechnique and the École des Mines de Paris, where his PhD was supervised by *Lévy. From 1954 to 1963 he worked for the French Geological Survey in Algeria and France, and then, after a period at the Nancy School of Mines, he returned in 1968 to the École des Mines. By 1986 this had been reorganized and Matheron was Director of the Centre de Géostatistique. In 1962, in his *Traité de Géostatistique Appliquée,* Matheron set out the fundamentals of *geostatistics, including the optimal spatial interpolation procedure that he called *kriging. He was created an Honorary Fellow of the *RSS in 1988.

SEE WEB LINKS

- Appreciation and photograph.

matrices *See* MATRIX.

matrix An $r \times c$ **matrix** consists of a rectangular array with r rows and c columns, in which the elements are either numbers or algebraic expressions. Example **matrices** (the plural form) are:

$$\begin{pmatrix} 5 & 3 & 2 \\ 1 & 0 & 4 \\ 2 & -1 & 0 \\ 3 & 1 & 1 \end{pmatrix} \qquad \begin{pmatrix} x_{11} & x_{12} & x_{13} & x_{14} \\ x_{21} & x_{22} & x_{23} & x_{24} \end{pmatrix}.$$

When the array is not written out in full, a matrix is usually denoted by a bold-face capital letter, e.g. **X**, or by a typical **element** (or **entry**) from the array, shown in curly brackets, e.g. $\{x_{jk}\}$, where x_{jk} is the element in the jth row and kth column of the matrix. If $r=c$ the matrix is **square**.

If a matrix $\mathbf{X}=\{x_{jk}\}$ is multiplied by the real number s, then the result is the matrix $s\mathbf{X}$, in which the element in the jth row and kth column is sx_{jk}. In this context a real number s is often referred to as a ***scalar**.

Two matrices, **A** and **B**, can be multiplied together only if the number of columns of one matrix is equal to the number of rows of the other matrix. If **A** is an $m \times n$ matrix and **B** is an $n \times p$ matrix then the product **AB** is an $m \times p$ matrix. However, if $p \neq m$ then the product **BA** does not exist. The rule for the construction of the product is as follows. Let e_{jk} denote the element in the jth row and kth column of the product **AB**, with a_{jk} and b_{jk} denoting typical elements in **A** and **B**. Then e_{jk} is given by

$$e_{jk} = \sum_{l=1}^{n} a_{jl} b_{lk}.$$

If **A** and **B** have the same values of r and c and if $a_{jk}=b_{jk}$ for all j and k, then $\mathbf{A}=\mathbf{B}$.

A **diagonal matrix** is a square matrix with all elements equal to 0, except for those on the **leading diagonal** (which runs from top-left to bottom-right). This diagonal is also called the **main diagonal**. A matrix (not necessarily square) in which all the entries are equal on every negatively sloping diagonal is a **Toeplitz matrix**. For example:

$$\begin{pmatrix} 3 & 1 & 4 & 5 & 8 & 0 & 6 \\ 2 & 3 & 1 & 4 & 5 & 8 & 0 \\ 8 & 2 & 3 & 1 & 4 & 5 & 8 \end{pmatrix}.$$

An **identity matrix**, usually denoted by **I**, is a diagonal matrix with all leading diagonal elements equal to 1. The size of an identity matrix may be indicated using a suffix: $\mathbf{I}_n$ is an $n \times n$ identity matrix.

The **transpose** of an $m \times n$ matrix **M** is the $n \times m$ matrix formed by interchanging the elements of the rows and columns of **M**. It is denoted by $\mathbf{M}'$. The jth row of $\mathbf{M}'$ is the transpose of the jth column of **M** and vice versa

$$\mathbf{M} = \begin{pmatrix} 5 & 3 & 2 \\ 1 & 0 & 4 \end{pmatrix}, \quad \mathbf{M}' = \begin{pmatrix} 5 & 1 \\ 3 & 0 \\ 2 & 4 \end{pmatrix}.$$

If a square matrix **S**, with typical element s_{jk}, is equal to its transpose, $\mathbf{S}'$, then it is a **symmetric matrix** satisfying

$$s_{jk} = s_{kj}, \quad \text{for all } j, k.$$

A square matrix that is not symmetric is an **asymmetric matrix**. If a square matrix **S** satisfies the equation $\mathbf{SS}=\mathbf{S}$ then it is **idempotent**. The product **SS** may be written as $\mathbf{S}^2$. If it exists, the **inverse** of a square matrix, **S**, is denoted by $\mathbf{S}^{-1}$. It satisfies the relations that

$$\mathbf{S}\mathbf{S}^{-1} = \mathbf{S}^{-1}\mathbf{S} = \mathbf{I}.$$

Only square matrices can have an inverse (but see 'generalized inverse' below). If $\mathbf{S}^{-1}$ exists then it will be the same size as **S**. A matrix that has an inverse is said to be **non-singular** (or **regular**, or **invertible**). A square matrix without an inverse is said to be **singular**.

A square matrix is described as being an **upper triangular matrix** if all the elements below the leading diagonal are zero, or as a **lower triangular matrix** if all the elements above the leading diagonal are zero. The matrices **U** and **L** are examples:

$$\mathbf{U} = \begin{pmatrix} 1 & 2 & 3 & 4 \\ 0 & 5 & 6 & 0 \\ 0 & 0 & 8 & 9 \\ 0 & 0 & 0 & 10 \end{pmatrix} \quad \mathbf{L} = \begin{pmatrix} 1 & 0 & 0 & 0 \\ 2 & 3 & 0 & 0 \\ 4 & 0 & 0 & 0 \\ 7 & 8 & 9 & 10 \end{pmatrix}.$$

A **generalized inverse** (also called a **Moore–Penrose inverse**) of the $m \times n$ matrix **M** is any $n \times m$ matrix $\mathbf{M}^-$ satisfying

$$\mathbf{M}\mathbf{M}^-\mathbf{M} = \mathbf{M}.$$

If a matrix **M** is multiplied by its transpose (to give either $\mathbf{M}\mathbf{M}'$ or $\mathbf{M}'\mathbf{M}$) then the result is a symmetric matrix.

If **M** is square and the product $\mathbf{M}\mathbf{M}'$ is an identity matrix, then $\mathbf{M}' = \mathbf{M}^{-1}$ and **M** is said to be an **orthogonal matrix**.

A matrix with just one row is called a **row vector**. A matrix with just one column is called a **column vector**. Column vectors are usually denoted with a bold-face lower-case letter, e.g. **x**; row vectors are written as their transpose, e.g. $\mathbf{x}'$. A vector with a single element (i.e. a 1×1 matrix) is a scalar.

Vectors multiply together in the same way as matrices (see above). Thus, if **v** is an $n \times 1$ column vector, and $\mathbf{v}'$ is its transpose, then the product $\mathbf{v}\mathbf{v}'$ is an $n \times n$ symmetric matrix, and the product $\mathbf{v}'\mathbf{v}$ is a scalar.

The set of $n \times 1$ vectors $\mathbf{v}_1, \mathbf{v}_2, \ldots, \mathbf{v}_m$ is **linearly independent** if the only values of the scalars $a_1, a_2, \ldots, a_m$ for which

$$\sum_{j=1}^{m} a_j \mathbf{v}_j = \mathbf{0},$$

where **0** is an $n \times 1$ vector with every element equal to 0, is $a_1 = a_2 = \cdots = a_m = 0$. If the set is not linearly independent then it is **linearly dependent**, in which case there are values for the scalars a_1, $a_2, \ldots, a_m$, not all equal to 0, such that $\sum_{j=1}^{m} a_j \mathbf{v}_j = \mathbf{0}$. If a set of two or more vectors is linearly dependent then at least one of the vectors, $\mathbf{v}_k$, say, is a **linear combination** of the others, i.e.

$$\mathbf{v}_k = \sum_{j \neq k} b_j \mathbf{v}_j,$$

for some scalars $b_1, b_2, \ldots, b_{k-1}, b_{k+1}, \ldots, b_m$.

The **rank** of a matrix is the maximum number of linearly independent rows, which is the same as the maximum number of linearly independent columns. Thus the rank of a matrix is equal to that of its transpose. If a matrix has r rows and c columns, with $r \leq c$, then the rank is $\leq r$; if $r > c$ then the rank is $\leq c$. If the rank is equal to the smaller of r and c then the matrix is of **full rank.**

If **A** is a square matrix, **x** is a column vector not equal to **0**, and λ is a scalar such that

$$\mathbf{Ax} = \lambda \mathbf{x},$$

then **x** is an **eigenvector** of **A** and λ is the corresponding **eigenvalue**. Eigenvectors and eigenvalues are also referred to as **characteristic vectors** and **characteristic values**. If **x** is the column vector $(x_1\ x_2 \ldots x_n)'$ and **A** is an $n \times n$ symmetric matrix with typical element a_{jk}, then the product $\mathbf{x}'\mathbf{Ax}$, which is a scalar, is described as a **quadratic form** because it is equal to

m

$$\sum_{j=1}^{n} \sum_{k=1}^{n} a_{jk} x_j x_k,$$

which is a linear combination of all the squared terms (such as x_1^2) and cross-products (such as $x_1 x_2$).

A symmetric matrix **A** is a **positive definite matrix** if $\mathbf{x}'\mathbf{Ax} > 0$ for all non-zero **x**; it is a **positive semi-definite matrix** if $\mathbf{x}'\mathbf{Ax} \geq 0$ for all **x** and there is at least one non-zero **x** for which $\mathbf{x}'\mathbf{Ax} = 0$.

The **trace** of a square matrix is the sum of the terms on the leading diagonal.

The **determinant** of a 2×2 square matrix, **A**, is written as $|\mathbf{A}|$ or det(**A**), and is given by

$$|\mathbf{A}| = a_{11}a_{22} - a_{12}a_{21}.$$

The determinant of a larger matrix is defined recursively in terms of **cofactors**. The cofactor A_{jk} of the entry a_{jk} is equal to the product of $(-1)^{j+k}$ and the determinant of the matrix obtained by eliminating the jth row and kth column of **A**. The recursive definition is $|\mathbf{A}| = \sum_{j=1}^{n} a_{j1} A_{j1}$. In fact $\sum_{j=1}^{n} a_{jk} A_{jl} = |\mathbf{A}|$ if $k = l$ (otherwise the sum is 0). Similarly, $\sum_{k=1}^{n} a_{jk} A_{lk} = |\mathbf{A}|$ if $j = l$ and is otherwise 0. Thus, for a 3×3 matrix, **A**,

$$|\mathbf{A}| = a_{11}(a_{22}a_{33} - a_{23}a_{32}) - a_{12}(a_{21}a_{33} - a_{23}a_{31}) + a_{13}(a_{21}a_{32} - a_{22}a_{31}).$$

The eigenvalues of a square matrix **A** are the roots of the **characteristic equation**

$$\det(\mathbf{A} - \lambda\mathbf{I}) = 0.$$

Mauchly test *See* SPHERICITY TEST.

maximal information coefficient (MIC) A computationally intensive measure of the strength of the relationship between two *random variables. The measure, proposed in 2011, makes use of the mutual information (*see* BIVARIATE DISTRIBUTION) between the variables. A sequence of lattices is superimposed on a *scatter diagram showing the observed pairs of values. For each lattice, the mutual information value is scaled by division by the maximum possible mutual information value for that size of lattice. The MIC is then defined as the maximum over these scaled values. A strength of the procedure is that it can be used with any type of relationship (which might consist of a mixture of several functions), whereas the *correlation coefficient refers only to linear relationships.

maximum *See* STATIONARY POINT.

maximum flow/minimum cut theorem *See* NETWORK FLOW PROBLEM.

maximum likelihood estimate *See* METHOD OF MAXIMUM LIKELIHOOD.

MCAR *See* MISSING VALUES.

MCB *See* MULTIPLE COMPARISON TEST.

McFadden, Daniel Little (1937– ; b. Rayleigh, NC) American econometrician. McFadden studied at U Minnesota (BS physics, 1957; PhD behavioural science, 1962). From 1963 to 1979 he was on the economics faculty at UCB, leaving to join MIT, but returning to UCB in 1990. He was elected to the AAAS in 1977 and to the NAS in 1981. Together with *Heckman, he was awarded the Nobel Prize for Economics in 2000.

SEE WEB LINKS

- University website.

McFadden's R^2 *See* DEVIANCE.

MCMC *See* MARKOV CHAIN MONTE CARLO METHODS.

McNemar, Quinn (1900–86; b. West Virginia; d. Palo Alto, CA) American psychologist educated at Stanford U. The author of the influential text *Psychological Statistics,* he was President of the Psychometric Society in 1951 and President of the American Psychological Association in 1964.

SEE WEB LINKS

• Memorial.

McNemar test A test, introduced by *McNemar in 1947, for use with paired data when the observed variable is dichotomous (*see* CATEGORICAL VARIABLE). Suppose, for example, that competing drugs (A and B) are tested on pairs of patients. The outcome is either success or failure. Information about the relative merits of the drugs is provided only by occasions where just one is successful. Let a denote the number of pairs in which only drug A succeeds, and b denote the number of pairs in which only drug B succeeds. With a *continuity correction, McNemar's test statistic is

$$\frac{(|a-b|-1)^2}{a+b},$$

which, if the drugs are really equally effective, may be taken to be an observation from a *chi-squared distribution with one *degree of freedom. The generalization to more than two matched samples is provided by the **Cochran Q test.**

m

MD-plot *See* BLAND-ALTMAN PLOT.

mean Familiarly known as the *average, the word 'mean' is used as a shorthand for either the *population mean or the *sample mean, depending on context. *See also* EXPECTED VALUE.

mean absolute deviation (MAD) A *measure of spread. For *observations $x_1, x_2, \ldots, x_n$, with *sample mean $\overline{x}$ and *median m, the mean absolute deviation about the mean is

$$\frac{1}{n}\sum_{j=1}^{n}|x_j-\overline{x}|,$$

and the mean absolute deviation about the median is

$$\frac{1}{n}\sum_{j=1}^{n}|x_j-m|.$$

mean absolute error (MAE) If $y_1, y_2, \ldots, y_n$ are n observed values and $\hat{y}_1, \hat{y}_2, \ldots, \hat{y}_n$ are the corresponding values predicted (perhaps by some *model), then the MAE is

$$\frac{1}{n}\sum_{j=1}^{n}|y_j - \hat{y}_j|.$$

There are several similar statistics, all of which provide information about the extent of the agreement between the observed and predicted values. The **mean absolute percentage error (MAPE)** is

$$\frac{100}{n}\sum_{j=1}^{n}\frac{|y_j - \hat{y}_j|}{|y_j|},$$

the **mean squared error (MSE)** is given by

$$\text{MSE} = \frac{1}{n}\sum_{j=1}^{n}(y_j - \hat{y}_j)^2,$$

and the **root mean squared error (RMSE)** is $\sqrt{\text{MSE}}$.

mean absolute percentage error *See* MEAN ABSOLUTE ERROR.

mean chart *See* QUALITY CONTROL.

mean deviation *Alternative name for* MEAN ABSOLUTE DEVIATION.

mean square *See* ANOVA.

mean squared error (MSE) *See* ESTIMATOR; MEAN ABSOLUTE ERROR.

measurement The process of determining values for *numerical or *categorical variables.

measure of agreement A single statistic used to summarize the agreement between the rankings or classifications of objects made by two or more observers. Examples are the *coefficient of concordance, *Cohen's kappa, and a *rank correlation coefficient.

measure of association *See* ASSOCIATION.

measure of goodness-of-fit A statistic that compares the observed data with the *expected values estimated according to some proposed *model. The most common measures are the *chi-squared test statistic and the *likelihood-ratio goodness-of-fit statistic.

measure of location For a set of *data, or a *population, a single number, or data value, which is in some sense in the middle of the data, or the population. *See* EXPECTED VALUE; MEDIAN; MIDRANGE; MODE; POPULATION MEAN; SAMPLE MEAN.

measure of spread A measure of the extent to which the values of a *variable, in either a *sample or a *population, are spread out. The most commonly used measures of spread are *variance, *standard deviation, *mean deviation, *median absolute deviation, *range, interquartile range (*see* QUARTILE), and *semi-interquartile range. Of these, only the variance is not measured in the same units as the observations.

median If a set of numerical *data has n elements and is arranged in order so that either

$$x_1 \leq x_2 \leq \cdots \leq x_n \quad \text{or} \quad x_1 \geq x_2 \geq \cdots \geq x_n,$$

then the median is $x_{\frac{1}{2}(n+1)}$ if n is odd, and $\frac{1}{2}\left(x_{\frac{1}{2}n} + x_{\frac{1}{2}n+1}\right)$ if n is even.

For example, the time intervals (in minutes, to the nearest minute) between the eruptions of *Old Faithful on 1 August 1978 were

78, 74, 68, 76, 80, 84, 50, 93, 55, 76, 58, 74, 75.

Arranging these thirteen values in order, we get

50, 55, 58, 68, 74, 74, **75**, 76, 76, 78, 80, 84, 93,

to give a median of 75 minutes. For 4 August 1978, there were fourteen inter-eruption times, which arranged in order were

m

60, 66, 67, 68, 70, 72, **73**, **75**, 75, 75, 79, 84, 86, 86,

so that the median is $\frac{1}{2}(73+75)=74$.

Alternatively, an approximate value for the median can be read from a *cumulative frequency graph as the value of the variable corresponding to a *cumulative relative frequency of 50%.

For a *continuous random variable X, the median m of the distribution is such that $P(X \leq m) = \frac{1}{2}$. For a discrete random variable taking values $x_1 < x_2 < \ldots$, the median is x_i if $P(X < x_i) < \frac{1}{2}$ and $P(X > x_i) < \frac{1}{2}$, and it is $\frac{1}{2}(x_i + x_{i+1})$ if $P(X \leq x_i) = \frac{1}{2}$ and $P(X \geq x_{i+1}) = \frac{1}{2}$. *See also* PERCENTILE; QUARTILE.

median absolute deviation A *robust estimate of variability. For *observations $x_1, x_2, \ldots, x_n$, with *median m, the median absolute deviation is the median of the differences $|x_1 - m|, |x_2 - m|, \ldots, |x_n - m|$. *See also* MEAN ABSOLUTE DEVIATION.

median-median line *See* ROBUST REGRESSION.

median polish A *robust method suggested by *Tukey for estimating the *mean μ, row *parameters $r_1, r_2, \ldots, r_J$, column parameters $c_1, c_2, \ldots, c_K$, and residuals ε_{jk} for the *model

$$y_{jk} = \mu + r_j + c_k + \varepsilon_{jk},$$

with $j = 1, 2, \ldots, J$ and $k = 1, 2, \ldots, K$. This *iterative algorithm alternates row and column operations. Considering the rows first, for each row the row *median is subtracted from every element in that row. For each column, the median of the revised numbers is then subtracted from every element in that column. This continues until all medians are 0. The outcome may vary slightly depending on whether rows or columns are considered first. In the example, μ is estimated as 30, with $r_1 = -12$, $c_4 = -7$, and $\varepsilon_{1,4} = 7$ (so that $30 - 12 - 7 + 7 = 18$, the original value)

Original table

13	17	26	18	29
42	48	57	41	59
34	31	36	22	41

→

Parameter estimates

0	−1	0	7	0	−12
−1	0	1	0	0	18
9	1	−2	−1	0	0
−5	0	8	−7	11	30

median test *See* TEST FOR EQUALITY OF LOCATION.

Meier, Paul (1924–2011; b. New York City; d. New York City) American statistician. Meier was co-author with *Kaplan of the paper, published in 1958, that introduced the *Kaplan–Meier estimate of the survivor function (*see* HAZARD RATE). A graduate of Oberlin College in Ohio, Meier obtained his PhD at Princeton U (supervised by *Tukey) in 1951. His career took him to Johns Hopkins U (1952), U Chicago (1957), and Columbia U (1992). Meier was President of the *IMS in 1986. He was the *COPSS *Fisher Lecturer in 1992 and the *Wilks Award winner of the *ASA in 2004. He was elected a Fellow of the AAAS in 1980.

SEE WEB LINKS

- Obituary and photograph.

mesokurtic *See* KURTOSIS.

***M*-estimate** A *measure of location that is not as sensitive as the *mean to *outlier values. With *observations $x_1, x_2, \ldots, x_n$, the sample mean can be characterized as the value of θ that minimizes $\sum_{j=1}^{n} g(x_j - \theta)$, where $g(u) = u^2$. The sample *median can be characterized in a similar way, though now $g(u) = |u|$.

M-estimates can be characterized in this same way, but the functional forms for g are chosen to be less sensitive to *outlier values. One frequently used alternative is the **Huber function**:

$$g(u) = \begin{cases} \frac{1}{2}u^2, & |u| \le k, \\ k|u| - \frac{1}{2}k^2, & |u| > k, \end{cases}$$

where k is a *tuning constant (often set equal to twice the *median absolute deviation).

A second alternative is the **biweight function**:

$$g(u) = \begin{cases} \frac{1}{6}k^2\left[1 - \left\{1 - (u/k)^2\right\}^3\right], & |u| \le k, \\ \frac{1}{6}k^2, & |u| > k, \end{cases}$$

where k is again a tuning constant and is here often set equal to seven times the median absolute deviation. *See also* L-ESTIMATE.

meta-analysis A statistical methodology in which *data from previous tests are considered and analysed together. For example, a series of small experiments may all show only slight signs of the same effect (e.g. that one medicine is better than another), whereas the aggregation of the experiments provides overwhelming evidence. A difficulty with this approach is that experimental conditions and experimental protocols may vary, so that the aggregate outcome may not be a fair reflection of the true situation.

m

method of least squares A method originated by *Legendre, which refers to the process of estimating the unknown *parameters of a *model by minimizing the sum of squared differences between the observed values of a *random variable and the values predicted by the model. If every observation is given equal weight then this is **ordinary least squares (OLS)**.

For example, with n pairs of observations $(x_1, y_1), \dots, (x_n, y_n)$ and the *linear regression model

$$E(Y_j) = \alpha + \beta x_j + \varepsilon_j,$$

where α and β are unknown parameters, ε_j is an error of observation, and $E(Y_j)$ denotes the *expected value of Y_j, the ordinary least squares estimates are the values for α and β that minimize

$$\sum_{j=1}^{n} (y_j - \alpha - \beta x_j)^2.$$

See also GENERALIZED LEAST SQUARES; WEIGHTED LEAST SQUARES.

method of maximum likelihood A method for obtaining an estimate (*see* ESTIMATOR) of an unknown *parameter of an assumed population distribution. The *likelihood of a data set depends upon the parameter(s) of the *distribution (or *probability density function) from which the *observations have been taken. In cases where one or more of these parameters are unknown, a shrewd choice as an estimate would be the value that maximizes the likelihood. This is the **maximum likelihood estimate (mle)**. Expressions for maximum likelihood estimates are frequently obtained by maximizing the *natural logarithm of the likelihood rather than the likelihood itself (the result is the same). Sir Ronald *Fisher introduced the method in 1912.

method of moments An alternative to the *method of maximum likelihood as a method for the *estimation of the *parameters of a *distribution. Each *moment of a distribution can be expressed as a function of the parameters of the distribution, and often this implies that the parameters can be expressed as simple functions of the moments. In such cases, replacing the moments with their sample estimates provides estimates of the population parameters.

For example, the two-parameter *gamma distribution with *probability density function proportional to $x^{\alpha-1}e^{-x/\beta}$ has its first moment, μ_1' (equal to the mean, μ), given by $\mu = \alpha\beta$ and its second central moment, μ_2, given by $\mu_2 = \alpha\beta^2$. Solving these equations, we get $\alpha = \mu^2/\mu_2$ and $\beta = \mu_2/\mu$. The method of moments replaces the unknown quantities μ and μ_2 with the corresponding sample quantities $\bar{x} = \frac{1}{n}\sum_j x_j$ and $\frac{1}{n}\sum_j (x_j - \bar{x})^2$ so that, for example, the estimator of α is $\tilde{\alpha}$ given by

$$\tilde{\alpha} = \frac{\left(\sum_j x_j\right)^2}{n\sum_j (x_j - \bar{x})^2}.$$

method of steepest ascent (descent) *See* RESPONSE SURFACE.

metric *See* DISTANCE MEASURE.

Metropolis, Nicholas Constantine (1915–99; b. Chicago, IL; d. Los Alamos, NM) Greek-American mathematician. A graduate of U Chicago, where he studied experimental physics, gaining his PhD in 1941, Metropolis was one of the original recruits to the Manhattan project at the Los Alamos Laboratory. His career alternated between there and U Chicago. To statisticians he is best known for his work on *Monte Carlo methods. He coined this term, and he was also responsible for naming the elements astatine and technetium at the time of their discovery.

SEE WEB LINKS

- Fuller biography and photograph.

Metropolis–Hastings algorithm *See* MARKOV CHAIN MONTE CARLO METHODS.

mgf *Abbreviation for* MOMENT-GENERATING FUNCTION.

MIC *Abbreviation for* MAXIMAL INFORMATION COEFFICIENT.

Michaelis–Menten equation *See* GROWTH CURVE.

mid-P With a *continuous random variable, X, it is evident that the sum of the usual *upper-tail probability, $P(X \geq x)$, and the *lower-tail probability, $P(X \leq x)$, is 1. However, with a *discrete random variable this sum is equal to $1 + P(X = x)$. This suggests that, for example, the calculation of the upper-tail probability should then use mid-P, defined as

$$P(X > x) + \frac{1}{2}P(X = x).$$

midrange For a set of numerical data the midrange is the arithmetic *mean of the greatest and least values in the set. It is sometimes used as a *measure of location, but it is unreliable, since, by definition, it depends only on extreme values.

midspread *See* QUARTILE.

Mills ratio For a *continuous random variable, the ratio, $R(x)$, of the survivor function (*see* HAZARD RATE) to the *probability density function:

$$R(x) = \frac{1 - F(x)}{f(x)}.$$

m

Investigations of the ratio have mostly been undertaken with respect to the standard *normal distribution, for which, for $x > 2$,

$$R(x) \approx \frac{4}{3x + \sqrt{x^2 + 8}} - \frac{2}{x^7}.$$

minimal cut vector; minimal path vector *See* RELIABILITY THEORY.

minimax; minimum *See* STATIONARY POINT.

minimum chi-squared A method of *estimation in which the value chosen for the *parameter estimate is the value that minimizes the value of the test statistic of the *chi-squared test for goodness-of-fit.

MINITAB A statistical computer package particularly designed for teaching purposes.

SEE WEB LINKS

- Statistical package home page.

Minkowski, Hermann (1864–1909; b. Kaunas, Lithuania; d. Göttingen, Germany) Lithuanian-born mathematician. Minkowski moved to Germany in 1872 and was educated at U Königsberg, gaining his PhD in 1885. His academic career took him successively to posts at U Bonn (1885), U Zürich (1896), and U Göttingen (1902). Minkowski had a passion for pure mathematics and his work underpinned that of Einstein on relativity.

SEE WEB LINKS

- Fuller biography and photograph.

Minkowski distance *See* DISTANCE MEASURE.

Minkowski inequality For *random variables X and Y,

$$\{\mathrm{E}(|X+Y|^a)\}^{\frac{1}{a}} \leq \{\mathrm{E}(|X|^a)\}^{\frac{1}{a}} + \{\mathrm{E}(|Y|^a)\}^{\frac{1}{a}}$$

for any positive constant a. *See also* BERNSTEIN INEQUALITY; CHEBYSHEV INEQUALITY; HÖLDER INEQUALITY; KOLMOGOROV INEQUALITY; MARKOV INEQUALITY.

missing values Many sets of *data have missing values. If the *probability that an *observation is missing is independent of its value and of all the *variables of interest, then it is said to be **missing completely at random** (**MCAR**); an example would be if information on a sampled individual was accidentally lost. If the probability that an observation is missing depends on its value, but not on the values of other observations, then it is described as **missing at random** (**MAR**); an example would be the reluctance of individuals earning high incomes to reveal their incomes. Other missing values fall into the category **missing not at random** (**MNAR**). MCAR leads to a reduced *sample size, but no other problems. Methods for dealing with MAR and MNAR include the *EM algorithm and *imputation.

misspecified model A *model that provides an incorrect description of the *data. To some extent all models are misspecified. The consequences of using a misspecified model are of particular concern in the analysis of *time series, where *forecasting will take place and a misspecified model can lead to a highly inaccurate forecast.

Mitscherlich equation *See* GROWTH CURVE.

mixed effects design *See* EXPERIMENTAL DESIGN.

mixture distribution A *distribution made up of two or more component distributions. For example, suppose that light bulbs of type A have an *exponential lifetime with *parameter α and light bulbs of type B have an exponential lifetime with parameter β. A box contains a mixture of the two types of bulb, with a proportion p being of type A. Let X be the lifetime of a randomly selected bulb. The *probability density function f of X is given by

$$f(x) = p\frac{1}{\alpha}e^{-\alpha x} + (1-p)\frac{1}{\beta}e^{-\beta x}, \qquad x > 0.$$

mle Abbreviation for maximum likelihood estimate. *See* METHOD OF MAXIMUM LIKELIHOOD.

MNAR *See* MISSING VALUES.

mobility table A square *contingency table in which the rows and columns have equivalent classifications but the columns refer to a later time point than the rows. A **social mobility table** usually refers to the social classes (or occupations) of successive generations. **Voter mobility tables** cross-classify the votes of individuals in successive elections.

modal class *See* MODE.

modal frequency *See* MODE.

mode A data value, in a set of *categorical data, whose *frequency is not less than that of any other data value. For a set of numerical data, a mode is a data value whose frequency is not less than the frequency of neighbouring values. If the numerical data is in the form of *grouped data, a **modal class** is a class whose frequency is not less than the frequency in neighbouring classes. The **modal frequency** is the frequency with which the mode occurs, or the frequency in the modal class.

For example, the time intervals (in minutes, to the nearest minute) between the eruptions of *Old Faithful on 1 August 1978 were

78, 74, 68, 76, 80, 84, 50, 93, 55, 76, 58, 74, 75.

The values 74 and 76 both occur twice, and the remaining values occur just once. The values 74 and 76 are the two modes, both having a modal frequency of 2. Combining the data from 1–8 August, we have the following summary table:

Time interval	40–49	50–59	60–69	70–79	80–89	90–99
Frequency	5	21	10	37	30	4

The modal classes are 50–59 (with modal frequency 21) and 70–79 (with modal frequency 37).

For a *discrete random variable a mode is a value whose *probability is not less than that of its neighbours. For a *continuous random variable a mode is a value such that the *probability density function has a local maximum. If there is only one mode the distribution is **unimodal**. If there is more than one mode the distribution is **multimodal**. If there are two modes it is **bimodal**. The word 'mode' was coined by Karl *Pearson in 1895.

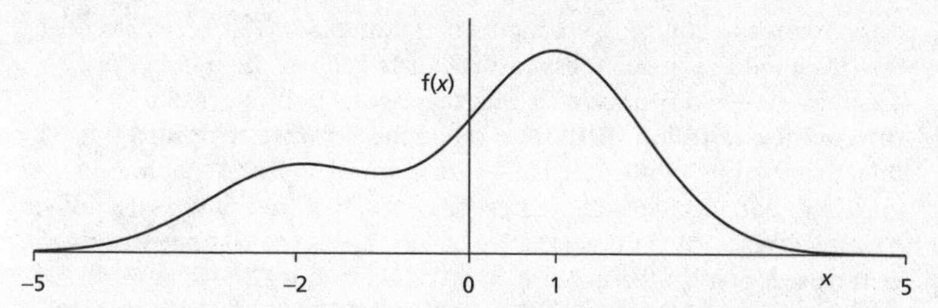

Mode. The multimodal probability density function illustrated is for a *mixture distribution of two normal distributions, each with unit variance. One is centred on $x = 1$ and the other on $x = -2$.

model A simple description of a probabilistic process that may have given rise to observed *data. For example, if the data consist of the numbers shown by a fair die during a game of Snakes and Ladders, then a simple model would state that for each roll, and *independent of the outcomes of other rolls, the *distribution of the number shown is a *discrete uniform distribution, on 1, 2, . . . , 6.

Models form the bedrock of Statistics. Specific distributions are often invoked. Many types of models are mentioned in this dictionary.

model selection procedure A procedure for choosing between competing *models that is based on balancing model complexity against the quality of that model's fit to the given *data.

For a *multiple regression model, one approach makes use of the **Mallows C_p** statistic, introduced by *Mallows in 1964. With n *observations and k explanatory variables (*see* REGRESSION), define s^2 as the estimate of the *experimental error *variance. Then, for a model using just p of the k variables,

$$C_p = \frac{1}{s^2}\sum_{j=1}^{n}(y_j - \hat{y}_j)^2 - n + 2p,$$

where y_j is an observation and $\hat{y}_j$ is the corresponding *fitted value. A model that fits well should have a C_p value close to p. An acceptable fit is provided by a model for which

$$C_p < (2p - k - 1) + (k - p + 1)F_{a,b}(\alpha),$$

where $a = k - p + 1$, $b = n - k - 1$, and $F_{a,\,b}(\alpha)$ is the value exceeded by chance on $100\alpha\%$ of occasions by a *random variable having an *F-distribution with a and b *degrees of freedom. Typically, $\alpha = 0.05$ or 0.01.

A more generally applicable alternative is based on **AIC (Akaike's information criterion)** proposed by *Akaike in 1969. For *categorical data this amounts to choosing the model that minimizes $G^2 - 2v$, where G^2 is the *likelihood-ratio goodness-of-fit statistic and v is the number of *degrees of freedom associated with the model. If the **Bayesian information criterion** (**BIC**) (also called the **Schwarz criterion**) is used, then the quantity minimized is $G^2 - v \ln n$, where ln is the *natural logarithm and n is the *sample size. This usually results in the selection of a simpler model. A third alternative of this type is the **Hannan-Quinn criterion**, for which the quantity to be minimized is $G^2 - 2v \ln(\ln n)$.

Whatever procedure is used for model selection, it is usually the case that the model fits less well (as measured by R^2, the coefficient of determination, *see* ANOVA) when it is applied to new data. The reduction in fit is described as **shrinkage**. *See also* STEPWISE PROCEDURE.

model validation The process of verifying that a fitted *model does indeed provide an acceptable description of the *data. Validation techniques usually involve the examination of plots such as a *scatter diagram of the residuals (*see* REGRESSION DIAGNOSTICS) against the values of the explanatory variable (*see* REGRESSION).

modulus *See* ABSOLUTE VALUE.

m

moment (uncorrected moment) For a *random variable X the rth moment (about the origin) is defined to be the *expected value of X^r, where r is a non-negative integer. It is usually denoted by μ'_r. So $\mu'_0 = 1$ and $\mu'_1 = \mu$, the mean of X.

The rth **moment about the mean** (or **central moment** or **corrected moment**) is defined to be the expected value of $(X - \mu)^r$ and is usually denoted by μ_r. Thus $\mu_1 = 0$ and $\mu_2 = \sigma^2$, the *variance of X. The moments about the mean can be expressed as *linear combinations of the uncorrected moments, for example:

$$\begin{aligned}\mu_2 &= \mu'_2 - \mu^2,\\ \mu_3 &= \mu'_3 - 3\mu'_2\mu + 2\mu^3,\\ \mu_4 &= \mu'_4 - 4\mu'_3\mu + 6\mu'_2\mu^2 - 3\mu^4.\end{aligned}$$

Either set of moments can also be expressed in terms of linear combinations of simple functions of the *cumulants. It should be noted that, for some distributions, μ'_r and μ_r may exist only for small values of r.

moment about the mean *See* MOMENT.

moment coefficient of skewness If the second and third *moments about the mean of a *distribution are μ_2 and μ_3, respectively, the moment coefficient of skewness is $\mu_3/\sqrt{\mu_2^3}$. It is invariant under change of scale and change of origin.

moment-generating function (mgf) The moment-generating function of a *random variable X involves a variable, usually denoted by t, and is defined to be the *expected value of e^{tX}. It is usually denoted by $M_X(t)$, or $M(X, t)$. Expanding $M_X(t)$ in powers of t gives

$$M_X(t) = \sum \mu_r' \frac{t^r}{r!},$$

where μ_r' is the rth *moment of X about the origin. By virtue of the *Maclaurin series, μ_r' is the value, when $t=0$, of the rth derivative with respect to t of $M_X(t)$. *See also* PROBABILITY-GENERATING FUNCTION.

Monte Carlo methods Methods that use *pseudo-random numbers in order to determine the properties of some function or set of functions. For example, the *acceptance–rejection algorithm can be used to evaluate integrals, *simulation makes use of pseudo-random numbers to study the outcome of some complex system, and *Monte Carlo tests use pseudo-random numbers to decide on the outcome of *hypothesis tests. The accuracy of the methods generally improves with an increase in the number of pseudo-random numbers used.

Monte Carlo test A procedure suggested by *Barnard in 1963, for testing whether a set of *data is consistent with a *null hypothesis. It is appropriate for a situation where the theoretical *distribution of the test statistic is unknown, although the distribution of the individual observations is known. The test procedure is to use *Monte Carlo methods to generate 99 (say) further data sets of the same size as the true data, under the conditions defined by the null hypothesis. The value for the test statistic is calculated for each data set and the distribution of these values is examined. If the value of the test statistic for the actual data is similar to the values obtained from the artificial data sets, then the null hypothesis is accepted, whereas if it is extreme the hypothesis is rejected.

Montmort, Pierre Rémond (1678–1719; b. Paris, France; d. Paris, France) French probabilist. Born Pierre Rémond, he bought the estate of Montmort in 1699, becoming 'Pierre Rémond de Montmort'. Originally trained as a lawyer, he developed an interest in mathematics, which he studied privately and through correspondence. He introduced the term *Pascal's triangle in his 1708 book *Essay d'analyse sur les jeux de hazard*, an early application of the ideas of *probability. He was elected FRS in 1715 on a visit to England. He died of smallpox.

SEE WEB LINKS

- Fuller biography.

Monty Hall problem Monty Hall was host of a TV show in which a contestant was faced by three doors. Behind two of the doors was a booby prize, and behind one was the real prize. The contestant was asked to choose a door. Another door was then opened to reveal a booby prize. The contestant was invited to change to the third door. Intuition suggests that changing would have no effect, yet actually it doubles the chance of winning the real prize.

SEE WEB LINKS

- Applet.

Mood, Alexander McFarlane (1913–2009; b. Amarillo, TX; d. Irvine, CA) American mathematical statistician. A physics undergraduate at U Texas, Mood obtained his statistics PhD (supervised by *Wilks) at Princeton U in 1940. He spent the Second World War at the National Defense Research Council, joining the faculty at ISU in 1945. His varied career included appointments at U Texas, Princeton U, Brown U, and U California at Irvine, as well as periods working for government organizations. He was co-author with *Graybill of the 1963 textbook *Introduction to the Theory of Statistics*, which for many years was the recommended introductory textbook worldwide. He was President of the *IMS in 1957 and of the Operational Research Society of America in 1963. He was the *ASA *Wilks Award winner in 1979.

SEE WEB LINKS

- Fuller biography.

Mood dispersion test *See* TEST FOR EQUALITY OF SCALE.

Mood median test *See* TEST FOR EQUALITY OF LOCATION.

Moore–Penrose inverse *See* MATRIX.

Moran, Patrick Alfred Pierce (1917–88; b. Sydney, Australia; d. Canberra, Australia) Australian statistician. Moran was educated at Sydney U and Cambridge U and was a researcher at Oxford U from 1946 to 1951, before returning to Australia. He was the founding Professor of Statistics at ANU, where his research students included *Hannan and *Heyde. He was elected to the AAS in 1962, and was elected FRS in 1975. He was the first President of the Australian Statistical Society (now the *SSAI) in 1963 and was awarded the Society's Pitman Medal (*see* PITMAN) in 1982. He was President of the Australian Mathematical Society in 1976.

SEE WEB LINKS

- Fuller biography and photograph.

Moran Medal A medal, awarded every two years by the AAS in commemoration of the work of *Moran, to recognize outstanding research by scientists aged 40 years and under.

Moran's *I* A *statistic, introduced by *Moran in 1950, that measures spatial *autocorrelation using information only from specified pairs of spatial *observations. Suppose there are n locations, the observed value at location j is x_j, and the overall mean value is $\bar{x}$. Let $w_{jk}=1$ if the comparison of the value at location j to the value at location k is of interest, and let $w_{jk}=0$ otherwise. By definition, $w_{jj}=0$, for all j. Usually, the only comparisons of interest are those between immediate neighbours. The statistic is given by

$$I = n\sum_j \sum_k w_{jk}(x_j - \bar{x})(x_k - \bar{x}) \bigg/ \left\{\sum_j \sum_k w_{jk}\right\} \left\{\sum_j (x_j - \bar{x})^2\right\}.$$

morbidity rate The *incidence rate of persons in a *population who become clinically ill during the period of time stated.

morphometrics The study of the mathematical and statistical properties of shape.

mortality rate (death rate) The number of deaths occurring in a *population during a given period of time, usually a year, as a *proportion of the number in the population. Usually the mortality rate includes deaths from all causes and is expressed as deaths per 1000. A disease-specific (or age-specific or sex-specific) mortality rate includes only deaths associated with one disease (or age or sex) and is reported as deaths per 1000 people of the specified type. The mortality rate may be standardized when comparing mortality rates over time, or between countries, to take account of differences in the population. *See also* AGE-SPECIFIC RATE; SEX-SPECIFIC RATE.

mosaic display A display that highlights departures from *independence ('residuals') in a *two-way table. The display, which is asymmetric, emphasizes the variations in the *conditional probability of the categories of one variable, given the category of the second variable (*see diagram overleaf*). For an alternative display that treats the variables in a symmetrical fashion, *see* COBWEB DIAGRAM.

Mosteller, Charles Frederick (1916–2006; b. Clarksburg, WV; d. Arlington, VA) American mathematical statistician. After his ScM at

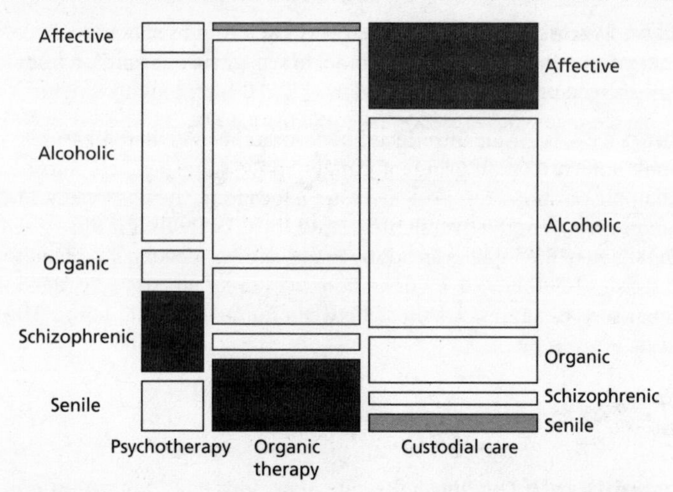

Mosaic display. The areas of the tiles are proportional to the cell frequencies. Dark shading indicates a large positive residual, pale shading a large negative residual.

m

Carnegie Mellon U in 1939, Mosteller was supervised by *Wilks for his 1946 PhD at Princeton U. That year he joined the faculty at Harvard U, where he spent his career. He served as President of the *Psychometric Society (1957), the *ASA (1967), and the *IMS (1975). He was awarded the *Wilks Award of the ASA in 1986 and the *COPSS *Fisher Lectureship in 1987. He was an Honorary Life Member of the *International Statistical Institute.

SEE WEB LINKS

- Biographical memoir with photographs.

most powerful test A *test of a *null hypothesis which has greater power than any other test for a given *alternative hypothesis. *See also* UNIFORMLY MOST POWERFUL TEST.

mother wavelet *See* WAVELET.

mover–stayer model A model for a *square table which classifies individuals into a group whose new location is *independent of their old location and a group who never move. The 'locations' may be geographical or they may refer to social class, occupation, or views on some subject. The model can provide a good fit to the data without actually making any sense when its implications are examined more closely.

moving average A method of smoothing a *time series to reduce the effects of *random variation and reveal any underlying trend or *seasonality. For the time series $x_1, x_2, \ldots, x_t$ the simple three-point moving average would replace the value of x_k, $k=2, 3, \ldots, t-1$, with

$$\frac{1}{3}(x_{k-1} + x_k + x_{k+1}).$$

Often, different weights are used, as in this five-point moving average which could be used for $k=3, 4, \ldots, t-2$:

$$\frac{1}{12}(x_{k-2} + 3x_{k-1} + 4x_k + 3x_{k+1} + x_{k+2}).$$

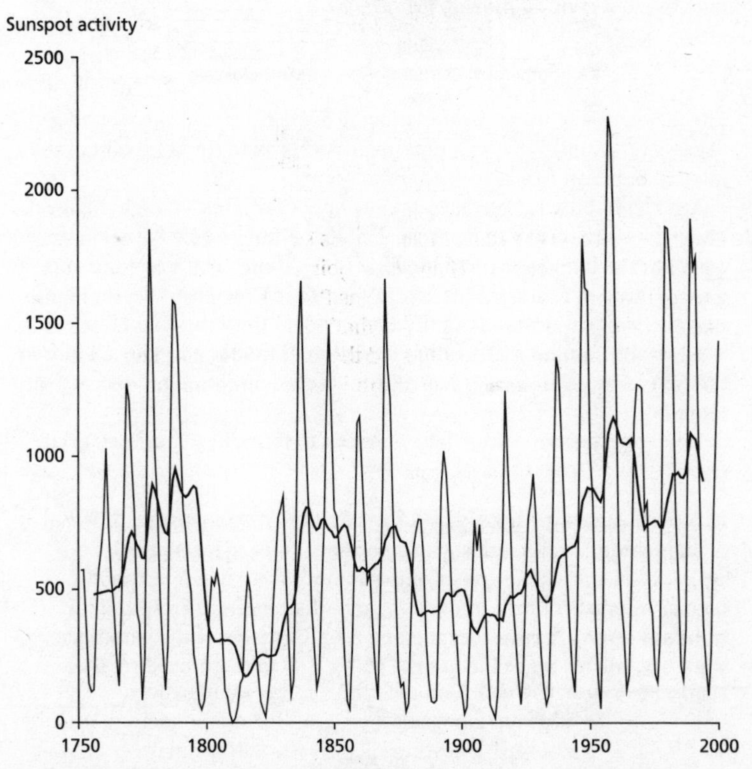

Moving average. The graph shows annual sunspot activity (in standardized units) from 1750 to 2000. There is a strong cycle with a period of around eleven years. Also shown is the eleven-year moving average. This removes the obvious cycle but reveals longer-scale fluctuations.

m

Another possibility is provided by **Daniell weights**: in the case of an average over m time points, the two end points are given weight $\frac{1}{2(m-1)}$, with the others each being given weight $\frac{1}{m-1}$.

The four-point moving averages (appropriate for quarterly data) are

$$\frac{1}{4}(x_1 + x_2 + x_3 + x_4), \quad \frac{1}{4}(x_2 + x_3 + x_4 + x_5), \ldots.$$

Twelve-point moving averages are similarly defined and are appropriate for monthly data.

For a cycle with an even period, e.g. quarterly or monthly data, the **centred moving averages** are the arithmetic means of the successive moving averages as defined above. For example, in the case of quarterly data the first centred moving average is

$$\frac{1}{8}(x_1 + 2x_2 + 2x_3 + 2x_4 + x_5).$$

The advantage of these centred moving averages is that the resulting values are associated with a time point rather than the midpoint of the interval between two successive time points.

A graph of moving averages against time may show changes against time which are obscured by cyclical effects. A line of best fit to the moving averages is a **trend line**, and its slope is the **trend**. The trend line may be used to forecast future values (in the short term). For example, for monthly data the average deviation of the January data from the trend line can be used as an estimate of the future deviation of the January deviation from the trend line. The deviation can be measured as either a difference or a ratio.

Note that the use of moving averages can introduce spurious *cycles (*see* SLUTSKY–YULE EFFECT).

moving average model (MA model; moving average process) A *model for a *time series with a constant *mean (taken as 0). Let $x_1, x_2, \ldots$ be successive values of the *random variable X, measured at regular intervals of time and let $\varepsilon_1, \varepsilon_2, \ldots$ denote the corresponding *random errors. A pth-order moving average model with *parameters $\alpha_1, \alpha_2, \ldots, \alpha_p$ relates the value at time j ($\geq p+1$) to the preceding p error values by

$$x_j = \alpha_p \varepsilon_{j-p} + \alpha_{p-1}\varepsilon_{j-p+1} + \cdots + \alpha_1 \varepsilon_{j-1} + \varepsilon_j.$$

Such a model is written in brief as MA(p). The errors are presumed to be *independent and to have mean 0 and hence the X-variables also have mean 0. Moving average models can also be expressed as *autoregressive models. Models combining both type of process include *ARMA models and *ARIMA models.

MSE Abbreviation for mean squared error. *See* ESTIMATOR; MEAN ABSOLUTE ERROR.

mu (μ) The Greek letter used to denote the *population mean.

multicollinearity *See* MULTIPLE REGRESSION MODEL.

multidimensional contingency table *See* CONTINGENCY TABLE.

multidimensional scaling A *multivariate method, resembling *principal components analysis, that gives a graphical representation of the main characteristics of a *proximity matrix using only a few dimensions. The success of the method is assessed by comparing the 'distances' between individuals as recorded by the proximity matrix with the (suitably scaled) Euclidean distances (*see* DISTANCE MEASURE) in the full-dimensional space; the resulting *measure of goodness-of-fit is called the **stress** (also called **Kruskal stress**) of the solution—a low value suggests a good solution.

	Great Britain	Ireland	Nether-Lands	Belgium	France	Italy	Spain
Great Britain	20	11	11	6	7	2	6
Ireland	11	20	11	10	11	10	10
Netherlands	11	11	20	12	11	6	6
Belgium	6	10	12	20	12	11	9
France	7	11	11	12	20	10	12
Italy	2	10	6	11	10	20	13
Spain	6	10	6	9	12	13	20

The proximity matrix shows, for seven European countries, the numbers of food products (out of twenty) that were found to similar extents (i.e. common in both countries or scarce in both countries) in each of the pairs of countries. The biggest difference is between Great Britain and Italy: this is reflected in a two-dimensional graph of the positions (in this respect) of the countries (*see diagram overleaf*).

Multidimensional scaling of countries and food products. The results may appear surprising, but they reflect the (very crude) information from which they were derived. The (unlabelled) axes reflect different aspects of eating behaviour.

m

multilevel model A multilevel model attempts to model all the components in a hierarchy of nested effects (*see* NESTED DESIGN). For example, the attitudes of schoolchildren, or the recovery rates of hospital patients, will be influenced at a number of levels. All children in the same class will have undergone almost identical teaching experiences. Their class will have had similar experiences to other classes in that school. The school may be similar to other schools in the neighbourhood.

multimodal *See* MODE.

multinomial distribution The extension of the *binomial distribution to the case of m (>2) possible outcomes. Suppose that the *probabilities of outcomes 1, 2, ..., m are $p_1, p_2, \ldots, p_m$ with $\sum_{j=1}^{m} p_j = 1$. Let X_j denote the number, in a sample of size n, of occurrences of outcome j, for $j = 1, 2, \ldots, m$. The *random variables $X_1, X_2, \ldots, X_m$ have a *multivariate distribution given by

$$P(X_1 = n_1,\ X_2 = n_2, \ldots,\ X_m = n_m) = \frac{n!}{n_1!n_2!\ldots n_m!} p_1^{n_1} p_2^{n_2} \ldots p_m^{n_m},$$

where $0 \leq n_1, n_2, \ldots, n_m \leq n$ and with $\sum_{j=1}^{m} n_j = n$. The random variable X_j has *expected value np_j and *variance $np_j(1 - p_j)$ and the *covariance of X_j and X_k $(j \neq k)$ is $-np_jp_k$. The term 'multinomial distribution' was introduced by Sir Ronald *Fisher in 1925.

multiple bar chart A *bar chart for comparing the *frequencies of a *categorical variable in two or more situations.

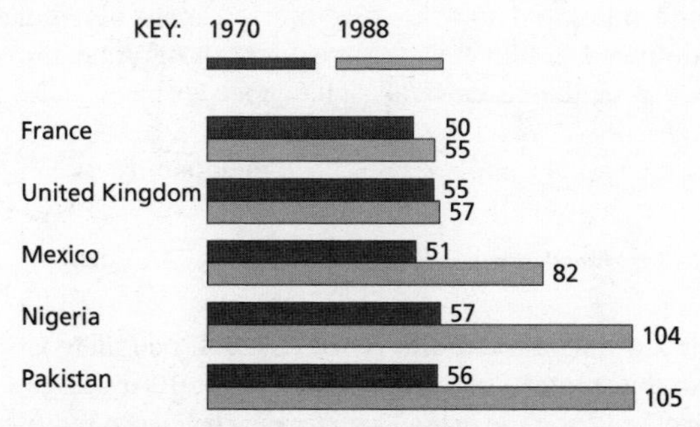

Multiple bar chart. This chart compares the population sizes in 1970 and 1988 (figures are in millions) for five countries. The small increases in the two European countries contrast sharply with the increases in countries on other continents.

multiple-choice question *See* CLOSED QUESTION.

multiple comparison test A *test suitable for the simultaneous testing of hypotheses concerning the equality of three or more *population means. When *samples have been taken from several *populations, a question of interest is whether the populations all have the same *mean. In the case of m populations, with the mean of population j denoted by μ_j, the *null hypothesis is

$$H_0: \mu_1 = \mu_2 = \cdots = \mu_m,$$

with the alternative being that H_0 is false.

In the simpler case $m = 2$, an appropriate test statistic (assuming the populations have the same *variance) is T given by

$$T = \frac{\bar{y}_1 - \bar{y}_2}{\sqrt{s^2\left(\frac{1}{n_1} + \frac{1}{n_2}\right)}},$$

where $\bar{y}_j$ is the mean of the n_j values sampled from population j, and s^2 is the pooled estimate of the common variance (*see* POOLED ESTIMATE OF COMMON MEAN). The statistic T has an approximate *t-distribution with

$v = (n_1 + n_2 - 2)$ *degrees of freedom (the approximation is exact for samples from *normal distributions). Denoting the upper 100α % point of a t-distribution with v degrees of freedom by $t(\alpha, v)$, H_0 is rejected at the 200α % level if $|T| > t(\alpha, v)$.

In the case of m populations, the null hypothesis can be rewritten in the form:

$$H_0\colon \ \mu_1 = \mu_2, \quad \mu_1 = \mu_3, \ldots, \mu_{m-1} = \mu_m,$$

which demonstrates that there are $c = \frac{1}{2}m(m-1)$ pairs of populations that could be compared. However, if c independent t-tests are performed, each at the 100α% level, then the overall significance level is $1 - (1 - \alpha)^c$ and is not α.

In the case of equal sample sizes (all n), the quantity

$$t(\alpha, v)\sqrt{\left(\frac{2}{n}\right)s^2}$$

is called the **least significant difference** (**LSD**). If no differences are greater than this, then H_0 may be accepted at the 100α% level.

One way of reducing the overall significance level is to reduce the value of α for the individual tests. The *Bonferroni inequality leads to the replacement of α by α/c : the resulting test is variously known as the **Dunn test** or as the **Bonferroni *t*-test**. A preferable alternative uses the **Sidak correction**, in which α is replaced by $1 - (1 - \alpha)^{1/c}$. However, both tests have rather low power (*see* HYPOTHESIS TEST) when m is large. A more relaxed approach involves controlling the *false discovery rate, rather than the overall significance level.

*Tukey suggested using the *Studentized range distribution in place of the t-distribution. The resulting test is familiarly called either the **Tukey test**, the **honestly significant difference test**, or the **HSD** test. This test assumes equal sample sizes; modifications for unequal sizes are the **Tukey–Kramer test** which uses $(1/n_i + 1/n_j)$ when comparing populations i and j, and the **Spjotvoll–Stoline test** which uses $2/n_s$, where n_s is the smallest of the m sample sizes. The Tukey tests are probably the best choices of all the multiple comparison tests. Similar in spirit to the Tukey tests are the **Hochberg test** and the **Gabriel test**; their test statistics are compared with the distribution of the maximum absolute value rather than with that of the Studentized range. The **Waller–Duncan test** is a test based on the F-test (*see* TEST FOR EQUALITY OF VARIANCE) for overall differences between treatments.

An alternative to comparing all pairs simultaneously is to use a **multistage test**. Suppose that the samples are labelled in order of their means, so that sample 1 has the least mean and sample m the greatest mean. Initially all m samples are compared. If H_0 is accepted, then testing ceases. However, if it is rejected, then the hypotheses

$\mu_1 = \mu_2 = \cdots = \mu_{m-1}$ and $\mu_2 = \mu_3 = \cdots = \mu_m$ are considered, using the Studentized range values for the comparison of $m-1$ populations. If a hypothesis is rejected, then comparisons of $m-2$ populations are made. Successive reductions are made until acceptable hypotheses are found. Examples of this type are **Duncan's test** (which uses the significance level $1-(1-\alpha)^{l-1}$ when l means are compared), the **Newman-Keuls test** (which uses α throughout), and the **Ryan-Einot-Gabriel-Welsch (R-E-G-W) test** which uses $1-(1-\alpha)^{l/m}$ for $l < m-1$ and α otherwise. A compromise between the Newman-Keuls test and the HSD test is the **Tukey wholly significant difference test**, which is also called the **WSD test** or **Tukey *b*-test**.

When one of the m populations under comparison is different to the remainder (for example, it refers to the use of a *control treatment) then interest focuses on the $(m-1)$ comparisons involving this population. In this case the **Dunnett test** is appropriate. The usual *t*-statistic is used, but with special tables of critical values. When the remaining $m-1$ treatments are ordered (for example, they represent different concentrations of some new substance) then the successive *T*-values will generally also be ordered and the number of tests reduced. This is known as the **Williams test**; revised tables of critical values are required. In yet another approach (the **Hsu MCB test**) attention is restricted to comparisons involving the best treatment.

If the comparisons of interest are contrasts (*see* ANOVA) of more than two population means then the **Scheffé test**, which is based on the *F-distribution, is appropriate.

In cases where the variances differ from one population to another, variants on the above tests are required. For example, the **Tamhane test** uses the *Welch statistic in place of *T*, together with the Sidak correction, while the **Games-Howell** test replaces the denominator of *T* by $\sqrt{\dfrac{s_i^2}{n_i} + \dfrac{s_j^2}{n_j}}$ when comparing populations *i* and *j* and also modifies the number of degrees of freedom.

multiple regression model The extension of the *linear regression model to the case where there is more than one explanatory variable (*see* REGRESSION). For the case of *p* *X*-variables and *n* *observations the model is

$$\mathrm{E}(Y_j) = \beta_0 + \beta_1 x_{1j} + \beta_2 x_{2j} + \cdots + \beta_p x_{pj}, \quad j = 1, 2, \ldots, n,$$

where $\beta_0, \beta_1, \ldots, \beta_p$ are unknown *parameters. An equivalent presentation is

$$Y_j = \beta_0 + \beta_1 x_{1j} + \beta_2 x_{2j} + \cdots + \beta_p x_{pj} + \varepsilon_j, \quad j = 1, 2, \ldots, n,$$

where $\varepsilon_1, \varepsilon_2, \ldots, \varepsilon_n$ are *random errors.

In practice the explanatory variables may be related as in the **quadratic regression model**

$$Y_j = \beta_0 + \beta_1 x_{1j} + \beta_2 x_{1j}^2 + \varepsilon_j, \quad j = 1, 2, \ldots, n,$$

the **cubic regression model**

$$Y_j = \beta_0 + \beta_1 x_{1j} + \beta_2 x_{1j}^2 + \beta_3 x_{1j}^3 + \varepsilon_j, \quad j = 1, 2, \ldots, n,$$

and the general **polynomial regression model** (also termed a **curvilinear regression model**)

$$Y_j = \beta_0 + \beta_1 x_{1j} + \beta_2 x_{1j}^2 + \cdots + \beta_p x_{1j}^p + \varepsilon_j, \quad j = 1, 2, \ldots, n.$$

In *matrix terms the model is written as

$$\mathrm{E}(\mathbf{Y}) = \mathbf{X}\boldsymbol{\beta},$$

where $\mathbf{Y}$ is the $n \times 1$ column *vector of *random variables, $\boldsymbol{\beta}$ is the $(p+1) \times 1$ column vector of unknown parameters, and $\mathbf{X}$ is the $n \times (p+1)$ **design matrix** given by

$$\mathbf{X} = \begin{pmatrix} 1 & x_{11} & x_{21} & \cdots & x_{p1} \\ 1 & x_{12} & x_{22} & \cdots & x_{p2} \\ \vdots & \vdots & & & \vdots \\ 1 & x_{1n} & x_{2n} & \cdots & x_{pn} \end{pmatrix}.$$

Equivalently,

$$\mathbf{Y} = \mathbf{X}\boldsymbol{\beta} + \boldsymbol{\varepsilon},$$

where $\boldsymbol{\varepsilon}$ is an $n \times 1$ vector of random errors.

Usually it is assumed that the random errors, and hence the Y-variables, are *independent and have common *variance σ^2. In this case, the ordinary least squares (*see* METHOD OF LEAST SQUARES) estimates (*see* ESTIMATOR) of the β-parameters are obtained by solving the set of simultaneous equations (the **normal equations**) which, in matrix form, are written as

$$\mathbf{X}'\mathbf{X}\,\hat{\boldsymbol{\beta}} = \mathbf{X}'\mathbf{y},$$

where $\mathbf{X}'$ is the transpose (*see* MATRIX) of $\mathbf{X}$ and $\mathbf{y}$ is the $n \times 1$ column vector of observations $(y_1\, y_2 \ldots y_n)'$. The matrix $\mathbf{X}'\mathbf{X}$ is a symmetric matrix. If it has an inverse then the solution is

$$\hat{\boldsymbol{\beta}} = (\mathbf{X}'\mathbf{X})^{-1}\mathbf{X}'\mathbf{y}.$$

The *variance-covariance matrix of the *estimators of the β-parameters is $\sigma^2(\mathbf{X}'\mathbf{X})^{-1}$. The **Gauss–Markov theorem** shows that $\boldsymbol{\beta}$ is the minimum variance linear *unbiased estimate of $\hat{\boldsymbol{\beta}}$.

If the random errors are not independent or have unequal variances then ordinary least squares is inappropriate and *weighted least squares may be appropriate. If the random errors are influenced by the X-variables, then a possible approach is to identify (if possible) W-variables that are highly correlated with the X-variables, but not with the errors. These W-variables are called **instrumental variables**. The subsequent analysis, commonly used in *econometrics, uses **two-stage least squares**, in which the first stage involves the regression of the X-variables on the W-variables to obtain fitted values:

$$\hat{\mathbf{X}} = \mathbf{W}(\mathbf{W}'\mathbf{W})^{-1}\mathbf{W}'\mathbf{X}.$$

These values are then used in place of the X-values in the regression for $\mathbf{Y}$ to give the estimate $\hat{\boldsymbol{\beta}}_2$:

$$\hat{\beta}_2 = (\hat{\mathbf{X}}'\hat{\mathbf{X}})^{-1}\,\hat{\mathbf{X}}'\mathbf{y}.$$

A perennial problem with multiple regression models is deciding which of the X-variables are really needed and which are redundant.

A related problem is that of **collinearity** (or **multicollinearity**). Suppose, for example, that the two explanatory variables X_j and X_k approximately satisfy the relation $X_j = a + bX_k$. In this case a multiple regression model that involves both X_j and X_k will run into problems, since either variable would be nearly as effective on its own. The situation can arise with other numbers of X-variables: in all cases the result is that the matrix $\mathbf{X}'\mathbf{X}$ is nearly singular. The equation $\mathbf{X}'\mathbf{X}\hat{\beta} = \mathbf{X}'\mathbf{y}$ is then said to be **ill-conditioned**. One suggestion is to replace the usual parameter estimate by

$$\hat{\beta}^* = (\mathbf{X}'\mathbf{X} + k\mathbf{I})^{-1}\mathbf{X}'\mathbf{y},$$

where k is a constant and $\mathbf{I}$ is an identity matrix. This technique is called **ridge regression**. A plot of the elements of $\hat{\beta}^*$ against k is called a **ridge trace** and can be used to determine the appropriate value for k. The estimates of the β-parameters obtained using this method are biased (*see* ESTIMATOR) but have smaller variance than the ordinary least squares estimates.

When there are many X-variables interpretation of the fitted model can be difficult and there may be several variables whose importance is questionable by reason of the small values of the corresponding β-parameters. A procedure that usually results in many of these β-parameters being set to zero (so that the corresponding explanatory variable is removed) is the **lasso**, in which the usual estimation procedure (*see* METHOD OF LEAST SQUARES) is modified by the restriction that

$$\sum_{j=1}^{p} |\beta_j| < t,$$

where t is a tuning constant (*see* M-ESTIMATE).

For methods of evaluating the fit of a multiple regression model, *see* REGRESSION DIAGNOSTICS. *See also* MODEL SELECTION PROCEDURE; STEPWISE PROCEDURE.

multiplication law for probabilities Law for probabilities stating that if A and B are *independent events then

$$P(A \cap B) = P(A) \times P(B),$$

and, in the case of n independent events, $A_1, A_2, \ldots, A_n$,

$$P(A_1 \cap A_2 \cap \cdots \cap A_n) = P(A_1) \times P(A_2) \times \cdots \times P(A_n).$$

This is a special case of the more general **law of compound probability**, which holds for events that may not be independent. In the case of two events, A and B, this law states that

$$P(A \cap B) = P(A) \times P(B|A) = P(B) \times P(A|B).$$

For three events, A, B, and C, this becomes

$$P(A \cap B \cap C) = P(A) \times P(B|A) \times P(C|A \cap B).$$

There are six (= 3!) alternative right-hand sides, for example $P(C) \times P(A|C) \times P(B|C \cap A)$. The generalization to more than three events can be inferred. For definitions of symbols, *see* CONDITIONAL PROBABILITY; INTERSECTION.

m

multistage test *See* MULTIPLE COMPARISON TEST.

multivariate analysis of variance (MANOVA) *ANOVA with a *multivariate distribution for the response variable. With g groups and n_j *observations in the jth group, the three basic *matrix quantities involve various sums of squares and cross-products. As with ANOVA the idea is to divide the total variation into a contribution explained by differences between the groups and a residual contribution resulting from variation within groups. The corresponding matrices are **T**, **B**, and **W**, defined by

$$\mathbf{T} = \sum_{j=1}^{g} \sum_{k=1}^{n_j} (\mathbf{x}_{jk} - \bar{\mathbf{x}})(\mathbf{x}_{jk} - \bar{\mathbf{x}})',$$

$$\mathbf{B} = \sum_{j=1}^{g} n_j (\bar{\mathbf{x}}_j - \bar{\mathbf{x}})(\bar{\mathbf{x}}_j - \bar{\mathbf{x}})',$$

$$\mathbf{W} = \sum_{j=1}^{g} \sum_{k=1}^{n_j} (\mathbf{x}_{jk} - \bar{\mathbf{x}}_j)(\mathbf{x}_{jk} - \bar{\mathbf{x}}_j)',$$

where $\bar{\mathbf{x}}_j$ is the column *vector of *means for the jth group, and $\bar{\mathbf{x}}$ is the column vector of overall means.

The most usual test of the hypothesis that the groups come from a single population uses **Wilks's lambda** (Λ), given by

$$\Lambda = \frac{\det(\mathbf{W})}{\det(\mathbf{T})},$$

where $\det(\mathbf{M})$ denotes the determinant of the matrix $\mathbf{M}$. Alternative tests use the **Hotelling–Lawley trace** (the sum of the eigenvalues of $\mathbf{BW}^{-1}$), the **Pillai–Bartlett trace** (the sum of the eigenvalues of $\mathbf{BT}^{-1}$), and **Roy's maximum root** (the largest eigenvalue of $\mathbf{BW}^{-1}$).

multivariate data Data collected on several *variables for each sampling unit. For example, if we collect information on weight (w), height (h), and shoe size (s) from each of a random sample of individuals, then we would refer to the triples (w_1, h_1, s_1), $(w_2, h_2, s_2), \ldots$ as a set of multivariate data.

multivariate distribution A description of the possible values, and corresponding *probabilities, of three or more *random variables. Examples include the *multinomial and *multivariate normal distributions. For related distributional ideas such as marginal distributions and conditional expected values in the simpler two-variable case, *see* BIVARIATE DISTRIBUTION.

m

multivariate distribution function For *random variables $X_1, X_2, \ldots, X_n$, the multivariate distribution function $F(x_1, x_2, \ldots, x_n)$ is the *joint probability that $X_1 \le x_1$ and $X_2 \le x_2$ and ... and $X_n \le x_n$.

multivariate method A method for analysing *multivariate data. Probably the most used procedure is *multiple regression. Other methods include *cluster analysis, *discriminant analysis, *factor analysis, *multidimensional scaling, and *principal components analysis.

multivariate normal distribution *Random variables $X_1, X_2, \ldots, X_n$ (each with ranges from $-\infty$ to ∞) have a multivariate normal distribution if their *joint probability density function (*compare* BIVARIATE DISTRIBUTION) f is given by

$$f(\mathbf{x}) = \frac{\exp\{-\frac{1}{2}(\mathbf{x}-\boldsymbol{\mu})'\boldsymbol{\Sigma}^{-1}(\mathbf{x}-\boldsymbol{\mu})\}}{\sqrt{(2\pi)^n \det(\boldsymbol{\Sigma})}},$$

where $\mathbf{x}$ is the $n \times 1$ *vector of values, $\boldsymbol{\mu}$ is the $n \times 1$ vector of *means, $\boldsymbol{\Sigma}$ is the $n \times n$ *variance-covariance matrix, and $\det(\boldsymbol{\Sigma})$ is the determinant of $\boldsymbol{\Sigma}$.

For the special case of the **bivariate normal distribution**, with random variables X and Y, the joint probability density function f is given by

$$f(x,y)=\frac{1}{2\pi\sigma_x\sigma_y\sqrt{1-\rho^2}}\exp\left[-\frac{1}{2(1-\rho^2)}\left\{\frac{(x-\mu_x)^2}{\sigma_x^2}-2\rho\frac{(x-\mu_x)(y-\mu_y)}{\sigma_x\sigma_y}+\frac{(y-\mu_y)^2}{\sigma_y^2}\right\}\right],$$

where the mean and variance of X are μ_x and σ_x^2, the mean and variance of Y are μ_y and σ_y^2, and ρ is the *correlation coefficient between the two variables.

mutual information *See* BIVARIATE DISTRIBUTION.

mutually exclusive events Two events (*see* SAMPLE SPACE), A and B, are **exclusive** if they cannot both occur simultaneously, i.e. if $A \cap B = \phi$, the empty set. In this case $P(A \cap B) = 0$. The n events $A_1, A_2, \ldots, A_n$ are mutually exclusive if, for all $j \neq k$, $A_j \cap A_k = \phi$.

mutually independent events *See* INDEPENDENT EVENTS.

m

N (μ, σ^2) The *normal distribution with *mean μ and *variance σ^2.

Nadaraya-Watson estimator *See* NON-PARAMETRIC REGRESSION.

Nagelkerke's R^2 *See* ANOVA.

National Lottery A nationwide opportunity to gamble, in which the probabilities of wins of various types are known in advance. The current United Kingdom National Lottery offers prizes for correctly choosing three or more of the balls drawn from a pool of 49 individually numbered balls. Assuming that each ball is equally likely to be drawn—which appears to be the case at the time of writing—the *probability of winning the lowest prize (which requires three correctly identified balls) is

$$\frac{\binom{6}{3}\binom{43}{3}}{\binom{49}{6}} = \frac{8\,815}{499\,422} \approx \frac{1}{57}.$$

The probability of identifying all six balls is

$$\frac{\binom{6}{6}\binom{43}{0}}{\binom{49}{6}} = \frac{1}{13\,983\,816}.$$

If all six numbers are identified correctly, then the amount won depends on the number of other individuals who have selected the same six numbers. A good choice would therefore be an unpopular set of numbers: perhaps, 44, 45, 46, 47, 48, and 49.

natural logarithm The natural logarithm of a positive real number a is denoted by $\ln a$ and defined by

$$\ln a = \int_1^a \frac{1}{x}\,\mathrm{d}x,$$

so it is the (signed) area enclosed by the curve $y = 1/x$, the x-axis, and the lines $x = 1$ and $x = a$. If $a > 1$ then $\ln a > 0$, and if $a < 1$ then $\ln a < 0$.

Also $\ln 1 = 0$. The natural logarithm is related to the *exponential function by $\exp(\ln x) = x$ for all $x > 0$ and $\ln\{\exp(x)\} = x$ for all values of x. It is also referred to as the logarithm to base e. For historical reasons a natural logarithm is sometimes referred to as a Napierian logarithm, after the Scottish mathematician John Napier (1550–1617).

natural parameter *See* EXPONENTIAL FAMILY.

nearest-neighbour methods

1. In the context of *point processes, methods underlying *tests of randomness. If locations are positioned randomly in the plane, at a density λ per unit area, then the squared distance from a location to its nearest neighbour will have an *exponential distribution with *mean $1/(\pi\lambda)$, and the distance to its kth nearest neighbour will have a *gamma distribution with mean $k/(\pi\lambda)$.

2. In the context of *experimental design, methods used to improve estimates of plot yields by using neighbouring values as an indicator of the fertility of a plot of interest.

3. In the context of *cluster analysis and *discriminant analysis, methods used to determine from which *population, of a given set of populations, an observation has arisen.

negative binomial distribution (Pascal distribution) For trials classified as 'success' or 'failure', the *distribution of X, the number of trials required in order to obtain n successes. The trials are presumed to be *independent and it is assumed that each trial has the same *probability of success, p ($\neq 0$ or 1). The *probability function (*see diagram opposite*) is

$$\mathrm{P}(X = r) = \binom{r-1}{n-1} p^n (1-p)^{r-n}, \qquad r = n, n+1, \ldots.$$

The *mean of this distribution is n/p and the *variance is $n(1-p)/p^2$.

Special cases of the distribution were discussed by *Pascal in 1679. The name 'negative binomial' arises because the probabilities are successive terms in the *binomial expansion of $(P - Q)^{-n}$, where $P = 1/p$ and $Q = (1-p)/p$. Writing $Y = X - n$, an equivalent form for the distribution is

$$\mathrm{P}(Y = r) = \binom{n+r-1}{n-1} p^n (1-p)^r, \qquad r = 0, 1, 2, \ldots.$$

The variable X may be regarded as the sum of n independent *geometric variables, each with *parameter p. The case $n = 1$ therefore corresponds to the *geometric distribution.

The **arc-sinh transformation**, suggested by *Anscombe in 1948,

$$S = \left(\sqrt{n - \frac{1}{2}}\right) \sinh^{-1} \sqrt{\left(X + \frac{3}{8}\right) \Big/ \left(n - \frac{3}{4}\right)},$$

results in a *random variable S having an approximate standard *normal distribution.

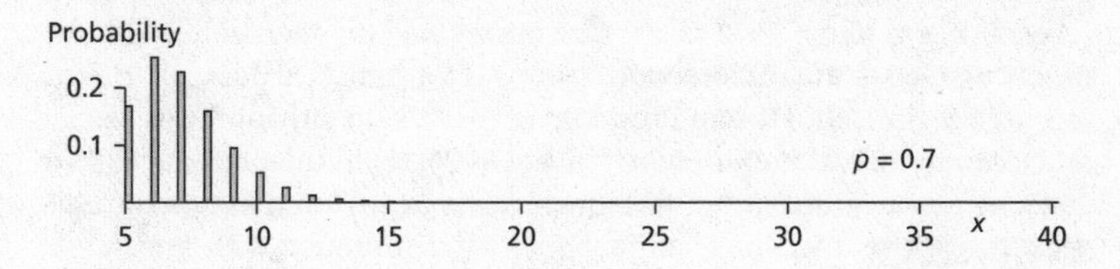

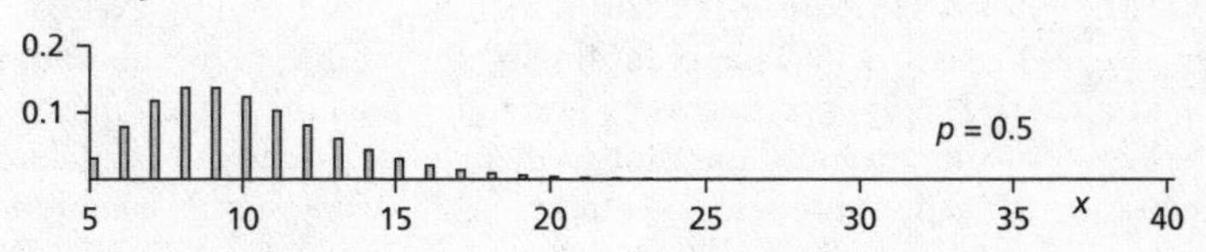

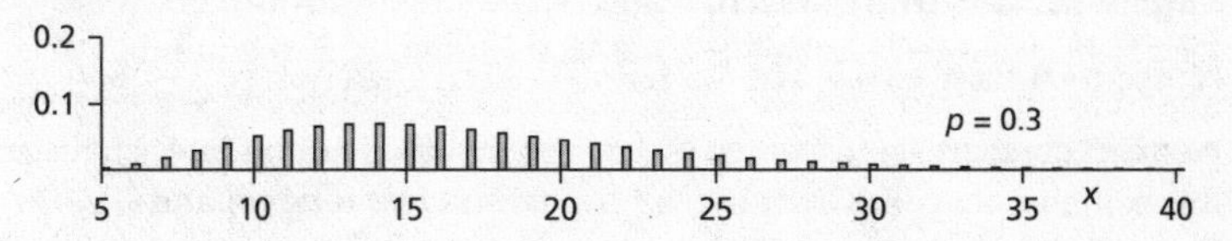

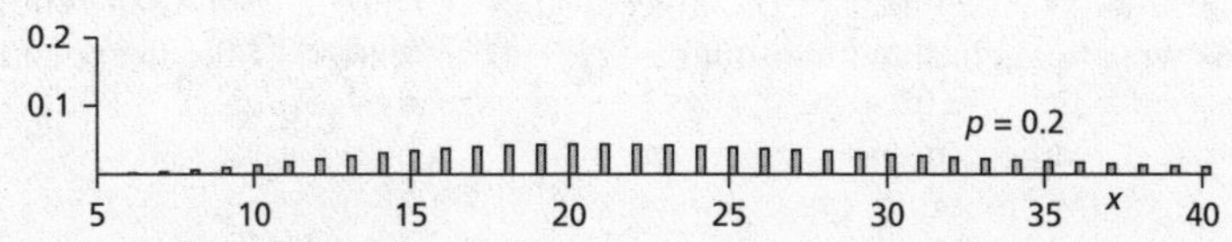

Negative binomial distribution. In each case $n = 5$; the shape is dependent upon the value of p.

negative correlation *See* SAMPLE CORRELATION COEFFICIENT.

negative exponential distribution *See* EXPONENTIAL DISTRIBUTION.

negative log-log link *See* GENERALIZED LINEAR MODEL.

negatively skewed; negative skewness *See* SKEWED.

Nelder, John Ashworth (1924–2010; b. Dulverton, England; d. Luton, England) English statistician. Nelder was a graduate of Cambridge U. In 1951 he joined the staff of the National Vegetable Research Station at Wellesbourne, becoming Head of Statistics. In 1968 he joined the staff of *Rothamsted, becoming Head of the Biomathematics Division. From 1972 to 2009 he was a visiting Professor at IC. He was co-author (with *Wedderburn) of the 1972 paper that introduced the *generalized linear model and led to the construction of the *GLIM statistical package. He was elected FRS in 1981. He was President of the *IBS in 1978 and was elected an Honorary Life Member of the Society in 2006. President of the *RSS in 1985, he was awarded its *Guy Medal in Silver in 1977 and in Gold in 2005.

SEE WEB LINKS

- Biography, interview, and photograph.

Nelder–Mead simplex method A method introduced by *Nelder and Roger Mead in a 1965 paper for finding the maximum (or minimum) of a function. The method begins by evaluating the function at $k+1$ vertices of a *simplex in k dimensions, where k is the number of variables. Subsequent locations are selected automatically, using simple geometric patterns that allow for both the extension of the search area (when, apparently, far from the maximum) and contraction (when close to the maximum). *See also* RESPONSE SURFACE.

Nelson–Aalen estimate *See* KAPLAN–MEIER ESTIMATE.

nested design An *experimental design in which the *variables have an implicit hierarchy. For example, a hospital has two wings (I and II). Patients in wing I are randomly assigned to either consultant A or consultant B. Patients in wing II are randomly assigned to either consultant C or consultant D. Thus consultants A and B are nested within the wing I patients, and consultants C and D are nested within the wing II

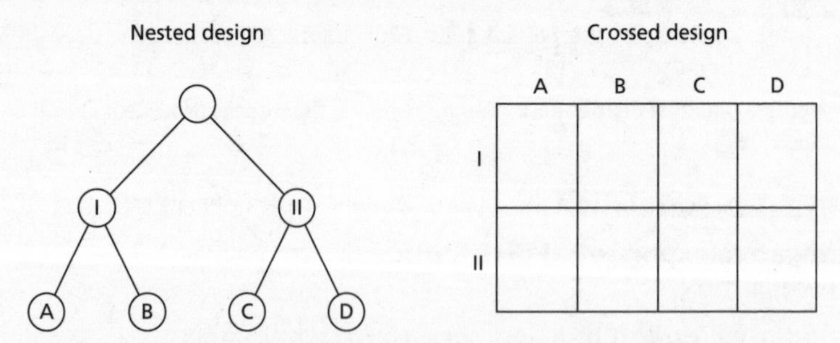

Nested design. The figure illustrates the difference between nested and crossed designs; in the latter, every combination of the crossed variables occurs.

patients. If, instead, all the patients in the hospital had been randomly assigned to one of the four consultants then this would have been a *crossed design.

Netherlands Society for Statistics and Operations Research The Society that has published the journals **Statistica Neerlandica* (since 1946) and *Kwantitatieve Methoden* (since 1982).

SEE WEB LINKS

- Society home page.

network A graphical representation of a problem by means of **nodes** connected by arcs of varying length or capacity. The arcs are usually directed. Examples include finding the shortest path, consisting of a connected set of arcs, that joins all the nodes, the *Chinese postman problem, and the *travelling salesman problem. In some applications it is the nodes that have associated values. *See also* CRITICAL PATH ANALYSIS; NETWORK FLOW PROBLEM; RELIABILITY THEORY.

network flow problem A *linear programming problem in which the objective is to maximize the overall flow from an initial **source** to a final **sink**. The *network consists of nodes connected by **directed arcs**, each arc having a given direction and a limited capacity. A **cut** is a line that breaks the network into two sections, one containing the source and the other containing the sink. The value of a cut is defined as the sum of the values on the broken arcs, ignoring arcs flowing from the sink section to the source section. The maximum flow can often be easily found using the **maximum flow/minimum cut theorem**, which states that these two quantities are equal.

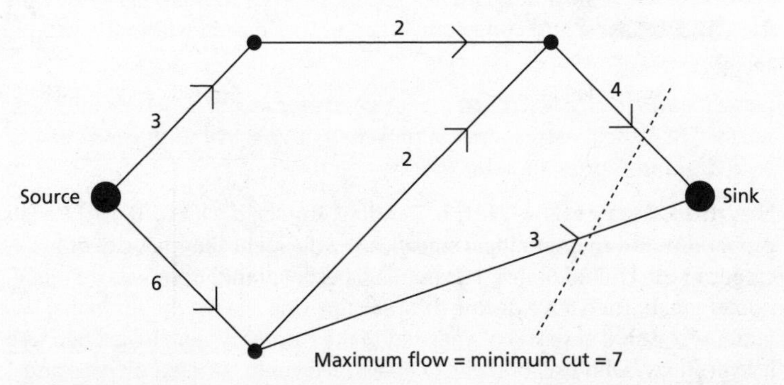

Network flow problem. This simple network illustrates the fact that the maximum flow equals the minimum cut.

neural net A mathematical structure that attempts to replicate the process by which the brain learns. The structure is conceived as consisting of a series of layers, the values of statistics computed in one layer being passed to processes ('neurons') in the next layer. The aim is that the output layer will provide accurate information about the process under investigation. The layers between the input and the output are called **hidden layers**. In order to be effective a neural network must be 'trained'—this is analogous to the *estimation of the *parameters of a *model. The *data used for this purpose are referred to as the **training sample**. Much of the process of statistical modelling can be viewed as the construction of a neural net.

Newman–Keuls test *See* MULTIPLE COMPARISON TEST.

Newton–Raphson method An algorithmic method for constructing a sequence of approximations to a root of an equation. Suppose it is desired to solve the equation $f(x)=0$ and that x_1 is an approximate value for the root of the equation. If we write $f'(x_n)$ for the value of the derivative of $f(x)$ (with respect to x) evaluated at the point $x=x_n$, the sequence defined by

$$x_{n+1} = x_n - \frac{f(x_n)}{f'(x_n)}$$

usually converges to a root of the equation. It is based on constructing the tangent to the curve $y=f(x)$ at the point $(x_n, f(x_n))$, and taking x_{n+1} to be the x-coordinate of the point where this tangent cuts the x-axis.

New Zealand Statistical Association This association was founded in 1948 and has about 400 members. Together with the *Statistical Society of Australia Inc. it publishes the **Australian and New Zealand Journal of Statistics*.

SEE WEB LINKS

- Association home page.

New Zealand Statistician The original journal published by the *New Zealand Statistical Association. Superseded in 1998 by the **Australian and New Zealand Journal of Statistics*.

Neyman, Jerzy (1894–1981; b. Bendery, Russia; d. Oakland, CA) Polish probabilist and mathematical statistician who spent the majority of his career in the United States. Neyman's paternal grandfather was a Polish nobleman, burned alive during the 1863 uprising against the Russians. His father was exiled as a twelve-year-old to the Crimea, where he trained as a lawyer. Jerzy, who was born Jerzy Splawa-Neyman, studied physics and mathematics at U Kharkov, where *Bernstein was an influential teacher. On the outbreak of war between Poland and Russia, Neyman was arrested as an enemy alien. On his release he got a job as a statistician although claiming to know no statistics! To train, he went first to Berlin and then to study under Karl *Pearson at UCL, where he met *Gosset and, crucially, Egon *Pearson. His work with Egon Pearson laid down the Neyman–Pearson theory of *hypothesis testing with the introduction of the ideas of an *alternative hypothesis and of power (*see* HYPOTHESIS TEST). At the onset of another war, fearing a further spell of imprisonment as an enemy alien, he moved to the USA, where he became Director of the Statistical Laboratory at UCB. In 1949 he was President of the *IMS and also its *Rietz Lecturer. He was elected to the NAS in 1963. He was awarded the *Guy Medal in Gold of the *RSS in 1966 and the *Wilks Award of the *ASA in 1968.

SEE WEB LINKS

- Biographical memoir by *Lehmann.

Neyman allocation *See* STRATIFIED SAMPLING.

Neyman Lecturer In honour of *Neyman, the Neyman Lectureship is awarded every three years by the *IMS to an individual who has made a major contribution to statistical theory.

Neyman–Pearson lemma A lemma, introduced in 1933 by *Neyman and Egon *Pearson, that gives a sufficient condition, in a *hypothesis test with *null hypothesis $\theta = \theta_0$ and *alternative hypothesis $\theta = \theta_1$, for choosing a *critical region, with given significance level, that maximizes the power of the test.

NH *See* HYPOTHESIS TEST; NULL HYPOTHESIS.

Nightingale, Florence (1820–1910; b. Florence, Italy; d. East Wellow, England) English pioneer of data analysis and graphical presentation (*see* CYCLIC DATA). Nightingale is best known for her work as a nurse at Scutari during the Crimean War, where the soldiers called her 'The Lady with the Lamp'. She was an efficient hospital administrator and compiled quantities of statistics in her drive for hospital reform: she standardized the reporting of deaths using *Miss Nightingale's Scheme for Uniform Hospital Statistics*. She has also been described as the 'the Passionate Statistician', and she wrote that Statistics is 'the most important science in the whole world'.

- Fuller biography and images.

NNT *Abbreviation for* NUMBER NEEDED TO TREAT.

node *See* NETWORK.

noise A term used by analysts of *time series to describe random fluctuations that may obscure the true *signal. A sequence of errors, in successive observations, that consists of *independent random values from a *normal distribution with zero *mean is termed **white noise**.

nominal variable *See* CATEGORICAL VARIABLE.

nomogram A graphical method for calculating the value of one *variable, given the values of other variables. A curve, or straight line, is drawn for each variable, and scales, usually non-linear, are marked on each curve in such a way that the points corresponding to any set of related values of the variables are collinear. Thus a straight line joining the points corresponding to the values of two of the variables will cross the curve corresponding to another variable at the point corresponding to the value of this variable.

non-ageing The property of a component whose *reliability is unaffected by age. *See also* FORGETFULNESS PROPERTY.

non-central chi-squared distribution If $Z_1, Z_2, \ldots, Z_\nu$ are *independent standard normal variables (*see* NORMAL DISTRIBUTION) and $\delta_1, \delta_2, \ldots, \delta_\nu$ are constants then X, given by

$$X = \sum_{k=1}^{\nu} (Z_k + \delta_k)^2,$$

is said to have a non-central chi-squared distribution with ν *degrees of freedom and **non-centrality parameter** λ, where

$$\lambda = \sum_{k=1}^{\nu} \delta_k^2.$$

The *probability density function f of X is given by

$$\mathrm{f}(x) = \frac{x^{(\frac{1}{2}\nu-1)}\mathrm{e}^{-\frac{1}{2}(x+\lambda)}}{2^{\frac{1}{2}\nu}} \sum_{j=0}^{\infty} \frac{x^j \lambda^j}{\Gamma(\frac{1}{2}\nu + j)2^{2j} j!}, \qquad x > 0,$$

where Γ is the *gamma function. The distribution, which is *unimodal, has *mean $(\nu + \lambda)$ and *variance $2(\nu + 2\lambda)$. If $\lambda = 0$ then the distribution becomes the *chi-squared distribution with ν degrees of freedom.

non-central *F*-distribution If Y_1 has a *non-central chi-squared distribution with ν_1 *degrees of freedom and non-centrality parameter λ, and if Y_2 is *independent of Y_1 and has a *chi-squared distribution with ν_2 degrees of freedom, then the ratio X, where

$$X = \frac{Y_1}{v_1} \Big/ \frac{Y_2}{v_2},$$

is said to have a non-central *F*-distribution with non-centrality parameter λ and with v_1 and v_2 degrees of freedom. The *probability density function f of X is given by

$$f(x) = \frac{e^{-\frac{1}{2}\lambda} v_1^{\frac{1}{2}v_1} v_2^{\frac{1}{2}v_2} x^{\left(\frac{1}{2}v_1 - 1\right)}}{B(\frac{1}{2}v_1, \frac{1}{2}v_2)(v_2 + v_1 x)^{\frac{1}{2}(v_1+v_2)}} \times$$

$$\sum_{j=0}^{\infty} \left[\left(\frac{\frac{1}{2}\lambda v_1 x}{v_2 + v_1 x} \right)^j \frac{(v_1 + v_2)(v_1 + v_2 + 2) \cdots (v_1 + v_2 + 2\{j-1\})}{j! v_1 (v_1 + 2) \cdots (v_1 + 2\{j-1\})} \right], x > 0,$$

where B is the *beta function. The distribution has *mean and *variance given by

$$\frac{v_2(v_1 + \lambda)}{v_1(v_2 - 2)} \quad \text{and} \quad \frac{2v_2^2\{(v_1 + \lambda)^2 + (v_1 + 2\lambda)(v_2 - 2)\}}{v_1^2(v_2 - 2)^2(v_2 - 4)},$$

respectively (with $v_2 > 4$). The distribution is *unimodal.

non-centrality parameter *See* NON-CENTRAL CHI-SQUARED DISTRIBUTION; NON-CENTRAL *F*-DISTRIBUTION; NON-CENTRAL *t*-DISTRIBUTION.

non-central *t*-distribution If Y_1 has a standard *normal distribution, if Y_2 is independent of Y_1 and has a *chi-squared distribution with v *degrees of freedom, and if λ is a non-zero constant, then X, given by

$$X = \frac{(Y_1 + \lambda)}{\sqrt{Y_2/v}},$$

is said to have a non-central *t*-distribution with v degrees of freedom and non-centrality parameter λ. The *probability density function f of X is given by

$$f(x) = \frac{e^{-\frac{1}{2}\lambda^2}}{\sqrt{\pi v}\,\Gamma(\frac{1}{2}v)} \left(\frac{v}{v + x^2} \right)^{\frac{1}{2}(v+1)} \sum_{j=0}^{\infty} \frac{\Gamma(\frac{1}{2}\{v + j + 1\})}{j!} \left(\frac{x\lambda\sqrt{2}}{\sqrt{v + x^2}} \right)^j, -\infty < x < \infty,$$

where Γ is the *gamma function. The *mean and *variance of X are

$$\frac{\Gamma(\frac{1}{2}\{v-1\})}{\Gamma(\frac{1}{2}v)} \lambda \sqrt{\frac{v}{2}} \quad \text{and} \quad \frac{v(1 + \lambda^2)}{(v - 2)} - \frac{1}{2} v\lambda^2 \left[\frac{\Gamma(\frac{1}{2}\{v-1\})}{\Gamma(\frac{1}{2}v)} \right]^2,$$

respectively (with $v > 2$). For large values of v the distribution of X is approximately a normal distribution with mean λ and variance 1.

non-ignorable non-response *See* NON-RESPONSE.

non-informative prior *See* BAYESIAN INFERENCE.

non-linear model A *model that involves a non-linear combination of its *parameters. For example,

$$E(Y) = \alpha + \beta e^{\gamma x},$$

where α, β, and γ are parameters and x and Y are *variables.

non-parametric regression The situation in which a *random variable, Y, has *expected value $g(x_1, x_2, \ldots, x_m)$, where $x_1, x_2, \ldots, x_m$ are m explanatory variables (*see* REGRESSION) and g is an unknown function. With $m = 1$ and n *observations, $y_1, y_2, \ldots, y_n$, *estimation may proceed using a *scatterplot smoother such as *loess, or by using **kernel regression**. In the latter case, to estimate the function g in the *regression equation

$$y_i = g(x) + \varepsilon_i \qquad (i = 1, \ldots, n),$$

a commonly used procedure employs the **Nadaraya-Watson estimator** given by

$$\widehat{g(x)} = \sum_{i=1}^{n} K(x, x_i) y_i \bigg/ \sum_{i=1}^{n} K(x, x_i),$$

where K is the chosen kernel (*see* KERNEL METHOD).

n

non-parametric test (distribution-free test) A test that makes no distributional assumptions about the *population under investigation. The adjective 'non-parametric' was introduced by *Wolfowitz in 1942. Non-parametric tests are usually very simple to perform (e.g. the *runs test and the *sign test) and often make use of *ranks (examples include the *Mann–Whitney and *Wilcoxon signed-rank tests). *See also* FRIEDMAN TEST; KENDALL'S TAU; KOLMOGOROV–SMIRNOV TEST; KRUSKAL–WALLIS TEST; SPEARMAN'S RHO; VAN DER WAERDEN TEST.

non-response A common problem for analysts of survey (*see* SAMPLE) *data. There are two types: complete non-response in which the individual selected for interview is absent, dead, or uncooperative; and **item non-response** where the interviewee fails to answer some (but not all) questions. Provided the lack of response is not a consequence of the questions being asked, neither type causes bias (*see* ESTIMATOR) in the subsequent analysis (this is called **ignorable non-response**). However, if the reason for the non-response is the question being asked (for example, the very rich may not wish to disclose exactly how rich they are), then this is described as **non-ignorable non-response**. *Imputation provides a partial solution to item non-response.

nonsense correlation A term used to describe a situation where two variables (X and Y, say) are *correlated without being causally related to one another. The usual explanation is that they are both related to a third variable, Z. Often the third variable is time. For example, if we compare the price of a detached house in Edinburgh in 1920, 1930, . . . with the size of the population of India at those dates, a 'significant' positive correlation will be found, since both variables have increased markedly with time. The first comprehensive study of nonsense correlation was undertaken in 1926 by *Yule, who considered the apparent connection between the fall in Church of England marriages and the concurrent increase in life expectancy. *See also* GOOSEBERRY BUSHES; RUM CONSUMPTION.

non-singular matrix *See* MATRIX.

normal approximation to the binomial distribution *See* BINOMIAL DISTRIBUTION.

normal approximation to the Poisson distribution *See* POISSON DISTRIBUTION.

normal distribution (Gaussian distribution) The distribution of a *random variable X for which the *probability density function f is given by

$$\mathrm{f}(x) = \frac{1}{\sigma\sqrt{2\pi}} \exp\left(-\frac{(x-\mu)^2}{2\sigma^2}\right), \qquad -\infty < x < \infty.$$

The *parameters μ and σ^2 are, respectively, the *mean and *variance of the distribution. The distribution is denoted by $\mathrm{N}(\mu, \sigma^2)$. If the random variable X has such a distribution, then this is denoted by $X \sim \mathrm{N}(\mu, \sigma^2)$ and the random variable may be referred to as a **normal variable**.

The graph of $\mathrm{f}(x)$ approaches the x-axis extremely quickly, and is effectively zero if $|x-\mu| > 3\sigma$ (hence the *three-sigma rule). In fact, $\mathrm{P}(|X-\mu| < 2\sigma) \approx 95.5\%$ and $\mathrm{P}(|X-\mu| < 3\sigma) \approx 99.7\%$. The first derivation of the form of f is believed to be that of *de Moivre in 1733. The description 'normal distribution' was used by *Galton in 1889, whereas 'Gaussian distribution' was used by Karl *Pearson in 1905.

The normal distribution is the basis of a large proportion of statistical analysis. Its importance and ubiquity are largely a consequence of the *Central Limit Theorem, which implies that averaging almost always leads to a bell-shaped distribution (hence the name 'normal'). *See* BELL-CURVE.

The **standard normal distribution** has mean 0 and variance 1. A random variable with this distribution is a **standard normal variable**. It is often denoted by Z and we write $Z \sim \mathrm{N}(0, 1)$. Its probability density function is usually denoted by ϕ and is given by

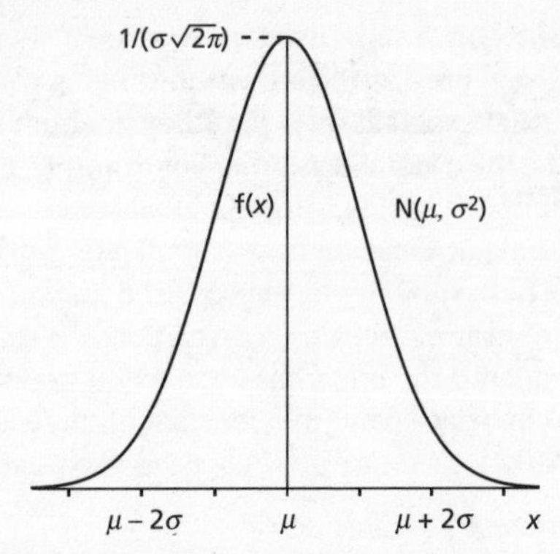

Normal distribution. The diagram illustrates the probability density function of a normal random variable X having expected value μ and variance σ^2. The distribution has mean, median, and mode at $x = \mu$, where the density function has value $1/(\sigma\sqrt{2\pi})$. Note that almost all the distribution (99.7%) lies within 3σ of the central value.

$$\phi(z) = \frac{1}{\sqrt{2\pi}} e^{-\frac{1}{2}z^2}, \qquad -\infty < z < \infty.$$

If X has a general normal distribution $N(\mu, \sigma^2)$ then Z, defined by the **standardizing transformation**

$$Z = \frac{X - \mu}{\sigma},$$

has a standard normal distribution. It follows that the graph of the probability density function of X is obtained from the corresponding graph for Z by a stretch parallel to the z-axis, with centre at the origin and scale-factor σ, followed by a translation along the z-axis by μ.

The *cumulative distribution function of Z is usually denoted by Φ and tables of values of $\Phi(z)$ are commonly available (*see* APPENDIX V; see also APPENDIX IV). These tables usually give $\Phi(z)$ only for $z > 0$, since values for negative values of z can be found using

$$\Phi(z) = 1 - \Phi(-z).$$

The tables can be used to find cumulative probabilities for $X \sim N(\mu, \sigma^2)$ via the standardizing transformation given above, since, for example,

$$P(X < x) = \Phi\left(\frac{x - \mu}{\sigma}\right).$$

As an example, if $X \sim N(7, 25)$ then the probability of X taking a value between 5 and 10 is given by

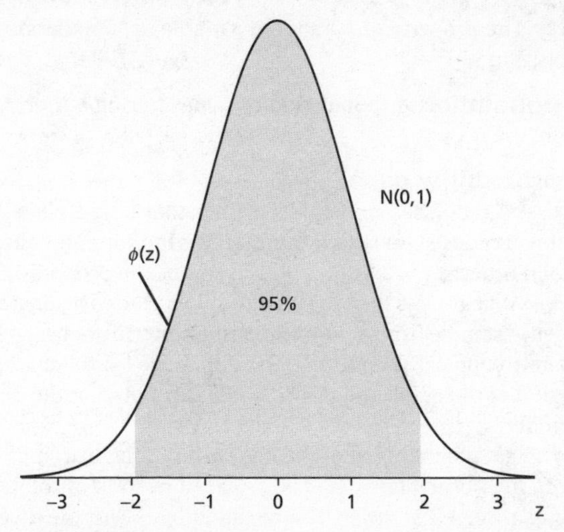

Standard normal distribution. The distribution is centred on 0, with 99.7% falling between − 3 and 3 and 95% falling between − 1.96 and 1.96.

$$\begin{aligned} P\left(\frac{5-7}{5} < Z < \frac{10-7}{5}\right) &= \Phi(0.6) - \Phi(-0.4) = \Phi(0.6) - \{1 - \Phi(0.4)\} \\ &\approx 0.7257 + 0.6554 - 1 \approx 0.38. \end{aligned}$$

The normal distribution plays a central part in the **theory of errors** that was developed by *Gauss. In the theory of errors, the **error function (erf)** is defined by

$$\text{erf}(x) = 2\Phi(x\sqrt{2}) - 1.$$

An important property of the normal distribution is that any *linear combination of *independent normal variables is normal: if $X_1 \sim N(\mu_1, \sigma_1^2)$ and $X_2 \sim N(\mu_2, \sigma_2^2)$ are independent, and a and b are constants, then

$$aX_1 + bX_2 \sim N(a\mu_1 + b\mu_2, a^2\sigma_1^2 + b^2\sigma_2^2),$$

with the obvious generalization to n independent normal variables. Many distributions can be approximated by a normal distribution for suitably large values of the relevant parameters. *See also* BINOMIAL DISTRIBUTION; CHI-SQUARED DISTRIBUTION; POISSON DISTRIBUTION; *t*-DISTRIBUTION.

normal equations *See* MULTIPLE REGRESSION MODEL.

normal inspection *See* QUALITY CONTROL.

normality The property of a *random variable or *population having a *normal distribution.

normal population A *population of values having a *normal distribution.

normal probability paper Specialized graph paper (*see diagram opposite*) that can be used to help decide whether a *sample of n *observations is consistent with a *normal distribution. The observations are ranked in order: $x_{(1)} \leq x_{(2)} \leq \cdots \leq x_{(n)}$ and p_j, the proportion of values in the sample that are less than or equal to $x_{(j)}$, is plotted against $x_{(j)}$ on the graph paper. A sample from a normal distribution will appear as a series of approximately collinear points. A strong departure from linearity in the central region of the graph will provide an indication that the distribution is not normal.

In order to get a better idea of the behaviour in the extremes of the distribution, the observations may be replaced by their absolute values before being arranged in order. The resulting plot, with the theoretical probabilities appropriately adjusted, is a **half-normal plot**. *See also* Q-Q PLOT.

normal random variable A *random variable having a *normal distribution.

normal score (rankit) The normal score corresponding to the kth largest of n *observations is the *expected value of the kth largest of n *independent observations from a standard *normal distribution. Several *non-parametric tests have been devised in which the ordered observations have their values replaced by the corresponding normal scores. These tests usually have greater power than the corresponding tests based on *ranks.

normal test *See* Z-TEST.

normal variable *See* NORMAL DISTRIBUTION.

Norwegian Statistical Association The Association was founded in 1919. It is one of the four societies that jointly publish the **Scandinavian Journal of Statistics*. It also publishes the Norwegian language *Tilfeldig Gang* (*Random Walk*).

SEE WEB LINKS

- Association home page.

not significant (n.s.) *See* HYPOTHESIS TEST.

nugget model A model of a *time series or *spatial process that allows for the *autocorrelation at lag 0 to be less than 1. This acknowledges the

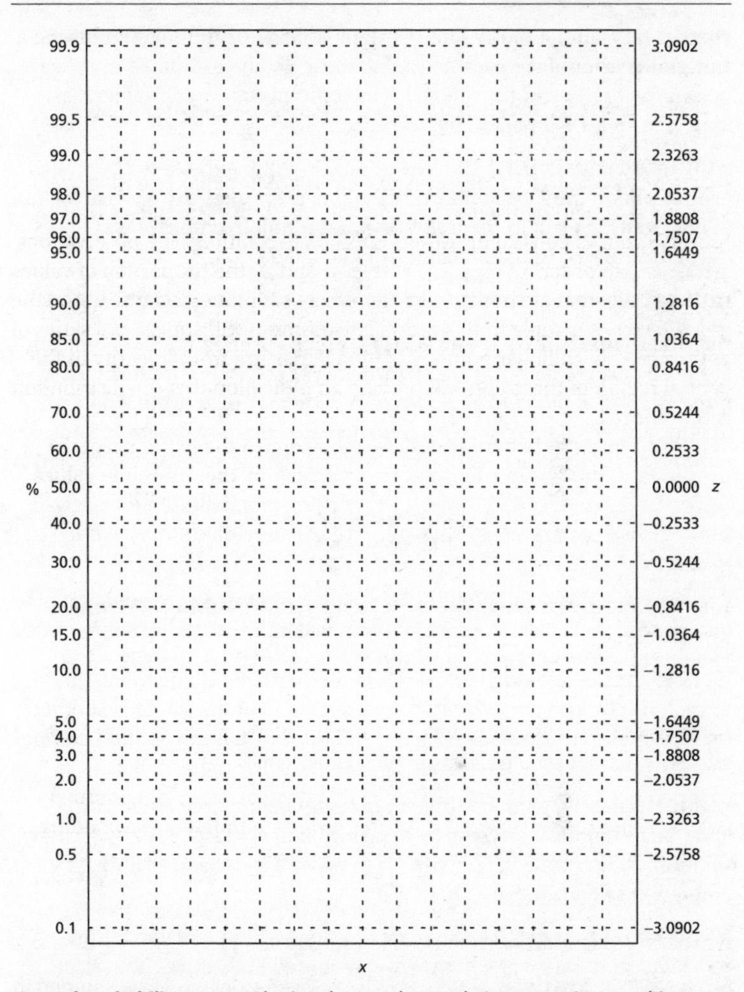

Normal probability paper. Plotting the sample cumulative proportions on this paper will result in an approximate straight line if the data have arisen from a normal distribution.

fact that, with real data, there are measurement errors such that repeat measurements at the 'same' time or place may actually vary.

nuisance parameter A *parameter that must be estimated, even though it is of no immediate interest. The most familiar example occurs in the construction of a *confidence interval for an estimate of a *population mean. The *sample mean is the obvious estimate, but the *population variance is a nuisance parameter that must also be estimated in order to determine the size of the interval. The term 'nuisance parameter' first appeared in a 1940 paper by *Hotelling.

null hypothesis (NH) The hypothesis, in a *hypothesis test, which is used to obtain the *probability distribution, and hence the *critical region, of the *statistic used in the test. The phrase 'null hypothesis' was introduced by Sir Ronald *Fisher in 1935.

null set An alternative name for the empty set (*see* SAMPLE SPACE).

number needed to treat (NNT) A phrase used in medical contexts. Suppose past results show that treatment j ($j = 1, 2$) is successful for a proportion p_j of patients. Suppose $p_1 > p_2$. On average, with N patients, treatment 1 will cure $N(p_1 - p_2)$ more patients than treatment 2. Thus the number needed to treat so that one extra patient is successfully treated is M, given by

$$M = \frac{1}{p_1 - p_2}.$$

If treatment 1 costs an amount k per patient more than treatment 2, then one interpretation of the above is that it will be worth using treatment 1 rather than treatment 2 only if treating a patient successfully has a value greater than kM.

numerical Usually referring to a *variable whose possible values are numbers (as opposed to, for example, categories).

numerical analysis The branch of mathematics concerned with obtaining numerical answers by approximations, rather than by analytic solution. An example is provided by the use of *Simpson's rule for the evaluation of an integral.

Nyquist frequency The shortest detectable *frequency in a *time series. All time series are, in practice, recorded at discrete time points (for example, one reading per second). Periodic variations that happen more rapidly than the shortest time interval cannot be detected. With time intervals of size t the Nyquist frequency is t^{-1} cycles per unit time.

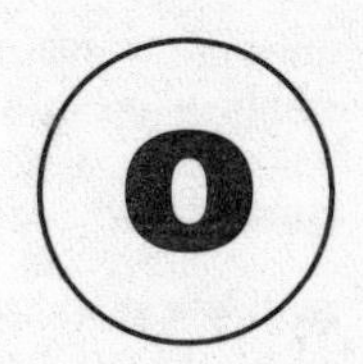

objective function *See* LINEAR PROGRAMMING.

O'Brien test In *regression, a test for *association between an explanatory variable and a dependent variable when the latter has been *censored. The usual application concerns the relation between the survival of a patient and a *variable describing some risk.

observation A result of an experiment or trial in which a variable, either *numerical or *categorical, is measured.

observed frequency *See* FREQUENCY.

occupancy distribution In a *queue, this is the long-term *probability distribution of the number of items in the system.

OC-curve *See* ACCEPTANCE SAMPLING.

Ockham's razor Essentially the **principle of parsimony** which states that if one is provided with a variety of explanations (e.g. a variety of statistical models) one should prefer the simplest. William of Ockham (*c.* 1285–1349) was an English philosopher who held that a complicated explanation should not be accepted without good reason, and wrote '*Frustra fit per plura, quod fieri potest per pauciora*'. ('It is vain to do with more what can be done with less.')

odds If the *probability of an event happening is p, then the odds on that event happening (as opposed to it not happening) are

$$\frac{p}{1-p}.$$

If the odds are expressed as a simple fraction m/n, where m and n are positive integers, then the probability of the event happening is

$$\frac{m}{m+n}.$$

In the case $m > n$ the odds are said to be 'm to n on'. In the case $m < n$ the odds are said to be 'n to m against'. Thus, for example, if $p = \frac{4}{5}$ then the odds are '4 to 1 on', and if $p=\frac{1}{5}$ then the odds are '4 to 1 against'. Confusingly, in the case $p=\frac{1}{2}$ the odds are said to be 'evens'!

odds ratio *See* TWO-BY-TWO TABLE.

ogive A diagram representing grouped numerical data in which *cumulative frequency is plotted against upper *class boundary, and a curve is fitted to the resulting points. *See* CUMULATIVE FREQUENCY POLYGON.

Old Faithful Geyser in the Yellowstone National Park which erupts in a fashion that has challenged analysts of *time series. The data in the table refer to the period 1–8 August in 1978. For each day, the table shows x, the duration (in minutes, to the nearest 0.1 minute) of the current eruption, and y, the time (in minutes, to the nearest minute) until the next eruption. Observations refer to daylight hours only.

DAY		DURATION (x) AND TIME INTERVAL (y) FOR SUCCESSIVE ERUPTIONS													
1	x	4.4	3.9	4.0	4.0	3.5	4.1	2.3	4.7	1.7	4.9	1.7	4.6	3.4	
	y	78	74	68	76	80	84	50	93	55	76	58	74	75	
2	x	4.3	1.7	3.9	3.7	3.1	4.0	1.8	4.1	1.8	3.2	1.9	4.6	2.0	
	y	80	56	80	69	57	90	42	91	51	79	53	82	51	
3	x	4.5	3.9	4.3	2.3	3.8	1.9	4.6	1.8	4.7	1.8	4.6	1.9	3.5	
	y	76	82	84	53	86	51	85	45	88	51	80	49	82	
4	x	4.0	3.7	3.7	4.3	3.6	3.8	3.8	3.8	2.5	4.5	4.1	3.7	3.8	3.4
	y	75	73	67	68	86	72	75	75	66	84	70	79	60	86
5	x	4.0	2.3	4.4	4.1	4.3	3.3	2.0	4.3	2.9	4.6	1.9	3.6	3.7	3.7
	y	71	67	81	76	83	76	55	73	56	83	57	71	72	77
6	x	1.8	4.6	3.5	4.0	3.7	1.7	4.6	1.7	4.0	1.8	4.4	1.9	4.6	2.9
	y	55	75	73	70	83	50	95	51	82	54	83	51	80	78
7	x	3.5	2.0	4.3	1.8	4.1	1.8	4.7	4.2	3.9	4.3	1.8	4.5	2.0	
	y	81	53	89	44	78	61	73	75	73	76	55	86	48	
8	x	4.2	4.4	4.1	4.1	4.0	4.1	2.7	4.6	1.9	4.5	4.5	4.8	4.1	
	y	77	73	70	88	75	83	61	78	61	81	81	80	79	

Old Faithful. Long time intervals between eruptions are generally followed by long eruptions, but predicting the future eruption pattern is a considerable challenge to time series experts.

OLS Abbreviation for ordinary least squares (*see* METHOD OF LEAST SQUARES).

one-sided confidence interval *See* CONFIDENCE INTERVAL.

one-sided hypothesis *See* HYPOTHESIS TEST.

one-tailed test *See* HYPOTHESIS TEST.

one-tail probability For the *random variable X, with *median m, the one-tail probability associated with the value x is $P(X \leq x)$ if $x < m$ and $P(X \geq x)$ if $x > m$. If $x < m$, then the corresponding **two-tail probability** is $P(X \leq x)+P\{X \geq (2m - x)\}$. For a *symmetric distribution this equals $2P(X \leq x)$.

open question *See* CLOSED QUESTION.

operating characteristic curve *See* ACCEPTANCE SAMPLING.

operational research; operations research The application of statistical and scientific methods to the management of commercial and industrial processes and of military and governmental activities.

opinion poll A survey of opinions, usually social or political. People in a *sample are interviewed or reply to a questionnaire. The questions asked usually have a list of possible responses, and the responses are analysed. The results are usually given as percentages of those expressing an opinion on the particular question.

opportunity sampling *See* CONVENIENCE SAMPLING.

optimization The minimization or maximization of some function, usually subject to restrictions (which are often on the values of the variables over which the optimization takes place).

optimum allocation *See* STRATIFIED SAMPLING.

optimum interpolation The term used by meteorologists for the procedure more generally known as *kriging.

order statistics Particular values in a data set that has been ordered by magnitude. The best-known statistics based on ordered values are the *median and the *range, which provide measures of location and spread. An example of naturally ordered data is provided by the lifetimes of failing light-bulbs, all initially new. *See also* NORMAL SCORES; Q-Q PLOT; TRIMMED MEAN.

ordinal variable A *categorical variable in which the categories have an obvious order (e.g. strongly disagree, disagree, neutral, agree, strongly agree). Specialist models are required for ordinal variables to take account of their ordered categories. *See also* ASSOCIATION.

ordinary least squares (OLS) *See* METHOD OF LEAST SQUARES.

ordinate *See* CARTESIAN COORDINATES.

origin *See* CARTESIAN COORDINATES.

Ornstein–Uhlenbeck process *See* BROWNIAN MOTION.

orthogonality *See* VARIANCE–COVARIANCE MATRIX.

orthogonal matrix *See* MATRIX.

orthogonal polynomials *Polynomials designed to simplify calculations when a response variable is believed to be related to a polynomial function of an explanatory variable. If $x_1, x_2, \ldots, x_n$ are n *observations on the *variable X, then the polynomials $\phi_1(x)$ and $\phi_2(x)$ are orthogonal if

$$\sum_{j=1}^{n} \phi_1(x_j)\phi_2(x_j) = 0.$$

If we wish to fit the model

$$\mathrm{E}(Y_j) = \beta_0 + \beta_1 x_j + \beta_2 x_j^2 + \cdots + \beta_m x_j^m, \qquad j = 1, 2, \ldots, n,$$

then we will need (*see* MULTIPLE REGRESSION MODEL) to invert the *matrix product $\mathbf{X}'\mathbf{X}$, where

$$\mathbf{X} = \begin{pmatrix} 1 & x_1 & x_1^2 & \ldots & x_1^m \\ 1 & x_2 & x_2^2 & \ldots & x_2^m \\ \vdots & \vdots & \vdots & \ddots & \vdots \\ 1 & x_n & x_n^2 & \ldots & x_n^m \end{pmatrix}.$$

This may not be easy, since $\mathbf{X}'\mathbf{X}$ will often be nearly singular. One solution is to rewrite the model in the form

$$\mathrm{E}(Y_j) = \gamma_0\phi_0(x_j) + \gamma_1\phi_1(x_j) + \cdots + \gamma_m\phi_m(x_j),$$

where $\phi_0(x) = 1$ and $\phi_k(x)$ is a polynomial of degree k ($k = 1, 2, \ldots, m$). With n observations on x, denoted by $x_1, x_2, \ldots, x_n$, the polynomials are chosen to be orthogonal, i.e. to satisfy the requirements that

$$\sum_{j=1}^{n} \phi_k(x_j)\phi_l(x_j) = 0, \qquad \text{for all } k \text{ and } l, k \neq l.$$

The revised $\mathbf{X}$ matrix has a typical element $\phi_k(x_j)$ and is such that the product $\mathbf{X}'\mathbf{X}$ is now a diagonal matrix and is therefore easily inverted.

Often the appropriate degree of dependence on x (in other words, the value of m) is unknown. The reformulation also makes it easy to select an appropriate model without the need for further iterations. In the *ANOVA table the contributions for each polynomial can be listed separately.

outcome *See* SAMPLE SPACE.

outcome variable *See* REGRESSION.

outer fence *See* QUARTILE.

outlier An observation that is very different to other observations in a set of data. Since the most common cause is recording error, it is sensible to search for outliers (by means of *summary statistics and plots of the data) before conducting any detailed statistical modelling.

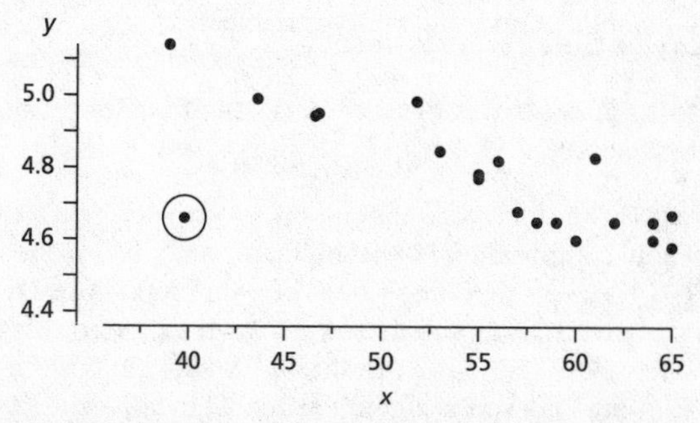

Outlier. A small data set showing an obvious outlier. Whenever feasible, data should be plotted, since including an outlier will usually make nonsense of any calculations.

Various indicators are used to identify outliers. One is that an observation has a value that is more than 2.5 standard deviations from the mean. Another is that an observation has a value that lies more than $1.5I$ beyond the upper or the lower quartile, where I is the interquartile range (*see* BOXPLOT).

If there is only a single outlier present, then an effective test for the presence of an outlier is the **Dixon test**. Denoting the ordered observations by $y_{(1)} \leq y_{(2)} \leq \cdots \leq y_{(n)}$ the test statistic (*see* HYPOTHESIS TEST) is either

$$\frac{y_{(n)} - y_{(n-1)}}{y_{(n)} - y_{(1)}} \quad \text{or} \quad \frac{y_{(2)} - y_{(1)}}{y_{(n)} - y_{(1)}},$$

depending on whether $y_{(n)}$ appears unusually large, or $y_{(1)}$ appears unusually small. Special tables are required in order to determine significance.

For data from a *normal distribution, the test statistic, G, of the **Grubbs test**, suggested by *Grubbs in 1969, is

$$G = \frac{1}{s}\max\{y_{(n)} - \bar{y}, \bar{y} - y_{(1)}\},$$

where $\bar{y}$ and s are the *sample mean and *standard deviation.

The **Rosner test** for multiple outliers relies on ordering the n observations in terms of their distance from $\bar{y}$. Let y_m be the observation that is the mth closest to $\bar{y}$ and let the mean and standard deviation of the $m-1$ observations closest to the mean be $\bar{y}_{m-1}$ and s_{m-1}. The decision as to whether y_m is an outlier is based on the value of

$$\frac{1}{s_{m-1}}|y_m - \bar{y}_{m-1}|.$$

out-of-bag sample *See* BOOSTING.

overdispersed; overdispersion *See* EXTRA-BINOMIAL VARIATION; INDEX OF DISPERSION.

overfitting This occurs when the *model used to describe a set of *data is unnecessarily complicated. For example, if, with x taking the values (1, 10, 40, 60), the corresponding values of y are found to be (2, 20, 80.1, 120), then the model $y=2x$ should suffice. Anything more complex would probably be described as overfitting. The term is used in *machine learning to describe a complex model that fits known data well but is less successful than a simpler model at fitting the values of a subsequent data set.

overparameterized model A *model having more *parameters than can be estimated from the *data. For example, suppose that the yields of two types of tomatoes are to be compared using the data $\{y_{jk}\}$, where $j\ (=1, 2)$ signifies the treatment and k is the number of the observation. Consider the model

$$y_{jk} = \mu + \tau_j + \varepsilon_{jk},$$

where the $\{\varepsilon_{jk}\}$ are *independent *random errors, each with *mean 0 and *variance σ^2. As it stands, this model is overparameterized, because for each value of j, only $(\mu + \tau_j)$ can be estimated. The value attributed to μ can be arbitrarily chosen. However, the problem can be solved either by rewriting the model

$$y_{jk} = \mu_j + \varepsilon_{jk},$$

or by retaining the original model and adding a constraint such as $\tau_1 + \tau_2 = 0$.

oversampling The deliberate selection of individuals of a rare type in order to obtain reasonably precise estimates of the properties of this type. In a *population which includes such a rare type, a random *sample of the entire population might result in very few (or none) of these individuals being selected. Oversampling implies the deliberate sampling of a much higher proportion of this type than of the rest of the population. This is a form of *stratified sampling.

Paasche, Hermann (1851–1925; b. Magdeburg, Germany; d. Detroit, MI) German economist remembered for suggesting the eponymous Paasche price index (*see* PRICE INDEX), though he was not the first to suggest it.

SEE WEB LINKS

- Photograph.

Paasche price index *See* PRICE INDEX.

PACF *Abbreviation for* PARTIAL AUTOCORRELATION FUNCTION.

paired comparisons *See* EXPERIMENTAL DESIGN.

paired-sample test *See* HYPOTHESIS TEST.

pairs plot *See* DRAFTSMAN'S PLOT.

pairwise independent Events (*see* SAMPLE SPACE) $A_1, A_2, \ldots, A_n$ are pairwise independent if every possible pair of these events are independent, i.e.

$$\mathrm{P}(A_j \cap A_k) = \mathrm{P}(A_j) \times \mathrm{P}(A_k) \qquad \text{for all } j, k, j \neq k.$$

It should be noted that pairwise independence is not sufficient to guarantee that

$$\mathrm{P}(A_1 \cap A_2 \cap \cdots \cap A_n) = \mathrm{P}(A_1) \times \mathrm{P}(A_2) \times \cdots \times \mathrm{P}(A_n).$$

See INDEPENDENT EVENTS; INTERSECTION.

panel study A *longitudinal study of the same group of people over time. By contrast with a *cohort study the panel study gathers information on the same people at each time point. In practice the participation is not constant over time, because of *attrition. A disadvantage compared with a *cross-sectional study is that the panel ceases to be representative of the entire population because of the absence of younger members in the later years of the study.

parallel *See* RELIABILITY THEORY.

parameter A constant appearing as part of the description of a *probability function or the *probability density function of a family of *distributions. The shape of the distribution depends on the value(s) given to the parameter(s). For example, the parameters of a *binomial distribution are n and p, the parameter of a *Poisson distribution is λ, the parameters of a normal distribution are μ and σ, and ν is the parameter of both a *chi-squared distribution and a *t-distribution. The term is also used more widely; for example it is used to describe the unknowns (α, β) in *linear regression and the corresponding terms in the *multiple regression model. The word 'parameter', in a statistical context, was first used by Sir Ronald *Fisher in 1922.

parent population The *population from which a *sample has been obtained.

Pareto, Marquis Wilfredo (1848–1923; b. Paris, France; d. Geneva, Switzerland) Italian economist famed for his studies of the distribution of income. After training at the Polytechnic Institute, Turin, Pareto began his career as a lecturer in mathematics and engineering at U Florence. However, he became increasingly concerned by what he saw as social injustices (*see* EIGHTY/TWENTY RULE) and subsequently moved to U Lausanne as Professor of Economics. He is best remembered for his *Trattato di Sociologia Generale*, completed in 1912 at his home, Angora Villa, which he shared with eighteen Angora cats.

Pareto distribution A *random variable x with the simplest type of Pareto distribution has *probability density function given by

$$f(x) = ak^a x^{-(a+1)}, \qquad x \geq k,$$

where a and k are positive constants (*see diagram overleaf*).

*Pareto proposed the distribution in terms of its *cumulative distribution function, $\{1 - (k/x)^a\}$ which he believed gave a good approximation to the *proportion of incomes that were less than x. According to Pareto the shape of the income distribution was the same for all countries, though the values of a and k varied from country to country. *See* EIGHTY/TWENTY RULE.

Pareto plot A diagram (*see overleaf*) similar to a *bar chart, with no gaps between the bars and with the bars ordered in decreasing order of the *frequencies they represent. The scale is usually *proportion rather than frequency. It is common to superimpose an indicator of cumulative proportion.

p

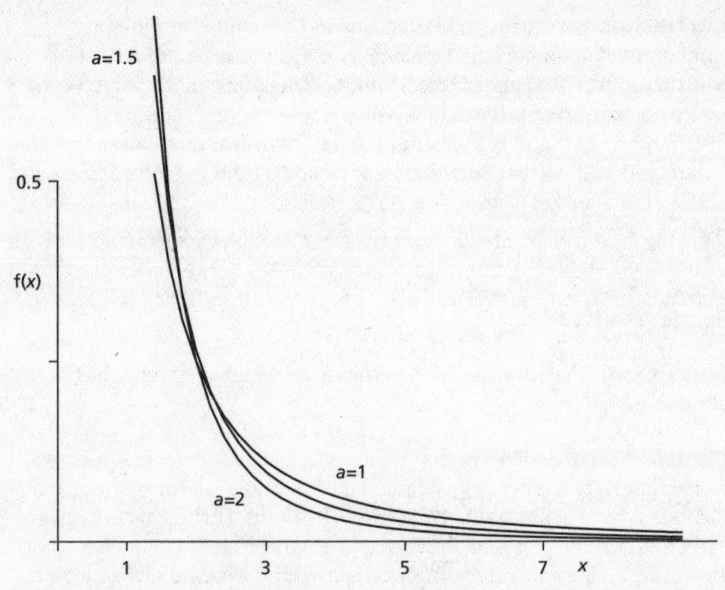

Pareto distribution. Pareto distributions were proposed to model the distribution of income. Those illustrated have $k = 1$.

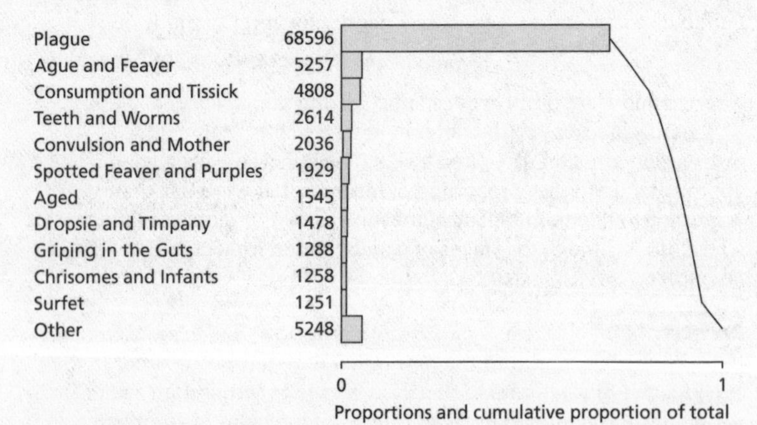

Pareto plot. This plot illustrates the causes of death of Londoners in 1665. The data and spelling are taken from the *Annual Bill of Mortality for London*.

partial autocorrelation function (PACF) A measure of *autocorrelation at lag k that takes account of the autocorrelation at lags less than k. The plot of the PACF can be a useful guide to the appropriate order of an autoregressive *time series.

partial confounding *See* FACTORIAL DESIGN.

partial correlation The *correlation between two *variables after allowing for the effect of other variables. For variables A, B, and C, with, for example, r_{AB} denoting the correlation coefficient between A and B, the partial correlation coefficient between A and B allowing for C is $r_{AB|C}$, given by

$$r_{AB|C} = \frac{r_{AB} - r_{AC}r_{BC}}{\sqrt{(1 - r_{AC}^2)(1 - r_{BC}^2)}}.$$

partial eta-squared *See* ANOVA.

partial least squares (PLS) A method for handling correlated explanatory variables in the context of a *multiple regression model. In PLS the first stage is to determine k *uncorrelated variables that are linear combinations of the explanatory variables. The combinations are chosen for their predictive ability. **Principal components regression analysis** uses a different technique to achieve the same objective.

partial likelihood Suppose that the *likelihood for a set of data depends on two *parameters θ and ϕ, both unknown. If the likelihood can be expressed as a product of two functions, one involving θ alone, and the second involving both parameters, then information about θ may be obtained by considering the first function (the 'partial likelihood') on its own. This idea was proposed by Sir David *Cox in 1972.

Parzen, Emanuel (1929-   ; b. New York City) American statistician specializing in the analysis of *time series. Parzen, a graduate of Harvard U, gained his doctorate in 1953 at UCB. He held posts at Columbia U (1953), Stanford U (1956), SUNY (1970), and Texas A & M U (1978). At Stanford U *Wahba was one of his research students. He was awarded the 1994 *Wilks Award of the *ASA.

SEE WEB LINKS

- Fuller biography, interview, and photographs.

Parzen Prize Established by *Parzen on his 65th birthday, this biennial prize is awarded by Texas A & M U for statistical innovation.

Parzen window *See* PERIODOGRAM.

Pascal, Blaise (1623–62; b. Clermont-Ferrand, France; d. Paris, France) French mathematical prodigy, whose first paper was published when he was seventeen. His father, also a mathematician, was employed at one time as a tax collector and Blaise, at the age of nineteen, invented a mechanical calculator capable of working with a currency that used multiples of both 12 and 20. In a correspondence with *Fermat concerning the *problem of points, Pascal laid the foundations for the modern treatment of *probability. Following that correspondence he wrote *Traité du triangle arithmetique* (*Treatise on the arithmetical triangle*) which showed the usefulness of *Pascal's triangle. He is quoted as saying 'It is not certain that everything is uncertain'. A street in Paris, a lunar crater, a computer language, and the S. I. unit of pressure are named after him.

SEE WEB LINKS

- Information and portrait

PASCAL A computer language named in honour of Blaise *Pascal.

Pascal distribution *Alternative name for* NEGATIVE BINOMIAL DISTRIBUTION.

Pascal's triangle A triangle of numbers in which the *r*th item on the *n*th row is the value of ${}^{n-1}C_{r-1}$, the number of different *combinations of $(r-1)$ objects chosen from $(n-1)$. The sum of the numbers on the *n*th row is 2^{n-1}.

```
          1
        1   1
      1   2   1
    1   3   3   1
  1   4   6   4   1
1   5  10  10   5   1
```

Thus ${}^{6-1}C_{4-1} = {}^{5}C_{3} = 10$. Apart from the 1 at the beginning and end of each row, each number is the sum of the two nearest numbers in the row above. For example $10 = 4 + 6$.

The relevance of the numbers in the context of *probability was noted by *Pascal and the description 'Pascal's triangle' was first used in a book on probability by *Montmort.

path analysis A procedure introduced by the geneticist *Wright that uses *correlation to study the relationships between *variables. It is now much used in the social sciences. The results may be displayed in a *causal diagram.

path vector *See* RELIABILITY THEORY.

pattern recognition A term used in *machine learning to describe the allocation of a class or a value following study of a set of *data.

PCA *Abbreviation for* PRINCIPAL COMPONENTS ANALYSIS.

pdf *Abbreviation for* PROBABILITY DENSITY FUNCTION.

Pearson, Egon Sharpe (1895–1980; b. London, England; d. Midhurst, England) English mathematical statistician and the only son of Karl *Pearson. Egon Pearson went up to Cambridge U in 1914, leaving after a year because of the First World War, when he worked at the Admiralty (his health was too poor for active combat). After the war, he started graduate work in astronomy. In 1921 he joined his father on the faculty at UCL, where his research students included *Box and *Johnson. Egon's association with *Neyman started in 1926 and led to the *Neyman–Pearson lemma and the development of a standard approach to *hypothesis testing. He was Editor of **Biometrika* from 1936 to 1965. He was President of the *RSS from 1955 to 1957 and was awarded its *Guy Medal in Gold in 1955. He was elected FRS in 1966.

SEE WEB LINKS

- Fuller biography and photograph.

Pearson, Karl (1857–1936; b. London, England; d. London, England) English mathematician, biometrician, and statistician. Pearson obtained his PhD (supervised by *Galton) from Cambridge U in 1879, joining the faculty at UCL, where he was appointed as Professor of Applied Mathematics and Mechanics in 1884. In 1890 he added the title of Gresham Lecturer in Geometry. It was not until 1893 that Pearson started publishing articles on statistics. By that time he already had a hundred publications to his name (including a number on German history and folklore). His first statistical work was entitled *The Chances of Death and Other Studies in Evolution*, and much of his subsequent work on statistical theory had a similar focus. During the period 1895–8 he presented a sequence of papers on *correlation and in 1900 he proposed the *chi-squared test. He founded the journal **Biometrika* in 1901 and was Editor until his death, when his son (*see* PEARSON, EGON SHARPE) took over. In 1911, at UCL, he was appointed Professor of Eugenics (the study of human evolution), a post he held until 1933. He was elected FRS in 1896 and FRSE in 1934.

SEE WEB LINKS

- Fuller biography and portrait.

Pearson coefficient of skewness A simple *statistic that uses the *mean, *mode, and *standard deviation:

$$\frac{\text{mean} - \text{mode}}{\text{standard deviation}}.$$

If the mode is unknown then the *median is used and the revised statistic is

$$3\,\frac{\text{mean} - \text{median}}{\text{standard deviation}}.$$

The coefficient is usually positive when the *distribution is positively *skewed, and negative when it is negatively skewed.

Pearson correlation coefficient *Alternative name for the* SAMPLE CORRELATION COEFFICIENT.

Pearson family of distributions A family of *distributions having *probability density functions of a variety of shapes. The family was proposed by Karl *Pearson in 1894 as a response to his recognition that not all *populations had *distributions that resembled the *normal distribution. He proposed twelve types of distribution which are variants of three basic distributions: Type I (the *beta distribution), Type VI (the *F-distribution), and Type IV (now little used).

Pearson goodness-of-fit test *See* CHI-SQUARED TEST.

Pearson residual *See* CHI-SQUARED TEST.

penalized likelihood An *estimation process in which the *likelihood is augmented by a function of the unknown *parameters, so as to ensure that the parameter estimates have desired characteristics. *See also* ROUGHNESS PENALTY.

penalized regression A method used so as to impose a desired relation on the unknown *parameters. For example, if the *data consist, for many locations, of annual tree growth and the twelve average monthly temperatures at those locations, then the dependence on the average April temperature should be very similar to the dependence on the average May temperature. In penalized regression the function minimized therefore includes a contribution from the parameter estimates that makes undesirable combinations unlikely.

penalty function A constraint specified in a *loss function which applies a penalty when an undesirable condition arises (e.g. an estimated value that is not feasible).

penetrance In the context of genetics, the *conditional probability of an individual possessing a particular characteristic, given knowledge of that individual's genotype. *See also* HARDY-WEINBERG LAW.

Penrose tiling *See* TESSELLATION.

percentage point (lower percentage point) For the *probability distribution of a *random variable X, the 100θ percentage point (or lower percentage point) of the distribution is x_1, such that $P(X < x_1) = \theta/100$. The corresponding **upper percentage point** of the distribution is x_2, such that $P(X > x_2) = \theta/100$. For example, the upper 5% point of a standard *normal distribution is 1.645.

percentile An approximate value for the rth percentile of a *data set can be read from a *cumulative frequency graph as the value of the *variable corresponding to a cumulative relative frequency of r%. So the lower *quartile is the 25th percentile and the *median is the 50th percentile. The term 'percentile' was introduced by *Galton in 1885. The term **centile** is also used.

perceptron A function used in supervised learning (*see* MACHINE LEARNING).

periodogram A periodogram is a useful graphical tool in the analysis of *time series. Consider a time series consisting of n *observations $x_1, x_2, \ldots, x_n$, corresponding to the times $t = 1, 2, \ldots, n$. The *mean of the n observations is denoted by $\bar{x}$. The series may be thought of as resulting from a mixture of cyclic variations of different periodicities so that, for n odd,

$$x_t = \bar{x} + \sum_{j=1}^{h} \left\{ a_j \cos\left(\frac{2\pi jt}{n}\right) + b_j \sin\left(\frac{2\pi jt}{n}\right) \right\},$$

where $h = \frac{1}{2}(n-1)$

$$a_j = \frac{2}{n}\sum_{t=1}^{n} x_t \cos\left(\frac{2\pi jt}{n}\right), \qquad b_j = \frac{2}{n}\sum_{t=1}^{n} x_t \sin\left(\frac{2\pi jt}{n}\right).$$

In the case where n is even the same expressions hold true, but now $h = \frac{1}{2}n - 1$ and there is one additional term in the equation for x_t namely, $\frac{1}{2}a_{\frac{1}{2}n}\cos(\pi t)$.

The dominating contributions are indicated by large values for a_j or b_j. The periodogram is a plot of $I(j)$ against j, for $j \le \frac{1}{2}(n-1)$, where

$$I(j) = a_j^2 + b_j^2 = \frac{1}{2}c_0 + \sum_{k=1}^{n-1} c_k \cos\left(\frac{2\pi jk}{n}\right), \quad j = 1, 2, \ldots,$$

and

$$c_k = \frac{1}{n}\sum_{t=1}^{n-k}(x_t - \bar{x})(x_{t+k} - \bar{x}), \quad k = 1, 2, \ldots, n-1.$$

Some authors define the periodogram using multiples (e.g. $2/\pi$) of some of these formulae.

The *population counterpart of the time series is the *stationary process $X(t)$, and the counterpart of the periodogram is the **spectral density function** (or **power spectrum** or **spectrum**).

The periodogram is not a consistent *estimator of the power spectrum and it is usual to give more *weight to the first m terms of the summation in the periodogram, using

$$I_m(j) = \frac{1}{2}c_0 + \sum_{k=1}^{m} \lambda_k c_k \cos\left(\frac{2\pi jk}{n}\right),$$

where $m < n$ and the weight λ_k is the **lag window**. A typical value for m is about $2\sqrt{n}$. Popular lag windows include:

Bartlett window	$\lambda_k = 1 - \frac{k}{m}$,	$k = 1, 2, \ldots, m$,
Daniell window	$\lambda_k = \frac{m}{k\pi}\sin(\frac{k\pi}{m})$,	$k = 1, 2, \ldots, m$,
Hamming window	$\lambda_k = 0.54 + 0.46\cos(\frac{k\pi}{m})$,	$k = 1, 2, \ldots, m$,
Parzen window	$\lambda_k = \begin{cases} 1 - 6(\frac{k}{m})^2 + 6(\frac{k}{m})^3, \\ 2(1 - \frac{k}{m})^3, \end{cases}$	$1 \le k \le \frac{1}{2}m$, $\frac{1}{2}m \le k \le m$,
Tukey window	$\lambda_k = \frac{1}{2}\{1 + \cos(\frac{k\pi}{m})\}$,	$k = 1, 2, \ldots, m$,

Direct analysis of the observed values $\{x_j\}$ is analysis in the **time domain** whereas analysis in terms of periodic functions is analysis in the **frequency domain**.

permutation An ordered arrangement of n different objects. The number of permutations (i.e. the number of different possible ordered arrangements) is $n!$.

For ordered selection of r objects from a set of $n(\ge r)$ different objects, the number of permutations of r from n, i.e. the number of different possible ordered selections, is usually denoted by ^nP_r. In fact,

$$^nP_r = n \times (n-1) \times (n-2) \times \cdots \times (n-r+1) = \frac{n!}{(n-r)!}.$$

Special values are $^nP_0 = 1$, $^nP_1 = n$, $^nP_n = n!$. For example, the 6 $(= 3!)$ permutations of $\{A, B, C\}$ are *ABC, ACB, BAC, BCA, CAB, CBA*, and there are 24 $(= 4 \times 3 \times 2)$ permutations of 3 letters from $\{A, B, C, D\}$.

If the n objects are not all different, and there are n_1 objects of type 1, n_2 objects of type 2, …, n_k objects of type k, where $n_1 + n_2 + \cdots + n_k = n$, then the number of different ordered arrangements is

$$\frac{n!}{n_1!n_2!\dots n_k!}.$$

For unordered selection, *see* COMBINATION.

permutation matrix A square *matrix in which the rows are a permutation of the rows of an identity matrix.

permutation test A simple type of *hypothesis test. Denote the value of some test statistic by T. The observed *data values are randomly redistributed amongst the experimental units. The test statistic is calculated for each such redistribution. Depending on the number of data values, either all possible permutations are made, or a random selection (of say 1000 permutations) is made. For each permutation the value of the test statistic is considered. The significance of the value T is determined by the proportion of permutations that lead to values greater than, or equal to, T.

Perron–Frobenius theorem The theorem stating that a square *matrix $\mathbf{A}$ with real non-negative elements has a positive real eigenvalue λ. The matrix is assumed to be irreducible, i.e. there is no *permutation matrix $\mathbf{P}$ such that $\mathbf{PAP}'$ has a zero submatrix in the bottom left-hand corner. Furthermore, every eigenvalue of $\mathbf{A}$ has modulus not exceeding λ, the eigenvalue λ is simple (i.e. λ is a non-multiple root of the characteristic equation (*see* MATRIX)), and there is a corresponding eigenvector with positive elements. Perron originally established the theorem for matrices with positive elements, which are necessarily irreducible, and Frobenius extended the result.

persistent state *See* MARKOV PROCESS.

personal probability *See* BAYESIAN INFERENCE.

Petersen estimator *See* CAPTURE–RECAPTURE METHODS.

pgf *Abbreviation for* PROBABILITY-GENERATING FUNCTION.

phase-type distribution The distribution of the time a finite *Markov process takes to reach a given state from its initial state distribution. Applications include *queues and *reliability.

Phillips curve A curve describing the relationship between the rate of inflation and the unemployment rate.

Phillips–Perron test *See* DICKEY–FULLER TEST.

pi (π) The ratio of the circumference of a circle to its diameter; $\pi = 3.141\ 592\ 65\dots$.

pictogram A diagram in which *frequency or quantity is represented by symbols that are small images of the objects or material being counted. The diagram should indicate exactly how many, or how much, each symbol represents.

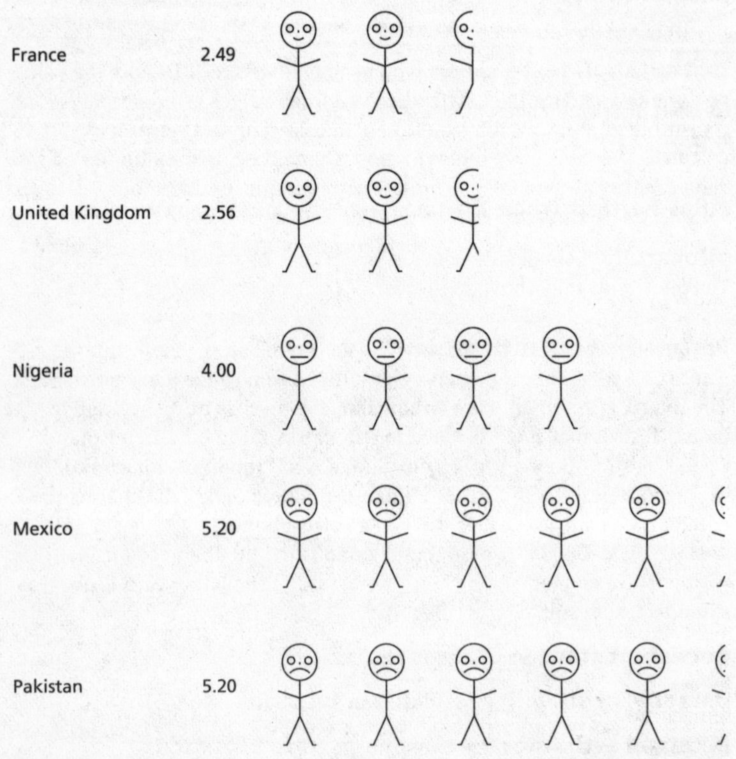

Pictogram. The diagram shows, for 1984, the average number of people per dwelling in five countries. In the diagram each extra stick person represents one extra person per dwelling. The two European countries are very similar and differ markedly from the other countries.

pie chart A diagram (*see opposite*) used when it is desired to emphasize the proportions of a set of *data when data items are grouped into classes according to the value of some *variable (usually *categorical). A circle is divided into sectors representing the classes. The area (or equivalently the angle) of a sector is proportional to the frequency of the corresponding class. If two pie charts are used to compare two populations, their areas can be made proportional to the sizes of the populations.

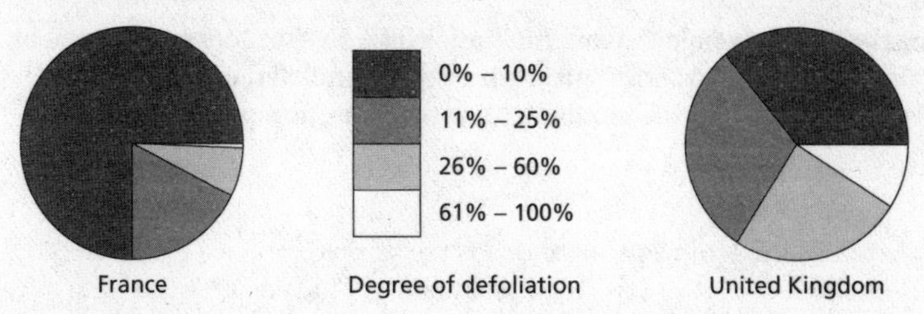

Pie chart. The charts contrast the amount of defoliation of conifers in France and the UK in 1989.

Pillai–Bartlett trace *See* MULTIVARIATE ANALYSIS OF VARIANCE.

pilot study A small survey taken in advance of a major investigation. The pilot study may show up problems in the organization of the intended major study. It can also give information about response variability that will help determine the size of the major study.

Pitman, Edwin James George (1897–1993; b. Melbourne, Australia; d. Kingston, Australia) Australian mathematical statistician. Pitman's parents, emigrating from England, met on the ship to Australia. He was educated at Melbourne U, graduating with an MA in mathematics in 1923 after his studies had been interrupted by two years of war service. Pitman was Professor of Mathematics at U Tasmania from 1926 to 1962. An early task there was to lecture in Statistics—a subject new to him. His subsequent research concerned the foundations of *statistical inference including pioneering work on *non-parametric tests. He was President of the Australian Mathematical Society in 1958. In 1978 the *SSAI instituted the award of the **Pitman Medal** for 'high distinction in statistics': Pitman himself was the first recipient.

SEE WEB LINKS

- Fuller biography and photograph.

Pitman efficiency A method for comparing two *estimators of the unknown value of a *population *parameter θ. Suppose that, with a *sample of n *observations, $\hat{\theta}$ and $\tilde{\theta}$ are two such estimators. The Pitman efficiency of $\hat{\theta}$ relative to $\tilde{\theta}$ is

$$\mathrm{P}(|\hat{\theta} - \theta| < |\tilde{\theta} - \theta|).$$

Pitman Medal *See* PITMAN.

pivotal quantity A function of the *observations having a *distribution that does not depend on any parameter other than the parameter of interest. For example, if a random sample of n observations is taken from a *normal distribution with unknown *mean μ and *variance σ^2 then a pivotal quantity for the parameter μ is the *statistic t, given by

$$t = \frac{\bar{x} - \mu}{s/\sqrt{n}},$$

where $\bar{x}$ is the sample mean and s^2 is the sample variance (calculated using the $(n-1)$ divisor). The distribution of t (a *t-distribution with $(n-1)$ degrees of freedom) does not depend on the value of σ^2.

pixel An area, often rectangular, for which a value has been recorded. The term was coined in connection with the analysis of pictures, such as satellite images of ground conditions, or medical scans of parts of the human body. Each dot in such a picture is a pixel conveying information. A **voxel** is the three-dimensional equivalent, representing a volume of space for which a value has been recorded.

placebo; placebo effect *See* BLINDING.

Plackett, Robert Lewis 'Robin' (1920–2009; b. Liverpool, England; d. Newcastle upon Tyne, England) English statistician. Plackett was a co-author of the 1946 paper that introduced Plackett and Burman designs (*see* FACTORIAL DESIGN). Plackett graduated from Cambridge U in 1942, joining *Barnard at the Ministry of Supply. In 1947 he joined the faculty at U Liverpool, moving, in 1962, to be the founding Professor of Statistics at U Newcastle upon Tyne. The *RSS awarded him its *Guy Medal in Bronze in 1968, in Silver in 1973, and in Gold in 1987. He was one of only three individuals to have the full set—the others being *Armitage and *Durbin.

Plackett and Burman design *See* FACTORIAL DESIGN.

platykurtic *See* KURTOSIS.

Playfair, William (1759–1823; b. Dundee, Scotland; d. London, England) Scottish political economist. At the age of thirteen he was apprenticed to a millwright. At twenty-one he became a draftsman to James Watt (of steam-engine fame) in Birmingham. His elder brother was Professor of Mathematics at Edinburgh U, so mathematical ideas ran in the family. However, William Playfair became interested in politics and the economy, writing several books (in particular, in 1786, *The Commercial and Political Atlas*). His books contain many effective graphical devices and he is credited with the introduction of *pie charts and *time series graphs. Playfair died in poverty in Covent Garden, London.

SEE WEB LINKS

- Fuller biography.

play-the-winner rule An allocation rule used in a medical context. The rule states that the treatment that should be allocated to the next patient should be that which has been most successful for past patients.

plot *See* EXPERIMENTAL UNIT.

PLS *Abbreviation for* PARTIAL LEAST SQUARES.

point-biserial correlation coefficient *See* BISERIAL CORRELATION.

point estimate *See* ESTIMATOR.

point kriging *See* KRIGING.

point process A *stochastic process concerned with the random positions of locations in space or of points in time. *See* POISSON PROCESS.

Poisson, Siméon Denis (1781–1840; b. Pithiviers, France; d. Sceaux, France) French mathematician. Poisson studied (under *Laplace) at the École Polytechnique in Paris. From 1802 to 1808 he taught at the École, attaining the chair in pure mathematics at the Faculté des Sciences. He published in many branches of mathematics. His major work on probability, published in 1837, was *Recherches sur la Probabilité des Jugements en Matière Criminelle et en Matière Civile* (*Researches on the Probability of Criminal and Civil Verdicts*). In this long book (over 400 pages) only about one page is devoted to the derivation of the distribution that now bears his name. He is alleged to have said that 'Life is good for only two things: to study mathematics and to teach it'. He was elected FRS in 1818 and was awarded the Society's Copley Medal in 1832. He was elected FRSE in 1820.

SEE WEB LINKS

- Information and portrait.

Poisson approximation to the binomial distribution *See* BINOMIAL DISTRIBUTION.

Poisson distribution A *random variable *X*, whose set of possible values consists of the non-negative integers, with *probability function given by

$$P(X = r) = \frac{e^{-\lambda}\lambda^r}{r!}, \qquad r = 0, 1, \ldots,$$

where λ is a positive constant, is said to have a Poisson distribution, or to be a **Poisson variable**, with parameter λ. (By convention, $\lambda^0 = 1$ and $0! = 1$.) The distribution was named after *Poisson, though the first derivation was by de Moivre in 1711. If we note that $P(X=0)=e^{-\lambda}$, successive *probabilities can be calculated by using the *recurrence relation

$$P(X=r)=\frac{\lambda}{r}P(X=r-1), \qquad r=1,2,\ldots.$$

The *mean and *variance of the *distribution are both λ. If λ is not an integer the *mode is the value of the integer r for which $r-1<\lambda<r$. If λ is an integer then $P(X=\lambda-1)=P(X=\lambda)$ and both $(\lambda-1)$ and λ are modes. If $\lambda<1$ the graph of the probability function decreases steadily, whereas if $\lambda>1$ the graph increases steadily to the value at the mode, then decreases steadily, tending to 0 as $r\to\infty$.

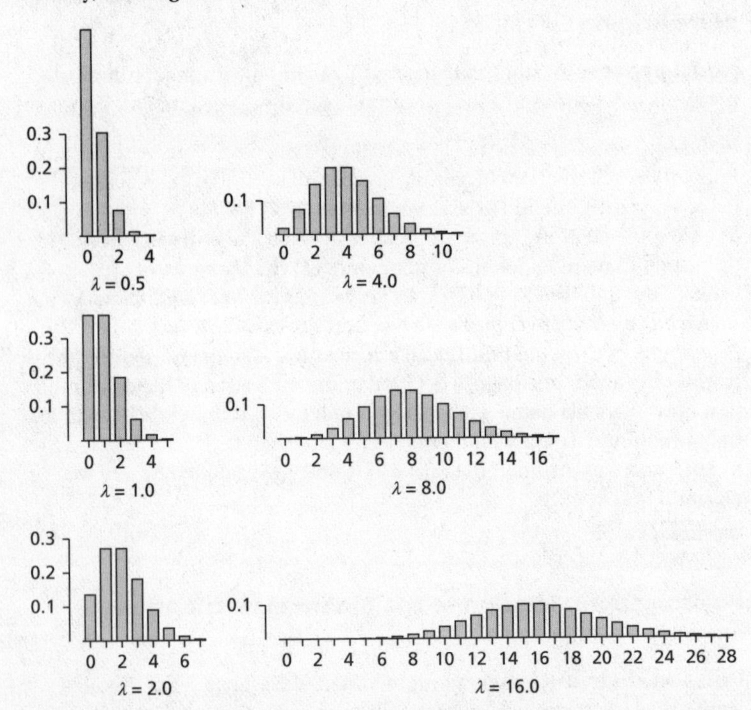

Poisson distribution. The outline of a Poisson distribution with parameter λ increasingly resembles that of a normal distribution as λ increases. Both the mean and the variance of a Poisson distribution with parameter λ are equal to λ.

For large values of λ a **normal approximation to the Poisson distribution** may be used:

$$P(X\le r)\approx\Phi(z), \qquad \text{where } z=\frac{r+\frac{1}{2}-\lambda}{\sqrt{\lambda}},$$

and Φ is the cumulative distribution function for a standard normal variable (*see* NORMAL DISTRIBUTION). The '$\frac{1}{2}$' is a *continuity correction. The approximation may be described as 'For large values of λ a Poisson variable with mean λ is approximately $N(\lambda, \lambda)$'.

In a *Poisson process the number of events in a given region, or a given time interval, has a Poisson distribution.

Poisson process Description of a situation where events occur randomly in time, or space, in such a way that for each small interval of time, or small region of space, the *probability that it contains exactly one event is proportional to the size of the interval or region. It is assumed that this probability is *independent of whether or not any other small intervals, or regions, contain an event, and it is also assumed that the probability of two events occurring in the same small interval or region is 0. This is a mathematical description of randomness. As the *diagram overleaf* shows, randomness usually gives rise to apparent clustering, despite the natural expectation that randomness would lead to regularity.

For events that occur at random instants in time, the formal defining properties of a Poisson process are

$$\frac{\text{P(one event in a time interval of length } \delta t)}{\delta t} \to k \text{ as } \delta t \to 0,$$
$$\frac{\text{P(two or more events in a time interval of length } \delta t)}{\delta t} \to 0 \text{ as } \delta t \to 0,$$

where k is a positive constant. The number of events in a time interval of length t has a *Poisson distribution with parameter kt.

The time intervals between events, and from the start of observations to the first event, have independent *exponential distributions with parameter k, *mean $1/k$, and *variance $1/k^2$. The time interval from the start of observations to the nth event has an Erlang distribution (*see* GAMMA DISTRIBUTION) with *probability density function f given by

$$f(x) = \frac{k^n x^{n-1} e^{-kx}}{(n-1)!}, \qquad x > 0.$$

The mean of this distribution is n/k and the variance is n/k^2.

An example of a naturally occurring Poisson process is the emission of α-particles from a radioactive source. A **compound Poisson process** is a process in which the events of interest occur in 'packets' (e.g. buses) that arrive randomly at a steady rate. The number in each packet is an observation from the same distribution (e.g. a *uniform distribution between zero and the capacity of a bus).

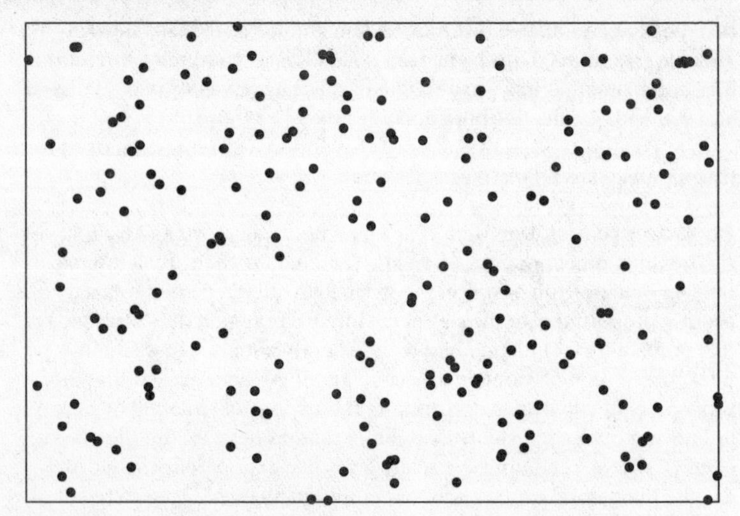

Poisson process. The diagram shows 200 points, randomly scattered such that each location was equally likely; the points were generated using pseudo-random numbers. The arrangement shows the typical characteristics of a spatial Poisson process with apparent clumps and blank spaces.

Poisson regression *Regression in which the dependent variable has a *Poisson distribution.

Poisson sampling This method can be used when events are happening in time at random at an unknown rate. In Poisson sampling the rate is estimated by noting the number of events that occur in a predesignated time. The alternative is **gamma sampling** in which the estimate is based on the time taken to record a predesignated number of events.

Poisson variable *See* POISSON DISTRIBUTION.

polar coordinates A coordinate system in a plane in which the position of a point *P* is specified by the distance $r = OP$, where *O* is a fixed point (the pole), and the angle θ between *OP* and a fixed line through *O* (the initial line). The polar coordinates are given as (r, θ), where $r \geq 0$, and $-180° < \theta \leq 180°$ or $0° \leq \theta < 360°$. Referred to *Cartesian coordinates *Oxy*, with *Ox* along the initial line, Cartesian and polar coordinates are related by

$$x = r\cos\theta, \qquad y = r\sin\theta, \qquad r = \sqrt{x^2 + y^2}, \qquad \tan\theta = \frac{y}{x}.$$

Pollaczek–Khintchine formula *See* QUEUE.

Pólya, György 'George' (1887–1985; b. Budapest, Hungary; d. Palo Alto, CA) Hungarian statistician who spent nearly half his life in the USA. Pólya obtained his PhD from U Budapest in 1912. From 1914 to 1940 he was a member of the faculty of the Swiss Federal Institute of Technology in Zürich. He then moved to the USA, occupying posts at Brown U and at Smith College. His final post, from 1946 to 1953, was as Professor at Stanford U. He was elected to membership of the NAS in 1976. He is quoted as saying that 'If you cannot solve a problem, then there is an easier problem you can solve: find it'.

SEE WEB LINKS

- Fuller biography and photograph.

Pólya distribution *See* BETA-BINOMIAL DISTRIBUTION.

Pólya's urn model A model that has been used to describe the spread of contagious diseases. In this *urn model the urn initially contains b black balls and w white balls. A ball is chosen at random and is then replaced by $c\ (>1)$ balls of the same colour. This process is repeated indefinitely.

polychoric correlation coefficient *See* TWO-BY-TWO TABLE.

polynomial A polynomial of **degree** n is an expression of the form $a_0+a_1x+a_2x^2+\cdots+a_nx^n$, where x is a variable, $a_0, a_1, a_2, \ldots, a_n$ are constants, and $a_n \neq 0$.

polynomial regression model *See* MULTIPLE REGRESSION MODEL.

polynomial-time complexity *See* COMPLEXITY.

polytomous *See* CATEGORICAL VARIABLE.

pooled estimate of common mean An estimate obtained by combining information from two or more *independent *samples taken from *populations believed to have the same mean. *Observations $x_{11}, x_{12}, \ldots, x_{1m}$ are randomly selected from a population. Their mean is $\bar{x}_1$, given by

$$\bar{x}_1 = \frac{1}{m}(x_{11}+x_{12}+\cdots+x_{1m}).$$

Random observations from a second population are denoted by $x_{21}, x_{22}, \ldots, x_{2n}$ and have mean $\bar{x}_2$. If the two populations are believed to have the same mean, then a pooled estimate of the common mean is $\bar{x}$, given by

$$\bar{x} = \frac{m\bar{x}_1+n\bar{x}_2}{m+n}.$$

With k samples of sizes $n_1, n_2, \ldots, n_k$ and with means $\bar{x}_1, \bar{x}_2, \ldots \bar{x}_k$, the pooled estimate is given by

$$\bar{x} = \frac{n_1\bar{x}_1 + n_2\bar{x}_2 + \cdots + n_k\bar{x}_k}{n_1 + n_2 + \cdots + n_k}.$$

The unbiased estimate of the variance of the first population is s_1^2, given by

$$s_1^2 = \frac{1}{m-1}\left\{\sum_{j=1}^{m} x_{1j}^2 - \frac{1}{m}\left(\sum_{j=1}^{m} x_{1j}\right)^2\right\}.$$

The corresponding estimate for the second population is s_2^2. If it is believed that the two populations have the same variance, but possibly different means, then the **pooled estimate of common variance** is s^2, given by

$$s^2 = \frac{(m-1)s_1^2 + (n-1)s_2^2}{m+n-2}.$$

In the case of k samples, with the estimate from sample j being s_j^2, the pooled estimate is given by

$$s^2 = \frac{(n_1-1)s_1^2 + (n_2-1)s_2^2 + \cdots + (n_k-1)s_k^2}{(n_1 + n_2 + \cdots + n_k) - k}.$$

See also HYPOTHESIS TEST.

pooled estimate of common variance *See* POOLED ESTIMATE OF COMMON MEAN.

population The complete set of all people in a country, or a town, or any region (or just the number of such people). By extension the term is used for the complete set of objects of interest; for example, all cars built by a particular company in the year 2001, all apples sold as Grade I by a particular supermarket, all students in a university, all smokers. These are all real populations and are finite, though they may be large. The term is also used for the infinite population of all possible results of a sequence of statistical trials; for example, tossing a coin.

population correlation coefficient A measure of the *linear dependence of one numerical *random variable on another. The phrase 'coefficient of correlation' was apparently originated by *Edgeworth in 1892. It is usually denoted by ρ (rho). The value of ρ, which lies between -1 and 1, inclusive, is defined as the ratio of the *covariance to the square root of the product of the *variances of the marginal distributions (*see* BIVARIATE DISTRIBUTION) of the individual variables:

$$\rho = \frac{\mathrm{Cov}(X,Y)}{\sqrt{\mathrm{Var}(X)\mathrm{Var}(Y)}}.$$

If the correlation coefficient between the random variables X and Y is equal to 1 or -1 then this implies that $Y=a+bX$, where a and b are constants. If b is positive then $\rho=1$ and if b is negative then $\rho=-1$. The converse statements are also true.

If X and Y are completely unrelated (i.e. are *independent) then $\rho=0$. If $\rho=0$ then X and Y are said to be *uncorrelated variables. However, ρ is concerned only with linear relationships, and the fact that $\rho=0$ does not imply that X and Y are independent.

population covariance For two *random variables X and Y, this is the difference between the *expected value of their product and the product of their separate expected values. It is denoted by Cov (X,Y):

$$\mathrm{Cov}(X, Y) = \mathrm{E}(XY) - \mathrm{E}(X) \times \mathrm{E}(Y).$$

If X and Y are *independent then $\mathrm{Cov}(X, Y)=0$. However, if $\mathrm{Cov}(X, Y)=0$ then X and Y may not be independent. A useful result is $\mathrm{Var}(aX+bY)=a^2\mathrm{Var}(X)+2ab\,\mathrm{Cov}(X, Y)+b^2\mathrm{Var}(Y)$, where Var denotes *variance, and a and b are constants. The term 'covariance' was used by Sir Ronald *Fisher in 1930. *See also* POPULATION CORRELATION COEFFICIENT.

population mean The *average value of some *variable that is measured for all members of a (possibly infinite) population. If the value of the variable for a randomly chosen member of the population is denoted by X, then the population mean is the *expected value of X and is usually denoted by μ (the notation μ was introduced in 1936 by Sir Ronald *Fisher in the sixth edition of his *Statistical Methods for Research Workers*). Similarly, the **population variance**, usually denoted by σ^2, is the mean of the squared differences between the values of the members of the population and the population mean: this is the expected value of $(X-\mu)^2$.

population median In a *population of values of the *variable X, the population median, m, is a value for which $\mathrm{P}(X\geq m)=\mathrm{P}(X\leq m)$.

population parameter A key quantity that determines the precise shape of a *distribution. For example, the shape of a *Poisson distribution is determined by the *parameter λ, and that of a *normal distribution by the parameters μ and σ.

population pyramid A diagram (*see overleaf*) for representing the age distribution of a population. It is really a *histogram in which age is plotted vertically and *frequency, or relative frequency (i.e. *proportion), is plotted horizontally. Often drawn as a back-to-back pyramid with one side for males and the other side for females. Paired pyramids can be used to compare two populations.

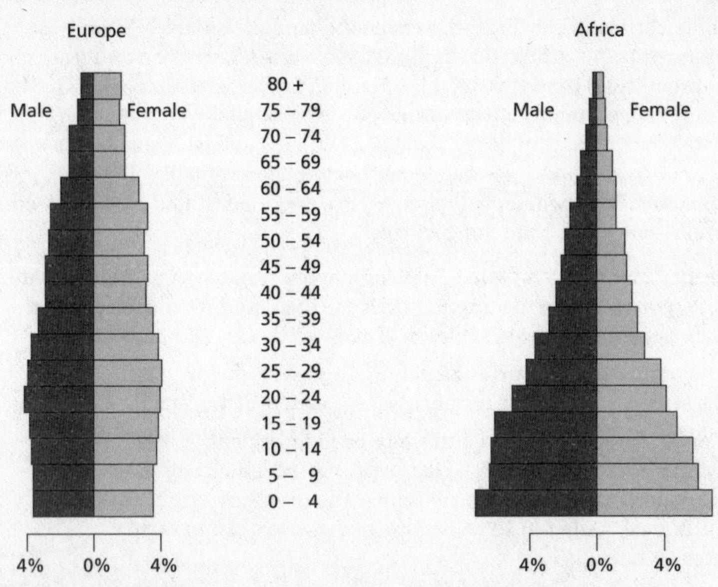

Population pyramid. The pyramids show the age distribution, by sex, for typical European and African nations. The pyramids have the same total area, so the wide base of the African pyramid reflects the high birth-rate, and its fast-tapering shape indicates the low expectation of life.

population standard deviation The square root of the *population variance.

population variance A measure of the variability in the values of a *random variable. In context the word 'population' is often omitted. It is defined as the *expected value of the squared difference between the random variable and its expected value:

$$\mathrm{Var}(X) = \mathrm{E}[\{X - \mathrm{E}(X)\}^2] = \mathrm{E}[X^2] - \{\mathrm{E}(X)\}^2,$$

where E(X) denotes the expected value of the random variable X. For a *discrete random variable X, taking values $x_1, x_2, \ldots, x_n$, the variance of X can be calculated as follows:

$$\begin{aligned}\mathrm{Var}(X) &= \sum_{j=1}^{n} \mathrm{P}(X = x_j)\{x_j - \mathrm{E}(X)\}^2, \\ &= \sum_{j=1}^{n} x_j^2 \mathrm{P}(X = x_j) - \{\mathrm{E}(X)\}^2,\end{aligned}$$

where

$$\mathrm{E}(X) = \sum_{j=1}^{n} x_j \mathrm{P}(X = x_j).$$

For a *continuous random variable X, with *probability density function f, the variance of X can be calculated as follows:

$$\begin{aligned}\mathrm{Var}(X) &= \int_{-\infty}^{\infty} \{x - \mathrm{E}(X)\}^2 \mathrm{f}(x)\mathrm{d}x \\ &= \int_{-\infty}^{\infty} x^2 \mathrm{f}(x)\mathrm{d}x - \{\mathrm{E}(X)\}^2,\end{aligned}$$

where

$$\mathrm{E}(X) = \int_{-\infty}^{\infty} x\mathrm{f}(x)\mathrm{d}x.$$

The term 'variance' was coined by Sir Ronald *Fisher in 1918. Fisher used the symbol σ^2, since the variance was the square of the *standard deviation that Karl *Pearson had denoted by σ in 1894. *See* COVARIANCE. *See also* POPULATION MEAN.

Porter, George Richardson (1792–1852; b. London, England; d. London, England) English economist and statistician. A founder of the *RSS, he was a pioneer of the use of *index numbers. In 1834 he was appointed as the first head of the Statistical Department of the Board of Trade. That department evolved into the present day Office for National Statistics. He was elected FRS in 1838.

positive correlation *See* SAMPLE CORRELATION COEFFICIENT.

positive definite; positive semi-definite *See* MATRIX.

positively skewed; positive skewed *See* SKEWED.

posterior distribution; posterior probability *See* BAYESIAN INFERENCE.

potential capability *See* CAPABILITY ANALYSIS.

Potts model *See* ISING MODEL.

power; power curve *See* HYPOTHESIS TEST.

power-divergence statistics Statistics, used in *goodness-of-fit tests, that have the form

$$\frac{2}{\lambda(\lambda+1)} \sum_{j=1}^{n} f_j \left\{ \left(\frac{f_j}{e_j} \right)^{\lambda} - 1 \right\},$$

where f_j is an observed *frequency, e_j is the corresponding *expected frequency, and λ is a constant. The case $\lambda = 1$ gives the *chi-squared test statistic, the limiting case as λ approaches 0 gives the *likelihood-ratio goodness-of-fit statistic, and the case $\lambda = -\frac{1}{2}$ gives the *Freeman–Tukey test statistic.

power spectrum *See* PERIODOGRAM.

p-p plot *See* Q-Q PLOT.

PRE *Abbreviation for* PROPORTIONAL REDUCTION IN ERROR.

prediction interval A statement about the likely value of a future *observation. It has a similar interpretation to a *confidence interval but the interval is wider because it allows for the *random error associated with the future observation.

For example, if a random sample of n observations is taken from a population with known variance σ^2 and unknown mean μ, then the natural estimate of μ is $\bar{x}$, the sample mean. A confidence interval for μ is based on the uncertainty in the estimate, which has variance σ^2/n. However, associated with a future observation from this distribution is the future random sampling variation with variance σ^2. A prediction interval for a future observation is based on the sum of the two variances, $\sigma^2 + \sigma^2/n$.

prediction model In a medical context, a *model used to predict an outcome or the probability of an outcome. Frequently the outcome is dichotomous (*see* CATEGORICAL VARIABLE) (e.g. tumour is benign, patient will survive for 5 years, etc.) and the model is then a *logistic regression model.

p

predictor variable *See* REGRESSION.

PRESS statistic A *statistic that provides an indication of the extent to which a *multiple regression model can be generalized. This is achieved by fitting the model to every subset of $n-1$ of the n *observations, $y_1, y_2, \ldots, y_n$, of the response variable (*see* REGRESSION). Let $\hat{y}_{-j}$ be the *fitted value for y_j based on the model that uses all the observed values except y_j. The corresponding residual $\hat{\varepsilon}_{-j}$ is given by

$$\hat{\varepsilon}_{-j} = y_j - \hat{y}_{-j}.$$

The PRESS statistic is

$$1 - \frac{\sum_{j=1}^{n} (\hat{\varepsilon})_{-j}^2}{\sum_{j=1}^{n} (y_j - \bar{y})^2},$$

where $\bar{y}$ is the *mean of the n observations. The abbreviation PRESS is derived from predicted residual sum of squares. The statistic is an analogue of the R^2 statistic (*see* ANOVA) and values close to 1 are preferable to those close to 0.

pre-whitening The process of removing unwanted *autocorrelations from a *time series, prior to the analysis of interest.

price index A measure of the value of money in which the cost of a standard collection, or basket, of goods and services at some particular date is compared with the cost of the same basket at a base date. The basket is chosen to represent the expenditure of a typical household at the date under consideration.

Writing p_{0j} and p_{nj} as the prices at times 0 and n of the jth item in the basket, and q_{0j} and q_{nj} as the corresponding quantities of that item, the **Laspeyres price index**, suggested by *Laspeyres in 1871, is the ratio

$$P_L = \sum_j q_{0j}p_{nj} \Big/ \sum_j q_{0j}p_{0j},$$

and the **Paasche price index**, suggested by *Paasche in 1874, is the ratio

$$P_P = \sum_j q_{nj}p_{nj} \Big/ \sum_j q_{nj}p_{0j}.$$

In 1922 Irving *Fisher suggested using the *geometric mean:

$$P_I = \sqrt{P_L P_P}.$$

This is now known either as the **Fisher index** or as the **ideal index**. An alternative, variously known as the **Bowley index**, or the **Marshall–Edgeworth index**, is

$$P_B = \sum_j (q_{0j} + q_{nj})p_{nj} \Big/ \sum_j (q_{0j} + q_{nj})p_{0j}.$$

See also RETAIL PRICE INDEX.

Priestley skill score *See* SKILL SCORE.

primary sampling unit (PSU) *See* CLUSTER SAMPLING.

principal components analysis (PCA) A technique for making *multivariate data easier to understand. Suppose that there is information on n *variables for each data item. These variables are unlikely to be *independent of one another; a change in one is likely to be accompanied by a change in another. The idea of PCA is to replace the original n *variables by m $(< n)$ *uncorrelated variables, each of which is a linear combination of the original variables, so that the bulk of the variation can be accounted for using just a few explanatory variables. In the *diagram overleaf*, the two original variables x_1 and x_2 can be replaced by the first **principal component**, y_1.

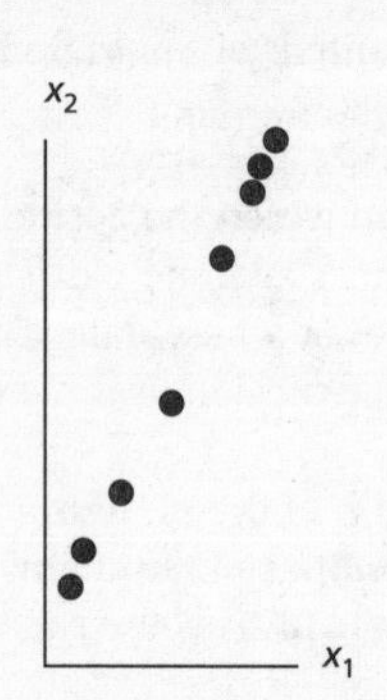

Explaining the variation between observations requires statements to be made about both original variables

Explaining the variation between observations requires a statement to be made about the principal component only

Principal components analysis. Principal components analysis aims to explain an n-dimensional situation using m ($< n$) uncorrelated variables. Here $n = 2$ and $m = 1$.

Let **R** denote the *correlation matrix for the case of p x-variables. The coefficients of the x-variables corresponding to the kth principal component are the elements of the eigenvector corresponding to the kth largest eigenvalue, λ_k, of **R**. All the eigenvalues are real and non-negative, since **R** is positive semi-definite. The proportion of the variation in the data explained by the kth principal component is $\lambda_k / \sum_{j=1}^{p} \lambda_j$. The first few principal components should account for the majority of the observed variation. The hope is that the linear combinations thus identified will have some natural interpretation. The **Kaiser rule** proposes that only components having eigenvalues greater than unity should be retained. *See also* SCREE PLOT.

principal components regression analysis *See* PARTIAL LEAST SQUARES.

principle of indifference; principle of insufficient reason *See* BAYESIAN INFERENCE.

principle of parsimony *See* OCKHAM'S RAZOR.

prior distribution; prior probability *See* BAYESIAN INFERENCE.

probability The probability of an event (*see* SAMPLE SPACE) is a number lying in the interval $0 \le p \le 1$, with 0 corresponding to an event that never occurs and 1 to an event that is certain to occur. For an experiment with N equally likely outcomes the probability of an event A is n/N, where n is the number of outcomes in which the event A occurs. For some experiments, such as throwing a drawing pin and seeing whether it lands point up, there is no possible set of equally likely outcomes. In the 'frequentist' view of probability, the probability of getting 'point up' is the limit, in some sense, of the relative *frequency as the number of experiments tends to infinity. In the context of *Bayesian inference, each observer has his or her own *a priori* *distribution for the probability, which is then modified *a posteriori* in the light of whatever results have been obtained. *Laplace claimed that 'probability theory is nothing but common sense reduced to calculation'.

probability density function (pdf); probability density For a *continuous random variable X the probability density function f is such that

$$P(x_1 < X < x_2) = \int_{x_1}^{x_2} f(x)dx,$$

for all $x_1 < x_2$. Because, for a continuous random variable, $P(X = x_j) = 0$ for any value x_j, either or both of the '$<$' signs in the left-hand side can be replaced with '$\leq$'. If the interval of possible values for X is (a, b), then

$$\int_a^b f(x)dx = 1.$$

It is often convenient to regard the function f as being defined for all real values of x. This can be achieved by taking $f(x)=0$ for x outside the interval (a, b), so that

$$\int_{-\infty}^{\infty} f(x)dx = 1.$$

This property, and the property that $f(x) \geq 0$ for all x, are essential properties of a probability density function. It should be noted that, although $f(x)$ is related to probability, $f(x)$ can exceed 1.

The probability density function is often referred to simply as the probability density. It is related to the *cumulative distribution function F by

$$F(x) = \int_{-\infty}^{x} f(t)dt \qquad \text{and} \qquad f(x) = \frac{dF(x)}{dx}.$$

The probability density function is not necessarily continuous. At a point of discontinuity the value assigned to $f(x)$ is immaterial.

probability distribution A description of the possible values of a *random variable, and of the *probabilities of occurrence of these values. For a *discrete random variable, *see* PROBABILITY FUNCTION. For a *continuous random variable, *see* PROBABILITY DENSITY FUNCTION.

probability function (probability mass function) For a *discrete random variable X, with possible values $x_1, x_2, \ldots,$ the function f, defined by

$$f(x_j) = P(X = x_j), \qquad j = 1, 2, \ldots$$

is the probability function of X. Of course, $\sum_j f(x_j) = 1$.

probability-generating function (pgf) For the *discrete random variable X, with *probability distribution $P(X = x_j)$, $j = 1, 2, 3, \ldots$, the probability-generating function G is defined by

$$G(t) = \sum_j t^{x_j} P(X = x_j),$$

where t is an arbitrary variable. Note that $G(t)$ is the *expected value of t^X and $G(1) = 1$. If the set of possible values of x is infinite, $|t|$ needs to be small enough for the series to converge.

If the first and second derivatives of $G(t)$ with respect to t are denoted by $G'(t)$ and $G''(t)$, respectively, the expected value and *variance of X are given by $G'(1)$ and $G''(1) + G'(1) - \{G'(1)\}^2$, where, for example, $G'(1)$ denotes the value of $G'(t)$ when $t = 1$.

Like the *moment-generating function, the probability-generating function can provide a useful alternative description of a probability distribution. For example, if Y denotes the sum of n *independent random variables, each having pgf $G(t)$, then $P(Y = y)$ is the coefficient of t^y in $\{G(t)\}^n$.

Another useful property is that, if X and Y are independent random variables with probability-generating functions $G_X(t)$ and $G_Y(t)$, respectively, then the probability-generating function of $Z = X + Y$ is $G_Z(t)$, where

$$G_Z(t) = G_X(t) \times G_Y(t).$$

*De Moivre used the probability-generating function technique in 1730. The term itself became common following its use by *Bartlett in 1940. *See also* MOMENT-GENERATING FUNCTION.

probability laws The basic properties of *probability. As well as the *addition law and the *multiplication law, these comprise $0 \le P(A) \le 1$ for every event A, and $P(\phi) = 0$ and $P(S) = 1$, where ϕ and S represent the empty set and the *sample space, respectively.

probability mass function *Alternative name for* PROBABILITY FUNCTION.

probability paper; probability plot *See* Q-Q PLOT.

probit The quantity $\Phi^{-1}(p)$, where p is a *proportion or *probability, and Φ is the *cumulative distribution function of the standard *normal distribution. When it is believed that the probability of success, p, is dependent upon the values of some predictor variables, it is usual to work with some function of p that has an infinite range in order to ensure that the estimate of p lies in the interval (0, 1). One such function is the *logit; another is the probit.

probit link *See* GENERALIZED LINEAR MODEL.

problem of points This problem was the subject of the correspondence between *Fermat and *Pascal that underpins the modern treatment of *probability. The problem is as follows.

Two gamblers are playing a series of fair games. The winner is to be the gambler who first wins n games. However, the sequence of games is interrupted at the point where player A requires a further a wins, and player B requires a further b wins. The question is how the stake should be fairly divided.

Fermat and Pascal attacked the problem in different ways, but agreed that the answer is that the proportion of the stake that player A should be awarded is

$$2^{-(a+b-1)}\sum_{j=0}^{b-1}\binom{a+b-1}{j}.$$

Procrustes analysis A method for matching two multidimensonal objects so as to match them (by stretching and rotation) as closely as possible. The name emanates from the character in Greek legend who 'adjusted' his guests to fit the bed provided.

Procrustes transformation A procedure used in *factor analysis to transform a *matrix to represent a target matrix as closely as possible.

producer's risk *See* ACCEPTANCE SAMPLING.

product-limit estimate *See* KAPLAN–MEIER ESTIMATE.

product-moment correlation coefficient *Alternative name for* SAMPLE CORRELATION COEFFICIENT.

profile likelihood The *likelihood maximized with respect to *nuisance parameters.

projection pursuit; exploratory projection pursuit (EPP) A simple alternative to *principal components analysis. To understand the underlying idea, visualize a data set as being a cloud of points in p-dimensional space. Now imagine shining a light on the cloud so that we see the projections of the points on some lower-dimensional 'wall'. If there is an obvious pattern that was not visible previously, then, indeed, light has been cast upon the problem! Typically, the patterns searched for are non-overlapping clusters.

propensity score A value used in the interpretation of the results of an *experimental design in which the allocation of *treatments to *experimental units was not made at random. A common example concerns a hospital patient, where the range of feasible treatments may be limited by aspects of the patient's condition. The propensity score is defined as the *conditional probability of the patient receiving a particular treatment given the values of relevant characteristics (e.g. age, sex) of the patient. The score is typically estimated using a *logistic regression model.

prophet inequality An onlooker has to choose from a sequence of values of a positive *random variable. The onlooker cannot choose a past value but wishes to choose a large value. The prophet inequality states that, on average, the ratio of the choice made by a prophet (the largest value) divided by the choice made by the onlooker (who may use any decision procedure) is at most 2.

proportion For a *population of size N, of which R have a particular characteristic, the population proportion P is given by $P = R/N$. For a *sample of size n, of which r have the characteristic, the sample proportion p is given by $p = r/n$.

proportional allocation *See* STRATIFIED SAMPLING.

proportional-hazards model *See* HAZARD RATE.

p

proportional odds model *See* CUMULATIVE ODDS RATIO.

proportional reduction in error (PRE) A criterion underlying some measures of association (*see* ASSOCIATION). The measures attempt to quantify the extent to which knowledge about one *variable helps with the prediction of another variable. Examples include R^2 (*see* ANOVA) and Goodman and Kruskal's lambda (*see* ASSOCIATION).

prosecutor's fallacy A mis-statement of a *probability as a result of a misunderstanding of *conditional probability. (*See also* DEFENDER'S FALLACY.) As an example, suppose a blood type possessed by only 1% of the population is found at a crime scene. The accused has blood of this type. The prosecutor argues that there is only a 1% chance that the accused would have blood of this type (event A, say) if innocent (event B, say) and concludes that the accused is guilty. The prosecutor has quoted $P(A|B)$ when it is $P(B|A)$ that is relevant.

prospective study An alternative name for a *cohort study or *longitudinal study. The term is used in epidemiology.

protocol The step-by-step procedure for carrying out an *experimental design.

proximity matrix A square *matrix in which the entry in cell (j, k) is some measure of the similarity (or distance) between the items to which row j and column k correspond. A simple example would be a standard mileage chart—the smaller the entry, the closer together are the two items. Proximity matrices form the data for *multidimensional scaling. Asymmetric matrices can occur (for example, if the measurement is time taken, then the journey from top to bottom of a hill will be shorter than the journey from bottom to top).

proxy variable A measurable variable that is used in place of a variable that cannot be measured. For example, since husbands and wives usually have similar views, an interviewer might use the view expressed by a wife who is present in place of the view that could not be expressed by an absent husband. *See also* SURROGATE VARIABLE.

pruning *See* DECISION TREE.

Prussian horse-kicks A set of *data (*see overleaf*) introduced by *Bortkiewicz. It illustrates the fact that the *frequencies of occurrence of unlikely events (*see* SAMPLE SPACE) follow a *Poisson distribution even when there may be variations in the *probabilities of the events. Bortkiewicz's data refer to the numbers of deaths in each corps of the Prussian Army during the period 1875–94. The data, given in the table, show significant variations from year to year and from corps to corps, but have mean (0.70) and variance (0.76) approximately equal.

pseudolikelihood If the values at n locations of a *spatial process are denoted by $x_1, x_2, \ldots, x_n$, the pseudolikelihood is the product

$$\prod_{j=1}^{n} \mathrm{P}(X = x_j | x_1, x_2, \ldots, x_n, \text{except for } x_j).$$

This product will depend on the *parameters defining the spatial process. Usually, these are unknown and are estimated by maximizing the pseudolikelihood.

pseudo-random numbers Numbers generated by a mathematical formula which has the property that the generated numbers appear to be independent observations from a uniform distribution. Depending on the context, the distribution is usually a continuous uniform distribution on the interval (0, 1), or a discrete uniform distribution on the integers 0, 1, . . . , 9. Pseudo-random numbers are often referred to simply as **random numbers** (or **random digits**). They form the basis for many

CORPS	1875	1876	1877	1878	1879	1880	1881	1882	1883	1884	1885	1886	1887	1888	1889	1890	1891	1892	1893	1894
G	0	2	2	1	0	0	1	1	0	3	0	2	1	0	0	1	0	1	0	1
I	0	0	0	2	0	3	0	2	0	0	0	1	1	1	0	2	0	3	1	0
II	0	0	0	2	0	2	0	0	1	1	0	0	2	1	1	0	0	2	0	0
III	0	0	0	1	1	1	2	0	2	0	0	0	1	0	1	2	1	0	0	0
IV	0	1	0	1	1	1	1	0	0	0	0	1	0	0	0	0	1	1	0	0
V	0	0	0	0	2	1	0	0	1	0	0	1	0	1	1	1	1	1	0	0
VI	0	0	1	0	2	0	0	1	2	0	1	1	3	1	1	1	0	3	0	0
VII	1	0	1	0	0	0	1	0	1	1	0	0	2	0	0	2	1	0	2	0
VIII	1	0	0	0	1	0	0	1	0	0	0	0	1	0	0	0	1	1	0	1
IX	0	0	0	0	0	2	1	1	1	0	2	1	1	0	1	2	0	1	0	0
X	0	0	1	1	0	1	0	2	0	2	0	0	0	0	2	1	3	0	1	1
XI	0	0	0	0	2	4	0	1	3	0	1	1	1	1	2	1	3	1	3	1
XIV	1	1	2	1	1	3	0	4	0	1	0	3	2	1	0	2	1	1	0	0
XV	0	1	0	0	0	0	0	1	0	1	1	0	0	0	2	2	0	0	0	0

Prussian horse-kicks. Numbers of deaths due to horse-kicks in each corps of the Prussian Army during the period 1875–94.

statistical methods including the *bootstrap, *Markov chain Monte Carlo methods and *simulation. A short table of pseudo-random numbers is given in Appendix IX. It is conjectured that the decimal digits of π (3, 1, 4, 1, 5, 9, . . .) may be a naturally occurring sequence of random digits.
See also ERNIE.

pseudovalue *See* JACKKNIFE.

PSI Abbreviation for Statisticians in the Pharmaceutical Industry, an organization founded in the United Kingdom in 1977. It has over 1000 members.

SEE WEB LINKS

- Organization home page.

PSU Abbreviation for primary sampling unit (*see* CLUSTER SAMPLING).

psychometrics A branch of psychology in which a numerical measurement is made of psychological factors; for example, an individual's or a group's preference in foods, or assessment of artistic or personal qualities.

Psychometric Society A society devoted to the advancement of quantitative measurement practices in psychology, education, and the social sciences. The Society, which has more than 850 members, publishes the journal **Psychometrika*. Past Presidents of the Society include *Thurstone (1935), *McNemar (1950), *Cronbach (1953), *Mosteller (1957), *Guttman (1970) and *Kruskal (1974).

SEE WEB LINKS

- Society home page.

Psychometrika The quarterly journal of the *Psychometric Society. It was first published in 1936.

SEE WEB LINKS

- Journal information.

publication bias The apparent tendency of journal editors to favour the publication of results showing rejection of the *null hypothesis as opposed to those that lead to acceptance of the hypothesis.

pure birth process *See* BIRTH-AND-DEATH PROCESS.

Puri, Madan Lal (1929– ; b. Sialkot, Pakistan) Indian mathematician and statistician whose career has been in the USA, where he arrived in 1957 to study mathematics at U Colorado. Puri obtained his PhD (supervised by *Lehmann) from UCB in 1962, joining the faculty at NYU before moving to Indiana U in 1968. His interests include *probability theory, *extreme-value distributions, *time series, and *non-parametric

p

tests. His 1971 book with *Sen entitled *Non-parametric Tests in Multivariate Analysis* remains a standard reference. *See* SEN–PURI TEST.

SEE WEB LINKS

- University website.

p-value *See* HYPOTHESIS TEST.

Pygmalion effect *See* BLINDING.

p

q-q plot A plot for comparing two *probability distributions, usually the *sample distribution function and a theoretical *distribution function. The sample values are ranked in order: $x_{(1)} \leq x_{(2)} \leq \cdots \leq x_{(n)}$. Define the sample cumulative *proportion as p_j, typically calculated as $(j - \frac{1}{2})/n$, and denote the theoretical distribution function by F. In a q-q plot, $F^{-1}(p_j)$ is plotted against $x_{(j)}$, for all j. If the sample has come from the theoretical distribution, the plotted values will lie on an approximate straight line. Specialized graph paper, that has the values of F^{-1} marked for interesting values of p is called **probability paper**.

The **p-p plot** is entirely equivalent: $F(x_{(j)})$ is plotted against p_j. Once again, if the sample has come from the theoretical distribution, the plotted values will lie on an approximate straight line. A plot of this type is called a **probability plot**. *See also* NORMAL PROBABILITY PAPER.

Quade test *See* FRIEDMAN TEST.

quadrant *See* CARTESIAN COORDINATES.

quadrat Originally, a square wooden frame thrown on to the ground by, for example, botanists who wished to count plant species at random locations. Now used more generally to refer to a sampled area of space.

quadratic form *See* MATRIX.

quadratic regression model *See* MULTIPLE REGRESSION MODEL.

qualitative variable A nominal or *categorical variable.

quality control The analysis of a sequence of small *samples taken at regular intervals from the output of an industrial production process, with the aim of ensuring that the output meets its required specification.

Standard statistical procedures expect large samples whereas here the samples are small. The solution suggested by *Shewhart was the use of a **control chart** (**Shewhart chart**). The basis of this chart is a simple plot of the successive sample means, though often a simultaneous parallel chart of some *measure of spread is maintained. The purpose of the chart is to give a quick visual indication of any trends in the production

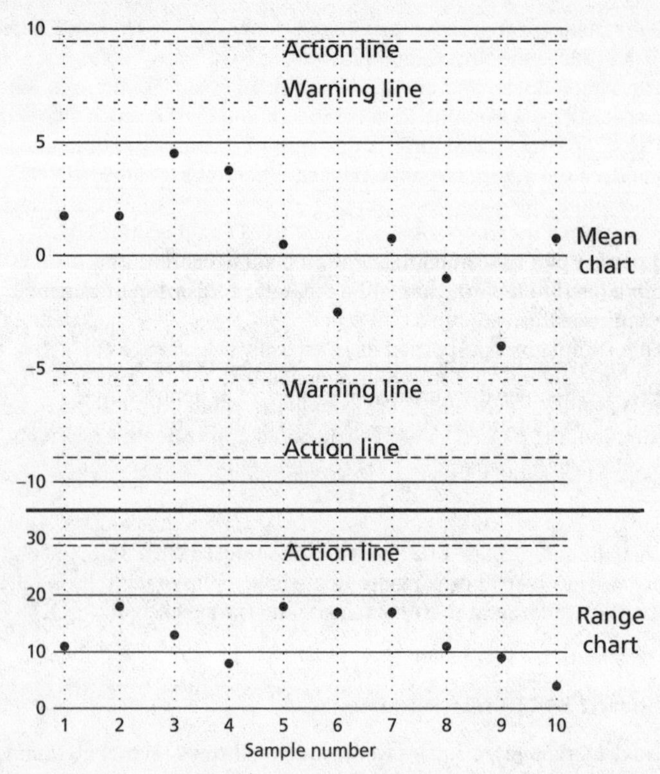

Quality control. This quality control chart plots the values of sample means (upper graph) and ranges (lower graph) as a function of time. If a value falls beyond an action line, or if a succession of sample means lie outside the warning lines, then production is stopped.

process (e.g. a drift in the mean), or an increase in variability (possibly a consequence of some problem in part of the process), without the need for advanced statistics. The diagram illustrates a combined **mean chart** (**$\bar{x}$- chart**) and **range chart** (**R-chart**).

On the $\bar{x}$-chart are two pairs of lines: **warning lines** and **action lines**. If any sample mean lies outside an action line, or if a succession of sample means lie outside the warning lines then production is stopped. One set of rules to have found favour are the **Western Electric Rules** which state that the process should be judged out of control if any of the following occur:

1. The sample mean is more than three *standard errors from the notional mean (this is the *three-sigma rule);
2. The sample mean and one of its two predecessors are more than two standard errors from the notional mean (and on the same side of it);
3. The sample mean and three of its four predecessors are more than one standard error from the notional mean (and on the same side of it);
4. The sample mean and its seven predecessors all lie on the same side of the notional mean. In some applications of quality control the discovery of a value outside the warning line will be a signal to inspect more often—this is **tightened inspection** (as opposed to **normal inspection**).

An alternative to the control chart is the **cumulative sum chart** (**cusum chart**). If $\bar{x}_j$ denotes the j th sample mean and m denotes the expected mean of the production process, then, on the cusum chart, $\Sigma_{j=1}^{k}(\bar{x}_j - m)$ is plotted against k. If production is normal then the plot will be roughly horizontal, whereas a trend indicates a departure from the expected mean. *See* ACCEPTANCE SAMPLING.

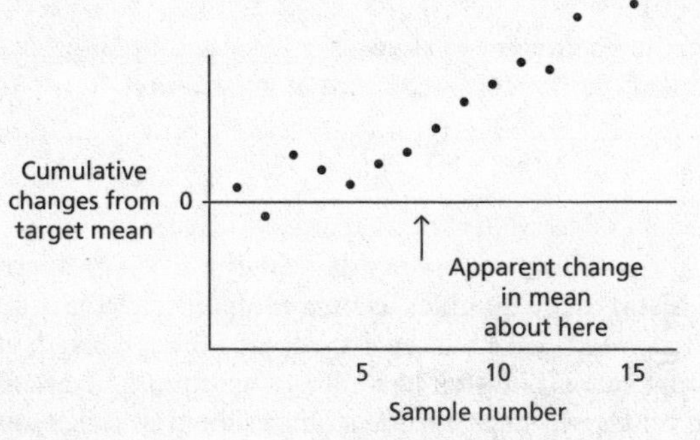

Cumulative sum chart. Small changes in mean show more clearly on a cusum chart than on a control chart.

quantile The qth quantile of the distribution of a random variable, X, is that value x such that $P(X<x)=q$. If $q=0.5$, the value is called the *median. The cases $q=0.25$ and $q=0.75$ correspond to the lower *quartile and upper quartile, respectively. *See also* DECILE; PERCENTILE.

quartile If a set of numerical *data has n elements and is arranged in increasing order

$$x_1 \leq x_2 \leq \cdots \leq x_n,$$

the **lower quartile** (Q_1) may be taken to be the *median of the lower half of the data, i.e. of $x_1, x_2, \ldots, x_{\frac{1}{2}(n-1)}$ if n is odd, and the median of $x_1, x_2, \ldots, x_{\frac{1}{2}n}$ if n is even. The **upper quartile** (Q_3) may be taken to be the median of the upper half of the data, i.e. of $x_{\frac{1}{2}(n+1)}, x_{\frac{1}{2}(n+3)}, \ldots, x_n$ if n is odd, and the median of $x_{\frac{1}{2}(n+2)}, x_{\frac{1}{2}(n+4)}, \ldots, x_n$ if n is even. The difference $Q_3 - Q_1$ is the **interquartile range**, a term introduced by *Galton in 1882. An alternative term is the **midspread**.

As an example, consider the ordered data:
101, 103, 104, 105, 106, 107, 108, 109, 111, 111, 111, 115, 118, 121, 124, 127, 130, 156, 199.

There are nineteen observations. The tenth largest is 111, the median. Within the lower nine values, the fifth largest is 106 ($= Q_1$). Within the upper nine values the fifth largest is 124 ($= Q_3$). The inter-quartile range is $124 - 106 = 18$.

When there are many observations it may be easier to read approximate values for the lower and upper quartiles from a *cumulative frequency graph. These will be the values of the variable corresponding to cumulative relative frequencies of 25% and 75%, respectively.

For a *continuous random variable X, the lower quartile of the distribution is such that $P(X < Q_1) = \frac{1}{4}$ and the upper quartile is such that $P(X < Q_3) = \frac{3}{4}$.

The **quartile coefficient of skewness** suggested by *Bowley in 1920 and also called the **Bowley coefficient of skewness** is

$$\frac{Q_3 + Q_1 - 2Q_2}{Q_3 - Q_1}.$$

The coefficient takes values in the interval $(-1, 1)$.

In his 1970 book on *exploratory data analysis, *Tukey referred (in the context of *data) to the quartiles as **hinges** and he called the interquartile range the **H-spread**. Tukey defined a **step** as $1.5 \times$ H-spread, and proposed that values one step beyond a hinge should be called **inner fences** and values two steps beyond a hinge should be called **outer fences**. Any data item beyond an outer fence would be called **far out**.

In the previous data the hinges are 106 and 124, thus the H-spread is $124 - 106 = 18$ and the step is $1.5 \times 18 = 27$. The inner fences are at $106 - 27 = 79$ and $124 + 27 = 151$. The outer fences are at $79 - 27 = 52$ and $151 + 27 = 178$. The observation 199 is greater than 178 and is therefore far out.

See also BOXPLOT; OUTLIER; QUANTILE; SKEWED; TRIMEAN.

quartile coefficient of skewness *See* QUARTILE.

quasi-independence model The situation, in a *contingency table, in which the usual *independence model holds for a subset of the cells (*see* CONTINGENCY TABLE) in the table.

108	12	18	22	7
12	24	36	44	14
18	36	54	66	21

In the table illustrated, all the cells except that at the top left display perfect independence: the quasi-independence model fits the remaining cells perfectly. The *mover–stayer model is a special case of a quasi-independence model.

quasi-likelihood A function of the *mean and *variance proposed by *Wedderburn as an alternative to the *likelihood when the complete *probability distribution is unknown.

quasi-symmetry model *See* SQUARE TABLE.

QUEST (Quick, Unbiased and Efficient Statistical Tree) A program for constructing *classification trees.

SEE WEB LINKS

- Outline of program.

questionnaire A list of questions—principally used to collect socio-economic or political data on individuals and households.

Quetelet, Lambert Adolphe Jacques (1796–1874; b. Ghent, Belgium; d. Brussels, Belgium) Belgian mathematician, astronomer, and statistician. Quetelet obtained his PhD (on conic sections) from U Ghent in 1819. By 1833 he was working as an astronomer and meteorologist at the Royal Observatory in Brussels, but his international fame was due to his work as a statistician on social science data. He spent much time constructing tables and diagrams to show relationships between variables. He was interested in the concept of an 'average man' as today we talk of the 'average family' and this was the subject of *Sur l'homme et le développement de ses facultés* (*A Treatise on Man and the Development of his Faculties*), published in 1835. He was one of the founders of the *RSS and, in 1853, he organized the first international Statistics conference. In 1835 he was elected FRSE and in 1838 FRS. A lunar crater is named after him.

SEE WEB LINKS

- Information and portrait.

Quetelet index (body mass index) An index of obesity, given as the ratio (weight in kg)/(height in m)^2. A person with an index greater than 30 is officially obese.

queue An example of a continuous-time Markov chain (*see* MARKOV PROCESS). The properties of queues are much studied by analysts of *stochastic processes. Three components of a queuing system are the inter-arrival times (a Markov process (M), a more general process (G), or a pre-determined process (D), the service times (also Markovian, general, or predetermined), and the number of servers (k). The standard nomenclature for queues describes them as M/M/1, M/D/1, M/G/1, or G/M/k queues as appropriate.

The basic quantities of interest are the *expected values of the number in the queue, the number waiting (i.e. not being served) in the queue, the queueing time, and the waiting time (the sum of the queueing and service times).

For an M/G/1 queue, with arrival rate λ and with $\frac{1}{\mu}$ and σ^2 denoting the *mean and *variance of the service time *distribution and with $\rho = \frac{\lambda}{\mu}$, then the *expected value of the number in the system (queueing or being served) is N given by the **Pollaczek-Khintchine formula**

$$N = \rho + \frac{\rho^2 + \lambda^2\sigma^2}{2(1-\rho)}.$$

A useful general result, with $\bar{t}$ denoting the average time spent in the system, is **Little's formula** which states that $\lambda\bar{t} = N$.

quincunx A simple arrangement of pegs on a board that can be used to illustrate the *binomial and *normal distributions. A funnel allows a ball to roll down and strike the single peg on the top line. The ball rolls to left or right (ideally, with equal probability) and then falls to strike a peg on the next row and the process is repeated on each row. At the bottom the ball is held in one of a number of channels. When many balls are fed through the system it is found that the central channels will contain more balls than the extreme ones. Sir Francis *Galton used a quincunx in his 1874 lecture on the normal distribution at the Royal Institution in London. *See diagram opposite.*

SEE WEB LINKS

- Animation.

quota sampling A method of *sampling in a survey in which the interviewer is instructed to include certain prescribed percentages of people who come from various identifiable subpopulations, e.g. men over 60, women under 30, unemployed men.

q

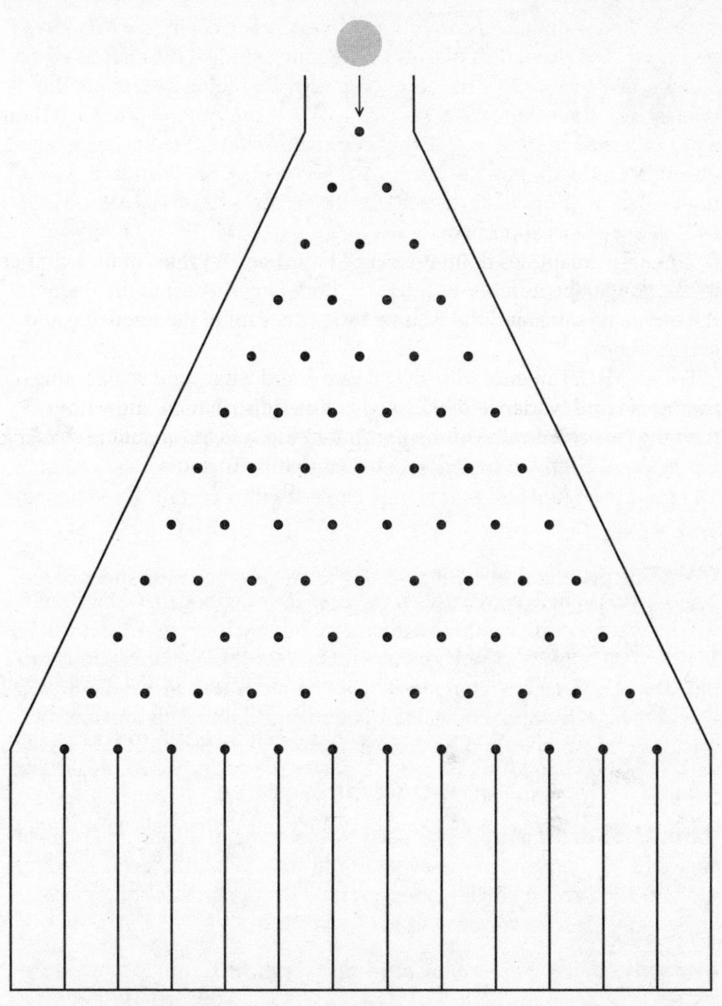

Quincunx. In the diagram each point represents a peg. A series of small balls is inserted at the top of the quincunx. Each ball hits a sequence of pegs before coming to rest in a channel at the bottom of the quincunx. The distribution of balls in channels will be a realization of a binomial distribution, with *n* being the number of rows of pegs.

R A freely available computer package based on *S-PLUS.

SEE WEB LINKS

- The R Project home page.

radar plot *See* SPIDER PLOT.

radial basis function (RBF) A class of function used in supervised learning (*see* MACHINE LEARNING). An RBF has a value at a point that is a function only of the distance, r, between that point and some fixed point. The most used function is, for some choice of the constant k, the function $\exp(-kr^2)$.

radian (rad) A unit of angle measurement. The radian measure of the angle enclosed between two lines OP and OQ is defined to be the length of the circular arc, with centre O and unit radius, enclosed between the lines. Thus a right angle is $\frac{1}{2}\pi$ rad, and a complete revolution corresponds to 2π rad. The relationship with degrees is that π rad $= 180°$ or 1 rad $\approx 57.296°$. Use of radians is important when trigonometrical functions are used in calculus. For example, in the relationship that the derivative of $\sin x$ is $\cos x$.

raking *See* DEMING–STEPHAN ALGORITHM.

Ramachandran plot A plot used by biologists to display the pairs of angles (ϕ, ψ) that govern the twists in the chain of amino-acids forming a protein. Each pair of angles is represented by a point in a square whose limits, on each axis, correspond to $(-180°, 180°)$.

random Adjective describing any process in which results may not be certain.

random cascade model A *model, based on the idea of *fractals, that seeks to model variablity in short time intervals (or over small areas) using information about the observed values and variability at longer time intervals (or larger areas).

random coefficient model In a random coefficient model the same *model holds for all individuals, but the *parameters of the model do not have fixed values. Instead, for each individual the value of every parameter is regarded as a random observation from its own specific distribution.

random digits *See* PSEUDO-RANDOM NUMBERS.

random effects *See* EXPERIMENTAL DESIGN.

random error An error of measurement as a consequence of recording the value $x+\varepsilon$ instead of the true value x, with ε being an *observation on a *random variable. The random error is often assumed to have a *normal distribution with *mean 0 and constant *variance (though this assumption should always be verified). In the case of errors introduced by rounding to a fixed number of decimal places, a *uniform distribution is appropriate. *See* SYSTEMATIC ERROR.

random forest In the *machine learning context, this is an *ensemble classifier consisting of a group of *decision trees. The classification chosen is that selected most frequently by the separate trees.

random graph A *graph constructed following rules governed by *probability. Let j and k denote two nodes (with $j=k$ being a possibility). With probability p_{jk}, construct an arc between these nodes. The value of p_{jk} might be the same for all pairs of nodes, or it might vary.

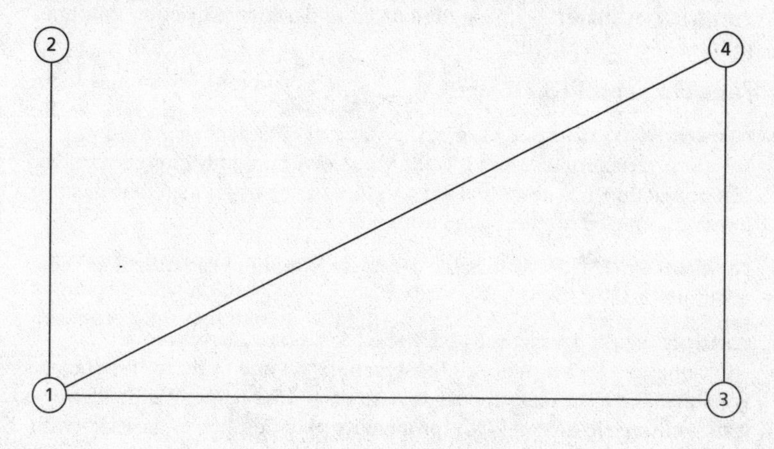

Random graph. The example shows a random graph that is not a connected graph, since node 5 is not connected to the other nodes.

randomized blocks design An *experimental design in which the *bt* *experimental units are divided into *b* *blocks each of size *t* units. Within each block the *t* *treatments under comparison are allocated at random. In this example, four treatments (A–D) have been randomly allocated within each of three blocks:

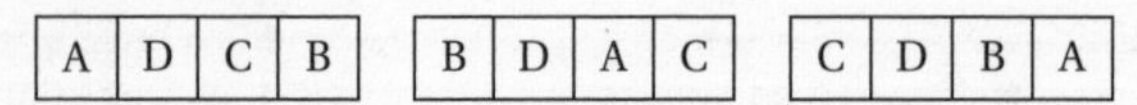

The phrase 'randomized blocks' was introduced by Sir Ronald *Fisher in 1926.

random man not excluded (RMNE) In the context of a crime scene, this is the estimated *probability that an individual, randomly sampled from the *population, would not be excluded by virtue of the DNA evidence at the scene. It is also known as the **combined probability of inclusion** (**CPI**).

random match probability (RMP) The *probability of obtaining a match between two distinct and unrelated individuals based on DNA evidence.

random number A phrase often used to describe a *pseudo-random number. *See also* ERNIE.

random sample *See* SAMPLE.

random variable (r.v.) When the value of a *variable is subject to *random variation, or when it is the value of a randomly chosen member of a population, it is described as a random variable—though the adjective 'random' may be omitted. *See* PROBABILITY DISTRIBUTION.

random variation A *variable is subject to random variation if its value is not predictable.

random walk A walk in which the walker's movements are a consequence of a sequence of *observations on one or more *random variables. For example, suppose that, at each time point an individual walks one step to the left (with *probability p) or one step to the right (with probability $1-p$). This simple *Markov process is a one-dimensional random walk. If the individual is allowed at each stage to move one step in any direction in two dimensions, then this is the **drunkard's walk**. The phrase 'random walk' was first used by Karl *Pearson in 1905. *See also* BROWNIAN MOTION.

In the particular case where $p=0.5$, it might be supposed that the proportion of time spent to the left of the starting point is likely to be close to a half. In fact the proportion is more likely to be near to 0 or 1.

The probability of the time proportion being less than α $(0 \leq \alpha \leq 1)$ is given by the approximate **arc-sine law**:

$$\frac{2}{\pi}\sin^{-1}(\sqrt{\alpha}).$$

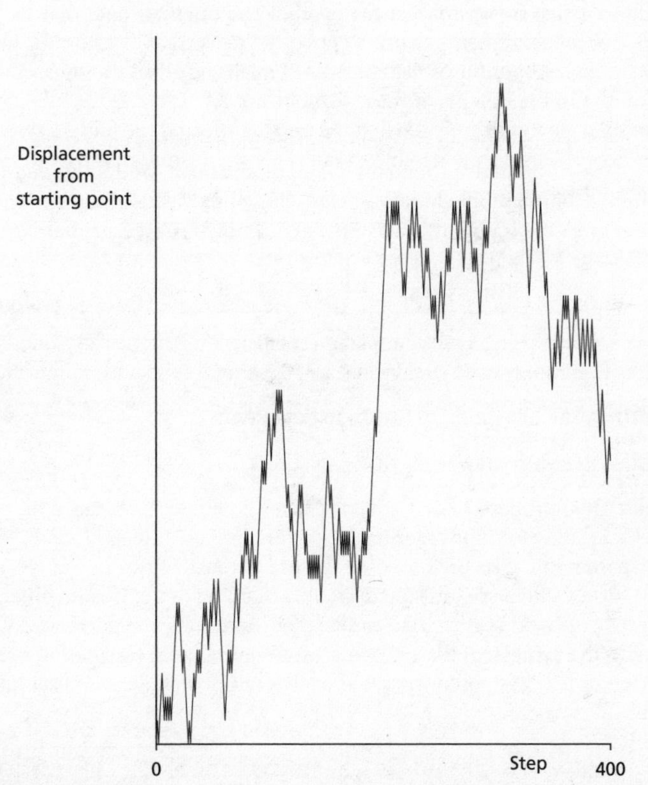

Random walk. The diagram shows the displacement from the starting point for the first 400 steps of a random walk in which each step has the same length, and moves up and down are equally likely. The *x*-axis represents the step number. The path is typical, in that it predominantly lies on one side of the starting line.

random zero *See* STRUCTURAL ZERO.

range The difference between the largest and the smallest items in a set of numerical *data.

range chart *See* QUALITY CONTROL.

rank (of a data item) *See* RANKS.

rank (of a matrix) *See* MATRIX.

rank correlation coefficient A *non-parametric measure of the extent to which two *variables are related. The original *data may be *ranks, or measurements that are converted to ranks. For example, in a flower show, six sunflower exhibits (A–F) might be given the ranks 1, 3, 2, 5, 6, and 4. If the heights of these exhibits are 2.4, 1.9, 2.2, 1.8, 1.85, and 2.0 m then we might suppose that the heights affected the judge's ranking, since on converting the heights to ranks we get a similar pattern to the judge's ranks:

Exhibit	A	C	B	F	D	E
Judge's rank	1	2	3	4	5	6
Rank of height	1	2	4	3	6	5

An attraction of using a rank correlation coefficient is that the calculations are simple. Commonly used coefficients are *Spearman's rho and *Kendall's tau.

ranking The allocation of *ranks to data items.

rankit *Alternative name for* NORMAL SCORE.

ranks The numbers 1, 2, . . . assigned to a set of objects arranged in order according to some criterion such as increasing numerical value. Many *non-parametric tests are based on the use of ranks. If two or more observations are indistinguishable according to the chosen criterion then they are said to have **tied ranks**; they are usually assigned a rank equal to the average of the ranks that would have been assigned in the absence of ties. The following set of twelve values includes two examples of ties:

Value	11	14	15	15	16	17	17	17	20	25	40	113
Rank	1	2	3.5	3.5	5	7	7	7	9	10	11	12

Rao, Calyampudi Radhakrishna (1920–   ; b. Huvina Hadagali, India) Indian statistician known for his fundamental insights into *statistical inference. After studying mathematics at Andhra U and statistics at Calcutta U, in 1944 Rao joined the *Indian Statistical Institute (ISI) (working under *Mahalanobis), where he worked on the Cramér–Rao inequality (*see* FISHER INFORMATION), which he allegedly proved overnight in response to a student inquiry, and on what is now known as the *Rao–Blackwell theorem. Mahalanobis sent Rao to Cambridge U to analyse data under the guidance of Sir Ronald *Fisher. On obtaining his PhD in 1948, Rao returned to the ISI first as head of the research section and subsequently as Director. *Basu and *Varadhan were two of his research students at the ISI. In 1982 Rao emigrated to the USA, first to U Pittsburgh, then, in 1990, to Penn State U as its founding Director of the Center for Multivariate Analysis. He has been President of the *IBS (1974), the *IMS (1977), and the *International Statistical Institute (1982). He is an Honorary Life Member of the latter. He was the *Wald Lecturer of the IMS in 1975. He was elected FRS in 1967 and is a Fellow of the AAAS. The *RSS awarded him its *Guy Medal in Silver in 1965, and in Gold in 2011. He was the recipient of the *COPSS *Fisher Lectureship in 1979, the *Wilks Award of the *ASA in 1989, and the *Parzen Prize in 2000. He was elected an Honorary Fellow of the RSS in 1969, an Honorary Life Member of the IBS in 1985, and member of the NAS in 1995. He was awarded the National Medal of Science in 2002.

SEE WEB LINKS

- Fuller biography and photograph.

Rao–Blackwell theorem A theorem, proved independently by *Rao in 1945 and *Blackwell in 1947. Let $\{X_i\}$ be independent identically distributed *random variables with θ being an unknown *parameter of their common probability distribution. The theorem states that, if T and S are two functions of the $\{X_i\}$, with T being an unbiased *estimator of θ, and S being a sufficient statistic, then, for all θ, the *expected value of T given S is an unbiased estimator of θ with *variance never greater than that of T.

Rao Prize A prize awarded in honour of *Rao by Penn State U. It is awarded biennially to 'recognize outstanding and influential innovations in the theory and practice of mathematical statistics'.

Rasch, Georg (1901–80; b. Odense, Denmark; d. Laesoe, Denmark) Danish statistician. Rasch was the son of a mathematics teacher who ran a mission high school for prospective seamen. He gained his first taste for mathematics through reading his father's trigonometry textbooks. He graduated in mathematics at U Copenhagen in 1925, obtaining his PhD in 1930. His subsequent work as a consultant brought him into contact with Statistics. In 1934 he joined Sir Ronald *Fisher in London. He returned to Denmark in 1935 to the Bio-Statistical Department of the Danish National Serum Laboratory. In 1947 he was a founder member of the *IBS. In 1961 he returned to U Copenhagen as Professor of Statistics. His work on the *Rasch model was published in 1960. He became a Danish knight in 1967.

SEE WEB LINKS

- Biographical PhD thesis with photograph.

Rasch model A simple model that describes the probability that, when a group of individuals are asked a number of questions, individual j makes a mistake in answering question k. The model, which was suggested by *Rasch in 1960, was originally applied to the answers given in an oral reading test and has subsequently been applied as a model of the responses to questions in many different tests (particularly psychological tests). The model is a particular type of *logit model and takes the form

$$\ln\left\{\frac{\text{P(Individual } j \text{ answers question } k \text{ correctly)}}{\text{P(Individual } j \text{ answers question } k \text{ incorrectly)}}\right\} = \mu + \alpha_j + \beta_k,$$

where $\Sigma_j \alpha_j = 0$ and $\Sigma_k \beta_k = 0$. *See also* ITEM RESPONSE FUNCTION.

ratio scale *See* INTERVAL SCALE.

Rayleigh, Lord (John William Strutt) (1842–1919; b. Maldon, England; d. Witham, England) English mathematical physicist awarded the Nobel Prize for Physics in 1904. Strutt graduated from Cambridge U in 1865 and joined the faculty. In 1871 he published his theory of scattering which explained why the sky was blue. In 1873, following his father's death, he became the third Baron Rayleigh. After a period managing his 7000-acre estates he returned to Cambridge as Head of the Cavendish Laboratory (1879–84). From 1887 to 1905 he was Professor of Natural Philosophy at the Royal Institution. His early work in mathematical physics extended to cover a wide range of physical problems. He was President of the LMS in 1876 and was awarded its de Morgan Medal in 1900. He was elected FRS in 1873 and was awarded the Royal Society's Royal Medal in 1882 and its Copley Medal in 1899, serving as the Society's President from 1905 to 1908. He was elected FRSE in 1886. He was awarded the Order of Merit in 1902 and was appointed Chancellor of Cambridge U in 1908. There are craters named after him on both the Moon and Mars.

SEE WEB LINKS

- Fuller biography and photograph.

Rayleigh distribution The distribution of the distance between a point and its nearest neighbour in a *spatial *Poisson process. Also the distribution of the distance from the origin in n-dimensional space to the point $(X_1, X_2, \ldots, X_n)$, where $X_1, X_2, \ldots, X_n$ are *independent normal variables (*see* NORMAL DISTRIBUTION), each with *expected value 0 and *variance σ^2. The *probability density function f is given by

$$\mathrm{f}(x) = \frac{2x^{n-1}\exp\left\{-\frac{1}{2}\left(\frac{x}{\sigma}\right)^2\right\}}{(2\sigma^2)^{\frac{1}{2}n}\Gamma\left(\frac{1}{2}n\right)}, \qquad x > 0,$$

where $\sigma > 0$ and Γ is the *gamma function. The distribution has *mean μ and variance V given by

$$\mu = \frac{\sigma\sqrt{2}\Gamma\{\frac{1}{2}(n+1)\}}{\Gamma(\frac{1}{2}n)}, \qquad \text{and} \quad V = n\sigma^2 - \mu^2.$$

The distribution has *mode $\sigma\sqrt{n-1}$. In the case $n=2$, the expressions for the mean and variance simplify to $\sigma\sqrt{\frac{1}{2}\pi}$ and $\frac{1}{2}\sigma^2(4-\pi)$ respectively.

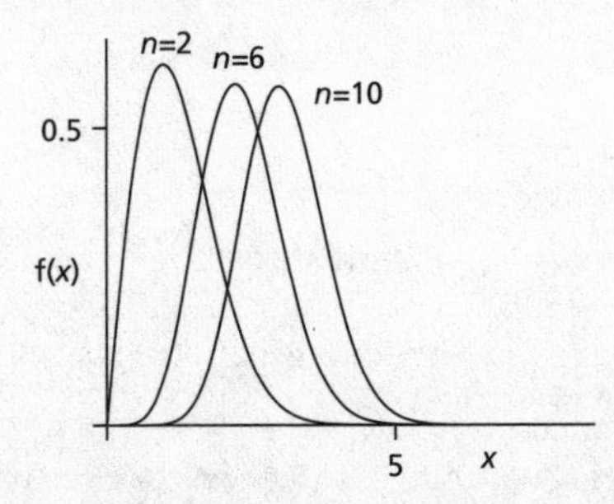

Rayleigh distribution. The shape of the distribution depends on the value of the parameters σ and n. The figure illustrates the dependence on n, with $\sigma = 1$.

Rayleigh test *See* CIRCULAR UNIFORM DISTRIBUTION.

RBF *See* RADIAL BASIS FUNCTION.

R-chart *See* QUALITY CONTROL.

recall data *Data provided by an individual remembering events in the past. Recall data are notoriously unreliable. For example, if voters are asked after a general election whether or not they voted, the replies will usually overestimate the true proportion voting. Similarly, when voters are asked which party they voted for, support for the winning party will usually be overestimated.

receiver operating characteristic curve *See* TWO-BY-TWO TABLE.

reciprocal The reciprocal of x is $1/x$.

recombinance; recombination probability *See* LINKAGE.

recommender systems *Algorithms and procedures that recommend items that may be suitable for a consumer. Much used in online sales, the techniques involved include *classification trees, *cluster analysis, *collaborative filtering, and *data mining.

rectangular cartogram *See* CARTOGRAM.

rectangular distribution Alternative name for a *uniform distribution of a *continuous variable.

recurrence relation An equation relating the *n*th term of a sequence to preceding terms. For example, the *probabilities in a *Poisson distribution with *parameter λ are related by

$$p_n = \frac{\lambda}{n} p_{n-1} \qquad \text{for } n = 1, 2, \ldots.$$

recurrent state *See* MARKOV PROCESS.

recursion A method of generating successive terms of a sequence when there is a *recurrence relation expressing the *n*th term in terms of some or all of the preceding terms. For example, $a_n = na_{n-1}$, with $a_1 = 1$, gives a recursive definition of $n!$ (*factorial *n*).

recursive model Let X_t and Y_t denote two *random variables measured at time *t*, with observed values x_t and y_t. A simple example of a recursive model is provided by

$$\mathrm{E}(Y_t) = a + bx_t, \qquad \mathrm{E}(X_t) = c + dy_{t-1},$$

where E denotes *expected value. The present value of *Y* depends upon the present value of *X*, but that value depends upon the previous value of *Y*.

refined boxplot *See* BOXPLOT.

regression A word introduced by *Galton that derives from his phrase '*regression towards the mean'. It is often used as shorthand for *linear regression or *multiple regression models. In these models the *expected value of one *variable *Y* is presumed to be dependent on one or more other variables ($x_1, x_2, \ldots$). The variable *Y* is variously known as the **response variable, dependent variable**, or **outcome variable.** The *x*-variables are variously known as **antecedent variables, background variables, predictor variables, explanatory variables, controlled variables**, or, potentially confusingly, **independent variables.** In the context of a factorial experiment (*see* FACTORIAL DESIGN) the *x*-variables are *factors. *See* LINEAR REGRESSION.

regression diagnostics Various *statistics that give information about the reliability of the estimates of the *multiple regression model

$$\mathrm{E}(\mathbf{Y}) = \mathbf{X}\boldsymbol{\beta},$$

where **Y** is an $n \times 1$ *vector of *independent and identically distributed response variables, $\boldsymbol{\beta}$ is a $p \times 1$ vector of unknown *parameters, and **X** is an

$n \times p$ matrix. If β is replaced by its least squares estimate, $\hat{\beta}$, the estimated column vector of *fitted values, $\hat{\mathbf{y}}$, is given by

$$\hat{\mathbf{y}} = \mathbf{H}\mathbf{y},$$

where the $n \times n$ matrix $\mathbf{H}$, the **hat matrix**, is given by

$$\mathbf{H} = \mathbf{X}(\mathbf{X}'\mathbf{X})^{-1}\mathbf{X}',$$

$\mathbf{X}'$ is the transpose of $\mathbf{X}$, $(\mathbf{X}'\mathbf{X})^{-1}$ is the inverse of the matrix $\mathbf{X}'\mathbf{X}$, and $\mathbf{y}$ is the column vector of observed values. Denote the element in the jth row and kth column of $\mathbf{H}$ by h_{jk}. The fitted value, $\hat{y}_j$, for the jth observation, y_j, is given by

$$\hat{y}_j = h_{jj}y_j + \sum_{k \neq j} h_{jk}y_k.$$

Thus there is a direct link between the fitted and observed values in the form of h_{jj}. This is the **leverage**: a large value (e.g. $> 2p/n$) indicates an observation having a large **influence** on the form of the fitted model.

The most obvious guide to the fit of a model are the **residuals**, $e_1, e_2, \ldots,$ where e_j is given by

$$e_j = y_j - \hat{y}_j.$$

If the random variables have common *variance σ^2 and if s^2 is an unbiased estimate of σ^2, then the **standardized residual** is sometimes defined as e_j/s. However, an unbiased estimate of the variance of e_j is not s^2 but $s^2(1-h_{jj})$ and a more appropriate residual (having unit variance if the model is correct) is given by r_j, where

$$r_j = \frac{e_j}{s\sqrt{1-h_{jj}}}.$$

This is sometimes called the **standardized residual** and sometimes the **Studentized residual**.

The *deletion residual is given by

$$d_j = y_j - \hat{y}_{j,-j},$$

where $\hat{y}_{j,-j}$ is the fitted value for observation j based on the fit of the model to all the observations except the observation y_j. Dividing the deletion residual by its estimated *standard error, we get the **Studentized deletion residual** which can be written as

$$r_{-j} = d_j/s_{-j}\sqrt{1-h_{jj}},$$

where s_{-j}^2 is the unbiased estimate of σ^2 obtained when observation j is omitted. Confusingly, this may also be called the **Studentized residual**. *See also* ANSCOMBE RESIDUAL; DEVIANCE RESIDUAL.

A related influence statistic is **DFFITS**, which is an abbreviation for difference in fits. For observation j, DFFITS$_j$ is

$$\frac{\hat{y}_j - \hat{y}_{j,-j}}{s_{-j}\sqrt{h_{jj}}}.$$

The influence statistic **DFBETA** (difference in beta values) applies the idea embodied in DFFITS to the parameter estimates rather than the fitted values. For β_k, DFBETA$_{k,-j}$ is

$$\frac{\hat{\beta}_k - \hat{\beta}_{k,-j}}{s_{-j}\sqrt{m_{kk}}},$$

where $\hat{\beta}_k$ is the estimate of β_k from the complete data, $\hat{\beta}_{k,-j}$ is the estimate when observation j is omitted, and m_{kk} is the corresponding diagonal element of the $p \times p$ matrix $(\mathbf{X}'\mathbf{X})^{-1}$.

A statistic that usefully combines information about leverage and influence is **Cook's statistic**, D_j, given by

$$D_j = \frac{h_{jj}r_j^2}{(p+1)(1-h_{jj})}.$$

This statistic (introduced by *Cook in 1977) can also be interpreted as measuring the effect on the parameter estimates of omitting the jth observation. Large values point to possible *outliers.

regression through the origin A situation in which there is one explanatory variable x and it is known that the *linear regression line must pass through the origin (because it is known that Y must be zero when x is zero). In this case the estimate of the slope using the *method of least squares is

$$\frac{\sum_{j=1}^n x_j y_j}{\sum_{j=1}^n x_j^2}$$

for data points $(x_1, y_1), \ldots, (x_n, y_n)$.

regression towards the mean An expression of the observation that, when we take pairs of related measurements, the more extreme values of one *variable will, on average, be paired with less extreme values of the other variable. In his early work on inheritance *Galton included a study of the heights of successive generations of people. Some of his data are summarized below.

Mean height of parents (inches)	72.5	70.5	68.5	66.5	64.5
Mean height of their adult children (inches)	72.2	69.5	68.2	67.2	65.8

Galton noted that, on average, the children of tall parents are shorter than their parents (72.2 < 72.5, etc.), whereas the children of short parents are taller than their parents (65.8 > 64.5, etc.): there is a regression towards the mean. These findings led Galton (in a talk entitled 'Regression towards Mediocrity in Hereditary Stature' given in 1885 to the British Association for the Advancement of Science) to refer to the summary line as a line of *regression.

SEE WEB LINKS

- Applet.

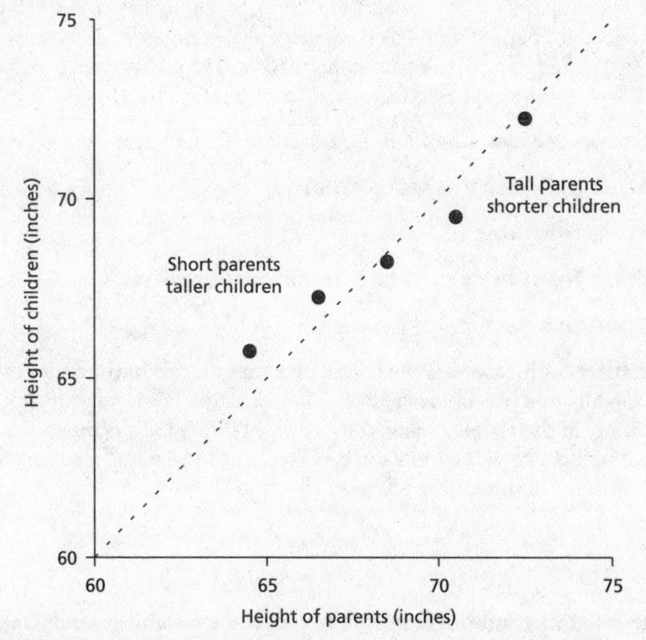

Regression towards the mean. The diagram illustrates Galton's data. The dotted line is the regression line.

regular *See* MATRIX.

regularity *See* INDEX OF DISPERSION.

R–E–G–W test *See* MULTIPLE COMPARISON TEST.

Reid, Nancy Margaret (1953- ; b. Niagara Falls, Canada) Canadian statistician specializing in *statistical inference. A graduate of U Waterloo and U British Columbia, she obtained her doctorate at Stanford U. She has been a faculty member at U Toronto since 1986. She was the 2000 *IMS *Wald Lecturer, was elected FRSC in 2001, and was awarded the *Parzen Prize in 2008. She was President of the *IMS in 1996 and of the *SSC in 2004. She was awarded the Gold Medal of the SSC in 2009.

SEE WEB LINKS

- University webpage.

rejection method A method for *simulation of a *random variable X. Suppose, for example, that X is *continuous, with *probability density function (pdf) f, and suppose that it is easy to simulate a random variable, Y, which has a pdf g that satisfies $\mathrm{f}(y) \leq c\mathrm{g}(y)$, for all y, where c is a constant. The rejection method then has three stages: (i) generate y, an observation of Y; (ii) generate a *pseudo-random number u in the interval $0 < u < 1$; (iii) if $\mathrm{f}(y) \geq uc\mathrm{g}(y)$, set x equal to y; otherwise return to (i).

rejection region *Alternative name for* CRITICAL REGION.

relative efficiency *See* ESTIMATOR.

relative frequency *See* FREQUENCY.

relative frequency polygon *See* FREQUENCY POLYGON.

relative risk *See* TWO-BY-TWO TABLE.

reliability A measure of the confidence that we can have in the results obtained from a psychological test. A key question is whether the variability in the scores obtained by different individuals is due to real differences between the individuals or to chance variations resulting from inadequacies in the testing process. The ratio

$$\frac{\mathrm{Var(true\ scores)}}{\mathrm{Var(test\ scores)}}$$

is the **reliability index** and its square root is the **reliability coefficient**. The true scores are unknown, but the coefficient can be estimated by using repeat tests. Suppose n individuals are given k similar tests. Let x_m

be the total score obtained by individual m over the k tests and let $\bar{x}$ and s^2 be the mean and variance, respectively, of $x_1, x_2, \ldots, x_n$. If each test consists of a single question for which an answer is either correct or incorrect, then let p_j represent the proportion of correct answers to question j. Alternatively, if a variety of scores are possible on test j, let s_j^2 be the variance of those obtained. An approximation to the reliability coefficient is provided by **Cronbach's alpha**, given by

$$\alpha = \frac{k}{k-1}\left(1 - \frac{1}{s^2}\sum_{j=1}^{k} s_j^2\right),$$

and other approximations are provided by the **Kuder–Richardson formulae** KR_{20} and KR_{21} (named after the equation numbers in Kuder and Richardson's 1937 paper):

$$KR_{20} = \frac{k}{k-1}\left\{1 - \frac{1}{s^2}\sum_{j=1}^{k} p_j(1-p_j)\right\},$$

$$KR_{21} = \frac{k}{k-1}\left\{1 - \frac{1}{s^2}\left(\bar{x} - \frac{1}{k}\bar{x}^2\right)\right\}.$$

An alternative approach is the **split-half method**, in which each test provides two scores (for example, the score on the even questions and the score on the corresponding odd questions). Let r be the *correlation coefficient between the scores obtained on the odd questions and the scores obtained on the even questions. The **Spearman–Brown formula** measures the reliability as

$$\frac{kr}{1+(k-1)r}.$$

See also TEST-RETEST RELIABILITY.

reliability coefficient *See* RELIABILITY.

reliability function *See* HAZARD RATE.

reliability index *See* RELIABILITY.

reliability theory Theory concerned with determining the *probability that a system (with n components) is working. Let $x_j = 1$ if the jth component is working and let $x_j = 0$ if it has failed. The *vector $\mathbf{x} = (x_1\ x_2 \ldots x_n)$ is called the **state vector**. The function $\phi(\mathbf{x})$, which takes the value 1 when the system is working and 0 when it has failed, is called the **structure function**. For n components in **series**,

$$\phi(\mathbf{x}) = \min\{x_1, x_2, \ldots, x_n\} = x_1 \times x_2 \times \cdots \times x_n.$$

For n components in **parallel**,

$$\phi(\mathbf{x}) = \max\{x_1, x_2, \ldots, x_n\} = 1 - (1 - x_1)(1 - x_2)\cdots(1 - x_n).$$

If $\phi(\mathbf{x}) = 1$ then $\mathbf{x}$ is a **path vector**: it traces a set of connected working components. If failure of any of its working components results in system failure, the vector is a **minimal path vector**. Correspondingly, if $\phi(\mathbf{x}) = 0$ then $\mathbf{x}$ is a **cut vector** and, if it is the case that repair of any of the failed components in $\mathbf{x}$ leads to the system working, then the vector is a **minimal cut vector**.

If p_j denotes the probability that the jth component continues to work during the next unit of time, then the probability that a structure consisting of n components in series continues to work is $p_1 \times p_2 \times \cdots \times p_n$ with the corresponding probability for n components in parallel being $1-(1-p_1)(1-p_2)\cdots(1-p_n)$.

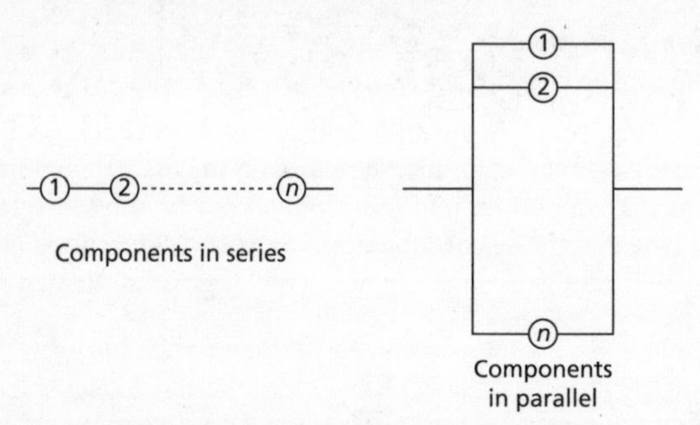

Reliability theory. Complicated systems can always be subdivided into combinations of components arranged either in series or in parallel.

renewal process A *stochastic process concerned with the times of replacement of components that are replaced as soon as they break down. Suppose a system has a single component, which is replaced immediately when it breaks down. Suppose the jth component has a lifetime X_j. The time, T_j, at which this component breaks down is given by

$$T_j = \sum_{k=1}^{j} X_k.$$

X_1 X_2 X_3

0 T_1 T_2 T_3 t

The sequence T_1, T_2, . . . is a renewal process. A useful result is that, if N(t) is the number of breakdowns by time t then

$$P(N(t) \geq n) = P(T_n \leq t).$$

renewal theory The study of *renewal processes.

repeated measures An *experimental design in which the measurements are taken at two or more points in time on the same set of *experimental units.

replicate; replication *See* EXPERIMENTAL DESIGN.

reproducibility The extent to which measurements made under one set of conditions (or by one observer) can be repeated under different conditions (or by another observer).

resampling The process of sampling from the *observations in a *sample, in order to obtain estimates and *confidence intervals for *population parameters without making assumptions about the form of the population *distribution. Suppose that we have taken a random sample of n observations, and assume, for simplicity, that all the sample values $x_1, x_2, \ldots, x_n$ are different. If we have no other information about the population then the obvious estimate of the *population mean, μ, is the *sample mean $\bar{x}$. This is not contentious. However, it is equally true that an unbiased estimate of the *probability of the value x_j is $\frac{1}{n}$. In a sense the sample is a surrogate for the population—if we want to know what other samples from the population might have looked like, we can find out by sampling from the sample. This is the process called resampling.

As an example, suppose that we wish to estimate the *median of a distribution. We take ten observations and obtain the values

3.1, 3.7, 3.8, 4.1, 4.4, 5.1, 5.2, 5.6, 5.9, 6.2.

A simple estimate of the median of the distribution is the median that we have observed, namely 4.75. Resampling enables us to derive an empirical *confidence interval for this estimate. Using the pseudo-random numbers in the first row of Appendix IX, which begins 07552 37078, we generate a new sample

6.2, 5.2, 4.4, 4.4, 3.7, 3.8, 5.2, 6.2, 5.2, 5.6,

which has median 5.2. Further resampling produces successive sets of ten 'observations' with medians 5.6, 4.4, 5.35, 4.4, 3.8, 4.25, 3.95, 4.75, 4.4, 4.75, 4.1. The fifteen resampled medians have mean 4.61 and *standard deviation 0.548 so that an approximate 95% confidence interval for the median is (3.5, 5.7). *See also* BOOTSTRAP; JACKKNIFE.

residual *See* REGRESSION DIAGNOSTICS.

residual sum of squares *See* ANOVA.

resistant line *See* ROBUST REGRESSION.

respondent A person who answers the questions posed by a *questionnaire.

response surface A surface in $(n+1)$ dimensions that represents the variations in the *expected value of a response variable (*see* REGRESSION) as the values of n explanatory variables are varied. Usually the interest is in finding the combination that gives a global maximum (or minimum). One interactive procedure is the **method of steepest ascent** (or **descent**), in which, in a sequence of experiments, the points corresponding to the successive values of the explanatory variables are collinear and lie on the estimated line of greatest (or least) slope that passes through the origin of the current *experimental design.

There are several specialist experimental designs for efficient experimentation near the supposed optimum. With n explanatory variables a **central composite design** consists of observations at vertices of a hypercube centred on the origin, together with repeated observations at the origin and observations on each axis at a distance c from the origin. If $c=\sqrt{n}$ then all the non-central points are at the same distance from the origin and the design is an example of a **rotatable design**. A **Box–Behnken design** uses fewer observations by replacing the observations at the vertices by observations at the mid-points of the edges. *See also* FACTORIAL DESIGN; NELDER–MEAD SIMPLEX METHOD.

response variable *See* REGRESSION.

resultant vector *See* CYCLIC DATA.

retail price index (RPI) A measure of the value of money in which the cost, C, of a standard collection, or basket, of goods and services at some particular date is compared with the cost, C_0, of an equivalent collection at a base date. The RPI is $100 \times C/C_0$, so that at the base date the RPI is 100. The basket is defined to represent the expenditure of a typical household, and the contents of the basket are changed periodically to accommodate changes in the pattern of expenditure. The RPI is a *weighted average of the prices of the goods and services in the basket. *See also* PRICE INDEX.

retrospective study A study that makes use of historical information. For example, if we are interested in the life expectancy of patients then it is convenient if all the patients have died, since only then will the lengths of their lives be known.

In a **case-controlled study** it is the outcomes (for example, death from a particular disease) that determine the sampling procedure. In effect, time is reversed and the early background of those with the outcome (the **cases**) is compared with those without the outcome (the **controls**). The approach enables *oversampling of the cases of interest.

rho (ρ) The symbol customarily used for the *population correlation coefficient.

Richards equation *See* GROWTH CURVE.

ridge regression; ridge trace *See* MULTIPLE REGRESSION MODEL.

ridit Ridits provide a method for replacing the categories of an *ordinal variable by scores between 0 and 1. Suppose that a variable has five ordered categories: 'very low', 'low', 'medium', 'high', and 'very high', with respective frequencies 16, 24, 36, 22, and 2 (totalling 100). The ridits would be $\frac{0.5\times16}{100}, \frac{16+0.5\times24}{100}, \frac{16+24+0.5\times36}{100}, \ldots$ (i.e. 0.08, 0.28, 0.58, 0.87, and 0.99). In this case the category 'very high' is given a more extreme score than the category 'very low' reflecting its comparative rarity.

Rietz, Henry Lewis (1875–1943; b. Gilmore, OH; d. Iowa City, IA) American mathematician and statistician. Rietz was a mathematics graduate at OSU in 1899, and obtained his PhD from Cornell U in 1902. After 15 years on the faculty at U Illinois, he moved to the U Iowa, where *Wilks was one of his research students. He was a founder of the *IMS and its first President (1935).

SEE WEB LINKS

- Fuller biography and photograph.

Rietz Lecturer; Rietz Lectureship The Rietz Lectureship is awarded every three years by the *IMS to honour an individual who has clarified the relationship of statistical methodology to another field.

right-censored *See* CENSORED DATA.

risk The estimated *probability of an undesirable outcome. Often a patient is subject to **competing risks**; for example, death due to a heart attack or death due to heart surgery.

RMNE *See* RANDOM MAN NOT EXCLUDED.

RMP *See* RANDOM MATCH PROBABILITY.

RMSE *Abbreviation for* root mean squared error. *See* ESTIMATOR; MEAN ABSOLUTE ERROR.

$\mathbb{R}^n$ *See* CARTESIAN COORDINATES.

Robbins, Herbert Ellis (1917–2001; b. Newcastle, PA; d. Princeton, NJ) American mathematical statistician. Robbins obtained his PhD from Harvard U in 1938. After a spell in the United States Navy, he joined *Hotelling at UNC. In 1952 he moved to Columbia U where he spent the rest of his career. He was the *Neyman Lecturer of the *IMS in 1982, having been its *Rietz Lecturer in 1963, its President in 1966, and its *Wald Lecturer in 1969. He was the *COPSS *Fisher Lecturer in 1993. He was elected to membership of both the NAS and the AAAS.

SEE WEB LINKS

- Obituary.

robust An adjective applied to a statistic or a statistical procedure which implies that the value of the statistic or the outcome of the procedure will be relatively unaffected by the presence of a small number of unusual or incorrect data values. Thus the *median is a robust *measure of location and the interquartile range (*see* QUARTILE) is a robust *measure of spread.

robust regression A method of *regression that is not greatly affected by discordant observations. The most usual method for obtaining regression estimates is ordinary least squares (OLS) (*see* METHOD OF LEAST SQUARES), which is quite sensitive to the presence of *outliers. One example of a robust alternative to OLS is *iteratively reweighted least squares.

When there is a single x-variable a simple alternative robust method is as follows.

1. Divide the data into three approximately equal-sized groups according to the sizes of the x-values. Call the groups L (low), M (medium), and H (High). Find the *medians of the x-values and y-values in each group: (x_L, y_L), (x_M, y_M), and (x_H, y_H).
2. Estimate the slope of the line as
$$\tilde{\beta} = \frac{y_H - y_L}{x_H - x_L}$$
3. Estimate the intercept as
$$\frac{1}{3}\left\{\left(y_H - \tilde{\beta}x_H\right) + \left(y_M - \tilde{\beta}x_M\right) + \left(y_L - \tilde{\beta}x_L\right)\right\}.$$

The resulting line is known as the **median-median line** or the **resistant line**.

ROC curve *See* TWO-BY-TWO TABLE.

root mean squared error (RMSE) *See* ESTIMATOR; MEAN ABSOLUTE ERROR.

rootogram A diagram, suggested by *Tukey in 1971, that resembles either a *histogram with equal-width categories, or a *bar chart. However, in the rootogram, it is the square-root of *frequency that is plotted on the vertical axis. The diagram gives increased prominence to the less frequent values. *See also* HANGING ROOTOGRAM.

rose diagram An alternative to the circular histogram (*see* CYCLIC DATA) as a method of displaying grouped *cyclic data. The areas of the sectors are proportional to the corresponding frequencies.

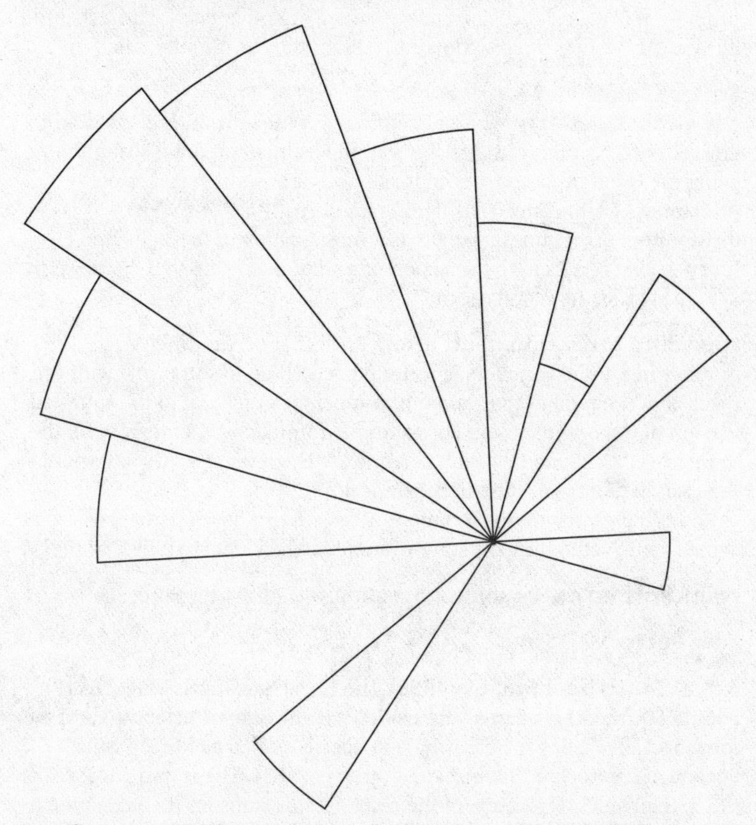

Rose diagram. A rose diagram illustrates the frequencies with which different directions occur. In this case the directions refer to the movements of Colorado beetles released in a wheat field.

Rosenbaum test *See* TEST FOR EQUALITY OF SCALE.

Rosenbrock pattern search A method for finding the maximum (or minimum) of a function of several variables. It is based on detecting and moving along ridges in the surface described by the function.

Rosner test *See* OUTLIER.

rotatable design *See* RESPONSE SURFACE.

Rothamsted Founded in 1843, it is the oldest agricultural research station in the world. It is situated in Harpenden, to the north of London. Some of the original field trials continue to this day. Many of the most prominent British statisticians have worked at Rothamsted. These include Sir Ronald *Fisher, *Yates, *Cochran, *Kempthorne, *Nelder, and *Wedderburn.

roughness penalty When a *sample of n *observations is taken from an unknown *continuous distribution, the estimate of the distribution using the *method of maximum likelihood consists of n spikes of probability (each of size $1/n$). This is an unrealistically 'rough' distribution. A roughness penalty is a modification of the *likelihood function that results in the selection of a smoother estimated distribution. *See also* PENALIZED LIKELIHOOD.

rounding error (round-off error) Error that occurs when each number in a set of numbers is rounded according to some rule with the result that the total of the rounded numbers is not equal to the rounded version of the original total. For example, rounded to the nearest 10, the *counts 21, 22, 23, and 24 each become 20. However, $4 \times 20 = 80$ whereas the total of the unrounded numbers is 90.

Rounding errors are most often noticed when a set of percentages that should sum to 100% actually sums to something different (e.g. 99% or 101%).

round-off error *See* ROUNDING ERROR.

row vector *See* MATRIX.

Royal Statistical Society (RSS) The Royal Statistical Society has about 6500 members across the world. It is the earliest established of all the world's Statistics societies, having been founded as the Statistical Society of London by *Babbage and others in 1834. The stated purposes of the RSS include: (i) to nurture the discipline of Statistics by publishing a journal, organizing meetings, setting and maintaining professional standards, accrediting university courses and operating examinations, (ii) to promote the discipline of Statistics by disseminating and encouraging statistical knowledge and good practice among producers and consumers of Statistics and in society at large. The Society publishes the **Journal of the Royal Statistical Society* (currently in three series, A, B, and C).

The Society awards the *Guy Medal in Gold to persons 'judged to have merited a signal mark of distinction by reason of their innovative contributions to the theory or application of Statistics'.

SEE WEB LINKS

- Society home page.

Roy's maximum root *See* MULTIVARIATE ANALYSIS OF VARIANCE.

RPI *Abbreviation for* RETAIL PRICE INDEX.

R^2 *See* COEFFICIENT OF DETERMINATION.

RSS *Abbreviation for* ROYAL STATISTICAL SOCIETY.

Rubin, Donald Bruce (1943– ; b. Washington, DC) American statistician specializing in *imputation. Rubin obtained his AB in psychology at Princeton U (1965) and then moved to Harvard U obtaining his MS in computer science in 1966 and his PhD in statistics (supervised by *Cochran) in 1970. After eleven years at the Educational Testing service at Princeton and two years on the faculty at U Chicago, he returned to Harvard U. He is a member of the AAAS. He was the winner of the *Wilks Award of the *ASA in 1995 and the *Parzen Prize in 1996. He was the *COPSS *Fisher Lecturer in 2004.

SEE WEB LINKS

- University website.

rugplot An arrangement of short parallel bars whose locations indicate the values of one or more *variables.

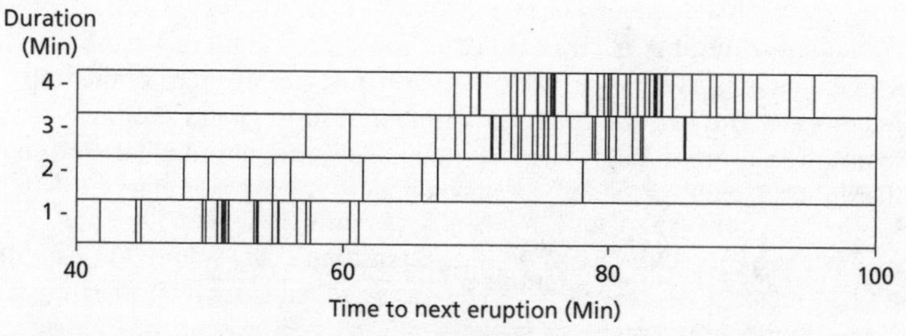

Rugplot. For the *Old Faithful data. Each bar indicates an inter-eruption time of the indicated length. The four rows correspond to eruptions of different durations.

ruin theory Theory concerned with aspects of reinsurance and with the probability of ultimate ruin.

rule of succession A term used by *Venn in discussing the proposition by *Laplace that, if an event has occurred on each of n successive *trials, then the *probability that it will occur on the next trial is $(n+1)/(n+2)$.

rum consumption A favourite example of *nonsense correlation. If the annual consumption of rum in Havana is plotted on a *scatter diagram against the salaries of ministers of religion in Massachusetts, then the graph will show a strong positive nonsense correlation that might suggest that it is the ministers who have been consuming the rum (and/or preaching in favour of its consumption).

run A run of a certain type of result is an uninterrupted sequence of results, all of the same type, bordered at each end either by the start or finish of the complete sequence or by a result of a different type.

running mean; running median *See* SCATTERPLOT SMOOTHER.

run-off triangle A presentation used by actuaries. The rows of the table correspond to years of interest and the columns to numbers of years subsequently elapsed. The entries in a run-off triangle for pay-outs are the accumulating amounts paid out over the years for events that occurred in the year of interest. Study of the past entries enables forecasts of future pay-outs.

runs test (Wald–Wolfowitz test) A versatile *non-parametric test. One application is as a test of the *null hypothesis that two *samples, of sizes m and n, have been taken from *random variables with the same *distribution. Arrange the two sets of *observations in joint order from least to greatest. Label the observations 1 or 2 according to their sample. This results in a sequence of 1s and 2s, e.g. 1-222-1-2-1111-2-1, containing a number of *runs (in this case, seven). If there are too few sequences (e.g. 1111111-22222), then this leads to rejection of the null hypothesis. The test statistic is the number of runs. Under the null hypothesis, for reasonably large m and n, this has an approximate *normal distribution, with mean and variance equal to

$$1+\frac{2mn}{m+n} \qquad \text{and} \qquad \frac{2mn(2mn-m-n)}{(m+n-1)(m+n)^2},$$

respectively. Since the count is an integer, a *continuity correction of 0.5 will be needed.

As an example, suppose that the reaction times, in ms, of 20 girls and 25 boys were:

Girls	428, 444, 446, 479, 492, 513, 522, 533, 544, 545, 560, 566, 581, 582, 590, 595, 599, 612, 634, 655
Boys	415, 439, 442, 477, 500, 512, 523, 532, 577, 580, 613, 614, 622, 633, 670, 671, 680, 688, 701, 703, 722, 730, 744, 750, 777

The null hypothesis is that the reaction times are *independent samples from a common distribution, the alternative hypothesis being that this is not the case. Denoting the boys by 2s, we get the sequence

2–1–22–11–2–11–22–11–22–11111–22–111111–2222–11–22222222222.

There are therefore just fifteen runs. The resulting *z-value is

$$\left\{15 + 0.5 - \left(1 + \frac{1000}{45}\right)\right\} \Big/ \sqrt{(1000 \times 955)/(44 \times 45^2)} = -2.36.$$

The corresponding tail probability (*see* TAIL AREA) is about 0.009. Considering both tails, the probability is 0.018 and this provides strong evidence to reject the null hypothesis.

r.v. *See* RANDOM VARIABLE.

Ryan–Einot–Gabriel–Welsch test *See* MULTIPLE COMPARISON TEST.

St Petersburg paradox A paradox that was the source of much correspondence among eighteenth-century mathematicians. Originally posed by Nicolaus *Bernoulli to *Montmort, it became well known as a result of the solution given by Daniel *Bernoulli in 1738 in the journal of the St Petersburg Academy.

Bernoulli's scenario was essentially as follows. Two players, A and B, play the following game. Player A repeatedly tosses a coin, stopping when a head is obtained. If A has to toss the coin k times, then A pays £2^k to B. Bernoulli's question is 'How much should B pay A in order to make the game fair?' The answer is that B must pay to A the average amount that A pays B. Half the time (assuming a fair coin) this will be £2. Half the remaining time it will be £4, and so on. However,

$$\left(\frac{1}{2} \times £2\right) + \left(\frac{1}{4} \times £4\right) + \cdots = £1 + £1 + \cdots .$$

Since the required number of tosses has no upper limit, this sum is infinite. Thus, for this game to be fair, B must pay A an infinite amount of money, even though B is certain to receive only a finite amount of money in exchange (and less than £10 on 87.5% of occasions).

sample A subset of a *population usually chosen in such a way that it can be taken to represent the population with respect to some characteristic, for example, height, or cost, or gender, or make of car. A list of members of the population of interest is called the **sampling frame**. If each member of the sample is selected by the equivalent of drawing lots, the sample is a **simple random sample** or, commonly, a **random sample**. In this case each **sampling unit**, i.e. each member of the population, has the same probability of being in the sample, independently of whether any other object is in the sample, and so all possible samples are equally likely. When the sampling units are people, the sample is often referred to as a sample **survey**. Various modifications of the simple random sample are often used. *See* CLUSTER SAMPLING; QUOTA SAMPLING; STRATIFIED SAMPLING; SYSTEMATIC SAMPLING.

sample autocorrelation *See* AUTOCORRELATION.

sample correlation coefficient (product-moment correlation coefficient; Pearson correlation coefficient) If the n pairs of values of *random variables X and Y in a random sample are denoted by $(x_1, y_1), (x_2, y_1), \ldots, (x_n, y_n)$, the sample correlation coefficient r is given by

$$r = \frac{S_{xy}}{\sqrt{S_{xx}S_{yy}}},$$

where

$$S_{xy} = \sum_{j=1}^{n} x_j y_j - \frac{1}{n}\left(\sum_{j=1}^{n} x_j\right)\left(\sum_{j=1}^{n} y_j\right), \qquad S_{xx} = \sum_{j=1}^{n} x_j^2 - \frac{1}{n}\left(\sum_{j=1}^{n} x_j\right)^2,$$

and S_{yy} is defined analogously to S_{xx}. If the sample means are denoted by $\overline{x}$ and $\overline{y}$, alternative definitions are

$$S_{xy} = \sum_{j=1}^{n} x_j y_j - n\overline{x}\overline{y}, \qquad S_{xx} = \sum_{j=1}^{n} x_j^2 - n\overline{x}^2.$$

The coefficient r can take any value from -1 to 1, inclusive. When increasing values of one variable are accompanied by generally increasing values of the other variable then $r > 0$ and the variables are said to display **positive correlation.** If $r < 0$ then the variables display **negative correlation.**

The idea of correlation was put forward by *Galton in 1869, and it was Galton who was the first to denote it by the symbol r in 1888. The formulae given here were introduced by Karl *Pearson in 1896.

The sample correlation coefficient r is an estimate of the population *correlation coefficient ρ.

Correlation is closely linked to *linear regression. If the least squares regression lines of y on x and of x on y for the sample $(x_1, y_1), (x_2, y_2), \ldots, (x_n, y_n)$ are, respectively, $y = a + bx$ and $x = c + dy$ then $r^2 = bd$.

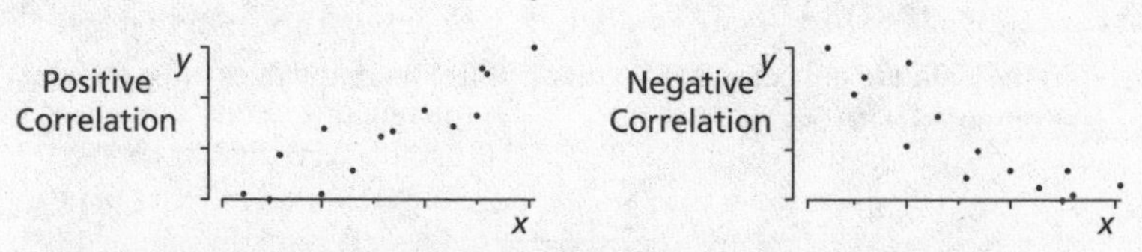

Sample correlation coefficient. When the correlation between two variables is positive, the values of one variable generally rise as the values of the other variable rise. The correlation is negative if the values of one variable generally rise as the values of the other fall.

In a *hypothesis test, to test for significant evidence of a linear relationship between X and Y, we compare the *null hypothesis that $\rho = 0$ with the *alternative hypothesis that $\rho \neq 0$, rejecting the null hypothesis if $|r|$ is too large. *See also* COEFFICIENT OF DETERMINATION; RANK CORRELATION COEFFICIENT.

SEE WEB LINKS

- Applet.

sample covariance Given the n pairs of *observations $(x_1, y_1), \ldots, (x_n, y_n)$, the sample covariance, c, is given by

$$c = \frac{1}{n-1}\left\{\sum_{j=1}^{n} x_j y_j - \frac{1}{n}\left(\sum_{j=1}^{n} x_j\right)\left(\sum_{j=1}^{n} y_j\right)\right\}.$$

sample distribution function The equivalent of the *distribution function for a *sample of *data. Let the ordered data be $x_{(1)} \leq x_{(2)} \leq \cdots \leq x_{(n)}$; then the sample distribution function $\mathrm{F}_n(x)$ is given by

$$\mathrm{F}_n(x) = \begin{cases} 0 & x < x_{(1)}, \\ j/n & x_{(j)} \leq x < x_{(j+1)}, \quad 1 \leq j \leq (n-1), \\ 1 & x_{(n)} \leq x. \end{cases}$$

sample mean The sample mean (or, simply, 'the mean') of a set of n items of *data $x_1, x_2, \ldots, x_n$ is $\left(\sum_{j=1}^{n} x_j\right)/n$, which is the arithmetic average of the numbers $x_1, x_2, \ldots, x_n$. The mean is usually denoted by placing a bar over the symbol for the variable being measured. If the variable is x the mean is denoted by $\bar{x}$. If the data constitute a *sample from a *population, then the sample mean is an unbiased estimate of the *population mean.

For example, the numbers of eruptions of the *Old Faithful geyser during the first eight days of August 1978 were 13, 13, 13, 14, 14, 14, 13, and 13. The mean is

$$(13 + 13 + 13 + 14 + 14 + 14 + 13 + 13)/8 = 13.375.$$

If the data are collected in *frequency form so that values $x_1, x_2, \ldots, x_n$ are obtained with frequencies $f_1, f_2, \ldots, f_n$ the mean is

$$\frac{\sum_{j=1}^{n} f_j x_j}{\sum_{j=1}^{n} f_j}.$$

For example, with the eruption data there are just two values, $x_1 = 13$ and $x_2 = 14$. Their respective frequencies are $f_1 = 5$ and $f_2 = 3$, so the mean is $\{(5 \times 13) + (3 \times 14)\}/(3 + 5) = 13.375$.

If the data are grouped into classes with mid-values $x_1, x_2, \ldots, x_c$ and corresponding *class frequencies $f_1, f_2, \ldots, f_c$, an approximate value for the mean of the original data is the **grouped mean**

$$\frac{\sum_{j=1}^{c} f_j x_j}{\sum_{j=1}^{c} f_j}.$$

The mean can be interpreted as the centre of gravity, or centre of mass, of a system of particles of masses $f_1, f_2, \ldots, f_n$ at points $x_1, x_1, \ldots, x_n$.

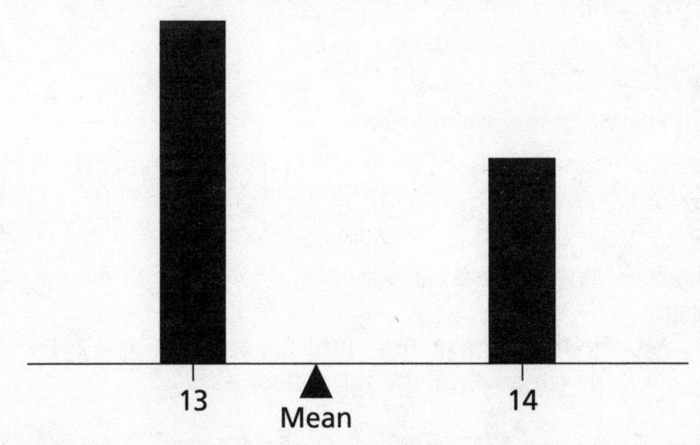

Sample mean. The data are the numbers of eruptions of the Old Faithful geyser during the first eight days of August 1978. The sample mean is seen to be the balance point of the observations.

sample size The number of *observations in a *sample.

sample space (universal set) A complete set of all possible results or **outcomes** for an experiment or observational procedure. The concept was introduced by *von Mises in 1931. The sample space is usually denoted by *S* or *E*.

An **event** is a particular collection of outcomes, and is a subset of the sample space. For example, when a die is thrown and the score observed, the sample space is {1, 2, 3, 4, 5, 6}, and a possible event is 'the score is even' i.e. {2, 4, 6}. If all the possible outcomes are equally likely, then the probability of an event *A* is given by

$$\mathrm{P}(A) = \frac{\text{Number of events in subset of sample space corresponding to } A}{\text{Number of events in sample space}}.$$

The word 'event' was used in this context by *de Moivre in 1718.

The subset of the sample space for the event 'the score is both even and odd' is an example of the **empty set**, usually denoted by ϕ, and $\mathrm{P}(\phi) = 0$.

S

The subset of the sample space for the event 'the score is less than 10', is the whole sample space, S, and $P(S) = 1$.

See also BOOLEAN ALGEBRA; COMPLEMENTARY EVENT; INTERSECTION; UNION.

sample standard deviation The square root of the *sample variance.

sample variance A measure of the variability of a set of *data. For data $x_1, x_2, \ldots, x_n$, with *sample mean $\bar{x}$ given by

$$\bar{x} = \frac{1}{n}\sum_{j=1}^{n} x_j,$$

the sample variance is defined to be

$$\frac{1}{n}\sum_{j=1}^{n}(x_j - \bar{x})^2 = \frac{1}{n}\left(\sum_{j=1}^{n} x_j^2 - n\bar{x}^2\right) = \frac{1}{n}\left\{\sum_{j=1}^{n} x_j^2 - \frac{1}{n}\left(\sum_{j=1}^{n} x_j\right)^2\right\}.$$

The variance is never negative and can be zero only if all the data values are the same.

In the case where the *frequency of the *observation x_j is f_j, for $j = 1, 2, \ldots, m$, the variance can be calculated using

$$\frac{1}{n}\sum_{j=1}^{m} f_j(x_j - \bar{x})^2 = \frac{1}{n}\left(\sum_{j=1}^{m} f_j x_j^2 - n\bar{x}^2\right) = \frac{1}{n}\left\{\sum_{j=1}^{m} f_j x_j^2 - \frac{1}{n}\left(\sum_{j=1}^{m} f_j x_j\right)^2\right\},$$

where n, the total *sample size, is given by

$$n = \sum_{j=1}^{m} f_j \quad \text{and} \quad \bar{x} = \frac{1}{n}\sum_{j=1}^{m} f_j x_j.$$

In these expressions for variance the divisor n is used. This is correct if the data set effectively constitutes the entire *population; for example, if the values $x_1, x_2, \ldots$ are the diameters of the planets of the solar system, or the lifetimes of all known patients with a rare disease. However, if the data constitute a random *sample from a population, and we are interested in the variance of the values in the population, as opposed to the variance of the values in the sample, then it is appropriate to use the divisor $(n-1)$, since this leads to an *unbiased estimate of the population variance. This sample variance is given by

$$s^2 = \frac{1}{n-1}\sum_{j=1}^{n}(x_j - \bar{x})^2 \quad \text{or} \quad \frac{1}{n-1}\sum_{j=1}^{m} f_j(x_j - \bar{x})^2,$$

as appropriate. The factor $\frac{n}{n-1}$ linking the population variance formulae and the sample variance formulae is known as the **Bessel correction**.

sampling distribution *Distribution that describes the variation in the values of a *statistic over all possible *samples. For example, if n values are sampled from a *population and if $X_1, X_2, \ldots, X_n$, are the *random variables representing the individual sample values, then $\bar{X}$, given by

$$\bar{X} = \frac{1}{n}(X_1 + X_2 + \cdots + X_n),$$

is a random variable. The variability of the n values about their mean,

$$V^2 = \frac{1}{n}\left\{(X_1 - \bar{X})^2 + (X_2 - \bar{X})^2 + \cdots + (X_n - \bar{X})^2\right\},$$

is also a random variable. The form of the sampling distributions of $\bar{X}$ and V^2 will depend on the population, but statements can nevertheless be made about their *moments. If the population has *mean μ and *variance σ^2, then, for an infinite population, or for *sampling with replacement from a finite population, each of $X_1, X_2, \ldots$ has mean μ and variance σ^2. Consequently the *expected value of $\bar{X}$ is μ and the expected value of V^2 is $\frac{n-1}{n}\sigma^2$. This shows that $\bar{X}$ and $S^2 = \frac{n}{n-1}V^2$ are unbiased *estimators of μ and σ^2, respectively. The variance of the sample mean $\bar{X}$ is σ^2/n.

The **sample variance** is usually taken to be the value of S^2, though the value of V^2 is sometimes used.

sampling fraction The *sample size, n, divided by N, the size of the finite *population from which it has been drawn. Its importance, when sampling without replacement is that the *variance of the *sample mean is not σ^2/n (*see* SAMPLING DISTRIBUTION), but

$$\frac{\sigma^2}{n}\left(1 - \frac{n}{N}\right).$$

The term $(1 - \frac{n}{N})$ is the **finite population correction.**

sampling frame; sampling unit *See* SAMPLE.

sampling with replacement The case where *observations are taken one at a time from a *population. The sampled value is returned to the population before the next value is selected. For example, in a competition with several prizes it is possible for one lucky person to be selected as the winner of several prizes. By contrast, **sampling without replacement** implies that each member of the population can be chosen only once.

Sankhya The journal of the *Indian Statistical Institute, first published in 1933. There are two series (both published six-monthly): *Series A* concentrates on probability and mathematical statistics, whereas the emphasis of *Series B* is on data analysis and statistical methodology.

SEE WEB LINKS

- Journal home page.

SARIMA models *ARIMA models for *time series that display *seasonality.

SAS (Statistical Analysis System) A powerful statistical package permitting many types of analyses via user-written commands.

SEE WEB LINKS

- SAS Software home page.

Satterthwaite's formula A formula for finding the approximate *distribution of a *linear combination of *independent *chi-squared variables. Define the *random variable S^2 by

$$S^2 = \sum_j c_j S_j^2,$$

where $c_1, c_2, \ldots$ are known positive constants, and the chi-squared random variable S_j^2 has ν_j *degrees of freedom. Satterthwaite's suggestion, made in 1946, was that the distribution of $\nu S^2/\sigma^2$ is approximately chi-squared with ν degrees of freedom, where $\sigma^2 = \sum_j c_j \nu_j$, and ν is the nearest integer to

$$\frac{\left(\sum_j c_j s_j^2\right)^2}{\left\{\sum_j \frac{1}{\nu_j}\left(c_j s_j^2\right)^2\right\}},$$

where s_j^2 is the observed value of S_j^2.

saturated model A *model that perfectly fits the *data because it has as many *parameters as there are values to be fitted. It can provide useful information in the search for a simpler **unsaturated model** in which some of the parameters of the saturated model are set to 0.

S

Savage test *See* TEST FOR EQUALITY OF LOCATION.

SBR *Abbreviation for* STATISTICS IN BIOPHARMACEUTICAL RESEARCH.

scalar A number. The term is often used to distinguish a number from a *vector or *matrix.

scaling function *See* WAVELET.

Scandinavian Journal of Statistics A quarterly journal, published in English, concerned with advances in statistical theory and its applications. It is published on behalf of the *Danish Society for Theoretical Statistics, the *Finnish Statistical Society, the *Norwegian

Statistical Association, and the *Swedish Statistical Association. The first volume was published in 1974.

SEE WEB LINKS

- Journal home page.

scan statistic A statistic used to identify clusters of events (*see* SAMPLE SPACE) in space or time. A window of length or radius *h* traverses the *time series or spatial region. The value of the scan statistic is the maximum number of events that fall within this window.

scatter diagram (scattergram; scatter plot) The simplest display when the *data consists of pairs of values. The data are plotted as points using *Cartesian coordinates. If the data are ordered (for example, in time) then it may be sensible to join the successive points with a line. If there are other *categorical variables, their values can be indicated using different plotting symbols or different colours. A quantitative third variable can be indicated by varying the size of the plotting symbol.

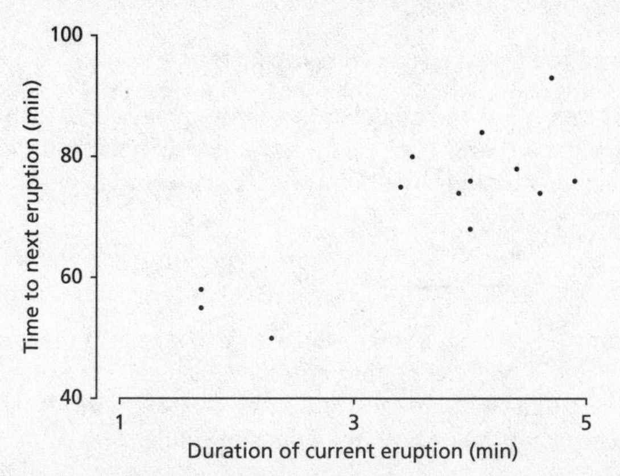

Scatter diagram. The diagram shows the *Old Faithful data for 1 August 1974. Long eruptions are followed by long intervals before the next eruption.

scatterplot smoother A procedure for revealing trends in *scatter diagrams by determining typical values based on observation of successive small intervals of the explanatory variable. Simple examples are provided by the **running mean** and **running median**. For a data set, $y_1, y_2, y_3, \ldots$, ordered by the corresponding values of the explanatory variable, the three-point running *means are given by

$$\frac{1}{3}(y_1 + y_2 + y_3), \frac{1}{3}(y_2 + y_3 + y_4), \frac{1}{3}(y_3 + y_4 + y_5), \ldots,$$

while, denoting the *median of y_{i-1}, y_i, and y_{i+1} by $M(y_i)$, for $i > 1$, the three-point running median values would be $M(y_2)$, $M(y_3)$, $M(y_4)$, ... These ideas can be extended by calculating, for example, $M(M(y_j))$. More sophisticated alternatives are *kernel methods, *loess, and the use of *splines. *See also* GENERALIZED ADDITIVE MODEL.

Scheffé, Henry (1907–77; b. New York City; d. Berkeley, CA) American statistician. Scheffé worked as a technical assistant at the Bell Telephone Laboratories for four years after leaving school. He then studied at U Wisconsin, Madison (BA 1931; PhD in differential equations, 1935), joining the faculty to teach pure mathematics. In 1941 he joined *Wilks at Princeton U and then took posts at Syracuse U (1944), UCLA (1946), and Columbia U (1948), where *Kruskal was his research student. In 1953 he became *Neyman's Assistant Director at UCB. He was President of the *IMS in 1955. In 1959 his classic book *The Analysis of Variance* was published. He died in a traffic accident while cycling on the UCB campus.

SEE WEB LINKS

- Fuller biography and photograph.

Scheffé test *See* MULTIPLE COMPARISON TEST.

Schmidt net A method of illustrating three-dimensional *spherical data in two dimensions. The point with *spherical polar coordinates $(1, \theta, \phi)$ is represented by the point with *polar coordinates (d, ϕ), where, for $0° \leq \theta \leq 90°$,

$$d = 2\sin\left(\frac{1}{2}\theta\right).$$

The quantity d is the (chordal) distance from the pole at $\theta = 0°$. Data having $\theta > 90°$ are plotted in the same way, using either a different diagram, or a different plotting symbol, and replacing $\sin(\frac{1}{2}\theta)$ with $\cos(\frac{1}{2}\theta)$; the distance is now taken from the opposite pole at $\theta = 180°$.

S

Schwarz criterion *See* MODEL SELECTION PROCEDURE.

score The derivative, with respect to some *parameter θ, of the logarithm of the *likelihood. The score is a sufficient statistic (*see* ESTIMATOR) for θ.

scree plot A plot, in descending order of magnitude, of the eigenvalues of a *correlation matrix. In the context of *factor analysis or *principal components analysis a scree plot helps the analyst visualize the relative importance of the factors—a sharp drop in the plot signals that subsequent factors are ignorable.

sd *Abbreviation for* STANDARD DEVIATION.

se *Abbreviation for* STANDARD ERROR.

seasonality A pattern in a *time series that repeats in a regular way—for example, daily average temperatures rise each summer.

seasonally adjusted A term applied to a *time series from which the *seasonality component has been removed.

second difference *See* DIFFERENCING.

second-order stationary *See* STATIONARY PROCESS.

secretary problem An interesting problem in *decision theory. There are n applicants for the post of secretary. The applicants are randomly ordered and each is interviewed in turn until an appointment is made. Before each interview, the employer must decide whether or not to appoint the previous applicant. The employer cannot subsequently decide to appoint a previously interviewed applicant. It turns out that the strategy that maximizes the probability of appointing the best secretary is to interview the first e^{-1} (approximately 37%) of the applicants, and then to appoint the first subsequent applicant who is superior to all those in the first group. If there are none, then the employer is left having to appoint the nth applicant. *See also* MATCHING PROBLEM.

self-organizing feature map (Kohonen map) A low-dimensional plot that provides a means of visualizing the proximities of multidimensional items. The plot is formed using a *self-organizing neural network. The procedure was first described by the Finnish computer scientist Teuvo Kohonen in 1982.

self-organizing neural network A *network consisting of three classes of nodes that might be termed input, computational, and output. The aim of the network is to assign (i.e. connect) the input nodes to the output nodes in an optimal fashion. Typically this involves a computer-intensive process involving an accumulating series of input-to-output paths. The aim is usually to assign the various items of input data to an appropriate class (i.e. to the correct output node).

semi-interquartile range If the lower and upper *quartiles of a *distribution are denoted by Q_1 and Q_3, respectively, the semi-interquartile range is $\frac{1}{2}(Q_3 - Q_1)$. The term can be used for either a set of *data or a *probability distribution.

semi-Markov process A *renewal process in which the *random variable X can take one of N states. When it is in state j it stays there for a random length of time, with *mean μ_j, and then moves to state k with *probability p_{jk}.

semi-parametric model A model that has both parametric and *non-parametric components.

semi-variogram *See* AUTOCORRELATION.

Sen, Pranab Kumar (1934– ; b. Calcutta, India) Indian statistician who has spent most of his career in the USA. Sen was educated at Calcutta U (BSc, 1955; MSc, 1957; PhD, 1962). He spent three years on the faculty at Calcutta U before moving to UCB in 1964. In the following year he moved to UNC, where he began the collaboration with *Puri that resulted in the publication of their influential 1971 book entitled *Non-parametric Tests in Multivariate Analysis*.

SEE WEB LINKS

- Fuller biography, interview, and photographs.

Sen–Puri test A *test, developed by *Sen and *Puri, which is most used in medical contexts, where individuals are re-examined at a sequence of time points. The individuals belong to one of a set of groups (usually corresponding to different treatments). The Sen–Puri test is a *non-parametric test of the *null hypothesis that the sequences of *observations have arisen from a common *distribution. The test uses scores, based on *ranks, accumulated over the time points.

sensitivity *See* TWO-BY-TWO TABLE.

sequential sampling Sampling in which *observations are taken one at a time, with an appraisal, after each observation, of whether there is any need for further sampling. This approach is useful when taking an observation is very expensive, or when there are ethical considerations. After each observation has been taken, the *data so far available are re-analysed and one of three possible decisions is taken: accept the *null hypothesis, reject the null hypothesis, or take another observation. The **average sample number** (**ASN**) is the average number of observations taken before a firm decision is made concerning the null hypothesis. *See also* DOUBLE SAMPLING; QUALITY CONTROL.

S

serial correlation *Autocorrelation between values that are at a constant distance apart in time or space. Suppose $x_1, x_2, \ldots, x_n$ is an ordered sequence of observations. The serial correlation at lag k is r_k, given by

$$r_k = \frac{\sum_{j=1}^{n-k}(x_j - \bar{x}_1)(x_{j+k} - \bar{x}_n)}{\sqrt{\left\{\sum_{j=1}^{n-k}(x_j - \bar{x}_1)^2\right\}\left\{\sum_{j=k+1}^{n}(x_j - \bar{x}_n)^2\right\}}}, \quad k = 1, 2, \ldots, n-1,$$

where

$$\bar{x}_1 = \frac{1}{n-k}\sum_{j=1}^{n-k} x_j, \quad \text{and} \quad \bar{x}_n = \frac{1}{n-k}\sum_{j=k+1}^{n} x_j.$$

In practice, a less complicated formula is routinely used. This formula, which involves $\bar{x}$, the overall *mean, is

$$r_k = \frac{n\sum_{j=1}^{n-k}(x_j - \bar{x})(x_{k+j} - \bar{x})}{(n-k)\sum_{j=1}^{n}(x_j - \bar{x})^2}, \quad k = 1, 2, \ldots, n-1.$$

The term 'serial correlation' was coined by *Yule in a 1926 paper.

series *See* RELIABILITY THEORY.

sex-specific rate When a rate such as a *birth rate, *incidence rate, or *mortality rate is calculated for individuals of a specified sex then the rate is described as being sex-specific.

Shannon index *See* DIVERSITY INDEX.

Shapiro, Samuel Sanford (1930– ; b. New York City) American statistician and engineer. A statistics graduate of City College (now City U), New York in 1952, Shapiro took an MS in industrial engineering at Columbia U in 1954. After a period working as a statistician in the Army Chemical Corps, he joined the General Electric Corporation, obtaining his MS (1960) and PhD (1963) in statistics at Rutgers U. He was co-author of the 1965 paper that introduced the *Shapiro–Wilk test and of the 1972 paper introducing the *Shapiro–Francia test. In 1972 he joined the faculty at Florida International U.

Shapiro–Francia test *See* SHAPIRO–WILK TEST.

Shapiro–Wilk test A *test that the *population being sampled has a specified *distribution. It was introduced by *Shapiro and *Wilk in 1965. The test compares the ordered *sample values with the corresponding *order statistics from the specified distribution. The test is most commonly used to test for a *normal distribution, in which case the test statistic, W, is given by

$$W = \frac{\left(\sum_{j=1}^{n} w_j x_{(j)}\right)^2}{(n-1)s^2},$$

where $x_{(j)}$ is the jth largest of n observations, s^2 is the *unbiased estimate of the population variance, and w_j is a function of the *means, *variances, and *covariances of the order statistics. The **Shapiro–Francia test** uses a simpler substitute for w_j.

In the case where the hypothesized distribution is *exponential the statistic takes the simple form

$$W = \frac{n(\bar{x} - x_{(1)})^2}{(n-1)^2 s^2},$$

where $\bar{x}$ is the sample mean. In this case, W is closely related to the inverse of the **Darling test** statistic, K, given by

$$K = \frac{(n-1)s^2}{\bar{x}^2},$$

which, for $n > 500$, has an approximate normal distribution with mean $\frac{n(n-1)}{n+1}$ and variance $\frac{4n^4(n-1)}{(n+1)^2(n+2)(n+3)}$.

SEE WEB LINKS

- The genesis of the test.

shell sort An efficient *algorithm for arranging a set of n numbers in order of magnitude. It starts by applying a succession of *bubble or *shuttle sorts to carefully chosen subsets of the data. A shuttle sort is then applied to the resulting partially ordered data.

Shepard diagram A plot of two measurements of the distances between objects. One measurement is the true distance, and the other measurement is the apparent distance in some representation of the objects. For example, the apparent distance between objects in a photograph (two dimensions) and the real three-dimensional distance. The diagram is used in *multidimensional scaling to assess the extent of any distortion. Zero distortion would correspond to a set of collinear points.

Sheppard, William Fleetwood (1863–1936; b. Sydney, Australia; d. Berkhampstead, England) Australian mathematician, barrister, and civil servant whose education and career was in England. Sheppard graduated in mathematics from Cambridge U in 1884 and trained as a barrister. He joined the Education Department of the Civil Service in 1896. It was at this time that Sheppard worked on *correlation and on the corrections that bear his name. Sheppard was President of the Mathematical Association in 1928 and was appointed FRSE in 1932.

Sheppard's corrections Adjustments suggested by *Sheppard for the values of the *moments of the *sample when these are calculated from grouped data. For example, with class intervals of width d, the correction to the calculated estimate of the *variance is to subtract $d^2/12$.

Shewhart, Walter Andrew (1891–1967; b. New Canton IL; d. Troy Hills, NJ) American physicist and statistician. Shewhart studied physics at U Illinois and U California. In 1918 he joined the Western Electric Company which made hardware for Bell Telephone. A major problem was the reliability of the transmission systems. Study of this problem led to Shewhart's introduction of the control chart (*see* QUALITY CONTROL), invented in 1924 and popularized in his 1931 book *Economic Control of Quality of Manufactured Products*. He was President of the *IMS in 1937 and again in 1944. He was President of the *ASA in 1945.

SEE WEB LINKS

- Fuller biography and photograph.

Shewhart chart *See* QUALITY CONTROL.

Shewhart Medal A medal awarded annually by the *American Society for Quality. Named in honour of *Shewhart, it is awarded to an individual who has displayed 'outstanding technical leadership in the field of modern *quality control'.

shrinkage *See* MODEL SELECTION PROCEDURE.

shrinkage estimator *See* STEIN EFFECT.

shuttle sort A simple, but not very efficient, *algorithm for arranging a set of n numbers in order of magnitude. The method starts with the left-hand pair of numbers, swapping them if necessary. The second and third numbers are now considered. If they are swapped then the first pair are reconsidered. Next the third and fourth numbers are considered. If swapped then previous pairs are again reconsidered, working from right to left. As with the *bubble sort, $\frac{1}{2}n(n-1)$ comparisons may be required. *See diagram overleaf.*

Sidak correction *See* MULTIPLE COMPARISON TEST.

Siegel, Sidney (1916–61; b. New York City; d. Stanford, CA) American psychometrician. Siegel obtained his PhD from Stanford U in 1953. From 1954 he taught at Penn State U. His *Nonparametric Statistics for the*

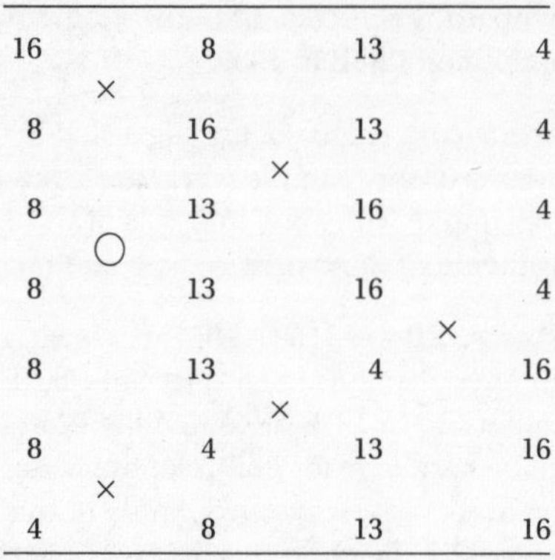

16		8		13		4
	×					
8		16		13		4
			×			
8		13		16		4
	○					
8		13		16		4
					×	
8		13		4		16
			×			
8		4		13		16
	×					
4		8		13		16

Shuttle sort. A sorting method based on swaps of pairs of numbers. In the example, × indicates a swap and ○ that no swap is required. The sort works from left to right, reconsidering earlier pairs when a swap is made.

Behavioral Sciences, first published in 1956, remains a standard introductory reference to the topic.

SEE WEB LINKS

- Web page with links.

Siegel–Tukey test *See* TEST FOR EQUALITY OF SCALE.

sigma (σ) The symbol usually used to denote the *standard deviation of a *population; σ^2 denoting the *variance.

sigma (Σ) Symbol used to denote a sum. Thus

$$\sum_{k=1}^{5} k = 1 + 2 + \cdots + 5,$$

$$\sum_{j=1}^{n} j^2 = 1^2 + 2^2 + \cdots + n^2,$$

$$\sum_{r=1}^{n} x_r = x_1 + x_2 + \cdots + x_n,$$

$$\sum_{k=1}^{3} a_{kj}x_k = a_{1j}x_1 + a_{2j}x_2 + a_{3j}x_3.$$

The *sample mean for n *observations $x_1, x_2, \ldots, x_n$ is

$$\frac{1}{n}\sum_{j=1}^{n} x_j$$

and the unbiased estimate (*see* ESTIMATOR) of the *population variance is

$$\frac{1}{n-1}\left\{\sum_{j=1}^{n} x_j^2 - \frac{1}{n}\left(\sum_{j=1}^{n} x_j\right)^2\right\}.$$

signal The value of a measurement that would be observed if the measurement were not contaminated by *random errors.

signed ranks *See* WILCOXON SIGNED-RANK TESTS.

significance level; significant *See* HYPOTHESIS TEST.

sign test A simple *non-parametric test that is used in two situations:

1. A random *sample of n *observations $x_1, x_2, \ldots, x_n$ is taken on the *random variable X; the *null hypothesis is that the *population has *median m_0.

2. A random sample of n observations $(x_1, y_1), (x_2, y_2), \ldots, (x_n, y_n)$ is taken on the pair of random variables (X, Y); the null hypothesis is that the distribution of $X - Y$ has median 0.

In both cases the analysis begins by noting the signs of the differences $d_1, d_2, \ldots, d_n$, where in case (1) $d_j = x_j - m_0$ and in case (2) $d_j = x_j - y_j$. In either case the test statistic, r, is the number of differences that have a positive value.

If the null hypothesis is correct and there are no zero differences, r is an observation from a *binomial distribution with *parameters n and 0.5. If the *one-tail or two-tail probability (as appropriate) is unusually low, then the null hypothesis will be rejected.

If there are k differences equal to 0, it is conventional to ignore the corresponding observations and to use a binomial distribution with parameters $(n - k)$ and 0.5.

As an example, suppose that the null hypothesis is that the distribution of the weights of 20-year-old males has median 77 kg, the alternative being that this is not the case. A random sample of thirteen 20-year-old males have the following weights (in kg): 59, 84, 99, 83, 65, 70, 77, 69, 85, 66, 76, 73, 81. The question is whether these data support the null hypothesis. In the sample there is one value equal to the hypothesized median. This is ignored. Of the remaining twelve values, five exceed 77 kg. Using the binomial distribution with $n = 12$ and $p = 0.5$, the probability of five or fewer is 0.387. The two-tail probability is therefore 0.774: we conclude that there is no significant evidence to refute the null hypothesis.

Silverman, Bernard Walter (1952–   ; b. London, England) English mathematical statistician known for his work with *splines and *wavelets. Silverman was educated at Cambridge U, gaining his BA in 1973 and his PhD (supervised by David *Kendall) in 1977. In 1978 he joined the faculty at U Bath, moving to U Bristol in 1993. From 2003 to 2009 he was Master of St Peter's College, Oxford. Silverman was awarded the *Guy Medal in Bronze of the *RSS in 1984 and the Guy Medal in Silver in 1995. He was the *COPSS President's Award winner in 1991 and was elected FRS in 1997. He was President of the *IMS in 2001 and, briefly, of the RSS in 2010 before his appointment as Chief Scientific Adviser to the UK Home Office.

SEE WEB LINKS

- University website.

similarity *See* DISSIMILARITY.

simple graph *See* GRAPH.

simple random sample *See* SAMPLE.

simplex A generalized triangle or tetrahedron. Suppose $\mathbf{a}_1, \mathbf{a}_2, \ldots, \mathbf{a}_n$ are n linearly independent (*see* MATRIX) *vectors, or points, in $\mathbb{R}^n$. The set of all points $\sum_{j=1}^{n} \lambda_j \mathbf{a}_j$, where each λ_j is non-negative and $\sum_{j=1}^{n} \lambda_j = 1$, is an $(n-1)$ simplex. *See* BARYCENTRIC COORDINATES.

simplex method An *algorithm introduced in 1947 by *Dantzig for the solution of a *linear programming problem. The algorithm works by moving from one vertex of the *feasible region to an adjacent one.

Simpson index *See* DIVERSITY INDEX.

Simpson's paradox An intriguing paradox illustrating how one may be misled when a relevant *variable is overlooked. The paradox is illustrated in the following example, which shows a cross-classification of three dichotomous (*see* CATEGORICAL VARIABLE) variables, A, B, and C (where, for example, A_1 and A_2 are the two categories of A):

C_1					C_2								
	B_1	B_2	Total			B_1	B_2	Total			B_1	B_2	Total
A_1	95	800	895	+	A_1	400	5	405	=	A_1	495	805	1300
A_2	5	100	105		A_2	400	195	595		A_2	405	295	700
Total	100	900	1000		Total	800	200	1000		Total	900	1100	2000

In the *subpopulation corresponding to C_1, there is a strong positive association between A and B. The same is true for the subpopulation corresponding to C_2. However, when the information on these two very dissimilar subpopulations is pooled, the association for the entire population is strongly negative. *See* ECOLOGICAL FALLACY for a diagram that shows how this type of result can occur with continuous data.

Simpson's rule A method for finding an approximate value for an integral, which is usually more accurate than the *trapezium rule. Suppose we wish to find an approximate value for $\int_a^b f(x)dx$. The interval $a \leq x \leq b$ is divided into an even number, $2n$, of sub-intervals, each of length $h = (b - a)/(2n)$, at points $a = x_0, x_1, x_2, \ldots, x_{2n-1}, x_{2n} = b$, where $x_r = (a + rh)$ for $r = 0, 1, 2, \ldots, 2n - 1, 2n$. Writing $y_r = f(x_r)$, the integral is approximated by

$$\frac{h}{6}\{y_0 + 4(y_1 + y_3 + \cdots + y_{2n-1}) + 2(y_2 + y_4 + \cdots + y_{2n-2}) + y_{2n}\}.$$

Simpson's rule gives the exact answer when f is a *polynomial of order 3 or less. In general, the error in using Simpson's Rule is approximately proportional to $1/n^4$, so that if the number of sub-intervals is doubled then the error is reduced by a factor of 16. Simpson's rule is based on fitting a quadratic graph in each adjacent pair of sub-intervals such that the graph passes through the points (x_{2r}, y_{2r}), (x_{2r+1}, y_{2r+1}) and (x_{2r+2}, y_{2r+2}).

simulated annealing A method for maximizing a function that permits an *algorithm to choose apparently suboptimal routes, but with decreasing *probability as the number of iterations increases (*see* ITERATIVE ALGORITHM). The initially fluid choice of values becomes increasingly set as the algorithm progresses. The object is to reduce the chance of ending at a *local maximum.

simulation A procedure used when there is no analytic solution available for a problem involving *random variables. *Pseudo-random numbers are used to mimic the random variables involved. Two general methods are the *inverse transformation method and the *rejection method. *See also* MONTE CARLO METHODS.

simultaneous equation model A collection of m *linear models in which each model involves k response variables ($k \leq m$) in addition to explanatory variables. Many *econometric models have this form. Often the explanatory variables are *latent variables, in which case the model is a **structural equation model**.

Sinclair, Sir John (1754–1835, b. Thurso, Scotland; d. Edinburgh, Scotland) Scottish politician and landowner, with a keen interest in agriculture. Between 1791 and 1799 he was responsible for the publication of the 21 volumes of the *Statistical Account of Scotland*, an assembly of largely non-numerical agricultural information. This introduced the term 'statistics' into the English language. He was knighted in 1780, appointed FRS in 1784 and FRSE in 1798. He was the oldest of the founders of the *RSS.

single linkage clustering A method of collecting *multivariate data into clusters (*see* CLUSTER ANALYSIS; CLUSTER SAMPLING). *See* AGGLOMERATIVE CLUSTERING METHODS.

singular *See* MATRIX.

sink *See* NETWORK FLOW PROBLEM.

size (of a sample) The number of *observations in the *sample.

size (of a test) The *probability of a Type I error, i.e. that in a *hypothesis test the *null hypothesis is rejected when it is true; a synonym for significance level.

skewed; skewness If the *distribution of a *variable is not symmetrical about the *median or the *mean it is said to be skewed. The distribution has **positive skewness** if, in some sense, the tail of high values is longer than the tail of low values, and **negative skewness** if the reverse is true. Skewness is quantified by the *Pearson coefficient of skewness, the quartile coefficient of skewness (*see* QUARTILE), or (preferably) the *moment coefficient of skewness. For another measure of the shape of a distribution, *see* KURTOSIS.

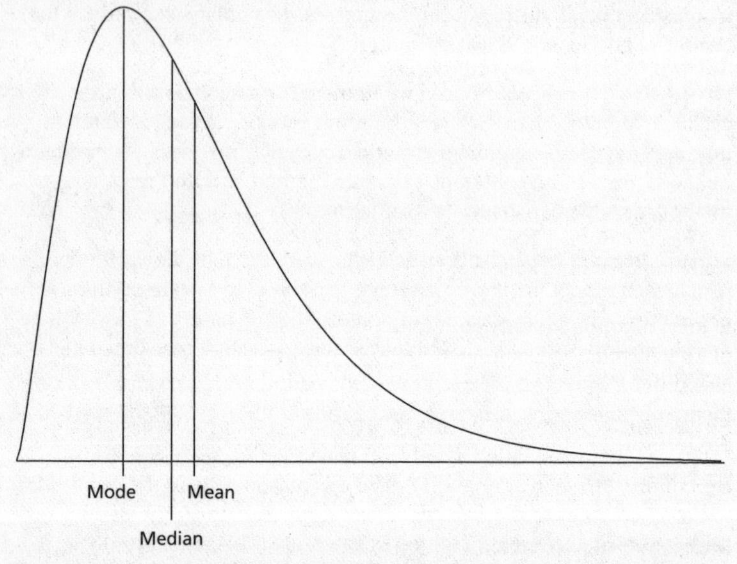

Skewed. Here it is supposed that the random variable has a finite lower bound but no upper bound. This typically results in the mean being bigger than the median and the median being bigger than the mode. The result is a positively skewed distribution.

skill score A measurement of the skill of a forecast. The **Priestley skill score**, P, applies to value forecasts (such as forecasts of temperature). Let f_i denote a forecast, and let o_i denote the subsequent observed value, with $\overline{o}$ denoting the mean of a set of n observed values. Then P is given by

$$P = 1 - \frac{\sum_{i=1}^{n}(o_i - f_i)^2}{\sum_{i=1}^{n}(o_i - \overline{o})^2}.$$

The **Brier score** was introduced in 1950 to assess the accuracy of probabilistic weather forecasts (e.g. 'There is a probability p_i of rain tomorrow'). Denoting the actual weather outcome by o_i (equals 1 if the event occurs, and 0 if it does not) the Brier score, B, is given (for n forecasts) by

$$B = \frac{1}{n}\sum_{i=1}^{n}(p_i - o_i)^2.$$

The smaller the value of B, the more skilful the forecasting procedure. If B_s denotes the Brier score for a standard forecasting system, then S, given by

$$S = \left(1 - \frac{B}{B_s}\right),$$

is called the **Brier skill score.**

Sklar's theorem *See* COPULA.

slack variable *See* LINEAR PROGRAMMING.

slippage test *See* TEST FOR EQUALITY OF LOCATION.

Slutsky, Evgeny Evgenievich (1880–1948; b. Novoe, Russia; d. Moscow, Russia) Russian statistician. Slutsky entered U Kiev in 1899 to study mathematics. However, after twice being involved in student unrest, he was expelled and completed his studies in Munich. In 1905 he returned to Kiev, this time to study political economics. In 1913 he joined the faculty of Kiev Institute of Commerce, moving to Moscow in 1926 to work in the government statistical offices. During his period there he worked on the theory of *time series and *stochastic processes. From 1938 he worked at the Institute of Mathematics at the USSR Academy of Sciences.

SEE WEB LINKS

- Fuller biography and photograph.

Slutsky–Yule effect An undesirable consequence (noted by *Slutsky and by *Yule) of applying a *moving average to a *time series. Suppose a time series consists of randomly chosen *observations from the same *population. We would therefore hope that any averaging would bring out

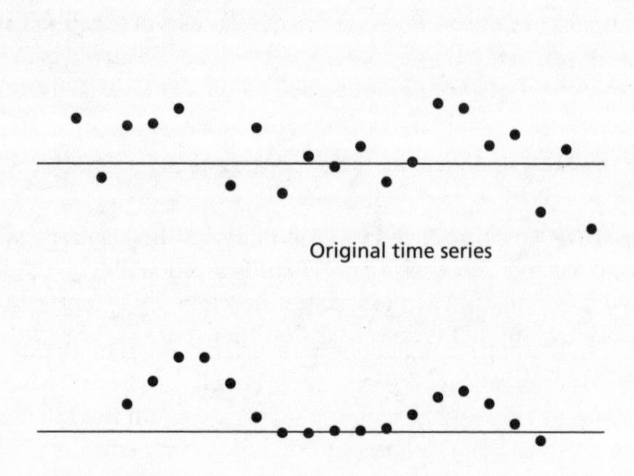

Slutsky–Yule effect. Chance variations in single values translate into apparent cycles in moving averages.

the fact that the *mean was constant. However, by chance some values will be larger than others. Let x_k be a particularly large value. When we apply a moving average, all the averages that involve x_k will be inflated. With most moving averages the inflation will be greatest for the average centred on the *k*th observation and will diminish on either side. Each extreme value will have a similar effect such that the series of averages will present oscillations that appear real but are due to chance.

Smirnov, Nikolai Visil'yevich (1900–66; b. Moscow, Russia; d. Moscow, Russia) Russian mathematician. Smirnov obtained his first degree from Moscow U in 1926. His doctoral thesis (1938) served as a foundation for his subsequent work on the theory of *non-parametric tests. His career was spent at the Steklov Mathematical Institute of the USSR Academy of Sciences. His forte lay in solving difficult computational problems using ingenious methods. He is remembered today for his 1939 work on the distribution of the test statistic proposed by *Kolmogorov in 1933.

Smith, Sir Adrian Frederick Melhuish (1946– ; b. Dawlish, England) English statistician specializing in *Bayesian inference. Smith obtained his first degree at Cambridge U in 1968 and his PhD (supervised by *Lindley) from UCL in 1971. He held faculty appointments at Oxford U (1971), UCL (1974), U Nottingham (1977), and IC (1980). From 1988 to

1998 he was Principal of QMUL. After a period as a civil servant he was appointed Vice-Chancellor of the University of London in 2012. President of the *RSS in 1995, Smith had been awarded its *Guy Medal in Bronze in 1977 and in Silver in 1993. He was elected FRS in 2001. He was the *COPSS *Fisher Lecturer in 2003. He was knighted in 2011.

SEE WEB LINKS

- University web page.

smoothing spline A curve through a set of *data. It consists of a sequence of cubic polynomial curves with no discontinuities.

Snedecor, George Waddel (1881–1974; b. Memphis, TN; d. Amherst, MA) American biometrician. Snedecor received a BS in mathematics and physics from U Alabama in 1905 and an AM in physics from U Michigan in 1913. In that year he was one of the first faculty to join the Mathematics Department at Iowa State College and one of his first courses was on statistics. He spent his entire career at Iowa, and the Department now occupies a building named in his honour. His first graduate student, in 1931, was Gertrude *Cox. His *Statistical Methods Applied to Experiments in Agriculture and Biology*, co-authored with *Cochran, sold more than 125 000 copies. He was the 1970 winner of the *Wilks Award of the *ASA and was made an Honorary Life Member of the *IBS in 1971.

SEE WEB LINKS

- Fuller biography and photograph.

Snedecor *F*-distribution *See* F-DISTRIBUTION.

Snow, Sir John English physician (1813–58; b. York, England; d. London, England) He examined the locations of victims of the 1854 London cholera outbreak and deduced that the cause was the Broad Street water pump (later found to be located too close to a cesspit). His arguments provide one of the first examples of *spatial statistics.

SEE WEB LINKS

- Biography, links, and photograph.

snowflake curve An example of a curve having *fractal dimension. Starting with an equilateral triangle, the middle third, PQ, say, of a side is replaced by the two lines PR and RP so that P, Q, and R form the vertices of a smaller equilateral triangle, with R outside the original enclosed region. This process is repeatedly applied to each line segment. The resulting 'curve' is the snowflake curve and has infinite length but encloses a finite area. Its fractal dimension is defined to be ln 4/ln 3, since each 'edge' of the curve contains four copies each of 1/3 size. *See diagram overleaf.*

social mobility table *See* MOBILITY TABLE.

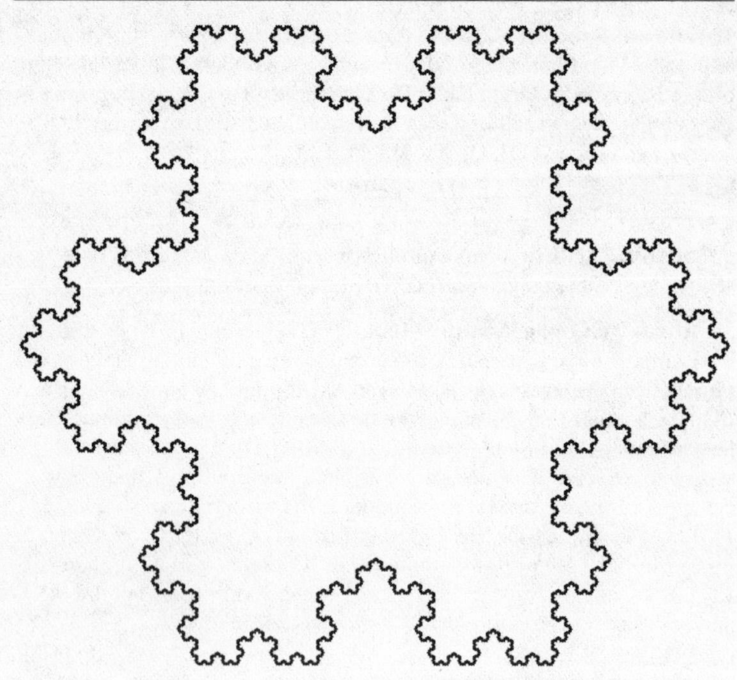

Snowflake curve. This is an example of a curve with fractal dimension $\neq 1$.

social statistics *Statistics applied to the social sciences, particularly sociology and demography. The *Lexis diagram and the *life table are two examples of statistical methods developed specifically for this branch of the subject.

Società Italiana di Statistica The Italian Statistical Society which publishes the English language **Statistical Methods and Applications*. The Society has about 1000 members.

SEE WEB LINKS

- Society home page.

Société Française de Statistique The French Statistical Society, formed from the amalgamation in 1997 of the Société de Statistique de Paris (founded in 1860) with two related organizations. It publishes the **Journal de la SFdS*. The Society has about 1000 members.

SEE WEB LINKS

- Society home page.

Somers's d_{BA} *See* ASSOCIATION.

Sorensen distance *See* DISTANCE MEASURE.

source *See* NETWORK FLOW PROBLEMS.

South African Statistical Association This association was founded in 1953 and has about 200 members.

SEE WEB LINKS
- Association home page.

spatial autocorrelation *See* AUTOCORRELATION.

spatial process The manner in which values change from one spatial location to another.

spatial statistics Statistical procedures informed by knowledge of the spatial location of *observations. An important example is *kriging. *Data may be collected using *quadrats or *transects.

SPC *Abbreviation for* STATISTICAL PROCESS CONTROL.

Spearman, Charles Edward (1863–1945; b. London, England; d. London, England) English psychologist who introduced the use of statistical methods into psychology. Spearman's initial choice of career was the army. He fought with distinction in the Burmese War and did not leave the army until 1897, when he went to Leipzig to study psychology, obtaining his PhD in 1904. His first two papers, introducing *Spearman's rho and laying the foundations for *factor analysis, appeared in that year. After posts in various German universities he returned to London in 1907 as Professor of Psychology at UCL. Spearman was elected FRS in 1924 and to the NAS in 1943.

SEE WEB LINKS
- Fuller biography and photograph.

S

Spearman–Brown formula *See* RELIABILITY.

Spearman rank correlation coefficient *Alternative name for* SPEARMAN'S RHO.

Spearman's rho (ρ) A *rank correlation coefficient that may be used as an alternative to *Kendall's tau. Individuals are arranged in order according to two different criteria (or by two different people). The *null hypothesis is that the two orderings are *independent of one another. It is based on the differences in the ranks given in two orderings. Suppose that the jth individual is given rank x_j in one ordering and rank y_j in the second ordering. Define d_j by $d_j = x_j - y_j$. Then ρ (which lies in the interval -1 to 1 inclusive) is given by

$$\rho = 1 - \frac{6\Sigma d_j^2}{n(n^2-1)},$$

where n is the number of individuals. In fact, ρ is the *sample correlation coefficient for the pairs $(x_1, y_1), (x_2, y_2), \ldots, (x_n, y_n)$.

As an example, suppose that someone is asked to arrange, in order of increasing mass, five similar boxes whose contents vary. The correct order is 1, 2, 3, 4, 5, but the order chosen is 2, 1, 3, 5, 4. The rank differences are $-1, 1, 0, -1, 1$, giving $\Sigma d_j^2 = 4$. The value of ρ is $1 - (6 \times 4)/(5 \times 24) = 0.8$. Comparing this value with a table of critical values we find that, at the 5% significance level, there is no significant evidence to reject the null hypothesis that the boxes were arranged in a random order.

specification limits *See* CAPABILITY ANALYSIS.

specificity *See* TWO-BY-TWO TABLE.

spectral density function; spectrum *See* PERIODOGRAM.

Speed, Terence Paul (1943- ; b. Melbourne, Australia) Australian statistician specializing in *bioinformatics who divides his time between Australia and the USA. Speed obtained his BSc at U Melbourne in 1965 and joined the faculty at Monash U where he gained his PhD in 1969. After spells at U Sheffield (1969–73) and U Western Australia (1974–82), he was chief of the CSIRO Division of Mathematics & Statistics (1983–87). In 1987 he was appointed Professor at UCB and since 1997 has divided his time between UCB and the Walter and Eliza Hall Institute of Medical Research in Melbourne. He was President of the *IMS in 2004 having been its *Rietz Lecturer in 1991 and its *Wald Lecturer in 2001. He was awarded the Pitman Medal (*see* PITMAN) of the *SSAI in 2004 and was the *COPSS *Fisher Lecturer in 2006.

SEE WEB LINKS

- Fuller biography and photograph

spherical data Directional data in three dimensions. The direction can be specified by the location of a point on a sphere of unit radius. Relative to axes at the centre of the sphere, the location can be expressed in either *Cartesian coordinates, (x, y, z) or *spherical polar coordinates $(1, \theta, \phi)$. These are related as follows:

$$x = \sin\theta\cos\phi, \quad y = \sin\theta\sin\phi, \quad z = \cos\theta, \qquad 0° \le \theta \le 180°, \quad 0° \le \phi < 360°.$$

Spherical data are often represented in two dimensions using a *Schmidt net. A *distribution commonly used to model spherical data is the *Langevin distribution.

spherical polar coordinates Coordinates that specify the location of a point in three dimensions. The basis of these coordinates is a set of three mutually perpendicular axes, *Ox*, *Oy*, and *Oz*, that intersect at *O*. Consider a point *P*, in three-dimensional space, at a distance *r* from *O*, the angle $\widehat{zOP}$ being equal to θ. Let *OM* be the projection of *OP* on the *xy*-plane and let the angle $\widehat{xOM}$ be equal to ϕ. The spherical polar coordinates of *P* are (r, θ, ϕ), where $0° \leq \theta \leq 180°, 0° \leq \phi < 360°$. These coordinates are related to the alternative Cartesian coordinates (x, y, z) by

$$x = r \sin\theta \cos\phi, \qquad y = r \sin\theta \sin\phi, \qquad z = r \cos\theta.$$

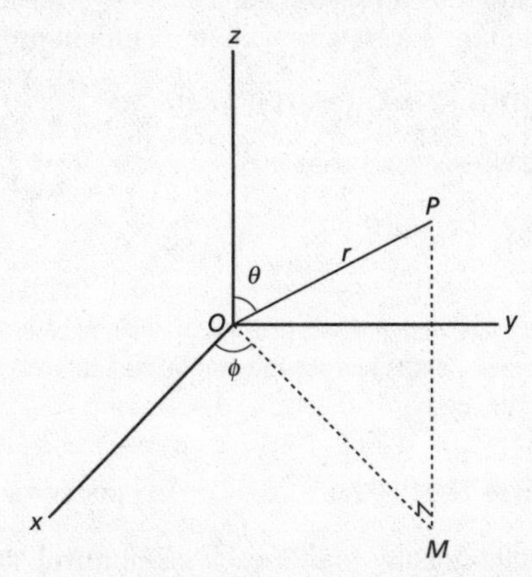

Spherical polar coordinates. Using Cartesian coordinates, the point *P* would be represented by (x, y, z), with $x = r \sin\theta \cos\phi$, $y = r \sin\theta \sin\phi$, $z = r \cos\theta$.

sphericity test A test of whether the null hypothesis of *independent observations with a constant *variance is valid. This hypothesis is of particular concern in the analysis of *longitudinal data, where the k *repeated measures of individuals may display a noticeable *correlation. The most common test is the **Mauchly test**, for which the test statistic is W, given by

$$W = \det(\mathbf{S})\left(\frac{k+1}{\mathrm{tr}(\mathbf{S})}\right)^{k+1},$$

where $\mathbf{S}$ is the $k \times k$ *variance-covariance matrix for the sample, and $\det(\mathbf{S})$ and $\mathrm{tr}(\mathbf{S})$ are, respectively, the determinant and the trace of $\mathbf{S}$.

spider plot Also referred to as **radar plot** or **star plot**, this is a plot that can be more effective than a *multiple bar chart for comparing small amounts of *multivariate data. *See diagram overleaf.*

S

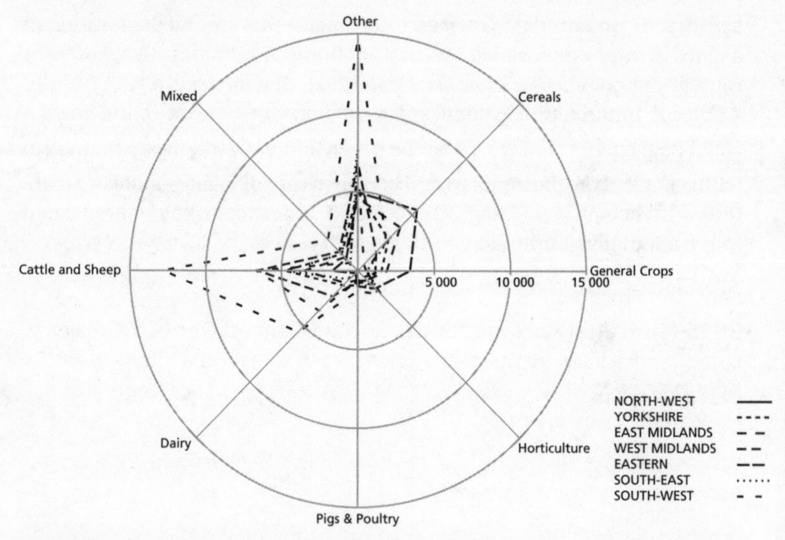

Spider plot. The plot shows the numbers of farms of different types in different regions of England. There are many cattle and sheep farms in the South-West, whereas cereal farms predominate in the East.

Spjotvoll–Stoline test *See* MULTIPLE COMPARISON TEST.

spline A set of polynomials, one for each sub-interval, that give an approximation to the function f(x), defined on some interval $a \leq x \leq b$, where $a = x_0 < x_1 < \cdots < x_n = b$ is a subdivision of the interval $a \leq x \leq b$. The polynomials are all of the same degree, d, and are chosen so that the values of the polynomials and their first $(d - 1)$ derivatives are continuous at the intermediate points of subdivision and the values of the polynomials agree with the value of f(x) at each of the points.

split-half method *See* RELIABILITY.

split plot An *experimental design used when comparisons of different types of treatments can be made at different scales. For example, after ploughing, different fields (the **whole plots**) might be treated with different fertilizers. Subsequently, within each field, **sub-plots** might be planted with different varieties of vegetable. The analysis of the variability of the varieties uses within-field sub-plot variation, and the analysis for the fertilizers uses between-fields whole-plot variation.

S-PLUS A computer language with extensive graphical and statistical capabilities.

SEE WEB LINKS

- Computer language home page.

spreadsheet Computer software for recording and analysing numerical *data. Arithmetic and statistical operations can be carried out by simple instructions. Any associated graphs and diagrams that are required can also be automatically drawn.

SPREE *See* DEMING–STEPHAN ALGORITHM.

SPSS (Statistical Package for the Social Sciences) One of the major computer packages permitting many types of statistical analysis.

SEE WEB LINKS

- Computer package home page.

spurious correlation The term used by Karl *Pearson in 1897 when describing a situation displaying a *nonsense correlation.

spurious precision A value stated with more precision than is actually possible, given the accuracy of the values from which it has been calculated. For example, if a three-metre length is cut off the top of a sixteen-metre tree, then, if the lengths had been quoted to the nearest metre, it would be spurious precision to describe this as being a reduction by 18.75%. The true percentage lies between $100 \times \frac{2.5}{16.5} \approx 15\%$ and $100 \times \frac{3.5}{15.5} \approx 23\%$. A reasonable precision would be provided by giving one significant figure in the percentage: 20%.

squared multiple correlation *See* ANOVA.

square matrix *See* MATRIX.

square table A *contingency table, having the same number of rows as columns, in which the definitions of the row and column categories are essentially the same, but differ, for example, by referring to different time points. With the cell (*see* CONTINGENCY TABLE) probability denoted by p_{jk}, and parameters relating to rows, columns, diagonals, and individual cells by r_j, c_k, d_l, and m_{jk}, respectively, specialist models include:

the **symmetry model**: $-p_{jk} = p_{kj}$ for all j, k;

the **mover-stayer model**: $-\, p_{jk} = \begin{cases} r_j c_k & \text{for } j \neq k, \\ m_{jj} & \text{otherwise;} \end{cases}$

the **diagonals model**: $-\, p_{jk} = r_j c_k d_l$, where $l = j - k$;

the **loyalty model**: $-\, p_{jk} = \begin{cases} r_j c_k & \text{for } j \neq k, \\ r_j c_j d_0 & \text{otherwise;} \end{cases}$

the **quasi-symmetry model**: $-\, p_{jk} = r_j c_k m_{jk}$, with $m_{jk} = m_{kj}$;

the **distance model**: $-\, p_{jk} = r_j c_k m_{jk}$, with m_{jk} equal to a product of terms corresponding to gaps between successive categories.

SSAI *Abbreviation for the* STATISTICAL SOCIETY OF AUSTRALIA INC.

SSC *Abbreviation for the* STATISTICAL SOCIETY OF CANADA.

stabilizing the variance A phrase that is used to describe the use of a transformation to equalize the *variance of a set of *observations initially measured with unequal accuracy.

standard deviation (sd) The square root of the *variance. Karl *Pearson introduced the term in 1893, using the symbol σ in the following year.

standard error (se) The square root of the *variance of a *statistic. For example, the standard error of the *sample mean of n *observations taken from a *population with variance σ^2 is $\sigma/\sqrt{n}$. The term is also used for $s/\sqrt{(n)}$ where s^2 is the *sample variance. The term 'standard error' was used by *Yule in 1897.

standardization The process of *standardizing.

standardized mortality ratio *See* MORTALITY RATE.

standardized residual *See* REGRESSION DIAGNOSTICS.

standardizing Converting a *random variable X with *expected value μ and *variance σ^2 to a random variable Y with expected value 0 and variance 1, using the transformation

$$Y = \frac{X - \mu}{\sigma}.$$

The same term is used for the process of converting a *data set $x_1, x_2, \ldots, x_n$, with *sample mean $\bar{x}$ and *sample variance s^2, into the data set $y_1, y_2, \ldots, y_n$, with mean 0 and variance 1, by the transformation

$$y_j = \frac{x_j - \bar{x}}{s}, \quad j = 1, 2, \ldots, n.$$

standard normal distribution; standard normal variable *See* NORMAL DISTRIBUTION.

STARIMA model; STARMA model The space-time equivalents of *ARIMA model and *ARMA model.

star plot *Alternative name for* SPIDER PLOT.

Stata A powerful and versatile command-line statistical package well suited to *robust data analysis.

SEE WEB LINKS

- Computer package home page.

state *See* MARKOV PROCESS.

state vector *See* RELIABILITY THEORY.

STATGRAPHICS A computer package designed for interactive statistical data analysis.

SEE WEB LINKS

- Computer package home page.

stationarity; stationary A *time series displays stationarity if the *expected value at all points in time is the same and if, additionally, the *correlation between the values at two time points, t and $t+\tau$, depends on the lag τ but not on t. The corresponding requirements hold for a *spatial process, with t replaced by the location of a point in space and τ replaced by a spatial lag. The time series or spatial process may be described as being stationary.

stationary chain *See* MARKOV PROCESS.

stationary point A point where a function has zero gradient. A function $f(x_1, x_2, \ldots, x_n)$ of one or more variables is said to have a **stationary value** at the point $(a_1, a_2, \ldots, a_n)$ if all the partial derivatives of f with respect to $x_1, x_2, \ldots, x_n$ vanish when $(x_1, x_2, \ldots, x_n) = (a_1, a_2, \ldots, a_n)$. The point $(a_1, a_2, \ldots, a_n)$ is a stationary point. The stationary point is a **maximum** (or **minimum**) if, for all neighbouring points, $f(x_1, x_2, \ldots, x_n)$ is less (or greater) than $f(a_1, a_2, \ldots, a_n)$. The stationary point is a **minimax** if there are points in the neighbourhood at which $f(x_1, x_2, \ldots, x_n) < f(a_1, a_2, \ldots, a_n)$, and points in the neighbourhood at which $f(x_1, x_2, \ldots, x_n) > f(a_1, a_2, \ldots, a_n)$. For example, x^2+3y^2 has a minimum (stationary) point at (0, 0), $-2x^2-5y^2$ has a maximum point at (0, 0), and $2x^2-5y^2$ has a minimax point at (0, 0).

stationary process A *stochastic process is said to be stationary (or **strictly stationary**) if the joint distribution of the sequence of measurements $x_{1+l}, x_{2+l}, \ldots, x_{k+l}$ is, for all k, *independent of l. A less restrictive requirement is that the *expected value of the x-values should be constant and that, for all k, the *covariance between the values of x_k and x_{k+l} should depend only on the lag l—such a process is described as **weakly stationary** (or **second-order stationary**). *See also* MARKOV PROCESS.

stationary value *See* STATIONARY POINT.

statistic A function of the set of *random variables corresponding to a set of *observations. Often used to refer to the corresponding function of the observations themselves. The word ‘statistic’ was introduced by Sir Ronald

*Fisher in 1922. Two simple examples are the *population mean and the *sample variance.

statistical inference The process of drawing conclusions about the nature of some system on the basis of data subject to random variation. There are several distinguishable and apparently irreconcilable approaches to the process of inference; comfortingly, there are rarely any gross differences in the inferences that result. Approaches include *Bayesian inference and *fiducial inference; the approach first met by a student of Statistics is usually that based on the *Neyman–Pearson lemma.

Statistical Methodology See *Journal of the Royal Statistical Society*.

Statistical Methods and Applications The English language journal of the *Società Italiana di Statistica. First published in 1994, there are four issues per year.

SEE WEB LINKS

- Journal home page.

statistical process control (SPC) The statistical methods used to monitor and improve the quality of the output of a manufacturing process. SPC includes the use of *quality control charts, *optimization, *reliability, and *experimental design.

Statistical Science A quarterly journal published by the *Institute of Mathematical Statistics. It was first published in 1985.

SEE WEB LINKS

- Journal home page.

Statistical Society of Australia Inc. (SSAI) This society, founded in 1962 as the Australian Statistical Society, has about 600 members. Together with the New Zealand Statistical Association it publishes the **Australian and New Zealand Journal of Statistics*.

SEE WEB LINKS

- Society home page.

Statistical Society of Canada (SSC) This Society, founded in 1972, has about 700 members. It publishes the **Canadian Journal of Statistics*.

SEE WEB LINKS

- Society home page.

Statistical Theory and Method Abstracts A publication of the *International Statistical Institute which provides comprehensive cross-referencing of new publications in Statistics by topic area. It is now a component of Zentralblatt MATH.

SEE WEB LINKS

- Information page.

Statistica Neerlandica An English language journal of the *Netherlands Society for Statistics and Operations Research. First published in 1946, with an emphasis on decision making, it appears quarterly.

SEE WEB LINKS

- Journal home page.

Statistica Sinica An English language journal of the *International Chinese Statistical Association. First published in 1991, it appears quarterly.

SEE WEB LINKS

- Journal home page.

Statistics The science of collecting, displaying, and analysing *data. The terms 'statistics' and 'statistical' appear to have been introduced by *Sinclair in his 21-volume *Statistical Account of Scotland*, published between 1791 and 1799.

statistics The plural of *statistic.

Statistics and Computing A quarterly journal, first published in 1990, that publishes articles that straddle the interface between Statistics and Computer Science.

SEE WEB LINKS

- Journal home page.

Statistics Education Research Journal The electronic journal published bi-annually by the *International Association for Statistical Education.

SEE WEB LINKS

- Journal home page.

Statistics in Biopharmaceutical Research **(*SBR*)** A quarterly journal first published in 2008 by the *American Statistical Association.

SEE WEB LINKS

- Journal home page.

Statistics in Medicine A journal, founded in 1982, that now appears twice monthly. It concentrates on practical applications of Statistics to Medicine.

SEE WEB LINKS

- Journal home page.

Statistics in Society *See* JOURNAL OF THE ROYAL STATISTICAL SOCIETY.

STATXACT A specialized statistical package for exact *non-parametric tests with *continuous variables or *categorical variables.

SEE WEB LINKS
- Package home page.

steepest ascent (hill-climbing); steepest descent A method for finding a maximum (or minimum) of a function. The idea is to proceed in the direction for which the slope of the corresponding surface is greatest. *See also* RESPONSE SURFACE.

Stein, Charles M. (1920– ; b. Brooklyn, NY) American probabilist. Having obtained his BS in mathematics from U Chicago in 1940, Stein joined the United States Army Air Force. After the Second World War he gained his PhD in mathematical statistics at Columbia U under the supervision of *Wald. He joined the faculty at Stanford U in 1953. He was the *IMS *Wald Lecturer in 1961, its *Rietz Lecturer in 1975, and its *Neyman Lecturer in 1984. He was elected to the NAS in 1975.

SEE WEB LINKS
- Fuller biography, interview, and photographs.

Stein effect A remarkable result presented by *Stein in 1956. Suppose n *vector *observations $\{\mathbf{x}_j\}$ are taken from a *multivariate normal distribution with $p\ (> 3)$ dimensions and unknown *mean $\boldsymbol{\mu}$. The most efficient *estimator of $\boldsymbol{\mu}$ is not $\bar{\mathbf{x}}$, the *sample mean, but is

$$\left(1 - \frac{p-2}{\sum_{j=1}^{n} \mathbf{x}_j' \mathbf{x}_j}\right)\bar{\mathbf{x}}.$$

The estimators of this type presented by Willard James and *Stein in 1961 are called **James–Stein estimators** or **shrinkage estimators** (since they 'shrink' $\bar{\mathbf{x}}$ towards the zero vector).

Stein method A procedure, introduced in 1972 by *Stein, for approximating complicated probability distributions by simpler ones, and providing estimates for the error in the approximation.

stem and leaf diagram A method of counting and ordering numerical *data without losing the detail of the individual data. The last significant digit for the whole data item is determined and each data item is represented by this digit (the leaf), with a stem consisting of the previous digits. For example, 58 is divided as 5|8, with a stem of 5 and a leaf of 8. Only the leaf is shown explicitly on the diagram. A **back to back stem and leaf plot** can be an effective method of comparing related populations.

5	8	Key: 5|8 means 58
4		
3	9	
2	2, 6	
1	0, 0, 1, 2, 3, 4, 7	
0	0, 0, 0, 1, 1, 1, 4, 6, 7	

Stem and leaf diagram. The data are 22, 58, 12, 17, 4, 26, 10, 13, 1, 39, 0, 1, 10, 6, 0, 11, 14, 1, 0, 7. The diagram retains all the information in the data, while also giving an idea of the underlying distribution.

Lengths (mm) of eggs laid by cuckoos

In dunnock nests									In reed-warbler nests				
							9	20					
							7	21	2	6	6	9	
						0	8	22	0	2	0	9	8
9	8	8	1	5	0	0	1	23	2				
							0	24					
KEY: 9|20 = 20.9							0	25	KEY: 20|9 = 20.9				

Back to back stem and leaf plot. The size of eggs laid by cuckoos varies according to the nest being used.

step *See* QUARTILE.

step diagram A diagram (*see overleaf*) showing *data values for a numerical *variable in which the *cumulative frequency of observations ≤ x is plotted against x. The diagram consists of horizontal segments, with jumps at the observed data values. The jump at x_j represents the frequency of x_j in the sample. *See also* CUMULATIVE FREQUENCY POLYGON; OGIVE; SAMPLE DISTRIBUTION FUNCTION.

Stephan, Frederick Franklin (1903–71; b. Chicago, IL; d. Princeton, NJ) American demographer. A graduate of U Illinois (BA, 1924), Stephan gained his MA from U Chicago (1926) and joined the faculty of U Pittsburgh in 1927. After a spell at Cornell U, he joined the faculty at Princeton U in 1947, initially as Professor of Social Statistics in the Sociology department, subsequently transferring to the Statistics department. He was Editor of the **Journal of the American Statistical Association* from 1935 to 1940. He collaborated with *Deming on the 1940 paper that introduced the *Deming–Stephan algorithm. He was President of the *ASA in 1966.

SEE WEB LINKS

- Obituary.

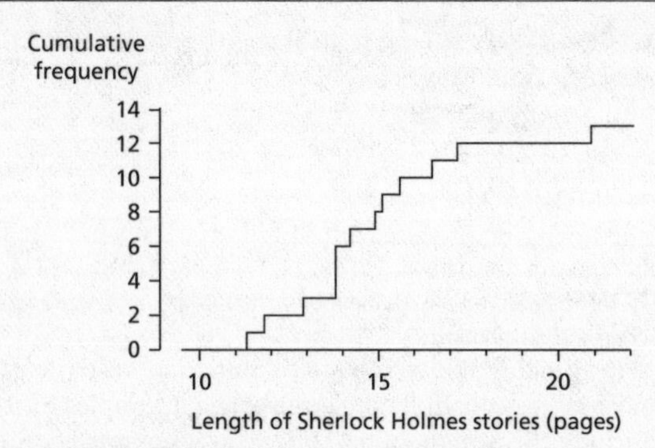

Step diagram. The data illustrated here are the lengths of the stories in *The Return of Sherlock Holmes*. Conan Doyle wrote the stories to fit regular slots in *Collier's Magazine* and the *Strand Magazine* between September 1903 and December 1904. The lengths of the stories are therefore very similar.

stepwise procedure A procedure for identifying an appropriate *linear model in the context of *multiple regression. The *expected value of the response variable, E(Y), is modelled as a *linear combination of many (p, say) explanatory X-variables. A natural question is whether all p of the X-variables are required.

Forward selection begins by determining which one of the X-variables is most highly correlated with Y. This variable is retained in all future models. At the second stage the procedure considers the remaining $(p-1)$ variables and determines which, in conjunction with the first variable, provides most additional information about Y.

Backward elimination mirrors forward selection by starting with the model containing all p X-variables and removing ineffective variables one by one (stepwise).

The order in which variables enter or leave the model may be determined using tests based on the *F-distribution, with the critical values for entry or removal being termed the ***F* to enter** and the ***F* to remove**.

An alternative stepwise selection criterion, suggested in 2004 by *Efron, *Johnstone, and co-workers, uses **least angle regression selection** (**LARS**), in which variables are selected on the basis of their *correlation with the currently unexplained variation in Y. A variant of LARS is the lasso (*see* MULTIPLE REGRESSION MODEL). *See also* MODEL SELECTION PROCEDURE.

Stirling's formula A formula providing an approximation, for large values of the positive integer n, to $n!$ (*see* FACTORIAL):

$$n! \approx n^{n+\frac{1}{2}} e^{-n} \sqrt{2\pi}.$$

stochastic calculus An extension, introduced by *Ito, of the methods of calculus to a *stochastic process.

stochastic process A finite collection of related *random variables, often ordered in time or space. The possible values of a random variable are referred to as **states** and the possible configurations of the states of all the random variables constitute the **state space**. Examples of stochastic processes include *Brownian motion, the *counting process, the Markov chain (*see* MARKOV PROCESS), the *Poisson process, and the *random walk. The phrase 'stochastic process' was used by *Kolmogorov in 1932.

strata *See* STRATIFIED SAMPLING.

stratified sampling When a *population contains easily recognizable subpopulations, or **strata**, of known sizes ($N_1, N_2, \ldots, N_s$), the method of stratified sampling will usually give better results than a simple random *sample from the whole population. With stratified sampling, simple random samples (of sizes $n_1, n_2, \ldots, n_s$) are taken from each stratum. The method was introduced by *Neyman in 1934.

With **proportional allocation** the sizes of these samples satisfy

$$\frac{n_j}{N_j} \approx \frac{n_k}{N_k},$$

for all j and k.

If the *standard deviations of the values of the items in the various strata are known to be $\sigma_1, \sigma_2, \ldots, \sigma_s$, then for a fixed sample size of n items, the **optimum allocation (Neyman allocation)** is obtained by choosing n_j so that

$$\frac{n_j}{n} \approx \frac{N_j \sigma_j}{\sum_{m=1}^{s} N_m \sigma_m}.$$

This allocation minimizes the *variance of the *estimator of the overall *population mean. If the standard deviations are not known then a *pilot study can be used to obtain estimates of their values.

stratum The singular of strata. *See* STRATIFIED SAMPLING.

stress *See* MULTIDIMENSIONAL SCALING.

strictly stationary *See* STATIONARY PROCESS.

strong law of large numbers *See* LAWS OF LARGE NUMBERS.

structural equation model *See* SIMULTANEOUS EQUATION MODEL.

structural zero An entry in a *contingency table that is certain to be zero, whatever the sample size, because it corresponds to an impossible outcome. The table shown contains three structural zeros (top right) as well as one **random zero**—a cell whose entry is zero by chance.

		Income of head of household		
		<£20 000	£20 000–£30 000	>£30 000
Total	<£20 000	5	0	0
household	£20 000–£30 000	2	8	0
income	>£30 000	0	5	12

structure function *See* RELIABILITY THEORY.

structure-preserving estimation *See* DEMING–STEPHAN ALGORITHM.

Stuart–Maxwell test A *test for *marginal homogeneity in a $k \times k$ *contingency table. When $k = 2$ the test reduces to the *McNemar test. The test statistic has an approximate *chi-squared distribution with $k - 1$ *degrees of freedom.

Student The pen-name used by *Gosset.

Studentized range distribution The *distribution of the *statistic

$$\frac{\bar{x}_{(k)} - \bar{x}_{(1)}}{s/\sqrt{n}},$$

where random *samples of size n have been taken from k *independent and identically distributed *normal populations, with $\bar{x}_{(1)}$ and $\bar{x}_{(k)}$ being, respectively, the smallest and largest of the k sample means, and s^2 being the pooled estimate of the common variance (*see* POOLED ESTIMATE OF COMMON MEAN). This statistic is particularly used in *multiple comparison tests.

Studentized residual; Studentized deletion residual *See* REGRESSION DIAGNOSTICS.

Student's *t*-distribution *See* *t*-DISTRIBUTION.

study population *See* TARGET POPULATION.

subjective probability *See* BAYESIAN INFERENCE.

sub-plot *See* SPLIT PLOT.

subpopulation A subset of the *population that has some characteristic in common.

subsample Part of a *sample.

SUDAAN A statistical package specifically designed for the analysis of correlated data from studies involving *longitudinal data, *repeated measures, and related complex surveys.

SEE WEB LINKS

- Package home page.

sufficient estimator; sufficient statistic *See* ESTIMATOR.

summary statistics For a set of *data, common summary statistics are *measures of location, *measures of spread, sums, sums of squares, and sums of cross-products. For a situation involving a dependent variable and explanatory variables, the entries in an *ANOVA table might also be referred to as summary statistics.

sum of squares *See* ANOVA.

supersaturated design *See* FACTORIAL DESIGN.

supervised learning *See* MACHINE LEARNING.

support vector machine A *machine learning procedure used to classify multidimensional *data.

surrogate variable A variable that can be measured (or is easy to measure) that is used in place of one that cannot be measured (or is difficult to measure). For example, whereas it may be difficult to assess the wealth of a household, it is relatively easy to assess the value of a house. *See also* PROXY VARIABLE.

survey *See* SAMPLE.

Survey Statistician The bi-annual journal of the *International Association of Survey Statisticians.

SEE WEB LINKS

- Journal home page.

survival curve; survival time *See* HAZARD RATE.

survivor function (survival function) *See* HAZARD RATE.

Swedish Statistical Association (Svenska Statiskersamfundet) One of the four associations that publishes the **Scandinavian Journal of Statistics*. It also publishes (in Swedish) the journal *Qvartilen*.

SEE WEB LINKS

- Association home page.

swing In the context of voting, if party A's share of the vote falls by $a\%$ and party B's share of the vote rises by $b\%$, then, according to one definition, the swing from A to B is $\frac{1}{2}(a+b)\%$. An alternative measure uses percentages of the total vote for parties A and B (so that the swing is $a\% = b\%$). When there are three major parties, however, neither definition is entirely satisfactory, since the majority vote can pass from party A to party B, while both parties are losing votes to party C.

symmetric confidence interval *See* CONFIDENCE INTERVAL.

symmetric distribution The *random variable X has a symmetric distribution if and only if there is a number c such that

$$\mathrm{P}(X<c-k) = \mathrm{P}(X> c+k), \text{for all} \quad k \geq 0.$$

In this case the distribution is symmetric about c. If X has a symmetric distribution and *expected value μ, then $c=\mu$. A distribution that is not symmetric is called an **asymmetric distribution**.

symmetric matrix *See* MATRIX.

symmetry model *See* SQUARE TABLE.

symmlet *See* WAVELET.

SYSTAT A computer package for statistical analysis which offers a wide range of graphical options.

SEE WEB LINKS

- Package home page.

systematic error An error that occurs when the result of measuring a variable whose actual value is x is $\mathrm{f}(x)$, where f is a fixed function. A simple example would be to consistently truncate (so that, for example, 11.9 is reported as 11 rather than 12). *See* RANDOM ERROR.

systematic sampling A method of choosing a *sample from a sampling frame of size N. It is assumed that a list exists of the individuals in the frame, the two ends of the list being notionally joined. One unit is chosen at random from the list and then every kth unit thereafter is chosen for the sample (continuing counting from the start of the list when the end is reached) until the desired sample size is reached. The number k can be chosen so that N/k is approximately equal to the desired sample size.

Taguchi, Genichi (1924–2012; b. Niigata, Japan; d. Tokyo, Japan) Japanese pioneer of modern industrial quality control. During the Second World War Taguchi learnt about *experimental design when working in the *Institute of Statistical Mathematics. From 1950 to 1962 he worked in the research and development section of the Nippon Telephone and Telegraph Company. During this period he met both Sir Ronald *Fisher and *Shewhart. In 1962 he was awarded a doctorate from Kyushu U. From 1964 to 1982 he was a Professor at Aoyama Gakuin U in Tokyo. Following his 1980 visit to Bell Laboratories, *Taguchi methods began to be applied in the United States. In 1986 he received the Indigo Ribbon from the Emperor of Japan for his contributions to Japanese economics and industry, and in 1995 he received the *Shewhart Medal of the ASQ.

SEE WEB LINKS

- Obituary and photograph.

Taguchi methods Methods concerned with the optimization of product and process before manufacture that concentrate on quality loss rather than quality. Optimization involves *experimental design using simple designs to estimate main effects. The success of Taguchi methods is partly a consequence of the experimentation being tailored to the application.

tail area; tail probability The probability of obtaining either a particular observed value, or a more extreme value. Suppose that a *null hypothesis specifies that the *random variable X has a *distribution with *median m. Denote the observed value of X by x. If $x \geq m$ then the tail area (or tail probability) is $\mathrm{P}(X \geq x)$, otherwise it is $\mathrm{P}(X \leq x)$.

tally chart A method of counting *frequencies, according to some classification, in a set of *data. One line on a sheet of paper is assigned to each category or number, in the case of a *discrete random variable, or class, in the case of *grouped data. The data set is then worked through, and each item is represented by a vertical stroke on the corresponding line. For ease of counting, every fifth observation is represented by a

Duration (completed minutes)	Tally
1	II
2	I
3	III
4	~~IIII~~ II

Tally chart. The chart is a tally of the eruption times of *Old Faithful on 1 August 1978. Note the five-barred gate that makes counting the last row easy.

diagonal line crossing the previous four to make a **five-barred gate**. Tally charts are often used with grouped observations.

Tamhane test *See* MULTIPLE COMPARISON TEST.

target population The *population about which information is desired. The population that is actually surveyed is the **study population**.

tau (τ) A Greek letter used to signify the measure of *association suggested by Sir Maurice *Kendall. Also used as a symbol for time.

Taylor, Brook (1685–1731; b. Edmonton, England; d. London, England) English mathematician. Taylor graduated from Cambridge U in 1709 and was elected FRS in 1712. His work on *Taylor series was published in 1715. A lunar crater is named after him.

SEE WEB LINKS

- Fuller biography.

Taylor expansion; Taylor series A series representation for a function f having continuous derivatives of all orders. The series is

$$f(a) + \frac{x}{1!}f'(a) + \frac{x^2}{2!}f''(a) + \frac{x^3}{3!}f'''(a) + \cdots.$$

It is the *Maclaurin series of F, where $F(x) = f(a+x)$, so in the case $a = 0$ the Taylor and Maclaurin series are identical. As in the case of Maclaurin series, the Taylor series may or may not be convergent to $f(a+x)$, and can vanish altogether.

Tchebycheff *See* CHEBYSHEV.

***t*-distribution (Student's *t*-distribution)** The *probability density function f for this *distribution is given by

$$f(t) = \frac{1}{\sqrt{\nu}B(\frac{1}{2}, \frac{1}{2}\nu)}\left(1 + \frac{t^2}{\nu}\right)^{-\frac{1}{2}(\nu+1)}, \quad -\infty < t < \infty,$$

where B is the *beta function and ν is a positive *parameter (usually an integer) known as the number of *degrees of freedom. The distribution is symmetrical about its *mode at 0, which is therefore (for $\nu > 1$) also its *mean. For $\nu > 2$ the distribution has *variance $\nu/(\nu - 2)$.

When $\nu = 1$ the distribution is a *Cauchy distribution. As ν increases, the distribution increasingly resembles the standard *normal distribution, which is its limit as $\nu \to \infty$. If X has a *t*-distribution with ν degrees of freedom, then X^2 has an **F*-distribution with 1 and ν degrees of freedom. A *t*-distribution with ν degrees of freedom may be described as a t_ν-distribution.

The form of the distribution was published in 1908 by *Gosset, writing under the pen-name 'Student', in the context of a random *sample of size n from a *population having a normal distribution. Gosset was finding the distribution of t, given by

$$t = \frac{\bar{x} - \mu}{s/\sqrt{n}},$$

where μ is the *population mean, and $\bar{x}$ and s are, respectively, the *sample mean and *sample standard deviation with divisor $(n - 1)$. In Gosset's case $\nu = (n - 1)$.

The *percentage points (*see* APPENDIX VI) of the *t*-distribution are used as *critical values in carrying out a *t*-test (*see* HYPOTHESIS TEST) based on the value of t when μ is replaced by μ_0, the value specified by the *null hypothesis.

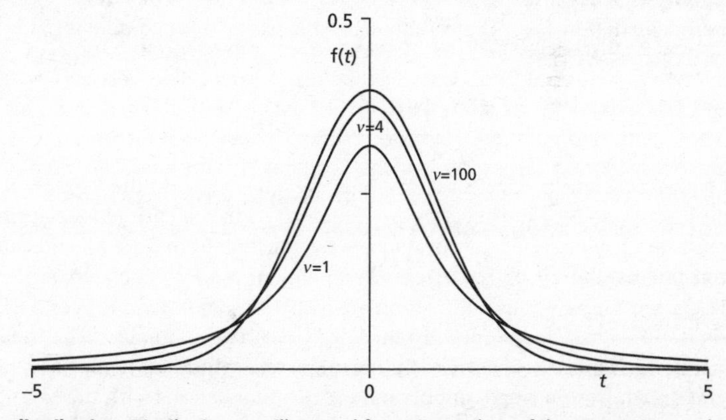

***t*-distribution.** Distributions are illustrated for various values of the parameter ν and all have mean 0. The case $\nu = \infty$ corresponds to the normal distribution, and the case $\nu = 1$ corresponds to the Cauchy distribution. The chance of a very extreme value is greater for a *t*-distribution than for the normal distribution, but decreases as ν increases.

t

Technometrics A quarterly journal established in 1959 as a joint publication of the *American Statistical Association and the *American Society for Quality. The journal publishes advances in the use of Statistics in the physical, chemical, and engineering sciences.

SEE WEB LINKS

- Journal home page.

ternary diagram *See* BARYCENTRIC COORDINATES.

tessellation A complete covering of a plane using a limited number of different shapes. Usually the shapes are polygons (as in the *Dirichlet tessellation). The plane can be tessellated with rectangles, or hexagons, or triangles (for example, using *Delaunay triangles). In a regular tessellation all the shapes are regular polygons (i.e. with all sides equal and all angles equal) of the same shape and size, and there are only three possible regular tessellations, using squares, equilateral triangles, or regular hexagons. Other semi-regular tessellations use two or more regular polygonal shapes, for example, squares and octagons. Many tessellations are periodic, i.e. the pattern repeats at regular intervals. A non-periodic tessellation, using two basic shapes, was invented by Sir Roger Penrose and is usually referred to as **Penrose tiling**.

test A means of determining whether a *hypothesis can be accepted. Typical hypotheses include 'the *distribution is *normal', 'the *population mean is 5', or 'the *sample was unbiased'. The question is whether a hypothesis is acceptable, or should be rejected in favour of some alternative hypothesis. *Data are then examined and a judgement is made. *See* HYPOTHESIS TEST.

test for equality of distribution A *test of the null hypothesis (*see* HYPOTHESIS TEST) that two (or more) *samples have been drawn from the same *distribution. An example is the *Van der Waerden test. *See also* CRAMÉR–VON MISES TEST; KOLMOGOROV–SMIRNOV TEST; TEST FOR EQUALITY OF LOCATION; TEST FOR EQUALITY OF SCALE; TUKEY POCKET TEST.

test for equality of location *Independent random *samples are taken from *populations with *cumulative distribution functions F_1 and F_2. It is assumed that, for some constant δ, $F_1(x) = F_2(x + \delta)$ for all x. The *null hypothesis is that $\delta = 0$, corresponding to the case where the samples have been drawn from a single population (or two populations with the same *distribution). The *alternative hypothesis is that $\delta \neq 0$, which corresponds to the case where one distribution has slipped relative to the other. This sort of test is called a **slippage test**; the most common example is the *Mann-Whitney test.

The matched-pair *Wilcoxon signed-rank test requires the distributions under comparison to be symmetric. The *Van der Waerden test is suitable for the comparison of *continuous distributions. For samples of size m and n, the test statistic (*see* HYPOTHESIS TEST) of the **Mood median test** is the number of observations in a sample that are greater than the *median of the combined set of *data. The **Savage test**, which is suited to comparing *skewed distributions, particularly the *exponential distribution, uses as its test statistic

$$\sum_{i=1}^{n}\sum_{j=1}^{R_i}(m+n-j+1)^{-1},$$

where $R_1, R_2, \cdots, R_n$ are the ranks of the sample of size n when the two samples are combined and jointly ordered. *See also* TEST FOR EQUALITY OF MEAN.

test for equality of mean When two *samples have been taken from *normal distributions that have the same (but unknown) *variance and possibly different means, then a t-test (*see* HYPOTHESIS TEST) is appropriate. In other cases a *non-parametric test may be used (*see* TEST FOR EQUALITY OF LOCATION). *See also* BEHRENS–FISHER PROBLEM; WELCH STATISTIC.

test for equality of scale A *non-parametric test of the *null hypothesis that *samples have been drawn from *populations with a common *distribution, with the alternative being that the distributions have the same *mean (or *median) but different scales (and thus different *variances).

The test statistic (*see* HYPOTHESIS TEST) of the 1953 **Rosenbaum test** is the number of *observations in one sample that exceed (or are less than) all the observations from the second sample.

The **Mood dispersion test**, introduced by *Mood in 1954, is suitable for symmetric populations. The test is based on replacing the original values with their *ranks in the combined sample. Suppose that the samples have sizes m and n, with $m+n=N$. The test statistic is M, given by

$$M=\sum_{j=1}^{m}\left\{r_j-\frac{1}{2}(N+1)\right\}^2,$$

where r_j is the overall rank in the combined sample of the jth observation in the sample of size m. If m and n are reasonably large and there are no tied ranks, then the transformed statistic z is an observation from an approximate standard *normal distribution, where

$$z=\left\{M-\frac{1}{12}m(N^2-1)\right\}\Big/\sqrt{\frac{1}{180}mn(N+1)(N^2-4)}.$$

The **Barton–David test** assumes a symmetric distribution, as does the **Levene test**, suggested by *Levene in 1960 for comparing several populations. Let the kth observation in the jth of m samples be denoted by y_{jk} and let the mean of the n_j observations in this sample be denoted by $\bar{y}_j$. Define z_{jk} by

$$z_{jk} = |y_{jk} - \bar{y}_j|,$$

with the mean of the z-values in the jth sample being denoted by $\bar{z}_j$ and the overall mean of the $n(=\sum_j n_j)$ z-values being $\bar{z}$. The Levene test statistic is W given by

$$W = \frac{n-m}{m-1}\left\{\sum_{j=1}^{m} n_j(\bar{z}_j - \bar{z})^2\right\} \Bigg/ \left\{\sum_{j=1}^{m}\sum_{k=1}^{n_j}(z_{jk} - \bar{z}_j)^2\right\}.$$

If the null hypothesis of equality of variance is correct, then the distribution of W is approximately an *F-distribution with $(m-1)$ and $(n-m)$ degrees of freedom.

A drawback of the Levene test is that the sample means may be affected by *outliers. The **Brown–Forsythe test**, suggested in 1974, avoids this problem by working with z'_{jk} instead of z_{jk}, where

$$z'_{jk} = |y_{jk} - M_j|,$$

and M_j is the median of the jth sample.

A related test is the **Fligner–Killeen test**, suggested in 1976. Let R_{jk} be the rank of z'_{jk} amongst the n ordered z'_{jk}-values and define a(j) by:

$$\mathrm{a}(j) = \Phi^{-1}\left(\frac{j}{2(n+1)} + \frac{1}{2}\right).$$

The test statistic, X^2, is given by:

$$X^2 = (n-1)\left\{\sum_{j=1}^{m} n_j(\bar{A}_j - \bar{a})^2\right\} \Bigg/ \left\{\sum_{j=1}^{n}\{\mathrm{a}(j) - \bar{a}\}^2\right\},$$

where

$$\bar{A}_j = \frac{1}{n_j}\sum_{k=1}^{n_j}\mathrm{a}(R_{jk}), \qquad \bar{a} = \frac{1}{n}\sum_{j=1}^{n}\mathrm{a}(j).$$

Under the null hypothesis of equality of variances, X^2 is an observation from a chi-squared distribution with $(m-1)$ degrees of freedom.

The **Siegel–Tukey test** introduced by *Siegel and *Tukey in 1960 uses alternatives to ranks that reflect the spread of the data. Denoting the N ordered values in the combined sample by $x_{(1)}, x_{(2)}, \ldots, x_{(N-1)}, x_{(N)}$, these replacement values are determined as follows:

$$x_{(1)} \to 1,\ x_{(N)} \to 2,\ x_{(N-1)} \to 3,\ x_{(2)} \to 4,\ x_{(3)} \to 5,\ x_{(N-2)} \to 6, \ldots.$$

The test statistic is the sum of these values corresponding to the observations in the smaller sample.

Similarly motivated, the **Ansari–Bradley test** introduced by Ansari and *Bradley in 1962 uses replacement values given by

$$x_{(1)} \to 1,\ x_{(N)} \to 1,\ x_{(N-1)} \to 2,\ x_{(2)} \to 2,\ x_{(3)} \to 3,\ x_{(N-2)} \to 3, \ldots,$$

while the 1962 **Klotz test** is based on the replacement of the value of the rth largest of the N observations by

$$\left\{\Phi^{-1}\left(\frac{r}{N+1}\right)\right\}^2.$$

Special tables (or exact calculation of tail probabilities) are required for most of these tests. *See also* TEST FOR EQUALITY OF VARIANCE.

test for equality of variance A test for equality of two or more *population variances. Suppose the *sample variance for the jth sample is s_j^2 based on v_j *degrees of freedom. In the case of *samples from two *populations having *normal distributions, the ***F*-test** compares the ratio

$$s_1^2/s_2^2$$

with the *critical values of an **F*-distribution with v_1 and v_2 *degrees of freedom. The F-test is encountered most frequently in the context of an analysis of variance table (*see* ANOVA), where it is often referred to as a **variance-ratio test.**

To test the hypothesis that m normal populations have the same variance, the **Bartlett test** (suggested by *Bartlett in 1937) has test statistic (*see* HYPOTHESIS TEST), B, defined by:

$$B = \left\{v \ln(s^2) + \sum_{j=1}^{m} v_j \ln(s_j^2)\right\} \Bigg/ \left\{1 + \frac{1}{3(m-1)}\left(\sum_{j=1}^{m}\frac{1}{v_j} - \frac{1}{v}\right)\right\},$$

where $v = v_1 + v_2 + \cdots + v_m$ and $s^2 = \frac{1}{v}\sum_{j=1}^{m} v_j s_j^2$. If the null hypothesis of equal variance is correct, then B has a *chi-squared distribution with $(m-1)$ degrees of freedom.

Alternatives in cases where the sample sizes are equal are the **Cochran *C* test**, introduced by *Cochran in 1941, and the **Hartley test** introduced by *Hartley in 1950. The Cochran test statistic, C, is given by

$$C = s_{\max}^2 \Bigg/ \sum_{j=1}^{m} s_j^2,$$

t

and the Hartley test statistic, H, is given by

$$H = s_{\max}^2 / s_{\min}^2,$$

where $s_{\max}^2$ and $s_{\min}^2$ are, respectively, the maximum and the minimum of $s_1^2, s_2^2, \ldots, s_m^2$. Unusually high values of C or H indicate unequal variances.

All of these tests are sensitive to departures from *normality. *See also* TEST FOR EQUALITY OF SCALE.

test for exponentiality A *test of whether a *sample may have been drawn from an *exponential distribution. Examples include adaptations of the *Anderson–Darling test and the *Shapiro–Wilk test.

test for independence *See* CHI-SQUARED TEST.

test for normality A test of the null hypothesis that a *sample has been drawn from a *normal distribution. An attractively simple test is that proposed by the Irish statistician Roy Geary in 1935. The **Geary test** compares the ratio of the mean deviation (*see* MEAN ABSOLUTE DEVIATION) divided by the *standard deviation with the theoretical ratio for a normal distribution, which is $\sqrt{2/\pi}$. An alternative is the **D'Agostino test** suggested by *D'Agostino in 1971. With ordered *observations $x_{(1)} \leq x_{(2)} \leq \cdots \leq x_{(n)}$, the test statistic (*see* HYPOTHESIS TEST) is D, given by

$$D = \frac{\sum_{j=1}^{n} \{j - \frac{1}{2}(n+1)\} x_{(j)}}{n\sqrt{n \sum_{j=1}^{n} (x_{(j)} - \bar{x})^2}},$$

where $\bar{x}$ is the *sample mean. Special tables are required.

Several more general tests of specified distributions have special cases for testing for normality. Examples include the *Anderson–Darling test and the *Shapiro–Wilk test. *See also* CRAMÉR–VON MISES TEST; KOLMOGOROV–SMIRNOV TEST.

test–retest reliability A measure of the trustworthiness of a test procedure. Usually measured by the *correlation between the values obtained in apparently identical tests of supposedly identical experimental units. For example, when individuals are presented with two different IQ tests, or with the same test applied at two points in time.

test statistic *See* HYPOTHESIS TEST.

tetrachoric correlation coefficient *See* TWO-BY-TWO TABLE.

The International Environmetrics Society (TIES) A society formed for the purpose of promoting statistical and other quantitative methods in the environmental sciences. It publishes the journal *Environmetrics*.

SEE WEB LINKS

• Society home page.

theory of errors *See* NORMAL DISTRIBUTION.

The Statistician *See* *JOURNAL OF THE ROYAL STATISTICAL SOCIETY*.

Thiele, Thorvald Nicolai (1838–1910; b. Copenhagen, Denmark; d. Copenhagen, Denmark) Danish astronomer, actuary, and mathematical statistician who developed the theory of *cumulants. Thiele studied astronomy at U Copenhagen (MSc, 1860; ScD, 1866). He was Professor of Astronomy at Copenhagen Observatory from 1875 to 1907, Rector of U Copenhagen from 1900 to 1906, and founder of the Danish Actuarial Society in 1901.

SEE WEB LINKS

• Photograph.

Thiessen, Alfred Henry (1872–1956; b. Troy, NY; d. Arlington, VA) American meteorologist. A graduate of Cornell U, he joined the US Weather Bureau at Pittsburgh in 1898. In 1900 he took part in the first use of the telegraph for reporting local weather. Thiessen was stationed on Cobb Island in the Potomac River. On receiving a telegraph command to report the local weather, he duly replied to base (just 1 km distant!) that it had just started snowing. He resigned from the Bureau in 1920 to join the US Army, retiring as a Major. He rejoined the Bureau for a year in 1941.

Thiessen polygon *See* DIRICHLET TESSELLATION.

three-sigma rule An empirical rule stating that, for many reasonably *symmetric *unimodal distributions, almost all of the *population lies within three *standard deviations of the *mean. For the *normal distribution about 99.7% of the population lies within three standard deviations of the mean. *See also* TWO-SIGMA RULE.

Thurstone, Louis Leon (1887–1955; b. Chicago, IL; d. Chapel Hill, NC) American psychometrician. Thurstone graduated from Cornell U with a BS in electrical engineering in 1912. This led to employment with Thomas Edison. Thurstone was interested in the way people learn, and this led to a PhD in psychology from U Chicago in 1917. He then joined the faculty at Carnegie Institute of Technology, moving in 1924 to a chair at U Chicago. In 1931 he introduced the method of *factor analysis. He was President of the American Psychological Association in 1933 and, in 1935, the first President of the *Psychometric Society.

SEE WEB LINKS

- Fuller biography and photograph.

tied ranks *See* RANKS.

TIES *Abbreviation for* THE INTERNATIONAL ENVIRONMETRICS SOCIETY.

tightened inspection *See* QUALITY CONTROL.

time domain *See* PERIODOGRAM.

time-homogeneous Markov chain *See* MARKOV PROCESS.

time-reversible stationary chain *See* MARKOV PROCESS.

time series A series of measurements over time, usually at regular intervals, of a *random variable. A prime concern is the *forecasting of future values using methods such as *exponential smoothing, *Holt–Winters forecasting, or *Box–Jenkins methods. Models fitted include *autoregressive models and *moving average models. It is often necessary to *deseasonalize the data and to remove any underlying trend (*see* MOVING AVERAGE) before undertaking the analysis. *See also* PERIODOGRAM.

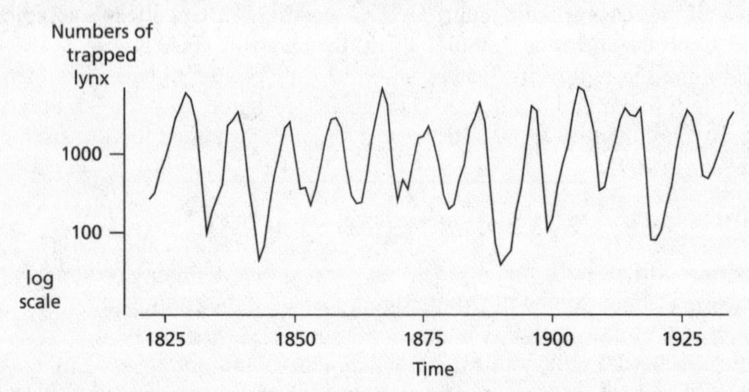

Time series. The famous series illustrated shows the variation in the numbers of lynx trapped in the neighbourhood of the Mackenzie River, in Canada, for the years 1821 to 1934. The numbers of lynx vary according to the amount of food available. A major component of this food is the snowshoe hare, which shows a similar cycle: a rise in the lynx population leads to a fall in the hare population, which leads to a fall in the lynx population, which leads to a rise in the hare population, and so on.

tobit analysis *See* CENSORED REGRESSION MODELS.

Toeplitz matrix *See* MATRIX.

total probability law *See* ADDITION LAW FOR PROBABILITIES.

total quality management (TQM) The monitoring of quality levels throughout a process, often through the use of simple graphs such as *Pareto plots, control charts (*see* QUALITY CONTROL), *scatter diagrams, and *histograms.

total sum of squares *See* ANOVA.

TQM *Abbreviation for* TOTAL QUALITY MANAGEMENT.

trace *See* MATRIX.

traffic intensity A measure used in describing the behaviour of a *queue. The traffic intensity, ρ, is defined by

$$\rho = \frac{\text{Mean rate of arrival at the queue}}{\text{Mean rate of service}} = \frac{\text{Mean service time}}{\text{Mean inter-arrival time}}.$$

training sample *See* NEURAL NET.

training set *See* DISCRIMINANT ANALYSIS.

transect In the context of *sampling natural *populations, a line drawn across the region of interest. Sampling may consist of examining the occurrence of organisms along the line, or visible from the line, or within a sequence of *quadrats centred on the line.

transformation A function of a *random variable (or of a *statistic). Examples are $\sqrt{x}$ or $\ln X$, where X is the variable. Using the transformed variable may, for example, simplify a *model or *stabilize the variance.

transient state *See* MARKOV PROCESS.

transition matrix A square *matrix in which the rows and columns correspond to categories defined in equivalent ways. Usually the row categories refer to one time period and the column categories to a subsequent time period. The entries may be *frequencies, *probabilities, or *conditional probabilities. *See also* MARKOV PROCESS.

transition probability *See* MARKOV PROCESS.

transportation problem A *linear programming problem concerned with identifying the cheapest method for moving commodities (or people) from one set of locations to another. At locations $L_1, L_2, \dots, L_l$ specified amounts of a commodity are available. There are demands for specified amounts of the commodity at locations $M_1, M_2, \dots, M_m$. The cost of transportation from L_j to M_k is known for all j and k. The problem is to satisfy the demands at minimum transportation cost.

t

transpose *See* MATRIX.

trapezium rule A method of finding an approximate value for an integral, based on finding the sum of the areas of trapezia. Suppose we wish to find an approximate value for $\int_a^b f(x)dx$. The interval $a \leq x \leq b$ is divided up into n sub-intervals, each of length $h = (b - a)/n$, and the integral is approximated by

$$\frac{1}{2}h(y_0 + 2y_1 + 2y_2 + \cdots + 2y_{n-1} + y_n),$$

where $y_r = f(a + rh)$. This is the sum of the areas of the individual trapezia, one of which is shown in the diagram. The error in using the trapezium rule is approximately proportional to $1/n^2$, so that if the number of sub-intervals is doubled, the error is reduced by a factor of 4. A more accurate method for approximating an integral is *Simpson's rule.

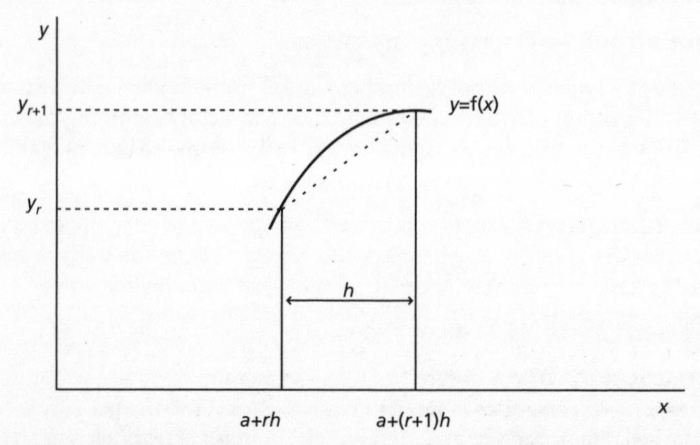

Trapezium rule. A method of approximate integration. A slight underestimate (as shown) will often be cancelled by a similar slight overestimate from another trapezium. Using narrower intervals will improve accuracy.

trapping state *See* MARKOV PROCESS.

travelling salesman problem A *network problem that can be formulated as a *combinatorial optimization problem. A salesman has to visit a number of interconnected destinations. The problem is to determine the route that minimizes the total distance travelled.

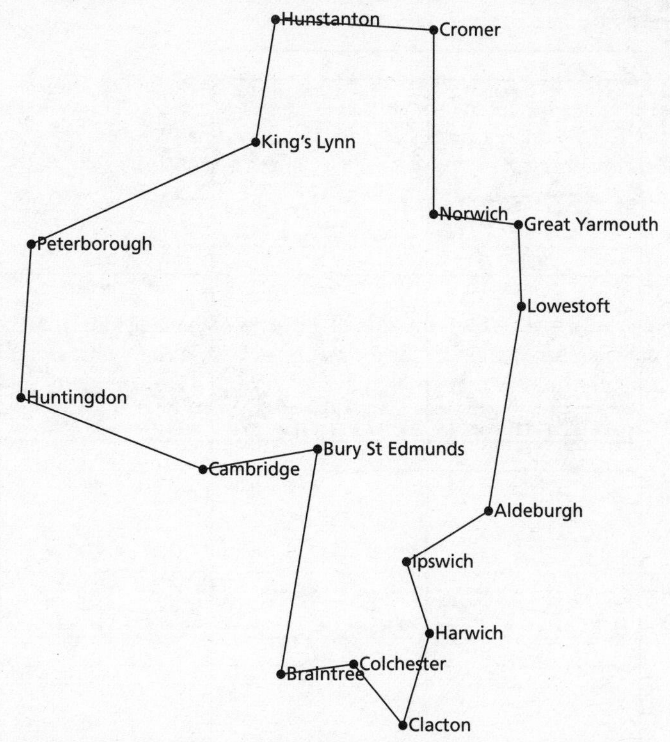

Travelling salesman problem. The route illustrated would be appropriate for a travelling salesman in East Anglia, England.

treatment A term used in the context of an *experimental design to refer to any prescribed combination of values of explanatory variables (*see* *regression). In the original agricultural context a treatment was, for example, the application of a particular fertilizer. The term is particularly used in the context of *balanced incomplete blocks, *Latin squares, and *randomized blocks. *See also* FRIEDMAN TEST.

trellis plot An array of plots, all having the same format, with each member of the array referring to a distinct subset of the entire data. *See diagram overleaf.*

trend; trend line *See* MOVING AVERAGE.

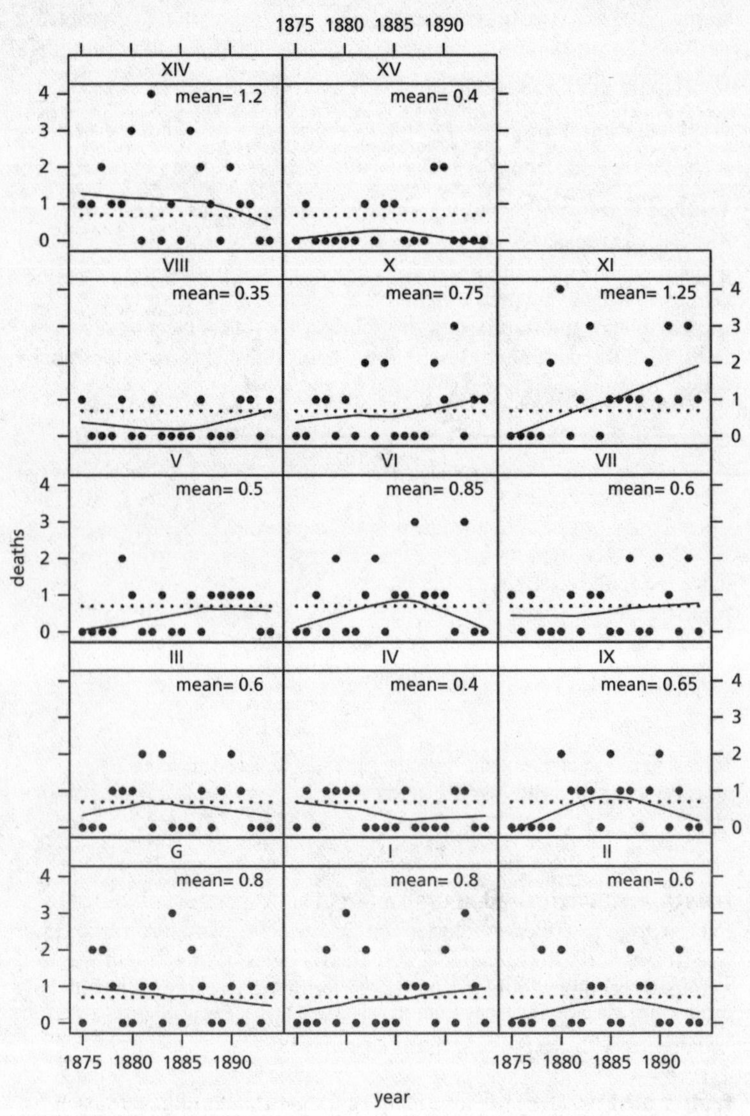

Trellis plot. The plot illustrates the data on *Prussian horse kicks. Each panel refers to a separate corps. The overall yearly mean is shown as a dotted line, with separate trends shown for each corps using *loess.

trend surface analysis Essentially multiple regression (*see* MULTIPLE REGRESSION MODEL) using physical location as the explanatory variable(s) and with random errors having non-zero *autocorrelation.

trial A statistical experiment, often leading to one of two outcomes: success or failure.

Trial of the Pyx A *sampling procedure conducted by the English Royal Mint. The ceremony dates back to the thirteenth century. Typically, 100 golden guineas were randomly sampled and placed in a ceremonial box (the Pyx), which was then weighed. The Master of the Mint was responsible for the standard of the coinage. He was subject to severe penalties if the combined weight of the coins differed from its nominal weight by more than one part in 400.

triangular distribution If X and Y are *independent *random variables each having the same *uniform distribution then $(X+Y)$ has a triangular distribution. In the case of a *continuous random variable the graph of the *probability density function is an isosceles triangle. In the case of a *discrete random variable the graph of the *probability function has a triangular shape. The diagram shows this for the case of throwing two ordinary dice. The most probable score is 7 (which results in a 'Chance' card when starting from 'Go' on a Monopoly board).

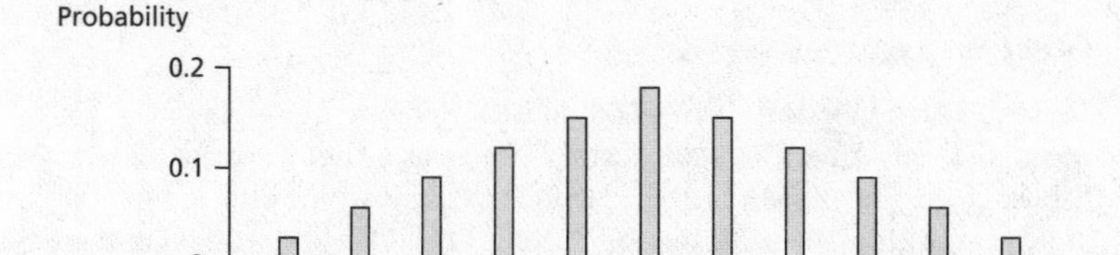

Triangular distribution. This is a graph of the distribution of the sum of the scores on two fair six-slide dice. The mode (corresponding to a probability of $\frac{6}{36}$) occurs when the sum is 7.

trimean A *robust *measure of location equal to $\frac{1}{4}(Q_1+2Q_2+Q_3)$, where Q_2 is the *median of the data and Q_1 and Q_3 are, respectively, the lower and upper *quartiles. It is an example of a *L-estimate.

trimmed mean A *robust estimate of the *population mean. For a *sample of size n, with *data ordered so that $x_{(1)} \leq x_{(2)} \leq \cdots \leq x_{(n)}$, the trimmed mean, $\bar{x}_k$, is the mean of the data if the k smallest values and the k largest values are discarded (with $1 \leq k < \frac{1}{2}n$). So

$$\bar{x}_k = \frac{1}{n-2k} \sum_{j=k+1}^{n-k} x_{(j)}.$$

The idea is to avoid the influence of extreme observations. The method was proposed by *Tukey in 1962.

Tukey also proposed the **Winsorized mean**, named after his former colleague Charles P. Winsor, who had died in 1951. In this case the k smallest values are replaced by $x_{(k)}$ and the k largest are replaced by $x_{(n-k+1)}$ to give the estimate w_k:

$$w_k = \frac{1}{n}\left(kx_{(k)} + \sum_{j=k+1}^{n-k} x_{(j)} + kx_{(n-k+1)}\right).$$

true-positive rate *See* TWO-BY-TWO TABLE.

truncated distribution A *distribution in which values less than some threshold, or more than some other threshold, or both, are not recorded. Another common situation is a distribution in which values of 0 are not reported. Compare CENSORED DATA.

T^2 *See* HOTELLING T^2.

***t*-test** *See* HYPOTHESIS TEST.

Tukey, John Wilder (1915–2000; b. New Bedford, MA; d. New Brunswick, NJ) American statistician. Tukey was a chemistry graduate at Brown U, gaining his MA in 1937. He followed this with a PhD in topology at Princeton U in 1939. During the Second World War he worked in the Fire Control Research Office alongside *Wilks and *Cochran. After the war he joined Wilks at Princeton U becoming a full Professor at the age of 35. His 55 research students included *Brillinger, *Dempster, *Goodman, *Meier, and *Mosteller. In 1946 he coined the word 'bit' as a shorthand for a 'binary digit' as used by computers, and, in 1958, the word 'software' to describe the programs used by computers. In 1962 he introduced the *trimmed mean as one of a number of *robust *summary statistics. In 1965, with John Cooley, he introduced the *fast Fourier transform. His 1970 book *Exploratory Data Analysis* introduced the *stem and leaf diagram and the *boxplot. He was elected to membership of the NAS in 1961 and was awarded the National Medal of Science in 1973. He was President of the *IMS in 1960 having been its *Wald Lecturer in 1958.

He was presented with the *Wilks Award of the *ASA in 1965 and the *Shewhart Medal of the ASQ in 1976. He was made an Honorary Fellow of the *RSS in 1986. Tukey once remarked that 'The best thing about being a statistician is that you get to play in everyone's backyard'.

SEE WEB LINKS

- Fuller biography, interview, and photographs.

Tukey *b*-test *See* MULTIPLE COMPARISON TEST.

Tukey depth *See* DATA DEPTH.

Tukey–Kramer test *See* MULTIPLE COMPARISON TEST.

Tukey pocket test (Tukey quick test) A simple slippage *test (*see* TEST FOR EQUALITY OF LOCATION) of the hypothesis that two *samples have been drawn from the same *distribution.

Suppose the largest observation is in sample 1. Tukey's test statistic, T, is calculated as follows:

1. If the smallest observation is also in sample 1, $T=0$.
2. Otherwise T is the sum of the number of observations in sample 1 that are larger than the largest observation in sample 2 and the number of observations in sample 2 that are smaller than the smallest observation in sample 1.

Not only is the statistic easy to calculate, but its distribution is remarkably unaffected by the sample sizes. As a guide, T is significant at the 5% level if $T > 7$. The corresponding values for the 1% and 0.1% levels are 9 and 13, respectively. *See* TEST FOR EQUALITY OF DISTRIBUTION.

Tukey's mean-difference plot. *See* BLAND-ALTMAN PLOT.

Tukey test *See* MULTIPLE COMPARISON TEST.

Tukey wholly significant difference test *See* MULTIPLE COMPARISON TEST.

Tukey window *See* PERIODOGRAM.

tuning constant A quantity whose value can be adjusted by the user so as to give a desirable outcome. A simple example is provided by the choice of the number of points over which to calculate a *moving average to give a *time series of the desired smoothness. *See also* *M*-ESTIMATE.

Turing, Alan Mathison (1912–54; b. London, England; d. Wilmslow, England) English mathematician. Turing graduated from Cambridge U in 1934. The following year he was elected to a Fellowship at King's College, Cambridge as a consequence of his novel proof of the *central limit theorem. In 1936 he introduced the concept of a *Turing machine. In 1938 he obtained his PhD from Princeton U. During the Second World War he was involved in the breaking of the German codes produced by the Enigma machine. After the war he held posts at Cambridge U and Manchester U. He was elected FRS in 1951.

SEE WEB LINKS

- Dedicated website.

Turing machine A theoretical model of a computer, in which the machine functions in a sequence of discrete operations. The machine can be in only one of a finite list of internal states at any given moment. It consists of an infinite tape carrying symbols, which represent instructions, and a mechanism that can move the tape and read from, or write to, the tape. The mechanism can also change the internal state of the machine in accordance with instructions read from the tape.

two-by-two table (fourfold table) A *two-way table with two rows and two columns. Let *a*, *b*, *c*, and *d* denote the *frequencies of the four (2 × 2) possible outcomes, with *m*, *n*, *r*, *s*, and *N* the totals of these frequencies as indicated in the table

a	b	m
c	d	n
r	s	N

A question of interest is whether the classifying variables are *independent. This is best tested using the **Fisher exact test** (introduced by Sir Ronald *Fisher in 1935). The test involves the use of the *hypergeometric distribution to calculate the *probability of the observed outcome, given the observed values of the *marginal totals *m*, *n*, *r*, and *s*, under the *null hypothesis of independence. The probability of the observed outcome is given by

$$\frac{m!n!r!s!}{a!b!c!d!N!}.$$

As an example, suppose that five out of six patients treated with drug A recover, whereas only three out of five patients treated with drug B recover. The null hypothesis is that the outcome was independent of the drug used. The possible outcomes (and their probabilities) are as follows:

t

Actual Outcome

	R	$\bar{R}$	
A	6	0	6
B	2	3	5
	8	3	11

$p = 2/33$

	R	$\bar{R}$	
A	5	1	6
B	3	2	5
	8	3	11

$p = 12/33$

	R	$\bar{R}$	
A	4	2	6
B	4	1	5
	8	3	11

$p = 15/33$

	R	$\bar{R}$	
A	3	3	6
B	5	0	5
	8	3	11

$p = 4/33$

where R denotes recovery and $\bar{\mathrm{R}}$ denotes non-recovery. Given the fixed marginal totals (6, 5, 8, 3), the probability of the observed outcome, or one in which drug A is more successful, is $\frac{12}{33} + \frac{2}{33} = \frac{14}{33}$. Similarly, the probability of the observed outcome, or one in which drug B is more successful, is $\frac{12}{33} + \frac{15}{33} + \frac{4}{33} = \frac{31}{33}$. Since neither $\frac{14}{33}$ nor $\frac{31}{33}$ is unusually small, the null hypothesis of independence is accepted.

The exact test is now routinely included in statistical packages, since the underlying theory is not restricted to the two-by-two table. However, for the two-by-two table, when the cell frequencies are large, a useful alternative is the **Yates-corrected chi-squared test** for which the test statistic is

$$X_c^2 = \frac{N\left(|ad - bc| - \frac{1}{2}N\right)^2}{mnrs},$$

where the $\frac{1}{2}N$ (which is in fact a *continuity correction) is called the **Yates correction**. The value of X_c^2 should be compared with a *chi-squared distribution with one *degree of freedom.

In some cases it may be reasonable to suppose that the classifying variables have a bivariate normal distribution (*see* MULTIVARIATE NORMAL DISTRIBUTION) but have been reported with respect to some cut-off values of interest (e.g. '< 1'), with the true values being unreported. A question of interest is the value of the *population correlation coefficient, ρ. One approximate estimate of ρ is provided by **Yule's *Q*** (suggested by *Yule in 1900), which is given by

$$Q = (ad - bc)/(ad + bc).$$

If the null hypothesis of independence between the classifying variables is correct and if the cell frequencies (*see* CONTINGENCY TABLE) are not too small, Q has an approximate *normal distribution with *mean 0 and *variance estimated by

$$\frac{1}{4}(1-Q^2)^2\left(\frac{1}{a}+\frac{1}{b}+\frac{1}{c}+\frac{1}{d}\right).$$

A better approximation to ρ, proposed by Karl *Pearson in 1901, is the **tetrachoric correlation coefficient**:

$$\sin\left\{\frac{\pi}{2}\left(\frac{\sqrt{ad}-\sqrt{bc}}{\sqrt{ad}+\sqrt{bc}}\right)\right\}.$$

If there are more than two rows or columns then the corresponding statistic is called the **polychoric correlation coefficient**.

An alternative assessment of association is provided by the **odds ratio**:

$$(a/b)/(c/d)=(a/c)/(b/d)=\frac{ad}{bc}.$$

This is the ratio of the *odds on something occurring in one situation to the odds of the same event occurring under a second situation. An odds ratio of 1 implies that the odds on an event occurring (and hence the probability of its occurrence) are unaffected by the change in situation: they are independent of the situation. The *interaction *parameters of *log-linear models can be interpreted in terms of odds ratios (or ratios of odds ratios). If any frequencies are zero then the **Haldane estimator** (proposed by *Haldane in 1955),

$$\frac{\left(a+\frac{1}{2}\right)\left(d+\frac{1}{2}\right)}{\left(b+\frac{1}{2}\right)\left(c+\frac{1}{2}\right)},$$

is more useful. An alternative is the **Jewell estimator**:

$$\frac{ad}{(b+1)(c+1)}.$$

Two-by-two tables are often used in a medical context. Typically, the rows of the table might refer to two different medicines and the columns to their success or failure. The question is whether there is a difference between the medicines. The **relative risk** is the ratio

$$\frac{a}{m}\Big/\frac{c}{n}.$$

A related context is that of screening patients for diseases. Each patient in a *sample, who may or may not have a disease, is tested for that disease. The test gives either a positive result or a negative result. The four possible outcomes are shown in the table, together with the corresponding frequencies a, b, c, and d:

		Disease	
		yes	no
Test	positive	a	b
	negative	c	d

Sensitivity is the *conditional probability of the test correctly giving a positive result, given that the patient does have the disease. An estimate is $a/(a+c)$. **Specificity** is the conditional probability of the test correctly giving a negative result, given that the patient does not have the disease. An estimate is $d/(b+d)$. **Youden's index**, J, proposed by *Youden in 1950, is given by

$$J = \text{sensitivity} + \text{specificity} - 1.$$

This provides a single summary measure of the efficiency of the screening procedure.

Sensitivity may be referred to as the **true-positive rate**, with (1 – specificity) being called the **false-positive rate** or the **false alarm rate**. Often the test involves calculation of a score and assessing whether the score exceeds a critical value. In that case the sensitivity and specificity of the test will depend on that critical value. As an example, consider a classification procedure that states that patients with scores greater than k (for some k) are diseased, and the remainder are normal. The scores, together with information on the true states of the patients is given in the following table:

Score	1–10	11–20	21–30	31–40	41–50	51–60
Normal	50	42	30	25	20	15
Diseased	2	3	10	11	18	25

Thus, with $k=30$, the procedure correctly classifies 54 diseased patients (sensitivity $=54/69$) with a specificity of 122/182.

A plot (*see diagram overleaf*), as the critical value changes, of the false-positive rate (on the x-axis) against sensitivity (on the y-axis) is known as an **ROC curve** (the acronym is of **receiver operating characteristic curve**, and has its roots in electrical engineering).

An alternative to the ROC curve is the **detection error tradeoff graph (DET graph)** which makes differences between alternative classifiers more apparent. The graph makes use of *probits and plots the false-positive rate against the false-negative rate. Denoting the *cumulative distribution function of a normal distribution by Φ, $\Phi^{-1}[b/(b+d)]$ is plotted on the x-axis against $\Phi^{-1}[c/(a+c)]$ on the y-axis (*see diagram opposite*).

In the context of significance tests the previous table could be labelled as shown:

		Null hypothesis (H_0) true	Null hypothesis (H_0) false
Result of test	H_0 accepted	a	b
	H_0 rejected	c	d

In this context the *expected value of the ratio $c/(c+d)$ is called the **false discovery rate (FDR)**.

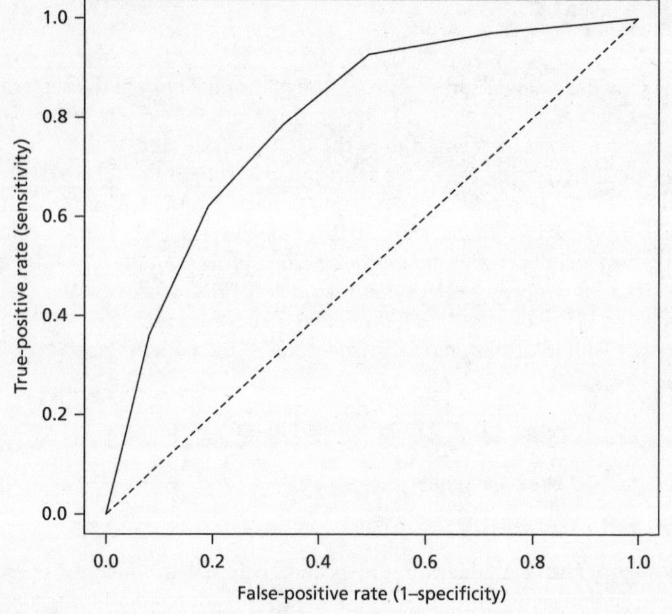

ROC curve. A plot of the false-positive rate against the true-positive rate for the tabulated data. In general, the greater the area below the ROC curve, the better the classification.

two-sample test *See* HYPOTHESIS TEST.

two-sided hypothesis *See* HYPOTHESIS TEST.

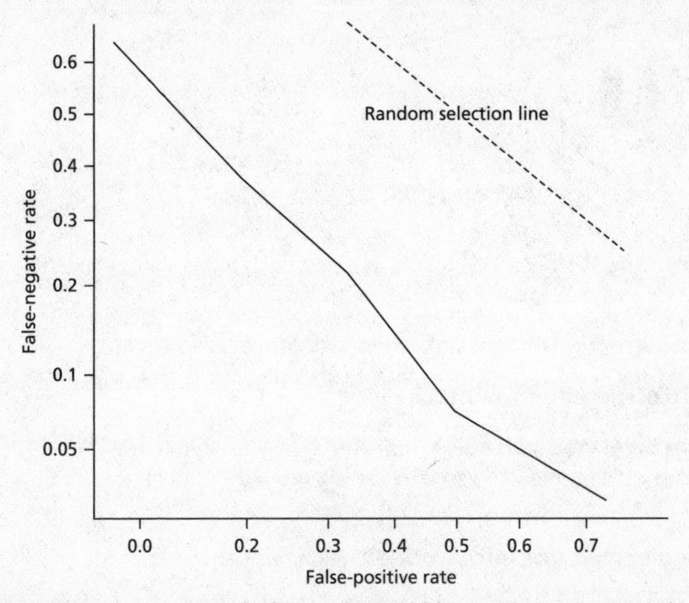

DET graph. A plot comparing the false-positive and false-negative rates, using probit transformations. An ROC curve is replaced by a near-straight line, making differences between alternative classifiers more apparent. The dotted line corresponds to classification at random.

two-sigma rule An empirical rule stating that, for many reasonably symmetric *unimodal distributions, approximately 95% of the *population lies within two *standard deviations of the *mean. *See also* THREE-SIGMA RULE.

two-stage least squares *See* MULTIPLE REGRESSION MODEL.

two-tailed test *See* HYPOTHESIS TEST.

two-tail probability *See* ONE-TAIL PROBABILITY.

two-way table A table with r rows and c columns in which the entry in cell (i, j) represents either the *frequency for that outcome (in the context of a *contingency table for *categorical variables) or a value resulting from that row and column combination in the context of *ANOVA. Such a table might be called an r-by-c table; if $r = c = 2$ then it would be called a *two-by-two table.

Type I error; Type II error *See* HYPOTHESIS TEST.

unbiased Fair. For example, a six-sided die is unbiased if, when thrown, it is equally likely to show any of its six sides.

unbiased estimate; unbiased estimator *See* ESTIMATOR.

uncorrected moment *See* MOMENT.

uncorrelated variables Variables displaying zero *correlation. *Independent random variables are uncorrelated, though the converse is not always true.

undirected graphical model *See* GRAPHICAL MODEL.

uniform association model A *model for a *contingency table that proposes that the odds ratio [*see* TWO-BY-TWO TABLE) for the categories in every component *two-by-two table comprising adjacent cells should be the same. Thus, if $p_{j,k}$ denotes the *probability of an individual belonging to cell (j, k) then the model is

$$\frac{p_{j,k}\ p_{j+1,k+1}}{p_{j,k+1}\ p_{j+1,k}} = c,$$

where c is constant. The case $c = 1$ corresponds to *independence.

uniform distribution, continuous (rectangular distribution) The *continuous distribution on the interval (a, b) with *probability density function f given by

$$f(x) = \frac{1}{b-a}, \qquad a < x < b,$$

where a and b are constants. The distribution has *mean $\frac{1}{2}(a+b)$ and *variance $\frac{1}{12}(b-a)^2$.

The error when a decimal number is rounded to the nearest whole number has a uniform distribution on $(-0.5, 0.5)$, with mean 0 and variance $\frac{1}{12}$.

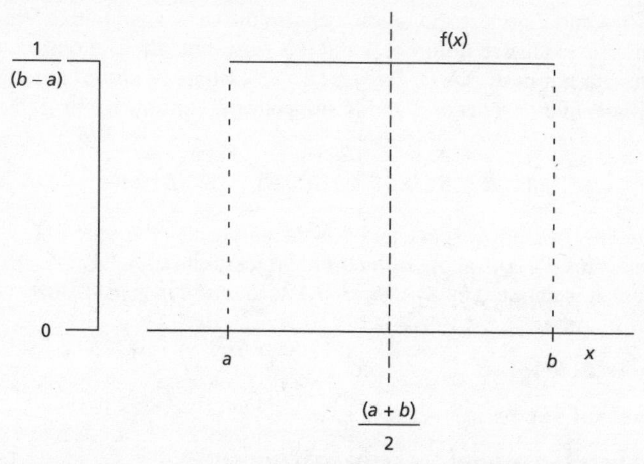

Uniform distribution, continuous. The density function is constant for all values of the random variable in the given range, which implies that, in this range, all intervals of the same width are equally probable.

uniform distribution, discrete A *distribution in which a *discrete random variable, X, say, can take a discrete set of possible values $x_1, x_2, \ldots, x_n$, with equal *probabilities, so that

$$P(X = x_j) = \frac{1}{n}, \qquad j = 1, 2, \ldots, n.$$

Often the set of possible values is the set of positive integers 1, 2, . . . , n, and in this case the *mean is $\frac{1}{2}(n+1)$ and the *variance is $\frac{1}{2}(n^2 - 1)$.

As an example, the score on a single throw of an *unbiased die has a uniform discrete distribution on 1, 2, 3, 4, 5, 6, with mean 7/2 and variance 35/12.

uniformly distributed Having a *uniform distribution.

uniformly most powerful test A test of a *null hypothesis which has power (*see* HYPOTHESIS TEST) which is at least as great as that of any alternative test for all values of the *parameter under test. The concept was introduced by Sir Ronald *Fisher in 1934.

unimodal An adjective used to describe a *distribution or set of *data having a single *mode. If there is more than one mode then the distribution is multimodal (*see* MODE).

union The union of two events (*see* SAMPLE SPACE) A and B is the event 'either A or B occurs', denoted by $A \cup B$. It should be noted that the 'or' is inclusive and therefore the union includes the case when both events occur. The **exclusive union** of A and B is the event 'either A occurs or B occurs but not both', i.e. $(A \cap B') \cup (A' \cap B)$, where A' and B' are the complementary events to A and B, respectively. For any events A, B, C,

$$A \cup A = A, \qquad A \cup S = S, \quad A \cup \phi = A,$$
$$A \cup B = B \cup A, \quad A \cup (B \cup C) = (A \cup B) \cup C,$$

where S is the sample space and ϕ is the empty set. The union of the n events $A_1, A_2, \ldots, A_n$ is the event 'at least one of $A_1, A_2, \ldots, A_n$ occurs'. It is denoted by $A_1 \cup A_2 \cup \cdots \cup A_n$. *See also* INTERSECTION; VENN DIAGRAM.

universal kriging *See* KRIGING.

universal set *See* SAMPLE SPACE.

unsaturated model *See* SATURATED MODEL.

unsupervised learning *See* MACHINE LEARNING.

upper percentage point *See* PERCENTAGE POINT.

upper quartile *See* QUARTILE.

upper-tail probability The *probability that the *random variable X takes values $\geq x$, where, usually, x is appreciably greater than the *median value of X.

upper triangular matrix *See* MATRIX.

urn model A model of a problem described in terms of balls being drawn from an urn. An example of a problem where a *hypergeometric distribution is appropriate is as follows. Three people are to be chosen, at random, from a group of twenty. In the group of twenty are four named Smith. The event of interest is that all those chosen are named Smith. The corresponding urn model has an urn containing four black balls and sixteen white balls. Three balls are to be selected at random and without replacement. The event of interest is that all are black.

A trivial example of a problem where a *binomial distribution is appropriate concerns the chance of getting three heads when tossing a fair coin four times. The corresponding urn model involves an urn containing one black ball and one white ball. A ball is selected at random and replaced on four occasions. The event of interest is that a black ball is chosen on three occasions.

The phrases **without replacement** and **with replacement** can best be understood by imagining that the balls are selected one at a time. In the first case, after each selection the ball is placed on one side, so that the number in the urn has reduced by one. In the second case the ball is replaced in the urn, so that before each draw the urn contains the same mixture of balls as before.

utility A function that takes a numerical value for each possible state of a system (usually an economic system) and is intended as a measure of the benefit or usefulness of that state.

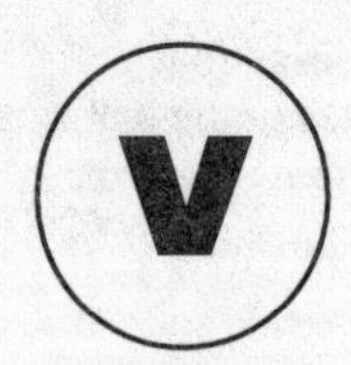

van Dantzig, David (1900–59; b. Rotterdam, Netherlands; d. Amsterdam, Netherlands) Dutch mathematician. From a poor home, van Dantzig worked for several years before entering U Amsterdam in 1923, where his interest was algebraic topology. In 1927 he joined the faculty at Delft Technical U, leaving in 1940 following the German invasion. In 1946, on his appointment at U Amsterdam, he changed his field of interest to mathematical statistics. The *Netherlands Society for Statistics and Operations Research awards a prize in his honour every five years.

SEE WEB LINKS

- Fuller biography.

van der Waerden, Bartel Leendert (1903–96; b. Amsterdam, Netherlands; d. Zürich, Switzerland) Dutch mathematician. Van der Waerden studied at U Amsterdam and U Göttingen, joining the faculty at Göttingen in 1928. In 1931 he was appointed Professor of Mathematics at U Leipzig. After the Second World War he briefly held posts at Johns Hopkins U and U Amsterdam, before moving in 1951 to Zürich U. His *non-parametric test for the equality of two *populations was published in 1952.

SEE WEB LINKS

- Fuller biography, interview, and photographs.

van der Waerden test A *non-parametric test of the null hypothesis that two *populations have the same distribution, in the case where the *distributions are *continuous. Two *independent random samples x_1, $x_2, \ldots, x_m$ and $y_1, y_2, \ldots, y_n$ are drawn and the test statistic is W, given by

$$W = \sum_{j=1}^{m} \Phi^{-1}\left(\frac{r_j}{m+n+1}\right),$$

where r_j is the *rank of x_j when the $(m+n)$ *observations are arranged in increasing order of size, and Φ is the *cumulative distribution function of the standard *normal distribution. The restriction to continuous variables ensures that there are no *ties. The test, proposed by *van der Waerden in 1952, requires special tables.

Varadhan, Sathamangalam Ranga Srinivasa 'Raghu' (1940– ; b. Madras (Chennai), India) Indian probabilist whose career has been in the USA. Varadhan obtained his BSc (in 1959 with record marks) and MSc (1960) from Madras U. His PhD supervisor at the Indian Statistical Institute was *Rao. In 1963 Varadhan joined the Courant Institute of Mathematical Sciences in New York, where he has remained. He is a member of the AAAS (1988), the NAS (1995) and was elected FRS in 1998. He was President of the *IMS in 2003, its *Rietz Lecturer in 1987 and its *Wald Lecturer in 2005. He was awarded the 2007 Abel Prize (the mathematics equivalent of a Nobel Prize).

SEE WEB LINKS

- Fuller biography.

variable The characteristic measured or observed when an experiment is carried out or an observation is made. Variables may be non-numerical (*see* CATEGORICAL VARIABLE) or numerical. Since a non-numerical observation can always be coded numerically, a variable is usually taken to be numerical. Statistics is concerned with *random variables and with variables whose measurement may involve *random errors.

variance The word 'variance' is used as a shorthand for either the *population variance or the *sample variance, depending on context. It is the square of the *standard deviation.

variance components model *See* COMPONENTS OF VARIANCE MODEL.

variance–covariance matrix (dispersion matrix) A square symmetric *matrix in which the elements on the main diagonal are *variances and the remaining elements are *covariances. Suppose $X_1, X_2, \ldots, X_p$ are *random variables and the variance of X_j is σ_j^2 and the covariance of X_j and X_k is $c_{jk} = c_{kj}$. Then the variance–covariance matrix is $\boldsymbol{\Sigma}$, given by

$$\boldsymbol{\Sigma} = \begin{pmatrix} \sigma_1^2 & c_{12} & \cdots & c_{1p} \\ c_{21} & \sigma_2^2 & \cdots & c_{2p} \\ \vdots & \vdots & \ddots & \vdots \\ c_{p1} & c_{p2} & \cdots & \sigma_p^2 \end{pmatrix}.$$

The same term is also used for the corresponding matrix based on sample values. If the matrix is diagonal then the variables display **orthogonality**.

variance-ratio test *See* TEST FOR EQUALITY OF VARIANCE.

variance reduction techniques Methods for reducing the number of observations required in a *simulation by efficient use of *pseudo-random numbers to reduce the *variance of the simulation estimates. Examples include the use of *antithetic variables and *stratified sampling.

variate A *random variable. A term used by Karl *Pearson in 1909.

varimax rotation A method used in *factor analysis. The aim is to rotate the *vector of factors so as to find a few key combinations that will simplify the analysis.

variogram *See* AUTOCORRELATION.

vector A *matrix with only one column (a column vector) or only one row (a row vector).

Venn, John (1834–1923; b. Hull, England; d. Cambridge, England) English logician. Following his graduation in mathematics from Cambridge U, Venn was elected to a fellowship of Gonville and Caius College, where a stained glass window now stands in his memory. In 1859 he was ordained as a priest and spent a year as a curate at Mortlake in Surrey. In 1862 he returned to Cambridge U as a lecturer specializing in logic. Venn had a general interest in all branches of Statistics, and a letter that he wrote in 1887 to the Editor of *Nature* stimulated an explosion of interest in the mathematical theory of Statistics. His interests were not confined to mathematics. In 1909 he constructed a bowling machine that was used by the visiting Australian cricket team. In retirement he compiled the first three volumes of a history of Cambridge U. He was elected FRS in 1883.

SEE WEB LINKS

- Fuller biography and portrait.

Venn diagram A simple diagram (*see opposite*) used to represent *unions and *intersections of sets. The diagram, described by *Venn in 1880 and popularized by his 1881 book *Symbolic Logic*, had been introduced by *Leibniz in the eighteenth century.

V

SEE WEB LINKS

- Applet.

violin plot A plot that combines the *kernel method with a type of *boxplot in a single diagram that presents the information provided by the two approaches. *See diagram opposite.*

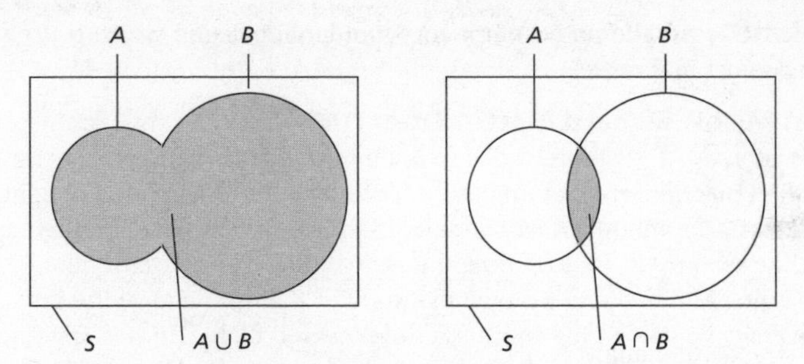

Venn diagram. These two diagrams illustrate the definitions of the union and intersection of the events A and B in the sample space S.

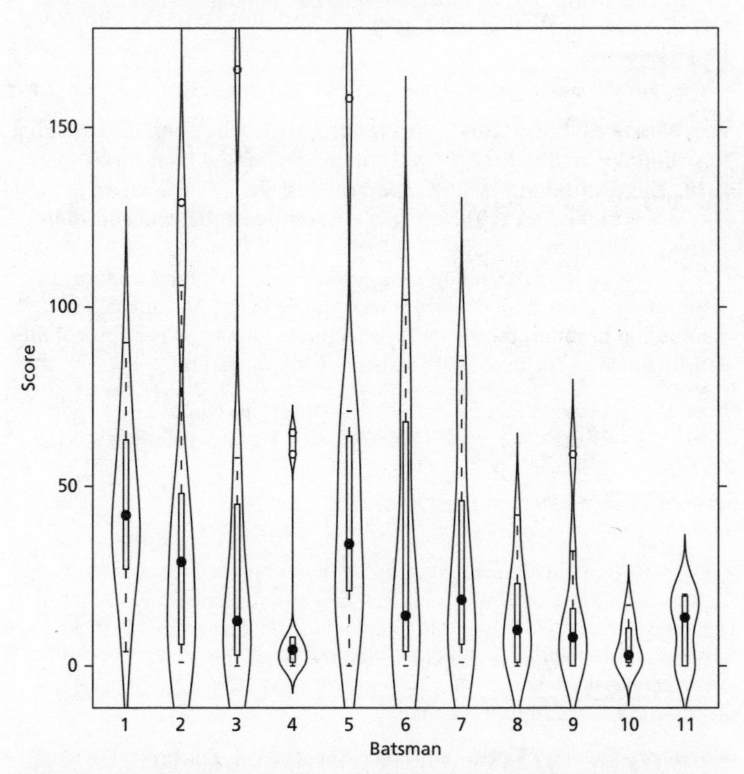

Violin plot. The plot refers to the scores obtained by the successful England cricket team in the 2005 Ashes series against Australia. The strength of the middle order compensated for the consistently poor performances of the number 4 batsman. The dots indicate the *median scores obtained.

volatility An alternative name for *standard deviation which is used in the context of finance.

von Mises, Richard Martin Edler (1883–1953; b. Lviv, Ukraine; d. Boston, MA) Austrian Jewish mathematician and engineer. Having studied machine engineering at the Technical U of Vienna (gaining his PhD in 1907), von Mises was appointed Professor of Applied Mathematics at U Strasbourg in 1909. During the First World War he was a pilot, designing and building his own plane. After the war he gave the first university course on the mechanics of powered flight. He was appointed Director of the Institute of Applied Mathematics in Berlin in 1920. However, in 1933, he left Hitler's Germany for Turkey. At the outbreak of the Second World War he moved to the USA, joining the faculty first at MIT and then, in 1943, at Harvard U.

SEE WEB LINKS

- Fuller biography and photograph.

von Mises distribution (circular normal distribution) The principal *distribution used to model *cyclic data; derived by *von Mises in 1918. The distribution has two *parameters: the circular mean μ, $(-\pi < \mu \leq \pi)$, and κ (≥ 0), which is a measure of the concentration of the distribution.

If $\kappa = 0$ then the distribution degenerates to the *circular uniform distribution in which all directions are equally likely. As κ increases, the distribution becomes increasingly concentrated about μ. The *probability density function f is given (with directions in radians) by

$$f(\theta) = \frac{1}{2\pi I_0(\kappa)} \exp\{\kappa \cos(\theta - \mu)\}, \quad -\pi < \theta \leq \pi,$$

where $I_0(\kappa)$ is a modified Bessel function given by

$$I_0(\kappa) = \sum_{s=0}^{\infty} \left(\frac{\kappa^s}{2^s s!} \right)^2.$$

The density function can either be pictured around a circle or it can be 'unwrapped' on to a line—in which case it resembles a *normal distribution. *See diagram opposite.*

Voronoi, Georgy Fedoseevich (1868–1908; b. Zhuravka, Ukraine; d. Warsaw, Poland) Russian algebraist. Voronoi was a graduate of U St Petersburg in 1889, and gained his PhD there in 1896 (supervised by

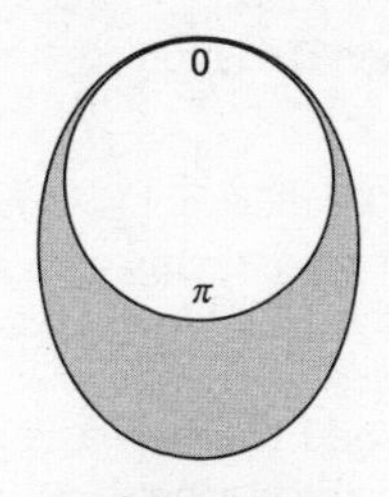

Von Mises distribution wrapped around the unit circle

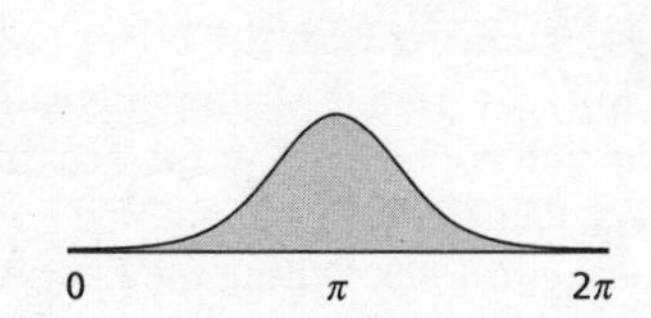

Unwrapped von Mises distribution

Von Mises distribution. The left diagram shows the density function wrapped around a central circle to give the continuous analogue of a *rose diagram. The right diagram gives a more conventional representation but fails to emphasize that the right-hand edge joins the left-hand edge of the diagram. In either diagram the area of a portion of the shaded region is proportional to probability.

---

*Markov). By that time he was on the faculty of U Warsaw, where he became Professor of Pure Mathematics and where he remained until his premature death. He is remembered by statisticians for his 1908 examination of the properties of the regions surrounding each of a finite set of points in many-dimensional space. In two dimensions these are now called Voronoi polygons.

SEE WEB LINKS

- Fuller biography and photographs.

Voronoi polygon *See* DIRICHLET TESSELLATION.

voter mobility table *See* MOBILITY TABLE.

voxel *See* PIXEL.

Wahba, Grace Goldsmith (1934– ; b. Washington, DC) American statistician specializing in the use of *splines and *machine learning methods. An undergraduate at Cornell U, her doctorate was supervised by *Parzen at Stanford U. Since 1967 she has been on the faculty at U Wisconsin, Madison. In 1994 she was the *IMS *Neyman Lecturer and also the *Parzen Prize winner. In 2003 she was the IMS *Wald Lecturer. She has been elected a member of the NAS and a fellow of the AAAS.

SEE WEB LINKS

- University webpage.

waiting time The time spent by an individual in a *queue waiting to be served.

Wald, Abraham (1902–50; b. Cluj, Romania; d. Travancore, India) Hungarian geometer and statistician. Wald gained his PhD in geometry in 1931 from U Vienna. In 1938, on the Nazi seizure of Austria, he emigrated to the USA and turned his attention to statistical *decision theory, making important advances in the theory of *sequential sampling. He held posts at Brown U (supervising *Chernoff) and Columbia U (where *Stein was one of his research students). He was President of the *IMS in 1948 and its *Rietz Lecturer in 1947. He died in a plane crash in India.

SEE WEB LINKS

- Fuller biography and photographs.

Wald distribution *See* INVERSE NORMAL DISTRIBUTION.

Wald equation Let $X_1, X_2, \ldots, X_N$ be a sequence of independent *random variables, each with *expected value μ, and let N itself be a random variable. Then the expected value of $\sum_{i=1}^{N} X_i$ is the product of μ and the expected value of N.

Wald Lecturer; Wald Lectureship This lectureship is awarded annually by the *IMS in memory of *Wald. The lecturer gives between two and four lectures on a single topic.

Wald statistic Any *statistic of the form

$$W = \{\mathbf{g}(\mathbf{T})\}'\mathbf{D}^{-1}\mathbf{g}(\mathbf{T}),$$

where $\mathbf{T}$ is an *estimator of a *vector parameter $\boldsymbol{\theta}$, $\mathbf{g}$ is some vector-valued function, and $\mathbf{D}$ is an estimator of the *variance-covariance matrix of the vector $\{\mathbf{g}(\mathbf{T}) - \mathbf{g}(\boldsymbol{\theta})\}$. The statistics are used to test the *null hypothesis that $\mathbf{g}(\boldsymbol{\theta}) = \mathbf{0}$, where $\mathbf{0}$ is a vector with all entries equal to 0. If $\mathbf{T}$ is the *maximum likelihood estimator then W has an approximate *chi-squared distribution with p *degrees of freedom (where p is the number of elements in $\boldsymbol{\theta}$).

Wald–Wolfowitz test *See* RUNS TEST.

Waller–Duncan test *See* MULTIPLE COMPARISON TEST.

Wallis, (Wilson) Allen (1912–98; b. Philadelphia, PA; d. Rochester, NY) American statistician. Wallis was a psychology graduate of U Minnesota in 1932. He then studied economics at Minnesota and at U Chicago. Subsequently he held posts in the economics departments at Columbia U, Yale U, and Stanford U. During the Second World War he headed a statistics think-tank. From 1946 to 1962 he was Professor of Statistics at the Business School of U Chicago. His paper with *Kruskal on the *Kruskal-Wallis test appeared in 1952. From 1951 to 1959 he was Editor of the **Journal of the American Statistical Association*. In 1962 he moved to U Rochester as President and then Chancellor (1975–82). On retirement from university life he was appointed Under-Secretary of State for Economic Affairs (until 1989). He was presented with the *Wilks Award of the *ASA in 1980.

SEE WEB LINKS

- Fuller biography, interview, and photographs.

Ward, Joe Henry Jnr (1926–2011; b. Denison, TX; d. San Antonio, TX) American mathematician and educational psychologist. Ward was educated at U Texas, gaining his BA in mathematics in 1947 and his PhD in educational psychology and mathematics in 1953. From 1951 to 1984 he was a personnel research psychologist with the US Air Force, working with U Texas in the development of computer-based instructional systems.

SEE WEB LINKS

- Fuller biography and photograph.

Ward's method An *agglomerative clustering method proposed by *Ward in 1963. The clustering criterion is based on the error sum of squares, E, which is defined as the sum of the squared distances of individuals from the centre of gravity of the cluster to which they have been assigned. Initially, E is 0, since every individual is in a cluster of its own. At each stage the link created is the one that makes the least increase to E.

warning line *See* QUALITY CONTROL.

wash-out period *See* CROSSOVER TRIAL.

Watson, Geoffrey Stuart (1922–98; b. Bendigo, Australia; d. Princeton, NJ) Australian statistician who spent most of his career in the USA. A mathematics graduate of U Melbourne (1942), Watson worked for five years as a teacher before travelling to the USA, where he obtained his PhD from North Carolina State U in 1951. After a succession of posts in Australia, England, and Canada, he moved to Princeton U in 1970. Watson is known for his part in the introduction of the 1950 *Durbin–Watson statistic for testing for *autocorrelation and in the 1964 *Wheeler–Watson test for *cyclic data. He was a talented artist.

SEE WEB LINKS

- Fuller biography, interview, and photographs.

Watson, Henry William (1827–1903; b. London, England; d. Berkswell, England) English mathematician. Watson was a graduate of KCL in 1846 and Cambridge U in 1850. In 1857 he became mathematics master at Harrow School and, in 1865, rector of Berkswell. During his time at Berkswell he became well known as an author of mathematics books. He collaborated with *Galton in developing the theory of *branching processes. He was elected FRS in 1881.

Watson's test A test of a specified *circular distribution, adapted from the *Cramér–von Mises test, and introduced by Geoffrey *Watson in 1961. Denote the n *observations by $\theta_1, \theta_2, \ldots, \theta_n$ and write

$$\bar{F} = \frac{1}{n}\sum_{j=1}^{n} \mathrm{F}(\theta_j),$$

where $\mathrm{F}(\theta)$ is the *probability of a value in the interval $(0, \theta)$ according to the *null hypothesis. The test statistic is U^2, given by

$$U^2 = \sum_{j=1}^{n} \{\mathrm{F}(\theta_j)\}^2 - \frac{1}{n}\sum_{j=1}^{n}\{(2j-1)\mathrm{F}(\theta_j)\} + n\left\{\frac{1}{3} - \left(\bar{F} - \frac{1}{2}\right)^2\right\}.$$

A large value of U^2 leads to rejection of the null hypothesis. A transformed version of U^2 is provided by U^*, given by

$$U^* = \left(U^2 - \frac{1}{10n} + \frac{1}{10n^2}\right)\left(1 + \frac{4}{5n}\right).$$

The distribution of U^* is approximately independent of n. The upper 10%, 5%, 2.5%, and 1% points of U^* are 0.152, 0.187, 0.222, and 0.268, respectively.

wavelet Functions that provide succinct and accurate representations of *time series and arrays of spatial *data. Wavelets have great potential as tools for data compression, when the data consist of, for example, images to be sent across computer networks. The wavelet representation of $g(t)$, a continuous function of time t, has the form

$$g(t) = a\phi(t) + b\psi(t) + \sum_{n,m} c_{n,m}\psi(2^n t - m),$$

where n and m are integers, with $n \geq 1$ and $m \geq 0$. The function ψ is the **mother wavelet** and ϕ is the **scaling function** or **father wavelet**. This wavelet representation of $g(t)$ has similarities to a *Fourier series representation but is more flexible, since both ϕ and ψ are chosen to take the value 0 outside finite intervals. For a time series measured at N equi-spaced time points, the functions ϕ and ψ are related by the equations

$$\phi(t) = \sum_{k=0}^{N} h_k \phi(2t - k),$$

$$\psi(t) = \sum_{k=0}^{N} (-1)^k h_{N-k} \phi(2t - k),$$

where $h_0, h_1, \ldots, h_N$ are constants, referred to as **filter coefficients**. The original N data values are represented by a sum of a weighted combination of the father and mother wavelets together with **daughter wavelets**. The daughter wavelet, ψ_{jk}, is related to the mother wavelet, ψ, by a simple formula that reflects a translation and dilation of the mother, so that the daughter looks like a compressed version of its mother. The relation is

$$\psi_{jk}(t) = 2^{\frac{1}{2}j}\psi(2^j t - k),$$

where $0 \leq t < 1$. The coefficient $2^{\frac{1}{2}j}$ simplifies subsequent analysis.

A summary of the wavelet decomposition for the case $N = 16$ is illustrated in the table:

<table>
<tr><td colspan="16">Father wavelet, ϕ</td></tr>
<tr><td colspan="16">Mother wavelet, ψ</td></tr>
<tr><td colspan="8">ψ_{10}</td><td colspan="8">ψ_{11}</td></tr>
<tr><td colspan="4">ψ_{20}</td><td colspan="4">ψ_{21}</td><td colspan="4">ψ_{22}</td><td colspan="4">ψ_{23}</td></tr>
<tr><td colspan="2">ψ_{30}</td><td colspan="2">ψ_{31}</td><td colspan="2">ψ_{32}</td><td colspan="2">ψ_{33}</td><td colspan="2">ψ_{34}</td><td colspan="2">ψ_{35}</td><td colspan="2">ψ_{36}</td><td colspan="2">ψ_{37}</td></tr>
<tr><td>1</td><td>2</td><td>3</td><td>4</td><td>5</td><td>6</td><td>7</td><td>8</td><td>9</td><td>10</td><td>11</td><td>12</td><td>13</td><td>14</td><td>15</td><td>16</td></tr>
</table>

Each of the ψ functions is a piecewise continuous function that takes the value 0 outside its range of influence. Thus, if observation 14 takes an unusually large value then this will be reflected in an unusually large coefficient of ψ_{36}; if observations 5 to 8 are unusually small then this will be reflected in an unusually small coefficient of ψ_{21}; and so on.

The first mother wavelet was given in the appendix of the 1909 PhD thesis by *Haar. This **Haar wavelet** is given by

$$\psi(t) = \begin{cases} -1, & 0 \le t < \frac{1}{2}, \\ 1, & \frac{1}{2} \le t < 1, \\ 0, & \text{otherwise.} \end{cases}$$

The family of wavelets that are currently most used were introduced by *Daubechies in 1988. These wavelets have *fractal properties. Other families of wavelets include **symmlets** and **coiflets**.

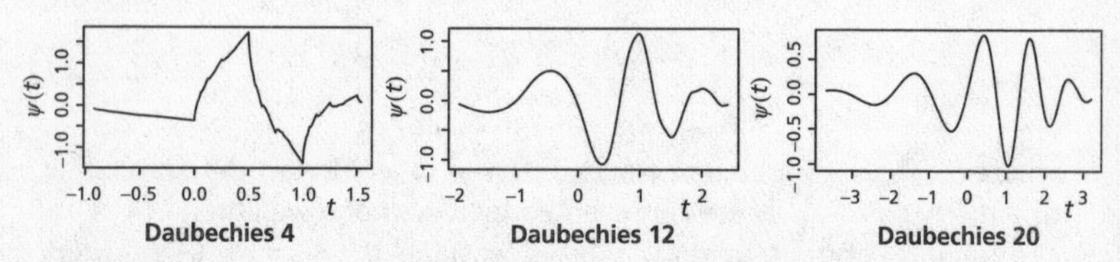

Mother wavelets. The D_2 wavelet is the Haar wavelet. The wavelets illustrated here are the D_4, D_{12}, and D_{20} wavelets. All are members of the Daubechies family.

weak law of large numbers *See* LAWS OF LARGE NUMBERS.

W

weakly stationary *See* STATIONARY PROCESS.

Wedderburn, Robert William Maclagan (1947–75; b. Edinburgh, Scotland; d. North Wales) Scottish statistician. Wedderburn was a graduate of Cambridge U. On graduation he joined the staff at *Rothamsted where he collaborated with *Nelder on the influential 1972 paper that unified the treatment of *generalized linear models. By the time of his early death (from anaphylactic shock following an insect bite) he had made valuable contributions to the *GENSTAT and *GLIM statistical packages.

Weibull, Ernst Hjalmar Wallodi (1887–1979; b. Vittskoevle, Sweden; d. Annecy, France) Swedish engineer. Weibull studied at the Royal Institute of Technology in Stockholm, graduating in 1924. He gained a doctorate from U Uppsala in 1932. He spent much of his career in the Royal Swedish Coast Guard, joining in 1904 as a midshipman and reaching the rank of major in 1940. Initially his job was concerned with studying the effects of underwater explosions, and subsequently with the lifetimes of components. Weibull is best known for his 1939 paper introducing the *Weibull distribution. In 1972 the American Society of Mechanical Engineers awarded him their Gold Medal citing him as 'a pioneer in the study of fracture, fatigue and reliability'.

SEE WEB LINKS

- Fuller biography and photograph.

Weibull distribution A *random variable, X, that has a Weibull distribution can take any positive value and has *probability density function f, given by

$$\mathrm{f}(x) = cx^{c-1}\exp(-x^c), \qquad x > 0,$$

where c is a positive constant. The distribution has *mean $\Gamma\left(\frac{1}{c}+1\right)$ and *variance $\Gamma\left(\frac{2}{c}+1\right) - \left\{\Gamma\left(\frac{1}{c}+1\right)\right\}^2$, where Γ is the *gamma function. The Weibull distribution has been found to be very useful for describing

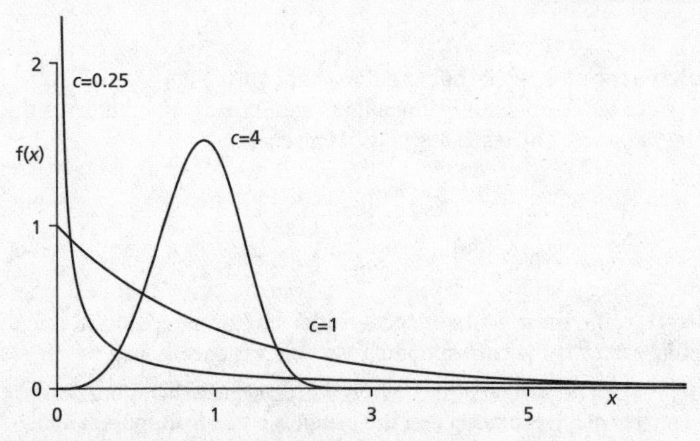

Weibull distribution. The shape depends on the parameter c. With $c < 1$ the distribution is often used to model extreme lifetimes.

the *distribution of the lifetimes of components and for analysing meteorological data—particularly in the context of extreme events.

weight *See* WEIGHTED AVERAGE.

weighted average (weighted mean) A weighted average is an average that can attach more importance (weight) to some *observations than to others. Suppose that the importance of the observation x_j is represented by the non-negative numerical coefficient w_j (the **weight**) ($j = 1, 2, \ldots, n$). The weighted average is

$$\frac{(w_1x_1 + w_2x_2 + \cdots + w_nx_n)}{(w_1 + w_2 + \cdots + w_n)}.$$

See also MOVING AVERAGE.

weighted least squares (WLS) A preferable alternative to ordinary least squares (*see* METHOD OF LEAST SQUARES) when estimating the *parameters of a *model using *independent *observations whose values are known to vary in accuracy in a specified way. *See also* GENERALIZED LEAST SQUARES.

weighted mean *See* WEIGHTED AVERAGE.

Weka A collection of *machine learning algorithms for *data mining tasks.

SEE WEB LINKS

- Website.

Welch statistic A *statistic used in a *test for the equality of the *means of m *normal populations that cannot be assumed to have the same *variance. The test statistic, W, is given by

$$W = \frac{(m+1)u^2\sum_{j=1}^{m} w_j(\bar{x}_j - \bar{x})^2}{\left\{(m^2-1)u^2 + 2(m-2)\sum_{j=1}^{m}(n_j-1)(u-w_j)^2\right\}},$$

where $\bar{x}_j$ is the mean of the n_j *observations in the sample from the jth population, s_j^2 is the corresponding *sample variance (using the $(n-1)$ divisor) of the jth sample, $w_j = n_j/s_j^2$, $u = \sum_{j=1}^{m} w_j$, and $\bar{x} = \frac{1}{u}\sum_{j=1}^{m} w_j\bar{x}_j$. Under the *null hypothesis that the samples come from populations with the same mean, W has an approximate *F-distribution with $(m-1)$ and v *degrees of freedom, where v is the nearest integer to

$$\frac{1}{3}u^2(m^2-1)\bigg/\sum_{j=1}^{m}\frac{(u-w_j)^2}{n_j-1}.$$

The special case $m=2$ corresponds to the *Behrens–Fisher problem.

Western Electric Rules *See* QUALITY CONTROL.

Wheeler–Watson test A *non-parametric test of the *null hypothesis that two *samples of *cyclic data have been drawn from the same *population. Suppose the samples contain n_1 and n_2 observations. These observations are first arranged in order of magnitude and then their actual values are replaced by coded values that (working in degrees) are multiples of $360/(n_1+n_2)$, so that the kth largest value becomes $360k/(n_1+n_2)$. Let the coded values for the first sample be $\theta_1, \theta_2, \ldots, \theta_{n_1}$, and calculate R^2 given by

$$R^2=\left(\sum_{j=1}^{n_1}\sin\theta_j\right)^2+\left(\sum_{j=1}^{n_1}\cos\theta_j\right)^2.$$

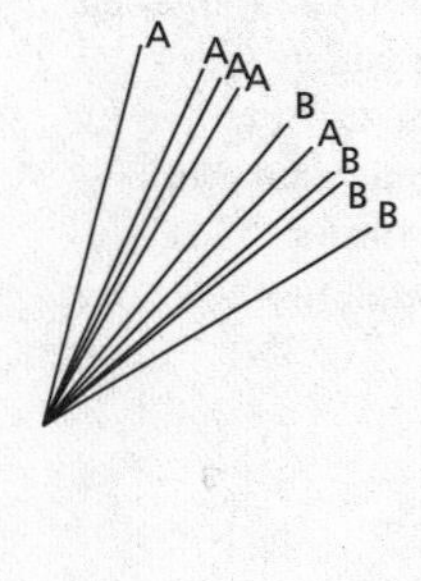

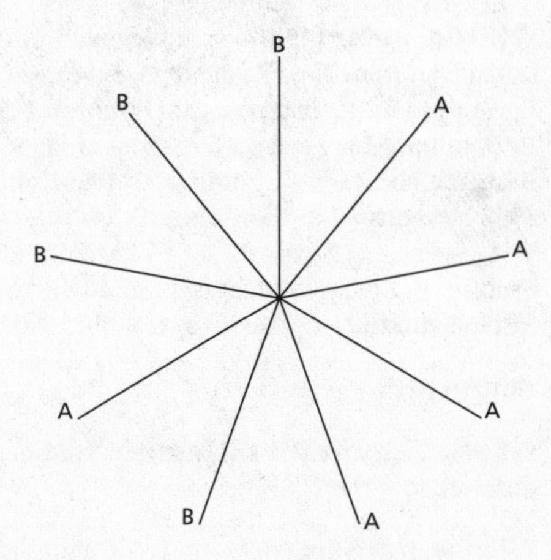

Original data Coded data

Wheeler–Watson test. Two sets of birds are released from the same location. The five birds in set A fly off in the directions 15°, 25°, 28°, 31°, and 45°. The four birds in set B fly off in the directions 40°, 50°, 52°, and 60°. Arranging these data in order, we have AAAABABBB. Since 360/9 = 40, the coded values for sample B are (in degrees) 200, 280, 320, and 360. The resulting value for R^2 is 4.88 and $T=3.90$. We conclude that there is no significant evidence that the samples have been drawn from different populations.

W

The value of R^2 for the second sample will be identical. The test statistic is T, given by

$$T = \frac{2(n_1 + n_2 - 1)R^2}{n_1 n_2}.$$

For large n_1 and n_2, T has an approximate *chi-squared distribution with two *degrees of freedom. The test was introduced by Stanley Wheeler and Geoffrey *Watson in 1964.

white noise *See* NOISE.

Whitney, Donald Ransom (1915–2001; b. E. Cleveland, OH; d. Monroe, NY) American statistician. Having obtained his MA at Princeton U, Whitney spent much of the Second World War teaching navigation to newly commissioned officers. After the war he obtained his doctorate at OSU under the supervision of Henry *Mann. The subject of his research was the *non-parametric Mann–Whitney test (published in 1947). Whitney remained at OSU throughout his career, retiring in 1982.

SEE WEB LINKS

- Biography and interview.

Whittle, Peter (1927– ; b. Wellington, New Zealand) Mathematical statistician from New Zealand whose career has been largely spent in England. Whittle held posts at U Uppsala (1950) where he obtained his PhD, in the New Zealand Department for Scientific and Industrial Research (1953), at Cambridge U (1959), and U Manchester (1961). In 1967 he returned to Cambridge U as Professor of Operational Research. He was elected FRS in 1978 and FRSNZ in 1981. He was the *IMS *Neyman Lecturer in 1990. Whittle was awarded the *Guy Medal in Silver of the *RSS in 1966 and the Guy Medal in Gold in 1996.

whole plot *See* SPLIT PLOT.

wholly significant difference test *See* MULTIPLE COMPARISON TEST.

Wiener, Norbert (1894–1964; b. Columbia, MO; d. Stockholm, Sweden) American mathematician. The son of Russian immigrants, Wiener was a child prodigy, obtaining his PhD (in mathematical logic) from Harvard U at the age of eighteen. He developed the mathematics of a **Wiener process**, the simplest type of *Brownian motion. His interests were wide-ranging: he became well known to the general scientific public for his philosophical discussion of cybernetics (a term he coined in 1945). Amongst his sayings is 'A professor is one who can speak on any subject—for precisely fifty minutes', though some might disagree with this observation! A lunar crater is named after him.

SEE WEB LINKS

- Fuller biography.

Wiener process *See* WIENER.

Wilcoxon, Frank (1892–1965; b. Glengarriffe Castle, Ireland; d. Tallahassee, FL) American chemist and self-taught statistician. Wilcoxon was a teenage rebel who was in turn a merchant seaman, a petrol-station attendant, and a tree surgeon. Subsequently conforming, he obtained his doctorate in physical chemistry from Cornell U in 1924. For the next 25 years he worked as a chemist in various of the larger chemical firms in the USA. Only for the last seven years of his working life was he officially employed as a statistician. However, his interest in the subject dated back to 1925, when he wished to devise statistical tests of the effectiveness of various types of insecticide and fungicide. He led the research group that worked on the development of various of the pyrethrin-based insecticides, including Malathion. His statistical work concentrated on devising methods of testing that were simple and easy to understand (the hallmarks of *non-parametric tests) and he is remembered now in the context of the *Wilcoxon signed-rank tests introduced in a 1945 paper.

SEE WEB LINKS

- Information and photograph.

Wilcoxon rank-sum test *See* MANN–WHITNEY TEST.

Wilcoxon signed-rank tests *Non-parametric tests that extend the *sign tests. The single-sample version (*observations $x_1, x_2, \ldots$), suitable for a *symmetric distribution, tests the *null hypothesis that the *population median has a specified value (m_0). The **matched-pair** (or **paired-sample**) version (observation pairs $(x_1, y_1), (x_2, y_2), \ldots$) is concerned with the differences $(x_1 - y_1), (x_2 - y_2), \ldots$. With the assumption that these differences are *independent observations from a symmetric distribution, the null hypothesis is that this distribution has median zero.

To determine the value of the test statistic, z, the first step is to calculate the differences $d_1, d_2, \ldots$, where $d_j = x_j - m_0$ (single sample) or $d_j = x_j - y_j$ (matched pairs). After zero differences have been discarded, the remaining n are arranged in ascending order of $|d_j|$. The magnitudes are replaced by the corresponding *ranks, with tied ranks where necessary. The signs of $d_1, d_2, \ldots$, are now attributed to the ranks, resulting in **signed**

ranks. Let P be the sum of the positive signed ranks and let T be the smaller of P and $\frac{1}{2}n(n+1)-P$. The test statistic, given by

$$z = \frac{\frac{1}{4}n(n+1) - T - \frac{1}{2}}{\sqrt{\frac{1}{24}n(n+1)(2n+1)}},$$

is an observation from the upper half of an approximate standard *normal distribution. The $\frac{1}{2}$ is a *continuity correction.

As an example, suppose that a symmetric distribution is believed to have median 100. A random sample of eight observations is reported as consisting of the values 92.3, 57.6, 88.8, 110.5, 100.0, 181.0, 96.0, 105.7. The supposed median is subtracted from each observation to give −7.7, −42.4, −11.2, 10.5, 0, 81.0, −4.0, 5.7. The value 0 is discarded and the remainder are arranged in order of ascending absolute magnitude: −4.0, 5.7, −7.7, 10.5, −11.2, −42.4, 81.0. Retaining the signs while replacing the values by ranks gives −1, 2, −3, 4, −5, −6, 7, so that $P=13$, $\frac{1}{2}n(n+1) - P=15$ (since $n=7$) and $T=13$. The test statistic is

$$z = \frac{14 - 13 - 0.5}{\sqrt{35}} = 0.085.$$

Comparing with the tables of upper-tail percentage points of the standard normal distribution (Appendix VI), we see that the null hypothesis that the median is 100 should not be rejected.

Wilk, Martin Bradbury (1922–2013; b. Montreal, Canada; d. Yorba Linda, CA) Canadian statistician. Wilk's first degree was a BS in chemical engineering from McGill U in 1945. For five years he worked as a chemical engineer. He then studied statistics at ISU, obtaining his MS in 1953 and his PhD in 1955 (supervised by *Kempthorne). Wilk was co-author of the 1965 paper that introduced the *Shapiro–Wilk test. After teaching at Rutgers U, he joined AT&T. In 1981 he was appointed Chief Statistician of Statistics Canada (the Canadian government's central statistics office). Subsequently he was an adjunct Professor at Carleton U, Ottawa. He was elected an Honorary Fellow of the *RSS in 1986. He was President of the *SSC in 1986 and was elected an Honorary Member in 1988.

SEE WEB LINKS

- Obituary.

Wilks, Samuel Stanley (1906–64; b. Little Elm, TX; d. Princeton, NJ) American mathematical statistician. Wilks initially studied architecture at the North Texas Teacher's College. He followed this with a BSc in mathematics at U Texas in 1928 and a PhD in statistics at ISU (supervised by *Rietz) in 1931. In 1933, after a year in England working with Sir Ronald

*Fisher and *Wishart, he joined the staff at Princeton U, where his research students included *Anderson, *Goodman, *Mood, and *Mosteller. He was a founder member of the *IMS, serving as its President in 1940, and was its *Rietz Lecturer in 1959. From 1938–49 he was Editor of the **Annals of Mathematical Statistics*. He was President of the *ASA in 1950.

SEE WEB LINKS

- Obituary.

Wilks Award An award established by the *ASA in 1964 to honour the memory and career of *Wilks by recognizing outstanding contributions to Statistics that continue the spirit of his work.

Wilks's lambda (Λ) *See* MULTIVARIATE ANALYSIS OF VARIANCE.

Williams test *See* MULTIPLE COMPARISON TEST.

Wilson, Edwin Bidwell (1879–1964; b. Hartford, CT; d. Brookline, MA) American mathematician. Wilson obtained his AB from Harvard U in 1899 and his PhD (supervised by *Gibbs) from Yale U in 1901. He left the Yale faculty in 1907 for MIT where he became in turn Professor of Mathematics (1911) and of Physics (1917). His interests centred on aerodynamics; his modelling of wind gusts led him to Statistics. He is credited with the original idea for *confidence intervals. In 1922 he was appointed Professor of Vital Statistics at the Harvard School of Public Health. He was elected to the NAS and the AAAS (President 1927–31). He was President of the *ASA in 1929.

SEE WEB LINKS

- Fuller biography and photograph.

WinBUGS Statistical software for implementing *Bayesian inference using the Gibbs sampler (*see* MARKOV CHAIN MONTE CARLO METHODS).

window width *See* KERNEL METHOD.

Winsorized mean *See* TRIMMED MEAN.

Wishart, John (1898–1956; b. Montrose, Scotland; d. Acapulco, Mexico) Scottish statistician. Either side of the First World War (when he served in the Black Watch regiment), Wishart studied mathematics and physics at Edinburgh U. In 1924 he joined Karl *Pearson at UCL, moving in 1927 to join Sir Ronald *Fisher at *Rothamsted. His work on the *Wishart distribution appeared in 1928. In 1931 he moved to Cambridge U, where he supervised *Bartlett, *Cochran, and *Hartley, and became Head of the Statistical Laboratory in 1953. He was elected FRSE in 1931.

SEE WEB LINKS

- Biography.

Wishart distribution The joint distribution (*see* BIVARIATE DISTRIBUTION) of the *variances and *covariances in *samples from a *multivariate normal distribution.

without replacement; with replacement *See* URN MODEL.

WLS *Abbreviation for* WEIGHTED LEAST SQUARES.

Wolfowitz, Jacob (1910–81; b. Warsaw, Poland; d. Tampa, FL) American statistician. Leaving Poland, Wolfowitz joined his father in New York when he was ten. He graduated in mathematics from City College (now City U) New York in 1932, spending the next ten years as a mathematics teacher. During this period he worked on *non-parametric tests (he coined the word 'non-parametric') and began his collaboration with *Wald; the Wald–Wolfowitz test (*see* RUNS TEST) was published in 1940. He also made pioneering contributions to *dynamic programming. He received his PhD in 1942 from NYU. After a period of war-related research at Columbia U, where *Levene was a research student, Wolfowitz held posts at several institutions, including Cornell U (1951), U Illinois at Urbana (1970), and U South Florida in Tampa (1978). Elected to membership of the NAS and the AAAS, he was President of the *IMS in 1959, and was its *Rietz Lecturer in 1957 and *Wald Lecturer in 1965.

SEE WEB LINKS

- Fuller biography and photograph.

working mean (assumed mean) A value, expected to be close to the *mean of a set of data, that is subtracted from each data item, so as to make calculations simpler. For example, suppose the data consist of the following values: 2001, 2004, 2007, 2007, 2005, 2006. Using a working mean of 2000, the data reduce to the simpler 1, 4, 7, 7, 5, 6, with mean 5. Thus the mean of the original data is $2000+5=2005$.

In other cases a more involved transformation helps. For example, the the x-values 19.42, 19.37, 19.88, and 19.62 are simplified by using the transformation $100(x-19)$ to give the **coded data** 42, 37, 88, and 62.

wrapped Cauchy distribution A distribution used to model circular data (*see* CYCLIC DATA). It can be regarded as wrapping a *Cauchy distribution around the circle. The *probability density function f is given by

$$\mathrm{f}(\theta)=\frac{1-\rho^2}{2\pi(1+\rho^2-2\rho\cos\,\theta)},\qquad -\pi<\theta\leq\pi,$$

where ρ $(0<\rho<1)$ is a constant. For many values of ρ the distribution closely resembles a *von Mises distribution.

Wright, Sewall Green (1889–1988; b. Melrose, MA; d. Madison, WI) American zoologist specializing in population genetics and responsible for the method of *path analysis. His mathematical theory of evolution ran counter to that proposed by Sir Ronald *Fisher, leading to a debate that continues to this day. His final paper appeared in 1988, 76 years after his first.

SEE WEB LINKS

- Biography with photograph.

Wright's inbreeding coefficient The coefficient, named after *Wright, is the *probability that at a specified location in an individual's DNA, the contribution provided by each parent stems from a common ancestor.

WSD test *See* MULTIPLE COMPARISON TEST.

X (design matrix) *See* MULTIPLE REGRESSION MODEL.

$\bar{x}$-chart *See* QUALITY CONTROL.

Yates, Frank (1902–94; b. Didsbury, England; d. Harpenden, England) English mathematician and statistician. In 1924 Yates left Cambridge U with first class results in mathematics. After two years as a mathematics teacher, he joined the Gold Coast (now Ghana) Survey as a mathematics advisor. However, the climate affected his health and he returned to England, joining the statistics staff at *Rothamsted in 1931. He was a keen user of computers, writing that 'to be a good theoretical statistician one must also compute'. He was elected FRS in 1948, being awarded the Society's Royal Medal in 1966. Although he retired in 1967 he maintained his links with Rothamsted, and his last paper was published in 1990, when he was 88. He was President of the *RSS in 1967 and was awarded its *Guy Medal in Gold in 1960. He was elected an Honorary Life Member of the *IBS in 1971.

SEE WEB LINKS

- Fuller biography and photographs.

Yates-corrected chi-squared test; Yates correction *See* TWO-BY-TWO TABLE.

Youden, William John (1900–71; b. Townsville, Australia; d. Washington, DC) American chemical engineer and statistician. Youden's parents took him to the USA when he was seven. His first degree, at Rochester U, was in chemical engineering and his MSc and PhD (in 1924) were in chemistry at Columbia U. In 1928, whilst working in an institute for plant research, he came across Sir Ronald *Fisher's *Statistical Methods*, and this book engendered his subsequent interest in *experimental design. During the Second World War he worked as a systems analyst. After the war, he joined the National Bureau of Standards, continuing to work on experimental design. He was the *COPSS *Fisher Lecturer in 1967. In 1969 he was awarded the *Shewhart Medal of the *ASQ and also received the *Wilks Award of the *ASA.

He produced the diagram below which neatly encapsulates the view of Statistics held in the middle of the twentieth century:

THE
NORMAL
LAW OF ERROR
STANDS OUT IN THE
EXPERIENCE OF MANKIND
AS ONE OF THE BROADEST
GENERALIZATIONS OF NATURAL
PHILOSOPHY ◇ IT SERVES AS THE
GUIDING INSTRUMENT IN RESEARCHES
IN THE PHYSICAL AND SOCIAL SCIENCES AND
IN MEDICINE AGRICULTURE AND ENGINEERING ◇
IT IS AN INDISPENSABLE TOOL FOR THE ANALYSIS AND THE
INTERPRETATION OF THE BASIC DATA OBTAINED BY OBSERVATION AND EXPERIMENT

SEE WEB LINKS

- Fuller biography and photograph.

Youden's index *See* TWO-BY-TWO TABLE.

Youden square *See* LATIN SQUARE.

Yule, George Udny (1871–1951; b. Morham, Scotland; d. Cambridge, England) Scottish statistician In 1890 Yule graduated from UCL where he studied engineering. After two years as an engineer, and a year researching in physics, he was appointed by Karl *Pearson to a demonstratorship at UCL. Yule's first paper on statistics appeared in 1895. In 1911 he published *Introduction to the Theory of Statistics* which was the standard reference book on mathematical statistics for the next forty years (the fourteenth edition, written jointly with Sir Maurice *Kendall, was published in 1950). In 1912 he moved to Cambridge U, where he spent most of the rest of his life, retiring in 1931 (when he learnt to fly and obtained his pilot's licence). During his time at Cambridge he worked on the theory of *time series, introducing the terms correlogram (*see* AUTOCORRELATION) and *autoregressive models. He was elected FRS in 1921. He was President of the *RSS in 1924, having been awarded its *Guy Medal in Gold in 1911.

- Fuller biography.

Yule's Q *See* TWO-BY-TWO TABLE.

Yule–Walker equations *See* AUTOREGRESSIVE MODEL.

Zelen, Marvin (1927– ; b. New York City) American medical statistician specializing in the design and analysis of *clinical trials. A graduate of City College (now City U) NY, his doctorate was obtained from The American U (Washington, DC) in 1957. Since 1977 Zelen has been on the faculty at the Harvard School of Public Health. He received the *Wilks Award of the *ASA in 2006 and the *Parzen Prize in 2008. He was the *COPSS Fisher Lecturer in 2007. In 2009 he was awarded the Medal of Honor of the American Cancer Society.

Zelen's design *See* BLINDING.

zero-sum game *See* GAME THEORY.

Zipf's law An empirical observation that states that, in a population consisting of many different types, the proportion belonging to the nth most common type is approximately proportional to $1/n$. The law was put forward by George Kingsley Zipf (1902–1950), a Harvard linguistics professor, in the context of the frequencies with which words appear in texts. It has been found to be a good approximation in many other contexts (for example, the incidence of family names in South Korea). *See also* BENFORD'S LAW.

z-test (normal test) A *hypothesis test, using a *sample of n independent *observations, of the *null hypothesis that a *normal distribution with *variance σ^2, has *mean μ. Writing the *sample mean as $\bar{x}$, the test statistic is z, given by

$$z = \frac{\bar{x} - \mu}{\sigma/\sqrt{n}}.$$

In the case where the *alternative hypothesis is that the mean is not μ, then, if $|z| > 1.96$, there is evidence to reject the null hypothesis at the 5% significance level in favour of the alternative hypothesis.

z-value A value, usually of a test statistic (*see* HYPOTHESIS TEST), that is hypothesized to have come from a *normal distribution with *mean 0 and *variance 1. The z-notation was introduced by Sir Ronald *Fisher in 1924.

Appendices

Appendix I: Statistical Notation

Symbol	Reference
$\sim$	has the same distribution as
$\|$	conditional, e.g. A occurs given that B occurs
$\mathrm{B}(n, p)$	binomial distribution with parameters n and p
$\mathrm{Cov}(X, Y)$	covariance of the random variables X and Y
$\mathrm{E}(X)$	expected value of the random variable X
F	cumulative distribution function
f	probability density function
G	probability-generating function
G^2	likelihood-ratio goodness-of-fit statistic
H_0, H_1	null and alternative hypotheses
M	moment-generating function
$\mathrm{N}(\mu, \sigma^2)$	normal distribution (mean μ, variance σ^2)
n	sample size
P	probability (of an event)
p	probability value
r	product-moment correlation coefficient; residual
s^2	sample variance given by $\sum_j (x_j - \overline{x})^2/(n-1)$.
$\mathrm{Var}(X)$	variance of the random variable X
X, Y	random variables
X^2	chi-squared test statistic
x	an observation
$\overline{x}$	sample mean
$x_{(j)}$	an ordered observation, e.g. $x_{(1)} \leq x_{(2)} \leq \cdots$
$\overline{x}_{-j}$, $\hat{x}_{-j}$	values calculated omitting observation j
$\hat{y}$	an estimate of y
Z	random variable (often standard normal random variable)

Appendix II: Mathematical Notation

Symbol	Reference
∞	infinity
!	*see* FACTORIAL
$'$	For a function f, $f'(x)$ is the derivative of $f(x)$ with respect to x
$'$	For a set, *see* COMPLEMENTARY EVENT
$'$	For a matrix, the transpose. *See* MATRIX
(a, b)	an open interval, i.e. the set of real numbers x such that $a<x<b$
Q	the set of rational numbers (numbers expressible as the ratio of two integers)
Z	the set of integers
$\mathbb{R}$	the set of all real numbers
$\mathbb{R}^n$	the set of all row vectors, or column vectors, with n real elements
$\cup$	union
$\cap$	intersection
$\approx$	approximately equal to
$<$	less than
$\leq$	less than or equal to
$>$	greater than
$\geq$	greater than or equal to
$\rightarrow$	tends to, e.g. $n \rightarrow \infty$
$\|x\|$	absolute value of x
a^b	a raised to the power b
$a^{\frac{1}{2}}$	$\sqrt{a}$, the (positive) square root of a
B	beta function, *see* BETA FUNCTION
$\partial y/\partial x$	partial derivative of y with respect to x. *See* HESSIAN MATRIX
$\max\{x_1, x_2, \ldots, x_n\}$	the greatest of $\{x_1, x_2, \ldots, x_n\}$
$\min\{x_1, x_2, \ldots, x_n\}$	the least of $\{x_1, x_2, \ldots, x_n\}$
$\text{sign}(x)$	the sign of x, i.e. $\text{sign}(x) = \begin{cases} +1 & \text{if } x > 0 \\ 0 & \text{if } x = 0 \\ -1 & \text{if } x < 0 \end{cases}$
$\mathbf{X}$	a matrix
$\mathbf{x}$	a vector
$\{r_j\}$	the set $r_1, r_2, \ldots$
$\binom{n}{r}$	${}^nC_r$; ${}_nC_r$ binomial coefficient
$\sum$	sum of terms
Π	product of terms

Appendix III: Greek Letters

Symbol		Reference
α	alpha	*See* ALPHA; LINEAR REGRESSION
β	beta	*See* BETA; LINEAR REGRESSION
$\boldsymbol{\beta}$		*See* MULTIPLE REGRESSION MODEL
γ	gamma	*See* ASSOCIATION
Γ	capital gamma	*See* GAMMA FUNCTION
δ	delta	
ε	epsilon	
η	eta	
θ	theta	an angle; a parameter
κ	kappa	*See* COHEN'S KAPPA; CUMULANT; KURTOSIS
λ	lambda	*See* ASSOCIATION
Λ	capital lambda	*See* MULTIVARIATE ANALYSIS OF VARIANCE
μ	mu	*See* POPULATION MEAN; MOMENT
ν	nu	*See* CHI-SQUARED DISTRIBUTION; DEGREES OF FREEDOM; *F*-DISTRIBUTION; *t*-DISTRIBUTION
π	pi	*See* PI
$\prod$	capital pi	a product e.g. $\prod_{n=1}^{N} n = 1 \times 2 \times \cdots \times N = N!$
ρ	rho	*See* AUTOCORRELATION; POPULATION CORRELATION COEFFICIENT; SPEARMAN'S RHO
σ	sigma	standard deviation
Σ	capital sigma	summation (*see* SIGMA)
$\boldsymbol{\Sigma}$		a variance–covariance matrix
τ	tau	*See* KENDALL'S TAU
ϕ	phi	probability density function of a standard normal distribution; also the empty set
Φ	capital phi	the cumulative distribution function of a standard normal distribution
χ	chi	For χ^2, *see* CHI-SQUARED DISTRIBUTION
ψ	psi	*See* WAVELET

Appendix IV: Upper-Tail Percentage Points for the Standard Normal Distribution

The table gives the values of z for which $P(Z > z) = q\%$, where Z has a normal distribution with mean 0 and variance 1.

$q(\%)$	z	$q(\%)$	z	$q(\%)$	z	$q(\%)$	z	$q(\%)$	z
50	0.000	15	1.036	2.5	1.960	1.0	2.326	0.04	3.353
45	0.126	14	1.080	2.4	1.977	0.9	2.366	0.03	3.432
40	0.253	13	1.126	2.3	1.995	0.8	2.409	0.02	3.540
35	0.385	12	1.175	2.2	2.014	0.7	2.457	0.01	3.719
30	0.524	11	1.227	2.1	2.034	0.6	2.512	$0.0^{2}5$	3.891
25	0.674	10	1.282	2.0	2.054	0.5	2.576	$0.0^{2}1$	4.265
24	0.706	9	1.341	1.9	2.075	0.4	2.652	$0.0^{3}5$	4.417
23	0.739	8	1.405	1.8	2.097	0.3	2.748	$0.0^{3}1$	4.753
22	0.772	7	1.476	1.7	2.120	0.2	2.878	$0.0^{4}5$	4.892
21	0.806	6	1.555	1.6	2.144	0.1	3.090	$0.0^{4}1$	5.199
20	0.842	5	1.645	1.5	2.170	0.09	3.121	$0.0^{5}5$	5.327
19	0.878	4.5	1.695	1.4	2.197	0.08	3.156	$0.0^{5}1$	5.612
18	0.915	4	1.751	1.3	2.226	0.07	3.195	$0.0^{6}5$	5.731
17	0.954	3.5	1.812	1.2	2.257	0.06	3.239	$0.0^{6}1$	5.998
16	0.994	3	1.881	1.1	2.290	0.05	3.291	$0.0^{7}5$	6.109

In the table, the notation $0.0^{5}1$ means 0.000001

Appendix V: The Standard Normal Distribution Function

The table gives the values of $\Phi(z) = P(Z < z)$, where Z has a normal distribution with mean 0 and variance 1.

	$\Phi(z)$																		
z	0	1	2	3	4	5	6	7	8	9	1	2	3	4	5	6	7	8	9
															ADD				
0.0	.5000	.5040	.5080	.5120	.5160	.5199	.5239	.5279	.5319	.5359	4	8	12	16	20	24	28	32	36
0.1	.5398	.5438	.5478	.5517	.5557	.5596	.5636	.5675	.5714	.5753	4	8	12	16	20	24	28	32	36
0.2	.5793	.5832	.5871	.5910	.5948	.5987	.6026	.6064	.6103	.6141	4	8	12	15	19	23	27	31	35
0.3	.6179	.6217	.6255	.6293	.6331	.6368	.6406	.6443	.6480	.6517	4	7	11	15	19	22	26	30	34
0.4	.6554	.6591	.6628	.6664	.6700	.6736	.6772	.6808	.6844	.6879	4	7	11	14	18	22	25	29	32
0.5	.6915	.6950	.6985	.7019	.7054	.7088	.7123	.7157	.7190	.7224	3	7	10	14	17	20	24	27	31
0.6	.7257	.7291	.7324	.7357	.7389	.7422	.7454	.7486	.7517	.7549	3	7	10	13	16	19	23	26	29
0.7	.7580	.7611	.7642	.7673	.7704	.7734	.7764	.7794	.7823	.7852	3	6	9	12	15	18	21	24	27
0.8	.7881	.7910	.7939	.7967	.7995	.8023	.8051	.8078	.8106	.8133	3	5	8	11	14	16	19	22	25
0.9	.8159	.8186	.8212	.8238	.8264	.8289	.8315	.8340	.8365	.8389	3	5	8	10	13	15	18	20	23
1.0	.8413	.8438	.8461	.8485	.8508	.8531	.8554	.8577	.8599	.8621	2	5	7	9	12	14	16	19	21
1.1	.8643	.8665	.8686	.8708	.8729	.8749	.8770	.8790	.8810	.8830	2	4	6	8	10	12	14	16	18
1.2	.8849	.8869	.8888	.8907	.8925	.8944	.8962	.8980	.8997	.9015	2	4	6	7	9	11	13	15	17
1.3	.9032	.9049	.9066	.9082	.9099	.9115	.9131	.9147	.9162	.9177	2	3	5	6	8	10	11	13	14

Example: $\Phi(0.657) = 0.7422 + 0.0023 = 0.7445$

	Φ(z)										ADD								
z	0	1	2	3	4	5	6	7	8	9	1	2	3	4	5	6	7	8	9
1.4	.9192	.9207	.9222	.9236	.9251	.9265	.9279	.9292	.9306	.9319	1	3	4	6	7	8	10	11	13
1.5	.9332	.9345	.9357	.9370	.9382	.9394	.9406	.9418	.9429	.9441	1	2	4	5	6	7	8	10	11
1.6	.9452	.9463	.9474	.9484	.9495	.9505	.9515	.9525	.9535	.9545	1	2	3	4	5	6	7	8	9
1.7	.9554	.9564	.9573	.9582	.9591	.9599	.9608	.9616	.9625	.9633	1	2	3	4	4	5	6	7	8
1.8	.9641	.9649	.9656	.9664	.9671	.9678	.9686	.9693	.9699	.9706	1	1	2	3	4	4	5	6	6
1.9	.9713	.9719	.9726	.9732	.9738	.9744	.9750	.9756	.9761	.9767	1	1	2	2	3	4	4	5	5
2.0	.9772	.9778	.9783	.9788	.9793	.9798	.9803	.9808	.9812	.9817	0	1	1	2	2	3	3	4	4
2.1	.9821	.9826	.9830	.9834	.9838	.9842	.9846	.9850	.9854	.9857	0	1	1	2	2	2	3	3	4
2.2	.9861	.9864	.9868	.9871	.9875	.9878	.9881	.9884	.9887	.9890	0	1	1	1	2	2	2	3	3
2.3	.9893	.9896	.9898	.9901	.9904	.9906	.9909	.9911	.9913	.9916	0	1	1	1	1	2	2	2	2
2.4	.9918	.9920	.9922	.9924	.9927	.9929	.9931	.9932	.9934	.9936	0	0	1	1	1	1	1	2	2
2.5	.9938	.9940	.9941	.9943	.9945	.9946	.9948	.9949	.9951	.9952	0	0	0	1	1	1	1	1	1
2.6	.9953	.9955	.9956	.9957	.9958	.9960	.9961	.9962	.9963	.9964	0	0	0	0	1	1	1	1	1
2.7	.9965	.9966	.9967	.9968	.9969	.9970	.9971	.9972	.9973	.9974	0	0	0	0	0	1	1	1	1
2.8	.9974	.9975	.9976	.9977	.9977	.9978	.9979	.9979	.9980	.9981	0	0	0	0	0	0	0	1	1
2.9	.9981	.9982	.9982	.9983	.9984	.9984	.9985	.9985	.9986	.9986	0	0	0	0	0	0	0	0	0

Example: $\Phi(1.846) = 0.9671 + 0.0004 = 0.9675$

Appendix VI: Percentage Points for the *t*-Distribution

If T has a t-distribution with ν degrees of freedom then a tabulated value, t, is such that $P(T < t) = p\%$

ν	$p(\%)$								
	75	90	95	97.5	99	99.5	99.75	99.9	99.95
1	1.000	3.078	6.314	12.71	31.82	63.66	127.3	318.3	636.6
2	0.816	1.886	2.920	4.303	6.965	9.925	14.09	22.33	31.60
3	0.765	1.638	2.353	3.182	4.541	5.841	7.453	10.21	12.92
4	0.741	1.533	2.132	2.776	3.747	4.604	5.598	7.173	8.610
5	0.727	1.476	2.015	2.571	3.365	4.032	4.773	5.893	6.869
6	0.718	1.440	1.943	2.447	3.143	3.707	4.317	5.208	5.959
7	0.711	1.415	1.895	2.365	2.998	3.499	4.029	4.785	5.408
8	0.706	1.397	1.860	2.306	2.896	3.355	3.833	4.501	5.041
9	0.703	1.383	1.833	2.262	2.821	3.250	3.690	4.297	4.781
10	0.700	1.372	1.812	2.228	2.764	3.169	3.581	4.144	4.587
11	0.697	1.363	1.796	2.201	2.718	3.106	3.497	4.025	4.437
12	0.695	1.356	1.782	2.179	2.681	3.055	3.428	3.930	4.318
13	0.694	1.350	1.771	2.160	2.650	3.012	3.372	3.852	4.221
14	0.692	1.345	1.761	2.145	2.624	2.977	3.326	3.787	4.140
15	0.691	1.341	1.753	2.131	2.602	2.947	3.286	3.733	4.073
16	0.690	1.337	1.746	2.120	2.583	2.921	3.252	3.686	4.015
17	0.689	1.333	1.740	2.110	2.567	2.898	3.222	3.646	3.965
18	0.688	1.330	1.734	2.101	2.552	2.878	3.197	3.610	3.922
19	0.688	1.328	1.729	2.093	2.539	2.861	3.174	3.579	3.883
20	0.687	1.325	1.725	2.086	2.528	2.845	3.153	3.552	3.850
21	0.686	1.323	1.721	2.080	2.518	2.831	3.135	3.527	3.819
22	0.686	1.321	1.717	2.074	2.508	2.819	3.119	3.505	3.792
23	0.685	1.319	1.714	2.069	2.500	2.807	3.104	3.485	3.768
24	0.685	1.318	1.711	2.064	2.492	2.797	3.091	3.467	3.745
25	0.684	1.316	1.708	2.060	2.485	2.787	3.078	3.450	3.725
26	0.684	1.315	1.706	2.056	2.479	2.779	3.067	3.435	3.707
27	0.684	1.314	1.703	2.052	2.473	2.771	3.057	3.421	3.690
28	0.683	1.313	1.701	2.048	2.467	2.763	3.047	3.408	3.674
29	0.683	1.311	1.699	2.045	2.462	2.756	3.038	3.396	3.659
30	0.683	1.310	1.697	2.042	2.457	2.750	3.030	3.385	3.646
40	0.681	1.303	1.684	2.021	2.423	2.704	2.971	3.307	3.551
60	0.679	1.296	1.671	2.000	2.390	2.660	2.915	3.232	3.460
120	0.677	1.289	1.658	1.980	2.358	2.617	2.860	3.160	3.373
∞	0.674	1.282	1.645	1.960	2.326	2.576	2.807	3.090	3.291

Appendix VII: Percentage Points for the *F*-Distribution

Upper 5% points

ν_2	ν_1 = 1	2	3	4	5	6	7	8	12	24	40	∞
1	161.4	199.5	215.7	224.6	230.2	234.0	236.8	238.9	243.9	249.1	251.1	254.3
2	18.51	19.00	19.16	19.25	19.30	19.33	19.35	19.37	19.41	19.45	19.47	19.50
3	10.13	9.55	9.28	9.12	9.01	8.94	8.89	8.85	8.74	8.64	8.59	8.53
4	7.71	6.94	6.59	6.39	6.26	6.16	6.09	6.04	5.91	5.77	5.72	5.63
5	6.61	5.79	5.41	5.19	5.05	4.95	4.88	4.82	4.68	4.53	4.46	4.36
6	5.99	5.14	4.76	4.53	4.39	4.28	4.21	4.15	4.00	3.84	3.77	3.67
7	5.59	4.74	4.35	4.12	3.97	3.87	3.79	3.73	3.57	3.41	3.34	3.23
8	5.32	4.46	4.07	3.84	3.69	3.58	3.50	3.44	3.28	3.12	3.04	2.93
9	5.12	4.26	3.86	3.63	3.48	3.37	3.29	3.23	3.07	2.90	2.83	2.71
10	4.96	4.10	3.71	3.48	3.33	3.22	3.14	3.07	2.91	2.74	2.66	2.54
12	4.75	3.89	3.49	3.26	3.11	3.00	2.91	2.85	2.69	2.51	2.43	2.30
15	4.54	3.68	3.29	3.06	2.90	2.79	2.71	2.64	2.48	2.29	2.20	2.07
18	4.41	3.55	3.16	2.93	2.77	2.66	2.58	2.51	2.34	2.15	2.06	1.92
20	4.35	3.49	3.10	2.87	2.71	2.60	2.51	2.45	2.28	2.08	1.99	1.84
25	4.24	3.39	2.99	2.76	2.60	2.49	2.40	2.34	2.16	1.96	1.87	1.71
30	4.17	3.32	2.92	2.69	2.53	2.42	2.33	2.27	2.09	1.89	1.79	1.62
40	4.08	3.23	2.84	2.61	2.45	2.34	2.25	2.18	2.00	1.79	1.69	1.51
60	4.00	3.15	2.76	2.53	2.37	2.25	2.17	2.10	1.92	1.70	1.59	1.39
∞	3.84	3.00	2.60	2.37	2.21	2.10	2.01	1.94	1.75	1.52	1.39	1.00

NB. The lower 5% point of an F_{ν_1,ν_2} distribution is the reciprocal of the upper 5% point of an F_{ν_2,ν_1} distribution.

Upper 2.5% points

ν_2	ν_1											
	1	2	3	4	5	6	7	8	12	24	40	∞
1	647.8	799.5	864.2	899.6	921.8	937.1	948.2	956.7	967.7	997.2	1006	1018
2	38.51	39.00	39.17	39.25	39.30	39.33	39.36	39.37	39.41	39.46	39.47	39.50
3	17.44	16.04	15.44	15.10	14.88	14.73	14.62	14.54	14.34	14.12	14.04	13.90
4	12.22	10.65	9.98	9.60	9.36	9.20	9.07	8.98	8.75	8.51	8.41	8.26
5	10.01	8.43	7.76	7.39	7.15	6.98	6.85	6.76	6.52	6.28	6.18	6.02
6	8.81	7.26	6.60	6.23	5.99	5.82	5.70	5.60	5.37	5.12	5.01	4.85
7	8.07	6.54	5.89	5.52	5.29	5.12	4.99	4.90	4.67	4.42	4.31	4.14
8	7.57	6.06	5.42	5.05	4.82	4.65	4.53	4.43	4.20	3.95	3.84	3.67
9	7.21	5.71	5.08	4.72	4.48	4.32	4.20	4.10	3.87	3.61	3.51	3.33
10	6.94	5.46	4.83	4.47	4.24	4.07	3.95	3.85	3.62	3.37	3.26	3.08
12	6.55	5.10	4.47	4.12	3.89	3.73	3.61	3.51	3.28	3.02	2.91	2.72
15	6.20	4.77	4.15	3.80	3.58	3.41	3.29	3.20	2.96	2.70	2.59	2.40
18	5.98	4.56	3.95	3.61	3.38	3.22	3.10	3.01	2.77	2.50	2.38	2.19
20	5.87	4.46	3.86	3.51	3.29	3.13	3.01	2.91	2.68	2.41	2.29	2.09
25	5.69	4.29	3.69	3.35	3.13	2.97	2.85	2.75	2.51	2.24	2.12	1.91
30	5.57	4.18	3.59	3.25	3.03	2.87	2.75	2.65	2.41	2.14	2.01	1.79
40	5.42	4.05	3.46	3.13	2.90	2.74	2.62	2.53	2.29	2.01	1.88	1.64
60	5.29	3.93	3.34	3.01	2.79	2.63	2.51	2.41	2.17	1.88	1.74	1.48
∞	5.02	3.69	3.12	2.79	2.57	2.41	2.29	2.19	1.94	1.64	1.48	1.00

NB. The lower 2.5% point of an F_{ν_1,ν_2} distribution is the reciprocal of the upper 2.5% point of an F_{ν_2,ν_1} distribution.

Upper 1% points

ν_2	ν_1 = 1	2	3	4	5	6	7	8	12	24	40	∞
1	4052	4999	5403	5625	5764	5859	5928	5981	6106	6235	6287	6366
2	98.50	99.00	99.17	99.25	99.30	99.33	99.36	99.37	99.42	99.46	99.47	99.50
3	34.12	30.82	29.46	28.71	28.24	27.91	27.67	27.49	27.05	26.60	26.41	26.13
4	21.20	18.00	16.69	15.98	15.52	15.21	14.98	14.80	14.37	13.93	13.75	13.46
5	16.26	13.27	12.06	11.39	10.97	10.67	10.46	10.29	9.89	9.47	9.29	9.02
6	13.75	10.92	9.78	9.15	8.75	8.47	8.26	8.10	7.72	7.31	7.14	6.88
7	12.25	9.55	8.45	7.85	7.46	7.19	6.99	6.84	6.47	6.07	5.91	5.65
8	11.26	8.65	7.59	7.01	6.63	6.37	6.18	6.03	5.67	5.28	5.12	4.86
9	10.56	8.02	6.99	6.42	6.06	5.80	5.61	5.47	5.11	4.73	4.57	4.31
10	10.04	7.56	6.55	5.99	5.64	5.39	5.20	5.06	4.71	4.33	4.17	3.91
12	9.33	6.93	5.95	5.41	5.06	4.82	4.64	4.50	4.16	3.78	3.62	3.36
15	8.68	6.36	5.42	4.89	4.56	4.32	4.14	4.00	3.67	3.29	3.13	2.87
18	8.29	6.01	5.09	4.58	4.25	4.01	3.84	3.71	3.37	3.00	2.84	2.57
20	8.10	5.85	4.94	4.43	4.10	3.87	3.70	3.56	3.23	2.86	2.69	2.42
25	7.77	5.57	4.68	4.18	3.85	3.63	3.46	3.32	2.99	2.62	2.45	2.17
30	7.56	5.39	4.51	4.02	3.70	3.47	3.30	3.17	2.84	2.47	2.30	2.01
40	7.31	5.18	4.31	3.83	3.51	3.29	3.12	2.99	2.66	2.29	2.11	1.80
60	7.08	4.98	4.13	3.65	3.34	3.12	2.95	2.82	2.50	2.12	1.94	1.60
∞	6.63	4.61	3.78	3.32	3.02	2.80	2.64	2.51	2.18	1.79	1.59	1.00

NB. The lower 1% point of an F_{ν_1,ν_2} distribution is the reciprocal of the upper 1% point of an F_{ν_2,ν_1} distribution.

Appendix VIII: Percentage Points for the Chi-Squared Distribution

If X has a χ^2 distribution with ν degrees of freedom, then a tabulated value, x, is such that $P(X < x) = p\%$.

ν	p(%)								
	Lower tail			Upper tail					
	0.5	2.5	5	90	95	97.5	99	99.5	99.9
1	$0.0^4 3927$	$0.0^3 9821$	$0.0^2 3932$	2.706	3.841	5.024	6.635	7.879	10.83
2	0.01003	0.05064	0.1026	4.605	5.991	7.378	9.210	10.60	13.82
3	0.07172	0.2158	0.3518	6.251	7.815	9.348	11.34	12.84	16.27
4	0.2070	0.4844	0.7107	7.779	9.488	11.14	13.28	14.86	18.47
5	0.4117	0.8312	1.145	9.236	11.07	12.83	15.09	16.75	20.52
6	0.6757	1.237	1.635	10.64	12.59	14.45	16.81	18.55	22.46
7	0.9893	1.690	2.167	12.02	14.07	16.01	18.48	20.28	24.32
8	1.344	2.180	2.733	13.36	15.51	17.53	20.09	21.95	26.12
9	1.735	2.700	3.325	14.68	16.92	19.02	21.67	23.59	27.88
10	2.156	3.247	3.940	15.99	18.31	20.48	23.21	25.19	29.59
11	2.603	3.816	4.575	17.28	19.68	21.92	24.72	26.76	31.26
12	3.074	4.404	5.226	18.55	21.03	23.34	26.22	28.30	32.91
13	3.565	5.009	5.892	19.81	22.36	24.74	27.69	29.82	34.53
14	4.075	5.629	6.571	21.06	23.68	26.12	29.14	31.32	36.12
15	4.601	6.262	7.261	22.31	25.00	27.49	30.58	32.80	37.70
16	5.142	6.908	7.962	23.54	26.30	28.85	32.00	34.27	39.25
17	5.697	7.564	8.672	24.77	27.59	30.19	33.41	35.72	40.79

In this table $0.0^3 1$ means 0.0001.

ν	p(%)								
	Lower tail			Upper tail					
	0.5	2.5	5	90	95	97.5	99	99.5	99.9
18	6.265	8.231	9.390	25.99	28.87	31.53	34.81	37.16	42.31
19	6.844	8.907	10.12	27.20	30.14	32.85	36.19	38.58	43.82
20	7.434	9.591	10.85	28.41	31.41	34.17	37.57	40.00	45.31
21	8.034	10.28	11.59	29.62	32.67	35.48	38.93	41.40	46.80
22	8.643	10.98	12.34	30.81	33.92	36.78	40.29	42.80	48.27
23	9.260	11.69	13.09	32.01	35.17	38.08	41.64	44.18	49.73
24	9.886	12.40	13.85	33.20	36.42	39.36	42.98	45.56	51.18
25	10.52	13.12	14.61	34.38	37.65	40.65	44.31	46.93	52.62
30	13.79	16.79	18.49	40.26	43.77	46.98	50.89	53.67	59.70
40	20.71	24.43	26.51	51.81	55.76	59.34	63.69	66.77	73.40
50	27.99	32.36	34.76	63.17	67.50	71.42	76.15	79.49	86.66
60	35.53	40.48	43.19	74.40	79.08	83.30	88.38	91.95	99.61
70	43.28	48.76	51.74	85.53	90.53	95.02	100.4	104.2	112.3
80	51.17	57.15	60.39	96.58	101.9	106.6	112.3	116.3	124.8
90	59.20	65.65	69.13	107.6	113.1	118.1	124.1	128.3	137.2
100	67.33	74.22	77.93	118.5	124.3	129.6	135.8	140.2	149.4

NB. For values of $\nu > 100$ use the result that $\sqrt{2X}$ has an approximate normal distribution with mean $\sqrt{2\nu - 1}$ and variance 1.

Appendix IX: Pseudo-Random Numbers

07552	37078	70487	39809	35705	42662
28859	92692	51960	51172	02339	94211
64473	62150	49273	29664	05698	05946
55434	20290	33414	26519	65317	47580
20131	05658	01643	17950	74442	30519
04287	26200	37224	23042	85793	50649
19631	42910	35954	88679	34461	45854
52646	83321	52538	41676	71829	00734
11107	55247	73970	67044	29864	72349
16311	04954	92332	51595	96460	77412
37057	83986	98419	76401	15412	68418
33724	28633	85953	82213	07827	48740
43737	15929	19659	52804	72335	25208
16929	84478	31341	60265	19404	27881
10131	98571	20877	34585	22353	54505
29998	48921	60361	12353	28334	84764
96525	74926	82302	97562	57805	40464
49955	60120	14557	04036	55397	54710
27936	70742	69960	69090	25800	53457
43045	75684	77671	70298	21292	27677
38782	35325	61068	64149	73456	06831
47347	47512	09263	83713	04450	31376
98561	93657	76725	55243	95540	31611
30674	43720	80477	82488	44328	55607
20293	63332	24626	56001	23528	85302

Appendix X: Selected Landmarks in the Development of Statistics

Date	Event
1657	The first treatise on *probability, *De Ratiociniis in Ludo Aleae* (*Calculation in Games of Chance*), written by *Huygens.
1662	*Graunt publishes *Natural and Political Observations Mentioned in a Following Index and Made upon the Bills of Mortality*, introducing the *life table.
1711	*De Moivre publishes a (largely overlooked) derivation of the *Poisson distribution (*Poisson's better-known derivation was published in 1837).
1713	Jacob *Bernoulli publishes *Ars Conjectandi* (*The Art of Conjecture*), containing a derivation of the *binomial distribution.
1733	*De Moivre published *The Doctrine of Chances* in 1718. The second (1738) edition contains a supplement dated 12 November 1733 which gives the formula for the *probability density function of the *normal distribution.
1763	*Bayes introduces the idea of a *prior distribution.
1791	First volume of Sir John *Sinclair's *Statistical Account of Scotland* appears, containing the first use of the word 'statistics'.
1805	*Legendre publishes the first account of the method of *least squares.
1812	*Laplace uses *generating functions in his *Théorie Analytique des Probabilités.*
1834	*Royal Statistical Society founded.
1835	*Quetelet applies the *normal distribution to describe the 'normal man'.
1839	*American Statistical Association founded in Boston, Massachusetts.
1847	*De Morgan publishes his laws of probability.
1863	*Abbe publishes a derivation of the *chi-squared distribution.
1869	*Galton uses the term *correlation in its statistical sense in his book *Hereditary Genius.*
1877	*Galton uses the term *regression in a lecture, entitled 'Typical Laws of Heredity in Man', on 9 February to the Royal Institution.
1880	*Venn introduces *Venn diagrams.
1885	Establishment of the *International Statistical Institute.
1896	Karl *Pearson introduces the product-moment correlation coefficient.
1900	Karl *Pearson introduces the *chi-squared test.
1901	First issue of **Biometrika*, edited by Karl *Pearson.
1902	Publication of *Elementary Principles of Statistical Mechanics*, by *Gibbs.
1908	Publication of *The Probable Error of a Mean*, by *Gossett. This introduces the *t-test.
1911	*Yule publishes *An Introduction to the Theory of Statistics*. The fourteenth edition was published in 1950.

Date	Event
1912	Sir Ronald *Fisher introduces the method of *maximum likelihood for parameter estimation.
1922	Sir Ronald *Fisher introduces the *F*-test for the comparison of variance estimates.
1924	*Shewhart introduces the *control chart.
1925	Sir Ronald *Fisher publishes the first edition of *Statistical Methods for Research Workers*, setting out *inter alia* *ANOVA tables. The thirteenth edition was published in 1970.
1928	*Neyman and Egon *Pearson introduce the idea of a *confidence interval.
1931	*Von Mises introduces the idea of sample space.
1933	*Kolmogorov publishes his axiomatic treatment of probability, *Foundations of the Theory of Probability*.
1933	*Kolmogorov introduces the *Kolmogorov–Smirnov test.
1933	*Neyman and Egon *Pearson introduce the procedure for *hypothesis testing.
1935	Foundation of the *Institute of Mathematical Statistics.
1938	*Kolmogorov publishes *Analytic Methods in Probability Theory*, which sets out the foundations of *Markov processes.
1947	*Dantzig introduces the *simplex method for constrained *optimization.
1947	Foundation of the *International Biometric Society and the journal **Biometrics*.
1948	*Wiener publishes *Cybernetics: or Control and Communication in the Animal and the Machine*.
1950	*Feller publishes the first volume of *An Introduction to Probability Theory and its Applications*, the definitive text on *stochastic processes.
1958	*Kaplan and *Meier introduce their method for estimating the *survivor function.
1963	*Barnard suggests the *Monte Carlo approach to *hypothesis testing.
1963	*Matheron publishes *Traité de Géostatistique Appliquée*, setting out the fundamentals of *geostatistics.
1965	*Tukey and Cooley introduce the *fast Fourier transform.
1969	*Akaike introduces his criterion for model comparison.
1970	*Box and *Jenkins publish *Time Series Analysis: Forecasting and Control*.
1970	*Tukey publishes *Exploratory Data Analysis*, introducing the *boxplot and the *stem and leaf diagram.
1972	*Nelder and *Wedderburn introduce the framework for *generalized linear models.
1972	The *Cox regression model is introduced.
1977	*Cook introduces new *regression diagnostics.
1977	*Dempster, *Laird, and *Rubin introduce the *EM algorithm for handling incomplete data.
1979	*Efron introduces the *bootstrap and other *resampling methods.
1988	*Daubechies introduces her family of *wavelets.
2011	Introduction of the *maximal information coefficient as a generalization of correlation.

Appendix XI: Honours and Awards

American Statistical Association: Wilks Award

1964	F. E. *Grubbs	1965	J. W. *Tukey	1966	L. E. Simon
1967	W. G. *Cochran	1968	J. *Neyman	1969	W. J. *Youden
1970	G. W. *Snedecor	1971	H. F. Dodge	1972	G. E. P. *Box
1973	H. O. *Hartley	1974	C. Daniel	1975	H. Solomon
1976	S. *Kullback	1977	C. Eisenhart	1978	W. H. *Kruskal
1979	A. M. *Mood	1980	W. A. *Wallis	1981	H. Working
1982	F. Proschan	1983	W. E. *Deming	1984	Z. W. *Birnbaum
1985	L. A. *Goodman	1986	C. F. *Mosteller	1987	H. *Chernoff
1988	T. W. *Anderson	1989	C. R. *Rao	1990	B. *Efron
1991	I. Olkin	1992	W. Dixon	1993	N. L. *Johnson
1994	E. *Parzen	1995	D. B. *Rubin	1996	E. L. *Lehmann
1997	L. Kish	1998	D. O. Siegmund	1999	L. Billard
2000	S. E. Fienberg	2001	G. C. Tiao	2002	L. D. Brown
2003	D. L. Wallace	2004	P. *Meier	2005	R. J. A. Little
2006	M. *Zelen	2007	C. L. *Mallows	2008	S. L. Zeger
2009	L-J. Wei	2010	P. K. Sen	2011	N. M. Laird
2012	P. G. *Hall	2013	K. Mardia		

Committee of Presidents of Statistical Societies: R. A. Fisher Lecturer

1964	M. S. *Bartlett	1965	O. *Kempthorne	1967	W. J. *Youden
1968	L. A. *Goodman	1970	L. J. Savage	1971	C. Daniel
1972	W. G. *Cochran	1973	J. Cornfield	1974	G. E. P. *Box
1975	H. *Chernoff	1976	G. A. *Barnard	1977	R. C. Bose
1978	W. H. *Kruskal	1979	C. R. *Rao	1982	F. J. *Anscombe
1983	I. R. Savage	1985	T. W. *Anderson	1986	D. *Blackwell
1987	C. F. *Mosteller	1988	E. L. *Lehmann	1989	D. R. *Cox
1990	D. A. S. Fraser	1991	D. R. *Brillinger	1992	P. *Meier
1993	H. E. *Robbins	1994	E. A. Thompson	1995	N. E. *Breslow
1996	B. *Efron	1997	C. L. *Mallows	1998	A. P. *Dempster
1999	J. D. Kalbfleisch	2000	I. Olkin	2001	J. O. *Berger
2002	R. Carroll	2003	A. F. M. *Smith	2004	D. B. *Rubin
2005	R. D. *Cook	2006	T. P. *Speed	2007	M. *Zelen
2008	R. L. Prentice	2009	N. A. Cressie	2010	B. G. Lindsay
2011	C. F. J. Wu	2012	R. J. A. Little	2013	P. J. *Bickel

Institute of Mathematical Statistics: Wald Lecturer

1957	S. Karlin	1958	J. W. *Tukey	1959	R. C. Bose
1961	C. M. *Stein	1962	J. C. Kiefer	1963	L. M. *Le Cam
1964	E. L. *Lehmann	1965	J. *Wolfowitz	1967	W. *Hoeffding
1968	H. *Chernoff	1969	H. E. *Robbins	1970	M. Rosenblatt

1971	D. L. Burkholder	1972	P. J. *Huber	1973	P. Billingsley
1974	R. R. *Bahadur	1975	C. R. *Rao	1976	D. H. *Blackwell
1977	J. F. C. *Kingman	1978	M. Kac	1979	F. Spitzer
1980	P. J. *Bickel	1981	B. *Efron	1982	T. W. *Anderson
1983	D. R. *Brillinger	1984	D. O. Siegmund	1985	L. H. Brown
1986	H. Kesten	1987	P. Diaconis	1988	D. V. *Lindley
1989	I. A. Ibragimov	1990	D. R. *Cox	1991	E. B. Dynkin
1992	W. R. van Zwet	1993	D. Aldous	1994	C. Stone
1995	U. Grenander	1996	T. Liggett	1997	D. L. *Donoho
1998	D. Freedman	1999	C. Newman	2000	N. M. *Reid
2001	T. P. *Speed	2002	L. Breiman	2003	G. *Wahba
2004	I. *Johnstone	2005	S. R. S. *Varadhan	2006	P. G. *Hall
2007	J. O. *Berger	2008	R. Durrett	2009	J. H. *Friedman
2010	J-F. Le Gall	2011	G. F. Lawler	2012	S. L. *Lauritzen
2013	P. Groeneboom				

Institute of Mathematical Statistics: Rietz (R)/Neyman (N)/ Le Cam (L) Lecturer

1947	A. *Wald (R)	1949	J. *Neyman (R)	1951	H. *Hotelling (R)
1953	C. H. *Cramér (R)	1955	W. *Feller (R)	1957	J. *Wolfowitz (R)
1959	S. S. *Wilks (R)	1961	D. H. *Blackwell (R)	1963	H. E. *Robbins (R)
1965	J. L. *Doob (R)	1967	W. G. *Cochran (R)	1969	A. Dvoretzky (R)
1971	H. Kesten (R)	1973	D. R. *Cox (R)	1975	C. M. *Stein (R)
1977	B. *Efron (R)	1979	J. B. *Kruskal (R)	1981	G. E. P. *Box (R)
1982	H. E. *Robbins (N)	1983	L. A. *Goodman (R)	1984	C. M. *Stein (N)
1985	U. Grenander (R)	1986	A. P. *Dempster (N)	1987	S. R. S. *Varadhan (R)
1988	S. M. Stigler (N)	1989	H. *Chernoff (R)	1990	P. *Whittle (N)
1991	T. P. *Speed (R)	1992	S. Johansen (N)	1993	O. Barndorff-Nielsen (R)
1994	G. G. *Wahba (N)	1995	M. Talagrand (R)	1996	E. A. Thompson (N)
1997	S. Geman (R)	1998	J. Chambers (N)	1999	J. H. *Friedman (R)
2000	M. E. Bock (N)	2001	P. Dawid (R)	2002	W. Wong (N)
2003	D. L. *Donoho (L)	2004	P. J. *Bickel (R)	2005	D. R. *Brillinger (N)
2006	S. M. Stigler (L)	2007	D. O. Siegmund (R)	2008	P. McCullagh (N)
2009	A. W. van der Vaart (L)	2010	M. L. Stein (R)	2011	M. I. Jordan (N)
2012	P. Massart (L)	2013	L. Wasserman (R)		

International Biometric Society: Honorary Life Member

1964	C. I. Bliss	1964	G. M. *Cox	1971	G. Barbensi
1971	G. W. *Snedecor	1971	F. *Yates	1975	A. Linder
1976	W. G. *Cochran	1978	E. Weber	1984	D. J. Finney
1985	C. R. *Rao	1998	P. *Armitage	1998	P. Dagnelie
2001	D. R. *Cox	2004	L. C. A. Corsten	2006	J. A. *Nelder
2006	N. E. *Breslow	2008	N. Keiding	2010	R. N. Curnow
2012	S. Wilson				

Penn State University: C. R. and Bhargavi Rao Prize

2003	B. *Efron	2005	J. Sethuraman	2007	L. D. Brown
2009	P. J. *Bickel	2011	J. O. *Berger	2013	H. *Chernoff

Royal Statistical Society: Guy Medal in Gold

1892	C. *Booth	1894	R. Giffen	1900	J. A. Baines
1907	F. Y. *Edgeworth	1908	P. G. Craigie	1911	G. U. *Yule
1920	T. H. C. Stevenson	1930	A. W. Flux	1935	A. L. *Bowley
1945	M. *Greenwood	1946	R. A. *Fisher	1953	A. B. *Hill
1955	E. S. *Pearson	1960	F. *Yates	1962	H. *Jeffreys
1966	J. *Neyman	1968	M. G. *Kendall	1969	M. S. *Bartlett
1972	C. H. *Cramér	1973	D. R. *Cox	1975	G. A. *Barnard
1978	R. G. D. *Allen	1981	D. G. *Kendall	1984	H. E. Daniels
1986	B. Benjamin	1987	R. L. *Plackett	1990	P. *Armitage
1993	G. E. P. *Box	1996	P. *Whittle	1999	M. J. Healy
2002	D. V. *Lindley	2005	J. A. *Nelder	2008	J. *Durbin
2011	C. R. *Rao				

Statistical Society of Australia, Inc: Pitman Medal

1978	E. J. G. Pitman	1980	H. O. *Lancaster	1982	P. A. P. *Moran
1986	E. J. *Hannan	1988	C. C. *Heyde	1990	P. G. *Hall
1992	A. T. James	1993	E. J. Williams	1994	J. M. Gani
1996	W. J. Ewens	1998	E. Seneta	2000	G. N. Wilkinson
2002	T. P. *Speed	2004	A. J. Baddeley	2005	J. Darroch
2006	D. J. Daley	2008	J. Robinson	2010	G. McLachlan
2012	A. Welsh				

Statistical Society of Canada: Gold Medal

1985	D. A. S. Fraser	1986	C. W. *Dunnett	1987	V. P. Godambe
1988	D. A. Sprott	1989	M. A. Stephens	1990	C. Van Eeden
1991	D. A. Dawson	1992	D. R. *Brillinger	1993	J. N. K. Rao
1994	J. D. Kalbfleisch	1995	I. Guttman	1996	M. Csörgö
1997	I. P. Fellegi	1998	J. O. Ramsay	1999	J. F. Lawless
2000	J. V. Zidek	2002	M. Srivastava	2003	M. E. Thompson
2004	K. J. Worsley	2005	D. F. Andrews	2006	C. A. Field
2007	D. L. McLeish	2008	L. Devroye	2009	N. M. *Reid
2010	L-P. Rivest	2011	C. Genes	2012	R. Tibshirani
2013	J. Rosenthal				

Texas A & M University: Emanuel and Carol Parzen Prize for Statistical Innovation

1994	G. G. *Wahba	1996	D. B. *Rubin	1998	B. *Efron
2000	C. R. *Rao	2002	D. R. *Brillinger	2004	J. H. *Friedman
2006	A. E. Gelfand	2008	N. M. *Reid	2008	M. *Zelen
2010	R. Koenker	2012	A. E. Raftery		

Appendix XII: Further Reference

Introductory

Upton, G. J. G., and Cook, I. T. (1996) *Understanding Statistics.* Oxford University Press, Oxford.

Source of data and algorithms

The StatLib Website. http://lib.stat.cmu.edu/index.php

Historical

Johnson, N. L., and Kotz, S. (eds) (1997) *Leading Personalities in Statistical Sciences.* Wiley, New York.

Stigler, S. (1986) *The History of Statistics.* Belknap Press, Cambridge, MA.

The MacTutor Website. http://www-history.mcs.st-andrews.ac.uk/history/index.html

The York Website. http://www.york.ac.uk/depts/maths/histstat/

The Mathematics Genealogy Project website. http://genealogy.math.ndsu.nodak.edu

Special topics

Agresti, A. (2013) *An Introduction to Categorical Data Analysis,* 3rd edn. Wiley, New York.

Armitage, P., Berry, G., and Matthews J. N. S. (2001) *Statistical Methods in Medical Research,* 4th edn. Blackwell, Oxford.

Balakrishnan, N., Read, C. B., Vidakovic, B., Kotz, S., and Johnson, N. L (2010) *Methods and Applications of Statistics in the Life and Health Sciences.* Wiley, New York.

Bertsekas, D. P. (1999) *Nonlinear Programming,* 2nd edn. Athena Scientific, Belmont, MA.

Bertsekas, D. P., and Tsitsiklis, J. (1997) *Introduction to Linear Optimization.* Athena Scientific: Belmont, MA.

Bishop, C. M. (2007) *Pattern Recognition and Machine Learning.* Springer, New York.

Chatfield, C. (2003) *The Analysis of Time Series: An Introduction,* 6th edn. Chapman & Hall/CRC Press, London.

Chatfield, C., and Collins, A. J. (2000) *Introduction to Multivariate Analysis,* rev. edn. Chapman & Hall/CRC Press, London.

Chetwynd, A., and Diggle, P. (1995) *Discrete Mathematics.* Arnold, London.

Cochran, W. G. (1977) *Sampling Techniques,* 3rd edn. Wiley, New York.

Cochran, W. G., and Cox, G. M. (1957) *Experimental Designs,* 2nd edn. Wiley, New York.

Collett, D. (2002) *Modelling Binary Data*, 2nd edn. Chapman & Hall/CRC Press, Florida.

Cox, D. R. (1992) *Planning of Experiments*. Wiley, New York.

Cox, D. R., and Isham, V. (1980) *Point Processes*. Chapman & Hall/CRC Press, London.

Cressie, N. A. C. (1993) *Statistics for Spatial Data*, rev. edn. Wiley, New York.

Diggle, P. J., Heagerty, P., Liang, K.-Y., and Zeger, S. L. (2002) *Analysis of Longitudinal Data*, 2nd edn. Oxford Science Publications, Oxford.

Draper, N. R., and Smith, H. (1998) *Applied Regression Analysis*, 3rd edn. Wiley, New York.

Everitt, B. S., Landau, S., Leese, M., and Stahl, D. (2011) *Cluster Analysis*, 5th edn. Edward Arnold, London.

Everitt, B. S., and Hothorn, T. (2010) *A Handbook of Statistical Analyses using R*, 2nd edn. Chapman & Hall/CRC Press, London.

Everitt, B. S., and Rabe-Hesketh, S. (2001) *Analyzing Medical Data using S-PLUS* (Statistics for Biology and Health) Springer, New York.

Feller, W. (1968) *An Introduction to Probability Theory and its Applications*, Volume 1, 3rd edn. Wiley, New York.

Fisher, N. I. (1993) *Statistical Analysis of Circular Data*. Cambridge University Press, Cambridge.

Gilks, W. R., Richardson, S., and Spiegelhalter, D. J., (eds) (1995) *Markov Chain Monte Carlo in Practice*. Chapman & Hall/CRC Press, London.

Gujarati, D. N. and Porter, D. (2008) *Basic Econometrics*, 5th edn. McGraw-Hill, New York.

Hogg, R. V., McKean, J. W., and Craig, A. J. (2012) *Introduction to Mathematical Statistics*, 7th edn. Macmillan, New York.

Larsen, R. J. and Marx, M. L. (2010) *An Introduction to Mathematical Statistics and Its Applications*, 5th edn. Pearson, New York.

Lindley, D. V. (1991) *Making Decisions*, 2nd edn. Wiley, New York.

Mardia, K. V., Kent, J. T., and Bibby, J. M. (1979) *Multivariate Analysis*. Academic Press, London.

McCullagh, P., and Nelder, J. A. (1989) *Generalized Linear Models*, 2nd edn. Chapman & Hall/CRC Press, London.

Miller, I., and Miller, M. (2012) *John E. Freund's Mathematical Statistics*, 8th edn. Prentice Hall, New Jersey.

Morgan, B. J. T. (1984) *The Elements of Simulation*. Chapman & Hall/CRC Press, London.

Mount, D. W. (2004) *Bionformatics: Sequence and Genome Analysis*, 2nd edn. Cold Spring Harbor Laboratory Press, New York.

Robert, C. P., and Casella, G. (1999) *Monte Carlo Statistical Methods*. Springer, Berlin.

Ross, S. M. (2010) *Introduction to Probability Models*, 10th edn. Academic Press, San Diego.

Taha, H. A. (2010) *Operations Research: An Introduction*, 9th edn. Macmillan, New York.

Tufte, E. R. (2001) *The Visual Display of Quantitative Information*, 2nd edn. Graphics Press, Cheshire, CT.

Venables, W. N., and Ripley, B. D. (2002) *Modern Applied Statistics with S-Plus*, 4th edn. Springer-Verlag, Berlin.

Advanced

Johnson, N. L., Kotz, S., and Balakrishnan, N. (1994) *Continuous Univariate Distributions*, Volume 1, 2nd edn. Wiley, New York.

Johnson, N. L., Kotz, S. and Balakrishnan, N. (1995) *Continuous Univariate Distributions*, Volume 2, 2nd edn. Wiley, New York.

Johnson, N. L., Kemp, A. W., and Kotz, S. (2005) *Univariate Discrete Distributions*, 3rd edn. Wiley, New York.

Kotz, S., Johnson, N. L., Read, C. B., and Vidakovic, B. (eds) (2006) *Encyclopaedia of Statistical Sciences*, 2nd edn. 16 volumes. Wiley, New York.

O'Hagan, A., and Forster, J. (2004) *Kendall's Advanced Theory of Statistics*, Volume 2B.: Bayesian Inference, 2nd edn. Edward Arnold, London.

Stuart, A., and Ord, J. K. (1994) *Kendall's Advanced Theory of Statistics*, Volume 1, 6th edn. Edward Arnold, London.

Stuart, A., Ord, J. K., and Arnold, S. (1998) *Kendall's Advanced Theory of Statistics*, Volume 2, 6th edn. Edward Arnold, London.